Molecular genetic analysis of populations

The Practical Approach Series

SERIES EDITOR

B. D. HAMES
Department of Biochemistry and Molecular Biology
University of Leeds, Leeds LS2 9JT, UK

★ indicates new and forthcoming titles

Affinity Chromatography
★ Affinity Separations
Anaerobic Microbiology
Animal Cell Culture (2nd edition)
Animal Virus Pathogenesis
Antibodies I and II
★ Antibody Engineering
★ Antisense Technology
★ Applied Microbial Physiology
Basic Cell Culture
Behavioural Neuroscience
Bioenergetics
Biological Data Analysis
Biomechanics—Materials
Biomechanics—Structures and Systems
Biosensors
Carbohydrate Analysis (2nd edition)
Cell–Cell Interactions
The Cell Cycle
Cell Growth and Apoptosis
Cellular Calcium
Cellular Interactions in Development
Cellular Neurobiology
★ Chromatin
Clinical Immunology
★ Complement
Crystallization of Nucleic Acids and Proteins
Cytokines (2nd edition)
The Cytoskeleton
Diagnostic Molecular Pathology I and II
★ DNA and Protein Sequence Analysis
DNA Cloning 1: Core Techniques (2nd edition)
DNA Cloning 2: Expression Systems (2nd edition)
DNA Cloning 3: Complex Genomes (2nd edition)
DNA Cloning 4: Mammalian Systems (2nd edition)
Electron Microscopy in Biology
Electron Microscopy in Molecular Biology

Electrophysiology
Enzyme Assays
★ Epithelial Cell Culture
Essential Developmental Biology
Essential Molecular Biology I and II
Experimental Neuroanatomy
Extracellular Matrix
Flow Cytometry (2nd edition)
★ Free Radicals
Gas Chromatography
Gel Electrophoresis of Nucleic Acids (2nd edition)
Gel Electrophoresis of Proteins (2nd edition)
Gene Probes 1 and 2
Gene Targeting
Gene Transcription
★ Genome Mapping
Glycobiology
Growth Factors
Haemopoiesis
Histocompatibility Testing
HIV Volumes 1 and 2
Human Cytogenetics I and II (2nd edition)
Human Genetic Disease Analysis
★ Immunochemistry 1
★ Immunochemistry 2
Immunocytochemistry
In Situ Hybridization
Iodinated Density Gradient Media
Ion Channels
Lipid Modification of Proteins
Lipoprotein Analysis
Liposomes
Mammalian Cell Biotechnology
Medical Parasitology
Medical Virology
★ MHC Volumes 1 and 2
★ Molecular Genetic Analysis of Populations (2nd edition)
Molecular Genetics of Yeast
Molecular Imaging in Neuroscience
Molecular Neurobiology
Molecular Plant Pathology I and II
Molecular Virology
Monitoring Neuronal Activity
Mutagenicity Testing
★ Mutation Detection
Neural Cell Culture
Neural Transplantation
★ Neurochemistry (2nd edition)
Neuronal Cell Lines
NMR of Biological Macromolecules
Non-isotopic Methods in Molecular Biology
Nucleic Acid Hybridization
Oligonucleotides and Analogues
Oligonucleotide Synthesis
PCR 1
PCR 2
★ PCR *In Situ* Hybridization
Peptide Antigens
Photosynthesis: Energy Transduction
Plant Cell Biology

- Plant Cell Culture (2nd edition)
- Plant Molecular Biology
- Plasmids (2nd edition)
- ★ Platelets
- Postimplantation Mammalian Embryos
- Preparative Centrifugation
- Prostaglandins and Related Substances
- Protein Blotting
- Protein Engineering
- ★ Protein Function (2nd edition)
- Protein Phosphorylation
- Protein Purification Applications
- Protein Purification Methods
- Protein Sequencing
- ★ Protein Structure (2nd edition)
- ★ Protein Structure Prediction
- Protein Targeting
- Proteolytic Enzymes
- Pulsed Field Gel Electrophoresis
- RNA Processing I and II
- ★ Signalling by Inositides
- ★ Subcellular Fractionation
- Signal Transduction
- Transcription Factors
- Tumour Immunobiology

Molecular Genetic Analysis of Populations

A Practical Approach

SECOND EDITION

Edited by

A. R. HOELZEL
Department of Biological Sciences
University of Durham

Oxford University Press, Great Clarendon Street, Oxford OX2 6DP

Oxford New York
Athens Auckland Bangkok Bogota Bombay Buenos Aires
Calcutta Cape Town Dar es Salaam Delhi Florence Hong Kong
Istanbul Karachi Kuala Lumpur Madras Madrid Melbourne
Mexico City Nairobi Paris Singapore Taipei Tokyo Toronto Warsaw

and associated companies in
Berlin Ibadan

Oxford is a trade mark of Oxford University Press

Published in the United States
by Oxford University Press Inc., New York
First edition published 1992
Second edition published 1998

A catalogue record for this book is available from the British Library

Library of Congress Cataloging in Publication Data
(Data available)

ISBN 0 19 963634 6 (Hbk)
ISBN 0 19 963635 4 (Pbk)

Typeset by Footnote Graphics, Warminster, Wilts
Printed in Great Britain by Information Press, Ltd, Eynsham, Oxon.

Preface

The first edition of this book brought together researchers from a diversity of interests to create a general guide to lab protocols useful for studies in population genetics and molecular ecology. All the techniques share the objective of assessing levels and patterns of genetic diversity, but differ in factors such as resolution, focus, difficulty, and expense. In the 1960s, R. C. Lewonton and J. L. Hubby described enzyme electrophoresis, the first technique to permit the assessment of genetic variation in populations at the molecular level. Although this method provides relatively low resolution, it is simple, fast, and inexpensive enough to still be of use today. The first chapter in both editions describes starch-gel electrophoresis of allozymes, by far the most widely used method for screening enzyme variation in natural populations. The remainder of the chapters focus on the analysis of DNA. The techniques that have advanced the direct analysis of DNA have been revolutionary in their impact, from RFLP to DNA sequencing to PCR and beyond. The first edition reflected the most modern methods in 1992, but this is a field that moves forward quickly, and there are now a number of new protocols finding wide applications in population genetic studies. In this volume, methodologies provided in the first edition have been updated to reflect new time-saving procedures, further use of non-radioactive methods, and new ways to maximize resolution and ease interpretation. In addition there are new chapters on microsatellite DNA, conformational and gradient gel electrophoretic methods, and automated sequencing using fluorescent markers. New features of the appendices include quick-reference tables and graphs and a directory of computer programs and internet addresses. The great facilitation availed by these advancing technologies is reflected in the breadth and diversity of studies in the literature over the last several years. I expect that the protocols brought together in this volume will contribute to this trend, and would like to thank the contributors for their efforts. The end result is forward-looking and highly accessible.

Durham A.R.H.
January 1998

Contents

Contributors

CHARLES F. AQUADRO
Section of Genetics and Development, Biotechnology Building, Cornell University, Ithaca, NY 14853, USA.

D. R. BANCROFT
Max-Planck Institute for Molecular Genetics, Ihnestrasse 73, Berlin-Dahlem, D-14195, Germany.

DAVID J. BEGUN
Department of Zoology, University of Texas, Austin, TX 78712, USA.

JOHN F. Y. BROOKFIELD
Department of Genetics, University of Nottingham, Nottingham N67 2UH, UK.

MICHAEL W. BRUFORD
Institute of Zoology, Zoological Society of London, Regent's Park, London NW1 4RY, UK.

TERRY BURKE
Department of Zoology, University of Leicester, University Road, Leicester LE1 7RH, UK.

BRYAN N. DANFORTH
Department of Entomology, Comstock Hall, Cornell University, Ithaca, NY 14853, USA.

VICTOR A. DAVID
Intramural Research and Support Program, SAIC Frederick, National Cancer Institute-Frederick Cancer Research and Development Center, Frederick, MD 21702, USA.

MICHAEL DEAN
Building 560, Room 21-18, National Cancer-Frederick Cancer Research and Development Center, Frederick, MD 21702, USA.

LLEWELLYN D. DENSMORE III
Department of Biological Sciences, Texas Tech University, M53131, Lubbock, Texas 79409, USA.

CHRISTINE A. ORENGO
Department of Biochemistry and Molecular Biology, University College London, Gower Street, London WC1E 6BT, UK.

A. GREEN
National Centre for Medical Genetics, Our Lady's Hospital for Sick Children, Crumlin, Dublin 12, Ireland.

OLIVIER HANOTTE
International Livestock Research Institute, PO Box 30709, Nairobi, Kenya.

A. R. HOELZEL
Department of Biological Sciences, University of Durham, South Road, Durham DH1 3LE, UK.

BERNIE MAY
Genomic Variation Laboratory, Department of Animal Science, 2237 Meyer Hall, University of California, Davis, CA 95616, USA.

MARILYN MENOTTI-RAYMOND
Laboratory of Genomic Diversity, National Cancer Institute–Frederick Cancer Research and Development Center, Frederick, MD 21702, USA.

BROOK G. MILLIGAN
Department of Biology, New Mexico State University, Las Cruces, NM 88003, USA.

WILLIAM A. NOON
Department of Soil, Crop and Atmospheric Sciences, Bradfield Hall, Cornell University, Ithaca, NY 14853, USA.

RICHARD J. ROBERTS
New England Biolabs, 32 Tozer Road, Beverly, MA 01915, USA.

CHRISTIAN SCHLÖTTERER
Institut für Tierzucht und Genetik, Veterinärmedizinische Universität Wien, 1210 Wien, Austria.

JILL PECON SLATTERY
Laboratory of Genomic Diversity, National Cancer Institute–Frederick Cancer Research and Development Center, Frederick, MD 21702, USA.

J. CLAIBORNE STEPHENS
Laboratory of Genomic Diversity, National Cancer Institute–Frederick Cancer Research and Development Center, Frederick, MD 21702, USA.

OWATHA L. TATUM
LS-3 Genomics, Mailstop M888, Life Sciences Division, Los Alamos National Laboratory, Los Alamos, NM, USA, and Department of Biological Sciences, Texas Tech University, M53131, Lubbock, Texas 79409, USA.

HÅKAN TEGELSTRÖM
Department of Genetics, Uppsala University, Box 7003, S-75007, Uppsala, Sweden.

J. P. WARNER
Molecular Genetics Service, Human Genetics Unit, Molecular Medicine Centre, Western General Hospital, Edinburgh, EH4 2XU, UK.

P. SCOTT WHITE
LS-3 Genomics, Mailstop M888, Life Sciences Division and Center for Human Genome Studies, Los Alamos National Laboratory, Los Alamos, NM 87545, USA.

Abbreviations

ACD	acid citrate dextrose
AFLP	amplified fragment length polymorphism
ALP	automated linkage preprocessor
AMPS	ammonium persulfate
2-BE	2-butoxyethanol
BSA	bovine serum albumin
CIP	calf intestinal alkaline phosphatase
CTAB	heaxadecyltrimethylammonium bromide
DAPI	4,6′-diamidino-2-phenylindole
DGGE	denaturing gradient gel electrophoresis
DIG	digoxigenin
DMSO	dimethylsulfoxide
DTT	dithiothreitol
EMC	enzyme mismatch cleavage
ETS	external transcribed spacer
FAMA	fluorescence-assisted mismatch analysis
IPTG	isopropyl-β-D-thiogalactopyranoside
LDH	lactate dehydrogenase
LMP	low melting point
MDE	mutation detection enhancer
MDH	malate dehydrogenase
MHC	major histocompatibility complex
mtDNA	mitochondrial DNA
MTT	thiazolyl blue
NBT	Nitroblue tetrazolium
NP-40	Nonidet P-40
NTP	nucleoside triphosphate
OD	optical density
OPC	oligonucleotide purification cartridge
p.f.u.	plaque-forming units
PCR	polymerase chain reaction
PEG	polyethylene glycol
PMS	phenazine methosulfate
PVP	polyvinylpyrrolidone
rDNA	ribosomal DNA
RAPD	random amplified polymorphic DNA
RFLP	restriction fragment length polymorphism
RNase	ribonuclease
SDS	sodium dodecyl sulfate
SSCP	single-strand conformation polymorphism

T_m	melting temperature
TCA	trichloroacetic acid
TEAA	triethylamine acetate
TEMED	*N*,*N*,*N*′,*N*′-tetramethylethylenediamine
TMAC	tetramethylammonium chloride
VNTR	variable number tandem repeat
X-Gal	5-bromo-4-chloro-3-indolyl-β-D-galactoside
YAC	yeast artificial chromosome

1

Starch gel electrophoresis of allozymes

BERNIE MAY

1. Introduction

Thirty years ago a novel, relatively simplistic methodology revolutionized our ability to study evolution (starch gel electrophoresis of allozymic proteins). Lewontin and Hubby (1) and Harris (2) illustrated convincingly how allozyme electrophoresis permitted a wealth of Mendelian data on populations to be gathered in a relatively short period of time. The availability of these data has revolutionized many fields of evolutionary biology, including systematics, sociobiology, genomic organization, and population genetics.

Prior to the appearance of allozyme electrophoresis only morphological and behavioural mutants were available to study single gene variation. This new technique brought the electrophoretic separation of proteins together with the specificity of histochemical detection of the protein products of single loci (3). Now dozens of simple Mendelian markers are available in most organisms. The utter simplicity of this methodology has permitted thousands of scientists to use this technique in their research.

Recent developments in laboratory procedures with DNA would seem to suggest that not only the reign but even the relevance of allozyme electrophoresis in evolutionary biology is over. The research limelight has increasingly shifted over the past decade to focus on studies which directly measure DNA variation. The measure of the truth of this statement can be seen in the 'Situations vacant' section in various journals that advertise for evolutionary biologists who use modern molecular techniques. However, while these newer technologies offer tremendous gain in resolution, they have yet to supplant the applicability of allozyme electrophoresis for the study of genetic variation in natural populations. Starch gel electrophoresis of proteins coupled with histochemical visualization of locus-specific allozymes offers a relatively cheap, fast method of analysing single locus variation in natural and artificial populations of any life form from bacteria to man (4–8).

In the past 15 years in my laboratory we have examined allozyme variability in over 100 species of plants, fungi, fish, mammals, birds, and a variety of invertebrates. In the sections that follow I will discuss the theoretical basis of the allozyme methodology, illustrate the horizontal starch gel methods we use in my laboratory, describe some of the other variations in method used by others, summarize the typical applications of the data, and close with a prospectus for the future of allozyme electrophoresis.

2. General methodology (basic principles)

This methodology is called allozyme (shortened version of alloenzyme) electrophoresis because it is used to examine variation at single discrete loci; 'allozymes' are the different protein forms encoded by the various alleles at one locus. The term 'isoenzyme' or 'isozyme' is used to refer to a larger subset of protein forms, including the different protein products (intralocus and interlocus protein products) from all the genomic loci encoding a single type of enzyme (e.g., all the different forms of the enzyme malate dehydrogenase (MDH)). We equate allozyme electrophoresis with the separation and detection of Mendelian genetic variation at single protein-coding loci even though a few non-enzymatic proteins are examined by this methodology (e.g., albumin, transferrin, and haemoglobin).

The entire process of allozyme electrophoresis can be divided into the following five steps: extraction, separation, staining, interpretation, and application. These are described in more detail below.

2.1 Extraction

Handling enzymatic proteins differs from handling DNA in two significant ways. Most enzymes are more temperature labile. Once an organism or tissue sample has been taken the sample must be frozen as quickly as possible. Storage for most enzymes must be below –70°C (ultracold freezers, dry ice, or liquid nitrogen). While some very stable enzymes like lactate dehydrogenase (LDH) may last for decades, most enzymes begin to degrade seriously in activity and resolvability after two years at even the lowest temperatures. Genomic DNA is quite stable in a regular freezer (–20°C) indefinitely and, when analysed by the polymerase chain reaction (PCR), DNA can be effectively recovered from a variety of museum specimens (e.g., skins, scales, feathers, or bones) or naturally preserved specimens (e.g., those in amber, or dried or frozen material) (9).

The second difference is that contamination is not nearly as great a problem with proteins as it is with DNA. In the special case of DNA work with PCR, any extraneous DNA must be prevented from entering the sample because the contaminating DNA may be amplified. More generally, contamination of a DNA sample with any form of DNase will degrade the sample. The risk of protein contamination, on the other hand, is negligible because the contaminating protein must be nearly equal in concentration to the sample protein to affect correct interpretation.

Most of the proteins analysed electrophoretically are soluble enzymes which catalyse steps in the major metabolic pathways (e.g., the enzymes of the glycolytic pathway and the citric acid cycle). Since these enzymes are soluble in the cytoplasm rather than bound in membrane structures, a simple rupturing of the cell membrane or cell wall will release them into solution. It should be noted that allozyme-coding loci are not usually expressed equivalently in all tissues. This situation matters little when the organism is small and is ground whole. However, for larger organisms (e.g., fish), a collection of different tissues may need to be taken to ensure that most commonly examined enzymes are expressed in the sample taken. These individual tissues may be extracted and examined separately or combined to form a pooled extract for most diploid (or haploid) organisms.

2.2 Separation

The separation of the allozymes takes place in an electric field. A solid medium such as starch, cellulose acetate, or acrylamide is necessary to maintain the separation after the electric field is turned off. The medium can serve as a sieve, separating molecules by size and shape. However, the primary determinant of separation is differential charge on the surface of the molecule. Most enzymes move towards the positive terminal, although a few move towards the negative terminal under standard electrophoretic conditions. In gels with a single pH the proteins move constantly through the gel. In gels with a pH gradient the proteins move until they reach their isoelectric point, and then stop. This latter phenomenon is the basis of isoelectric focusing. Gels can be run in the horizontal or vertical plane, with vertical gels providing slightly sharper resolution. This chapter focuses on horizontal starch gel electrophoresis; other media are discussed further in section 4.

2.3 Staining

When a general protein detection system is used, only those proteins present in large quantities (e.g., transferrin or haemoglobin) are detected, and it is often difficult to distinguish the products of one locus from another. A more productive approach is to take advantage of the specific pathway catalysed by enzymes. In the presence of a specific substrate and necessary cofactors only a single enzyme will be able to carry out its reaction. One of the products of the reaction can be linked to a dye reaction or, if there is a change in fluorescence in the course of the reaction, visualized directly under UV light. Thus, replicate slices of a single gel can be stained for a number of different enzymes.

2.4 Interpretation

While allozyme electrophoresis refers to the electrophoretic separation and histochemical recognition of differences in banding patterns for a particular protein between individuals, the power of this methodology lies in the ability to assign a genetic basis to the observed phenotype (the banding pattern). Once the genetic basis of differences in banding pattern is understood, it becomes possible to make and confirm predictions about the inheritance of that genetic variation in the population.

2.4.1 Genetic basis of electrophoretic variation

The electrophoretic methodology necessary to detect genetic variation is quite simple (see Section 3 below). Deciphering the genetic basis of the observed protein banding patterns is the most difficult part of the entire process.

Some of the amino acids that comprise protein molecules are charged. As a result, the protein molecule itself usually has a net charge based on the arrangement of the amino acids in its three-dimensional (tertiary) structure. The charge on protein molecules causes them to migrate in an electric field towards the oppositely charged terminal. The migration rate of a molecule is directly related to its charge. The size and shape of a protein molecule also affect the rate of migration in a starch gel, although to a lesser extent. If the protein products of two alleles at a particular locus have different charges or significantly different shapes, their rates of migration will be different and they will appear as separate bands on the gel when it is stained. The matrix nature of the starch maintains the position of the protein molecule after electrophoresis has been terminated.

Most of the proteins examined during electrophoresis are enzymes that catalyse specific biochemical reactions. Some of these enzymes are composed of more than one copy of the polypeptide chain or gene product (quaternary structure). The enzyme is a monomer if one polypeptide chain is the active protein, a dimer if formed of two subunits each consisting of a single polypeptide chain, or a tetramer if four subunits are required to form the active protein. Usually each enzyme has a constant number of subunits no matter how many loci encode the enzyme or what life form is examined; for example, LDH is always a tetramer.

The result for the simplest case of a single locus with two alleles for a monomeric enzyme is illustrated in the upper left-hand corner of *Figure 1*. Two types of single-banded homozygous individuals may be observed along with a two-banded heterozygote (i.e., one that possesses two different alleles at a locus). The two bands (a and a′) represent two different protein products coded by the two alleles (A and A′, respectively).

The gene products of the loci that code for a specific multimeric enzyme usually combine at random to produce the active form of the enzyme. For example, an individual that is heterozygous (AA′) for a dimeric enzyme (two protein subunits combine to form the active enzyme) encoded by a single locus with two alleles (A and A′) produces a three-banded phenotype on the gel (see middle pattern on the left-hand side of *Figure 1*). The protein products (designated by lower case letters a and a′) combine to give rise to three

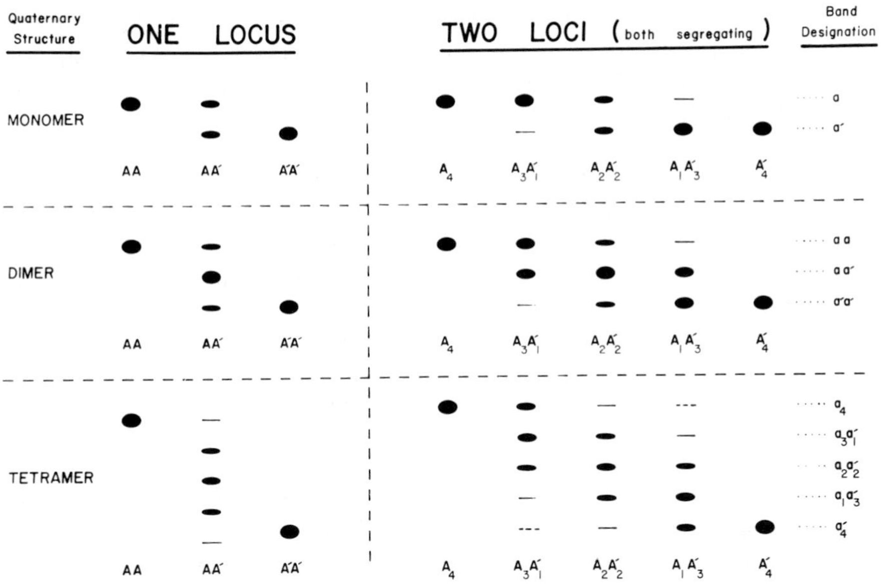

Figure 1. Possible banding patterns for one locus with two alleles (A and A′) and two loci which share the same two alleles for monomeric, dimeric, and tetrameric enzymes. The capital letters refer to alleles and the lower case letters to the protein products (subunits) coded by the alleles in a given genotype. Genotypes are listed below each banding pattern (phenotype) and the band designations (subunit compositions) are listed on the right. The expected ratios of band intensities for the single locus patterns are 1, 1:1, 1; 1, 1:2:1, 1; and 1, 1:4:6:4:1 for the monomeric, dimeric, and tetrameric cases, respectively. The expected ratios for the two loci patterns are 1, 3:1, 1:1, 1.3, 1; 1, 9:6:1, 1:2:1. 1:6:9, 1; and 1, 81:108:54:12:1, 1:4:6:4:1, 1:12:54:108:81, 1.

forms of the active enzyme (allozymes) represented by bands aa, aa′, and a′a′. These three bands appear in relative concentrations of 1:2:1 since random combination of subunits follows the binomial theorem. This heterozygous genotype can be contrasted with the single-banded phenotypes produced by the two possible homozygous genotypes AA and A′A′.

The aa′ band of the dimeric enzyme's heterozygous phenotype is often termed a 'hybrid band', but it is more aptly termed a 'heteromeric band' (i.e., one composed of different subunits). A heteromeric band cannot be formed in heterozygotes for monomeric enzymes since the active protein is only a single polypeptide. On the other hand, three heteromeric bands are found for a heterozygous phenotype derived from a tetrameric enzyme (aaaa′, aaa′a′, and aa′a′a′).

On the left-hand side of *Figure 1* are displayed the possible banding patterns for monomeric, dimeric, and tetrameric enzymes coded by one locus with two alleles which produce electrophoretically distinct protein products. It is only the heterozygous type which allows the distinction of quaternary structure;

this is an elegant yet simple method for determining such an important protein property. It should also be noted that the genotypic designation remains the same in all three cases and that it is only the number of bands, their relative intensities, and the individual band designations which change.

Assembly of subunits of multimeric enzymes encoded at a single locus tends to be random, while randomness of interlocus multimeric products depends on the amount of evolutionary time since gene duplication. The expected ratio of band intensities within an individual phenotype can be calculated with the multinomial expansion. A value is calculated for each band with the following formula:

$$[(a + b + c + ...)!/(a!b!c!...)][A^a B^b C^c...]$$

where capital letters refer to the number of copies of an allele (gene doses) in the genotype and lower case letters refer to the number of copies of the protein products of the particular alleles that give rise to each band. As an example, for the genotype AAA′A′ (listed as $A_2A'_2$) in the lower right-hand panel in *Figure 1*, the aaaa′ (or $a_3a'_1$) band would have an expected intensity of 64 where $A = 2$, B (number of A′ subunits) $= 2$, $a = 3$, and b (number of a′ subunits) $= 1$ in the multinomial formula. The use of this formula for each of the five bands in this balanced heterozygous phenotype would give, upon reduction, an expected ratio of 1:4:6:4:1 for the five bands in the electrophoretic phenotype. Alternatively, the two asymmetrical phenotypes on either side would have banding (staining) intensities of 81:108:54:12:1. The band of least intensity would barely be visible in either of these two phenotypes.

2.4.2 Electrophoretic complications

Some organisms are polyploid in origin and some specific genes have been duplicated in otherwise diploid organisms. Obviously, with more loci there are more gene copies coding for a protein; for two loci there are four gene copies, for three loci six gene copies, etc. This multigenicity adds complications to interpretations of the genetic bases of electrophoretic banding patterns; these complications may be represented in several ways.

On the right-hand side of *Figure 1* are displayed the banding patterns possible from a model of two loci both of which segregate for two electrophoretically indistinguishable alleles. The inherent features of this possibility are:

- each genotype is now designated by four letters (gene doses)
- it is not possible to assign the specific alleles to the respective loci
- there are three heterozygous types
- the genotypes have changed but not the band designations, as contrasted with the case for a single locus.

In *Figure 2* the first example illustrates a case where two loci have one identical allele (A) and one locus varies for an additional allele (A′). Only

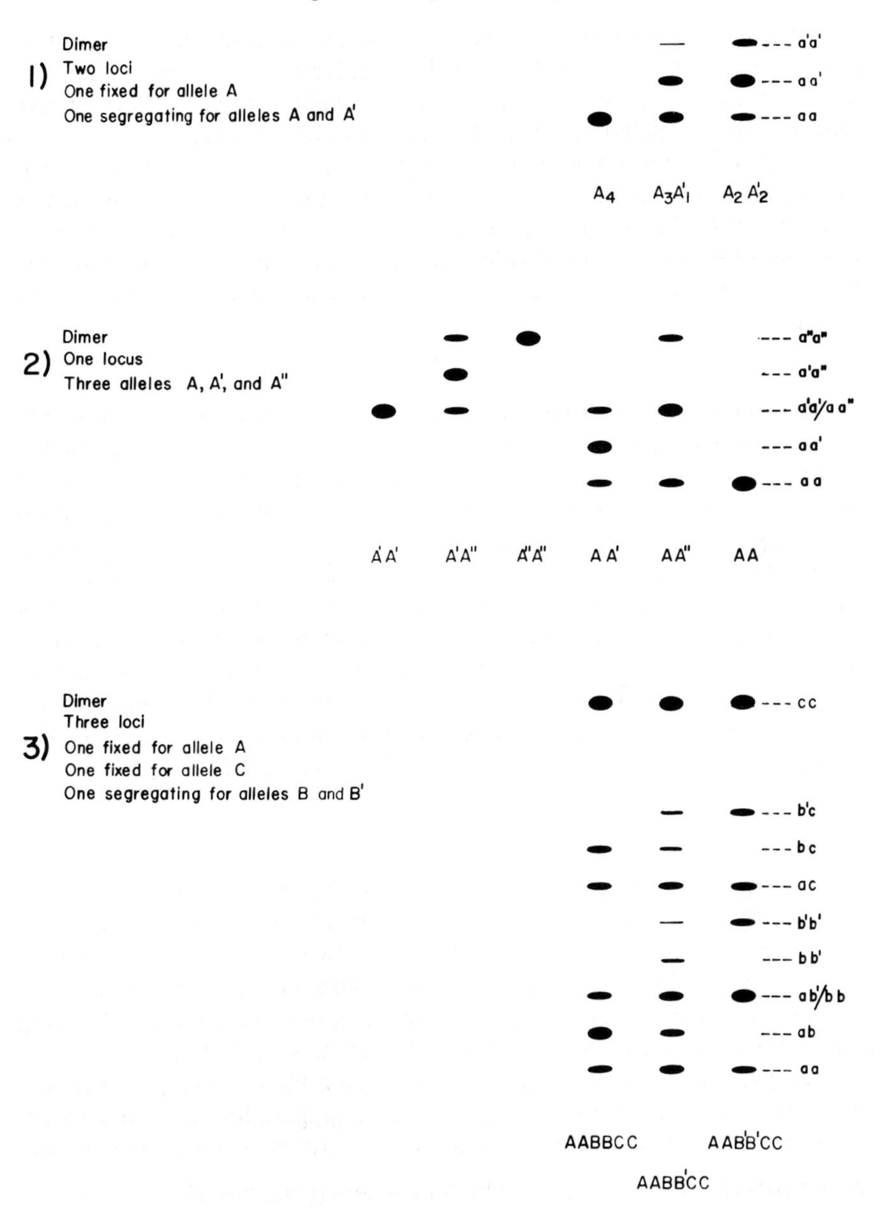

Figure 2. Examples of more complicated electrophoretic banding patterns. The overlap of bands a'a' and aa'' in the second example is entirely coincidental, although frequently seen due to the single change in charge of one amino acid being replaced by another. See text for more details about this figure.

these three phenotypes are ever seen for this enzyme system. If individuals exhibiting phenotypes of the $A_2A'_2$ type are crossed, all offspring will be also be $A_2A'_2$, indicating the homozygosity of this phenotype. This situation can be contrasted with a dimeric enzyme specified by two loci both segregating for the same two alleles (see middle panel on right-hand side of *Figure 1*). The second example in *Figure 2* illustrates the less than trivial point that a single locus can maintain more than two alleles in a breeding population. The third example typifies the situation where three loci code for a particular dimeric enzyme, the homomeric products of the three loci do not lie on top of one another (i.e., the zones of activity are sufficiently separated on the starch gel) and heteromers are formed between the products of the three loci. Represented in this example is a complex of bands (isozymes) with only three phenotypes (a single variable locus, the B locus). As illustrated, all bands can be identified as to their subunit composition (right hand side). Frequently in such situations the heterodimers formed between subunits a or b with c allow the variability in the a or b region (zone) to be correctly scored. Although the example states that only a single locus codes for c subunits, no information is available from these phenotypes to tell how many loci actually code for c.

Figure 3 shows a photograph of a gel and a diagrammatic representation of the phenotypes coding for supernatant MDH in brook trout. This is an example of two loci sharing common alleles for a dimeric enzyme.

One inherent weakness of the electrophoretic methodology is that it does not distinguish all genetic variability at a structural gene locus. This fact is due to:

- the multiplicity of codons for the same amino acid
- variation in the non-coding regions of the gene
- the fact that only approximately 30% of amino acid substitutions on the exterior of the molecule result in a charge shift.

Numerous studies have been done to overcome this conservative expression of genetic variability. Possible methods of altering the allozyme electrophoresis include heat denaturation, gel media (cellulose acetate instead of starch or one type of starch in place of another), buffer type, or buffer pH. Of course, the most direct and exhaustive method is to sequence the DNA itself (see Chapters 6 and 10).

Occasionally, the primary protein products of individual alleles or loci do not combine randomly, or do not combine at all in some instances. Utter *et al.* (10) showed that no heterodimers are formed between the protein products of alleles at *Ck-1* or between the products coded at *Ck-1* and those coded by *Ck-2* even though creatine kinase is known to be a dimeric enzyme in salmonids. Ferris and Whitt (11) suggested that this phenomenon may be due to some form of temporal or spatial isolation of RNA translation. This result implies that care needs to be taken when interpreting banding patterns and judicious

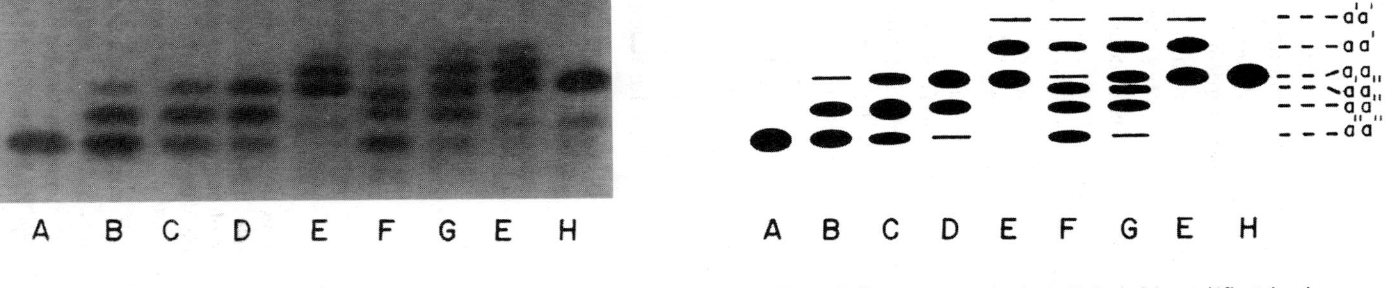

Figure 3. Banding patterns observed for *Mdh-(3,4)* among *Salvelinus fontinalis* and *S. namaycush*, their F_1 hybrid, and first backcross generation to either parental species. A is the common *S. fontinalis* allele, A′ a variant form, and A′′ is the common *S. namaycush* allele. Genotypes are as follows: A, A′′A′′A′′A′′; B, AA′′A′′A′′; C, AAA′′A′′; D, AAAA′′; E, AAAA′; F, AA′A′′A′′; G, AAA′A′′; H, AAAA. In the diagrammatic representation on the right the banding patterns have been expanded to show that the dimers aa and a′a′′ do not overlap in this case. The lighter bands which appear in the photograph at the bottom of phenotypes E and H are homodimers of the protein products of *Mdh-(1,2)* and heterodimers formed between the products of *Mdh-(1,2)* and *Mdh-(3,4)*.

use of prior knowledge about specific enzymatic subunit composition should be taken into account.

'Null alleles' describes those alleles where no active protein product is produced, whether the gene copy is not present, the gene copy is not transcribed, the message is not translated, or the protein product is inactive. The homomer may not be observed, or more likely, neither the homomer nor the heteromer containing the inactive protein product is observed. Null alleles are sometimes noted by the presence of an apparent excess of homozygous types.

A discussion of gel banding patterns (electrophoretic phenotypes) would not be complete without mention of so-called 'artefact bands' (*Figure 4*). These, by definition, are bands of dye deposition representing enzyme activity which do not fit simple Mendelian and biochemical models. They do indeed represent isozymes of the enzyme in question as documented by dye deposition, though their precise bases (whether conformational changes or addition or deletion of other moieties) in most instances are not known. The greatest difficulty in dismissing bands as artefactual is that they frequently occupy

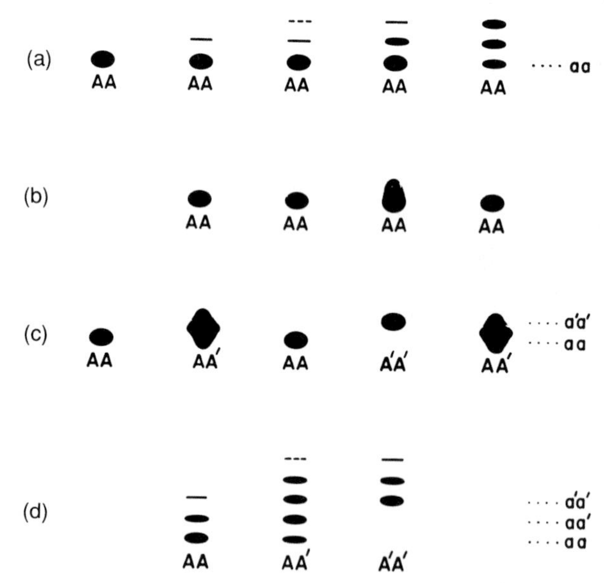

Figure 4. Example of artefact bands. (A) All phenotypes displayed are for one homozygous individual (AA). Additional bands of enzymatic activity are observed with increasing duration of frozen storage (left to right). (B) All phenotypes represent homozygous AA individuals. The aberrant pattern is due to a non-homogeneous gel. A re-run of this individual would show a sharp single band. Caution should be exercised in scoring low frequency variants. (C) Blurred phenotypes are probably heterozygous individuals (AA') because of their symmetry and the observance of AA and A'A' alternate homozygotes. (D) This represents a single locus with two alleles for a dimeric enzyme. The enzyme appears to have two states encoded by any allele. Thus a homozygote has three bands of activity; two of which might be regarded as artefactual

positions in line with the protein products of variant alleles. This is, of course, the greatest danger because they may be prematurely described as due to genetic variation. Two other categories of artefact bands include those from an included parasite or the expression of an alternative enzyme (e.g., LDH shows up on many NAD-dependent stains in vertebrates).

The types of observations which can be used to diagnose artefact bands include consistent variable expression among individuals on a single gel, variable expression in the same individual dependent on the length of storage time, and, most importantly, failure to fit classical electrophoretic phenotype expectations. Some typical kinds of artefact bands are displayed in *Figure 4*.

Several other overviews of the genetic basis of allozyme variation can be found elsewhere (6,7,12,13).

2.5 Application

The applications of the data follow directly from the nature of the data. If one interprets the genetic basis of the phenotypic banding pattern variation and assigns presumptive genotypes, one is left with a data set of single locus Mendelian genotypes, with co-dominant alleles. While quantifying the number and position of bands in a banding pattern provides sufficient information in some circumstances, it is the conclusions that can be drawn from knowledge of the functional Mendelian basis of the banding pattern that have made allozyme electrophoresis the powerful tool it is today.

Some of the types of application include, but are not limited to: systematics, population structure, inbreeding, mating systems, species boundaries, hybridization, origins of polyploids, ecological genetics, and genomic organization (see ref. 7 for a review of applications and Section 3.5 below)

3. Starch gel allozyme electrophoresis

In this section I focus on the details of starch gel electrophoresis as practised in my laboratory. These techniques have been optimized for cost-effectiveness, ease of implementation, and resolution. However, there are numerous variations on the theme and several viable alternatives can be found in other reviews (4,6,7,14–16). The beginner would be cautioned to choose one gel-running methodology and not to try to mix procedures from different alternatives, as each approach has many integrated features.

3.1 Extraction and sample handling

The majority of proteins examined are soluble enzymes, part of glycolysis or the citric acid cycle. Extraction of the proteins from the cells results from breaking the cell wall or membrane and releasing these proteins into an appropriate buffer solution. While a variety of hand-held tissue homogenizers and motorized homogenizers is available, extraction of most samples (vertebrate tissues, red blood cells, entire insects, fungi, and some plants like squash) can be done simply with a glass rod in a 12 mm × 75 mm disposable glass test tube in a mild Tris–HCl buffer (0.05 M, pH 7.0) or distilled water. About 20 strokes of the rod in the tube on the bench top break enough of the cells. The tube can then be spun in a low-speed centrifuge (~1000 × *g*) for 2–5 min. Always keep samples in iced water, or on an ice pack until they are loaded on to the gel. We often extract a day or two in advance, re-freeze, and then spin the samples on the day we run the gels to thaw them again and to remove cellular debris. Freezing extracts tends to increase general enzymatic activity, including that of some normally very weak zones of activity. This result can be attributed to the rupture of more cells by freezing and to the fractionation of mitochondria and the release of their specific isozymes (mitochondrial MDH (19) and mitochondrial malic enzyme (20)). Some plants require reducing agents to prevent the oxidation or inhibition of many of the

Table 1. Buffer types used[a]

Buffer[a]	Type	Recipe
A	Gel buffer	20.90 g Tris (T1503) (0.009 M), 11.53 g citrate (C-0759) (0.003 M), 7.01 g EDTA (ED2SS) (0.0012 M), 1.08 g NaOH (S-5881); pH 7.1
	Tray buffer	327.10 g Tris (0.135 M), 172.90 g citrate (0.045 M), 7.00 g EDTA (0.0012 M); pH 7.0
C	Gel buffer	Tray buffer diluted 1:9 with water
	Tray buffer	153.73 g citrate (0.04 M), 200 ml *N*-(3-aminopropyl)morpholine (A-9028); pH 6.1
H	Gel buffer	Tray buffer diluted 1:3 with water
	Tray buffer	201.76 g L-Histidine-free base (H-8000) to pH 6.5 with citrate
M	Gel buffer	Tray buffer diluted 1:3 with water; pH 8.7
	Tray buffer	436.10 g Tris (0.18 M); 123.66 g boric acid (B-0252) (0.1 M); 23.40 g EDTA (0.004 M); pH 8.7
R	Gel buffer	72.66 g Tris (0.03 M), 19.21 g citrate (0.005 M), 200 ml R tray buffer; pH 8.5
	Tray buffer	50.35 g lithium hydroxide (L-4256) (0.06 M), 370.98 g boric acid (0.3 M); pH 8.1
4	Gel buffer	19.4 g Tris (0.008 M), 12.6 g citrate (0.005 M), 2.2 g NaOH (0.003 M); pH 6.7
	Tray buffer	540 g Tris (0.223 M), 361.4 g citrate (0.094 M), 40.01 g NaOH (0.05 M); pH 6.3
9	Gel buffer	Tray buffer diluted 1:19 with water
	Tray buffer	242.00 g Tris (0.1 M), 232.00 g maleic acid (M-0375) (0.1 M), 67.10 EDTA (0.01 M), 40.60 g $MgCl_2$ (M-0250) (0.01 M), 126.75 g NaOH (0.12 M); pH 8.0

[a]All ingredients are mixed in 20 litres of distilled water. Sigma catalogue numbers are recorded for the first occurrence of each compound.
[b]As modified from the following references: buffer A, ref. 21; buffer C, ref. 22; buffer H, ref. 23; buffer M, ref. 24; buffer R, ref. 25; buffers 9 and 4, ref. 26.

enzymes. A number of different plant extraction buffers can be found in the literature (e.g., see review in ref. 17), and a few of these should be tried to identify the appropriate buffer for the plant being examined. The durable cells walls of some plant leaves may require the use of mortar and pestle. For a review of collection and storage of tissues see Dessauer *et al.* (18).

3.2 Preparing and running gels

Pouring gels is an art which is easily mastered (see *Protocol 1*). The gels may be made from a variety of buffers (see *Table 1*) which are tested empirically for effectiveness (enzyme activity and resolution) with each specific organism, tissue type, and enzyme. It should be noted that a few enzymes migrate towards the cathode while most migrate towards the anode. Also, some loci are specific as to tissue expression. Initial tests require keeping these two points in mind. Location of the cut for the wicks relative to the anode and cathode, thickness of the gel, temperature of the gel, type of starch, concentration of starch, and voltage also affect the resolving power of a particular tissue type/buffer/enzyme combination. Specific enzymes with which we frequently work are listed in *Table 2*. General results from our work on over 100 taxa are listed in *Tables 3* and *4*.

Table 2. Names, abbreviations, EC numbers, and subunit composition of enzymes used. Enzyme numbers are as recommended by IUBMB (27).

Abbreviation	Enzyme name	EC number	No. of subunits
AAT	Aspartate aminotransferase	2.6.1.1	2
AC	Aconitase	4.2.1.3	1
ACP	Acid phosphatase	3.1.3.2	2
ADA	Adenosine deaminase	3.5.4.4	1
ADH	Alcohol dehydrogenase	1.1.1.1	2
AK	Adenylate kinase	2.7.4.3	1
AKP	Alkaline phosphatase	3.1.3.1	2
ALD	Aldolase	4.1.2.13	4
CAT	Catalase	1.11.1.6	4
CK	Creatine kinase	2.7.3.2	–
DIA	Diaphorase	1.8.1.4	1
EST	Esterase	–	1,2
FBP	Fructose biphosphatase	3.1.3.11	–
FUM	Fumarase	4.2.1.2	4
α-GAL	α-Galactosidase	3.2.1.22	–
β-GAL	β-Galactosidase	3.2.1.23	–
GAM	Galactosaminidase	–	2
GAPDH	Glyceraldehyde-3-phosphate dehydrogenase	1.2.1.12	4
GDA	Guanine deaminase	3.5.4.3	2
GDH	Glutamic dehydrogenase	1.4.1.2–4	2
G2DH	Glycerate dehydrogenase	1.1.1.29	–
GK	Glucokinase	2.7.1.2	1
G3P	Glycerol-3-phosphate dehydrogenase	1.1.1.8	–
G6PDH	Glucose-6-phosphate dehydrogenase	1.1.1.49	2

Table 2. Continued

Abbreviation	Enzyme name	EC number	No. of subunits
GPI	Glucose-phosphate isomerase	5.3.1.9	2
α-GLU	α-Glucosidase	3.2.1.20	–
β-GLU	β-Glucosidase	3.2.1.21	–
GPT	Glutamic pyruvic transaminase	2.6.1.2	2
GR	Glutathione reductase	1.6.4.2	2
GUS	Glucoronidase	–	–
HA	Hexoseaminase	3.2.1.52	–
HBDH	Hydroxybutyric dehydrogenase	1.1.1.30	2
IDH	Isocitrate dehydrogenase	1.1.1.42	2
LAP	Leucine aminopeptidase	3.4.11.1	1
LDH	Lactate dehydrogenase	1.1.1.27	4
MADH	Mannitol dehydrogenase	1.1.1.67	–
α-MAN	α-Mannosidase	3.2.1.24	–
MDH	Malate dehydrogenase	1.1.1.37	2
ME	Malic enzyme	1.1.1.40	4
MPI	Mannose-phophate isomerase	5.3.1.8	1
MUP	Methylumbelliferyl phosphatase	–	–
NP	Nucleoside phosphorylase	2.4.2.1	3
ODH	Octanol dehydrogenase	1.1.1.73	2
PEP	Peptidase resolved with glycyl-leucine, leucyl-alanine, leucyl-glycyl-glycine, leucyl-leucyl-leucine, or phenyl-alanyl-proline	3.4.11–13	1,2
PER	Peroxidase	1.11.1.7	–
PFK	Phosphofructokinase	2.7.1.11	–
PGD	Phosphogluconate dehydrogenase	1.1.1.43	2
PGK	Phosphoglycerate kinase	2.7.2.3	1
PGM	Phosphoglucomutase	5.4.2.2	1
PK	Pyruvic kinase	2.7.1.40	4
PP	Inorganic pyrophosphatase	–	–
PRO	General protein	–	–
SDH	Sorbitol dehydrogenase	1.1.1.14	4
SK	Shikimic kinase	2.7.1.71	1
SOD	Superoxide dismutase	1.15.1.1	2
TPI	Triose-phosphate isomerase	5.3.1.1	2
XDH	Xanthine dehydrogenase	1.1.1.204	2

Table 3. Buffers found to be the most effective in resolving enzymes within and across taxa[a,b]

Enzyme[c]	Fish	Birds	Mammals	Fungi	Invertebrates	Plants	Generic
ACP	4	M	C,4	–	R	R,C,4	R
AC	R,4	–	C	C,4	4	–	4
ADA	M,C	R,C,4	C	4	R	–	R
AK	C,4	4	C	–	C	R,4	C
ADH	M,R,C	4	4	–	R	R,4	4
ALD	C	C	C	–	C	C	C
AKP	–	–	R,C	–	–	C,4	C
AAT	R	C	C	C,4	R	R	R
CAT	C	C	C	–	R,C,4	R	C
CK	R	C	R	–	–	–	R

Table 3. Continued

Enzyme[c]	Fish	Birds	Mammals	Fungi	Invertebrates	Plants	Generic
DIA	R	R	C	R	R	C	R
EST	R	R	M,R,C,4	C,4	R	R	R
FBP	M	4	C	C,4	R	R	R
FUM	C	4	M	–	4	–	4
α-GAL	–	–	–	–	–	9	9
β-GAL	R	–	–	–	C	R	R
GAM	R	M,R	4	C	R	C,4,9	R
GK	C,4	–	M,R,C	R,C,9	R	R	R
G6PDH	M,R,C	4	C	C	R	R,C,4	C
GPI	R	4	R,C	C	R	R	R
α-GLU	9	R	–	–	R,C,4	R	R
β-GLU	–	–	4	–	M	R,C,4,H	M,C
GDH	C	M	C,4	R,C	R	R	R
GPT	R,C	4	R,C,4	4	C	9	C
GR	C,4	R	C	4	R	R	4
GUS	R,9	–	–	–	R	–	R,9
GAPDH	C	C	C	–	C	CH	C
G2DH	C	–	–	–	R	R	R
G3P	C,A	R	–	–	4	–	C
GDA	M,R,C	4	4	R	M,4	–	4
HA	R	–	C	–	R	R	R
HBDH	–	–	C	–	R,4	–	R,C,4
IDH	C	C,4	C,4	C	4	4	4
LDH	R	4	4	M,C	R	M,R,C	R
LAP	4	R,C	4	R	R	R	R
MDH	C,4	4	C,4	M,C	C	C	C
ME	C	C	C	C	R,C	C	C
MADH	–	–	–	R	–	–	R
MPI	R,C	R	R	M,R	R	4,H	R
α-MAN	M	M	R	–	R	R	R
MUP	C	M,C,4	4	–	C,R	4	C
NP	C	4	4	C	4	M,4	4
ODH	–	–	–	–	R,C,4	–	RC
PEP	R,C	4	R	R	R	R	R
PER	–	–	–	–	–	C,4,9	C,4,9
PFK	9	–	–	–	–	–	4
PGM	4,A	R,C,4	R,C,4	R,4,9	R	R	R
PGD	C	C	M,C,4	C	C	4	C
PGK	C	C	M,C,4	4	C	4	C
PRO	M,R,4	R,4	R	–	R,C	R	R
PP	9	–	–	–	9	9	9
PK	M,C	C,4	M,C,4	–	M,R,C,4	–	C
SK	–	–	–	–	–	M,4	M,4
SDH	R	M,4	4	–	R	R	R
SOD	R	C	M,R	–	R	R	R
TPI	C	C	C,4	4	C	R	C
XDH	C	M,C	M,R,C	R	R	–	R

[a]The data are from complete screens of eight species of fish, six of birds, seven of mammals, four of fungi, 14 of invertebrates, and seven of plants.
[b]Buffers as in *Table 1*.
[c]Abbreviations as in *Table 2*.

Table 4. Variability found for each enzyme system among different taxa[a]

Enzyme[b]	Fish		Birds		Mammals		Fungi		Invertebrates		Plants	
	m	*p*	*m*	*p*	*m*	*p*	*m*	*p*	*m*	*p*	*m*	*p*
ACP	0	0	1	1	2	0	1	2	2	2	0	8
AC	1	0	0	0	1	0	0	0	0	0	0	0
ADA	1	3	9	7	9	6	0	3	5	3	0	0
AK	3	3	10	1	12	1	2	1	4	9	0	0
ADH	2	2	4	1	1	3	1	1	1	6	2	2
ALD	4	1	4	0	7	0	1	1	3	2	2	1
AKP	0	0	1	0	2	0	1	1	1	1	0	2
AAT	2	5	11	4	10	4	1	6	5	9	5	4
CAT	1	0	0	0	7	1	0	1	0	1	0	0
CK	3	5	12	4	6	1	0	0	0	0	0	0
DIA	3	2	3	0	10	4	0	9	2	3	2	7
EST	4	1	10	8	10	9	0	8	3	6	5	4
FBP	4	1	0	0	2	0	0	0	2	0	2	1
FUM	1	3	1	0	1	0	0	2	5	0	0	1
α-GAL	0	0	0	0	0	0	0	0	0	0	0	0
β-GAL	1	0	0	0	0	0	1	0	0	0	1	1
GAM	2	1	1	0	6	2	1	0	4	7	0	1
GK	1	0	0	0	1	0	2	3	0	5	1	0
G6PDH	1	0	0	1	8	0	0	4	1	1	0	0
GPI	0	8	11	9	14	8	0	14	2	19	2	7
α-GLU	0	2	1	0	0	0	1	0	0	1	0	0
β-GLU	0	0	0	0	0	1	0	4	2	2	0	2
GDH	0	1	2	0	0	0	2	6	1	0	2	2
GPT	5	0	7	0	4	0	0	6	0	2	2	0
GR	1	0	8	7	7	4	1	6	4	6	3	3
GUS	1	0	0	0	0	0	0	0	0	0	0	0
GAPDH	3	3	9	0	13	0	1	3	6	3	4	0
G2DH	2	2	1	0	1	0	0	0	0	0	0	0
G3P	1	3	4	4	2	1	0	0	1	3	0	0
GDA	1	0	8	1	3	0	0	1	0	2	0	0
HA	5	0	0	0	0	0	2	1	1	1	3	0
HBDH	0	0	0	0	0	0	0	0	2	6	0	0
IDH	3	5	9	9	11	6	0	4	7	13	3	3
LDH	0	8	16	3	19	3	2	3	4	6	1	0
LAP	0	0	2	0	11	0	3	1	3	3	4	2
MDH	0	6	17	4	18	1	1	9	6	18	3	4
ME	1	6	11	0	8	3	2	1	1	11	3	0
MADH	0	0	0	0	0	0	0	1	0	0	0	0
MPI	2	4	7	13	12	7	2	8	2	10	1	1
α-MAN	3	0	9	2	11	0	2	0	1	0	3	0
MUP	4	2	15	1	13	2	1	2	2	3	3	3
NP	2	1	2	13	6	7	0	4	1	0	2	0
ODH	0	1	0	0	0	0	0	0	1	0	0	0
PEP	2	6	10	11	11	11	2	12	3	16	6	2
PER	0	0	0	0	0	0	0	0	0	0	1	2
PFK	1	0	0	0	0	0	0	0	0	0	0	0
PGM	2	5	9	8	8	9	0	4	4	12	4	4

Table 4. Continued

Enzyme[b]	Fish		Birds		Mammals		Fungi		Invertebrates		Plants	
	m	*p*	*m*	*p*	*m*	*p*	*m*	*p*	*m*	*p*	*m*	*p*
PGD	4	2	10	11	12	7	3	7	2	13	3	4
PGK	2	4	10	0	7	0	1	4	2	1	0	1
PRO	4	0	10	6	8	11	0	0	4	0	1	1
PP	0	0	0	0	2	0	1	1	0	0	0	0
PK	3	0	3	0	1	0	0	0	0	1	0	0
SK	0	0	0	0	0	0	0	0	0	0	2	5
SDH	3	2	2	3	3	1	0	1	0	0	0	0
SOD	3	3	14	5	17	4	0	6	7	9	8	0
TPI	3	3	16	0	7	1	2	4	2	10	3	2
XDH	0	0	0	0	4	0	1	2	2	1	0	0

[a]*p* = number of species polymorphic for at least one locus in the enzyme system; *m* = number of species monomorphic at all loci in that system.
[b]Abbreviations in *Table 2*.

Protocol 1. Pouring gels

Equipment and reagents

- Starch, e.g. Connaught starch (Fisher Scientific.), or starch from Sigma or Starch Art. We use a 1:1 (w/w) mixture of Sigma and Connaught starches, sifted with an aquarium fish net (2 mm^2 nylon mesh) to remove lumps, and a 14% (w/v) starch/buffer formula. These starches are mixed to provide a starch with good resolving properties as well as high tensile strength for ease of handling after cutting
- Buffer (see *Table 1*)
- 1000 ml Erlenmeyer flask
- 500 ml volumetric flask
- Gas burner
- Glass plates (165 mm × 254 mm × 63 mm)
- Glass plate supports (50 mm high)
- Plexiglass strips (216 mm × 19 mm × 6.35 mm and 165 mm × 19 mm × 6.35 mm)

Method

1. Weigh dry starch into a 1000 ml Erlenmeyer flask:
   - to make one thin (60 mm) gel use 42 g starch and 300 ml buffer.
   - to make one thick (1 cm) gel use 56 g starch and 400 ml buffer.
   - to make two thin (60 mm) gels use 70 g starch and 500 ml buffer.
2. Measure buffer into a 500 ml volumetric flask.
3. Pour approximately one-third of the buffer into the dry starch and swirl to suspend.
4. Heat remaining buffer to a boil.
5. Add hot buffer to starch suspension while constantly swirling starch flask.(**Note**: this is the most important step in preparing gels. If starch is not fully suspended in cold buffer, lumps in gel will result.)

Protocol 1. *Continued*

6. Heat mixture to a light boil by placing on burner and removing to swirl every 5–10 sec.
7. When bubbles roll up the side of the flask, remove and de-gas by water aspiration (or vacuum pump) for approximately 30 sec. Do not overheat or de-gas for too long. The former results in a very runny gel that is difficult to de-gas and may be burned, while the latter produces a gel that is too stiff to pour evenly.
8. Pour mixture into gel form (see *Figure 5*) as quickly as possible to produce a uniform gel.
9. Remove lumps and bubbles larger than 1 mm within the first 15 sec.
10. Let gel room-cool for 1 h or cool in refrigerator or freezer (< 10 min) until cool to the touch of the back of your fingers and then cover with plastic wrap. Gels poured the previous day and left on the bench or in a refrigerator are too brittle to be used with our protocols.

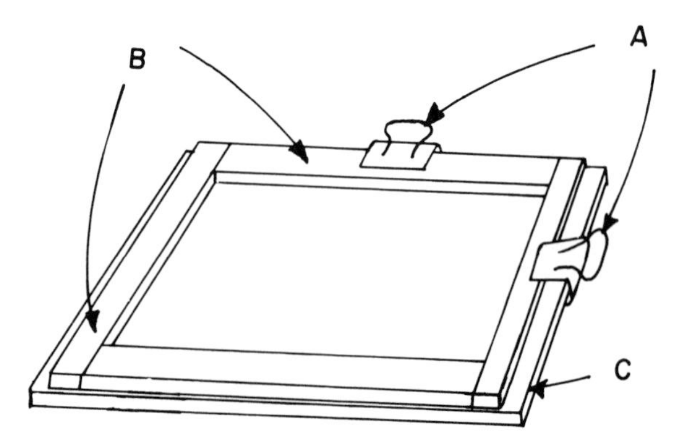

Figure 5. Form for starch gel. Plexiglass strips (B), two measuring 165 mm × 19 mm × 6.35 mm and two measuring 216 mm × 19 mm × 6.35 mm; are placed on a glass plate (C, 165 mm × 254 mm × 63 mm) with polished edges, and are held in place with clamps (A).

Once the gels have cooled the samples can be plated (see *Protocol 2*) and then the gels run (see *Protocol 3*).

Protocol 2. Plating gels

Equipment and reagents

- Gel prepared as described in *Protocol 1*
- Samples (40 per gel if using 4 mm wicks or 50 if using 3 mm wicks)
- Scalpel
- Forceps
- Marker dye (red food colouring)
- Wicks cut from filter paper (Schleicher & Schuell no. 470) or ordered pre-cut from Northfork Products (3 or 4 mm wide × 1.3 cm long)
- Plastic wrap (e.g. Saran Wrap)

Method

1. Remove the two long side pieces of the gel form.
2. Cut gel lengthwise 3.5 cm from one side, holding scalpel perpendicular to glass plate.
3. Separate the two pieces of gel by placing all fingers on the small piece and pulling via finger friction towards your body.
4. Using forceps, dip a wick into the supernatant of the first extracted sample and withdraw when fluid is half to two-thirds of the way up the wick.
5. Blot excess liquid on paper towel before plating (usually unnecessary except for blood samples).
6. Place the wick by touching it to the bottom of the glass plate and standing the wick vertically up against the edge of the large piece of gel (see *Figure 6*).
7. Repeat steps **4–6** for the remaining samples, keeping wicks straight and as close together as possible without touching. Allow a wick's width of space between each group of ten wicks, to permit easier reading later.
8. Plate one wick with marker dye (red food colouring) after all the samples have been plated.
9. Close gel by pushing the small slice of gel up against the wicks.
10. Fold plastic wrap down over gel.

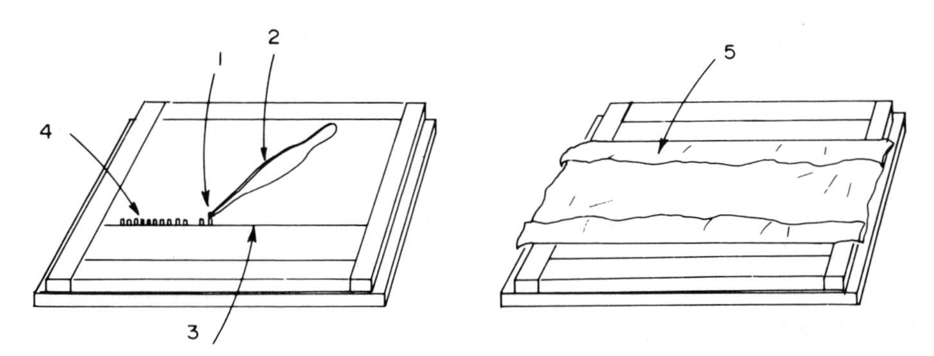

Figure 6. Placement of samples on starch gel. Filter paper wicks (1) are saturated with homogenates and inserted between the two pieces of the gel using forceps (2) after the gel has been cut lengthways (3) with a scalpel and the two pieces of starch separated by about 0.5 cm The wicks are placed 1 mm apart (4), with a wider gap of 4 mm after every ten samples. A marker dye is plated at the end of the gel. A layer of plastic wrap (5) is then placed over the gel and folded back to expose 1 cm of gel at each end. The longer strips are usually removed prior to insertion of the wicks.

Protocol 3. Running gels

Equipment and reagents

- Power pack (350 V, 150 mA)
- Buffer trays (e.g., lids of plastic butter dishes)
- Electrode (2 cm piece of 0.5–0.8 mm diameter platinum wire soldered to an alligator clip)
- Cloth towels
- Ice packs, stored in freezer at −15°C[a]
- Glass plates (165 mm × 254 mm) to insulate gels from ice packs
- Plexiglass strips (300 mm × 19 mm × 1.6 mm)
- Nylon thread (#2–4 test)
- Staining trays (preferably white or clear; e.g., Styrofoam, enamelled, glass, clear plastic, or white painted glass)

Method

1. Fold a cloth towel to give eight layers and place it in a buffer tray. Alternatively, thin cellulose sponges can be used.
2. Fold back plastic wrap about 0.5 cm from gel edge (see *Figure 7*).
3. Place paper towels over exposed gel on top and pinch in along edge of gel.
4. Fold plastic wrap back over paper towel.
5. Repeat steps **1–4** for the other side of the gel.
6. Insert electrode into buffer tray. Add buffer to within 0.5 cm of clip and turn on power. **Note**: caution should be exercised at all times to avoid electric shocks.
7. Run gel as follows:
   (a) Pre-run gel for 10 min (15 min for thick gels).
   (b) Turn off power to gel, lift Saran Wrap exposing cut in gel, separate blocks of starch, remove wicks, reassemble gel and add ice packs with glass plates underneath them straight from freezer.
   (c) After the gel has been running for 1 h; remove glass plates from ice packs.
   (d) After 1 h more, take new ice packs out of freezer and allow to warm up on bench.
   (e) After another 30 min, add new ice packs.
   (f) After another 1 h, cut the gels or remove glass plates if run is continuing.
8. Disconnect power when marker dye has migrated approximately 5–6 cm.
9. Remove paper towels, electrodes, buffer trays, and small side pieces from gel form.
10. Trim excess gel from the right side of the marker dye and above (anodal to) the marker dye.
11. Cut notch on both anodal and cathodal slices to identify end with first sample.

12. Slice gel (see *Figure 8*) by sequentially placing 1.6 mm Plexiglass strips along the side of the gel and drawing monofilament sewing thread or 2# fishing line along strips and through gel. Hold the thread by winding it around index finger and holding taut (near breaking strength). This procedure is superior to commercial or hand-made slicers employing steel wire because mistakes in mid-slice or uneven thickness gels can be accommodated easily.

13. Pick up cut slices using friction with fingers and place in staining trays or on glass plates (agar overlay or fluorescent stains).

[a] Alternatively, ice baths, circulating coolant below the gel or running the gel in a refrigeration unit can provide the necessary cooling.

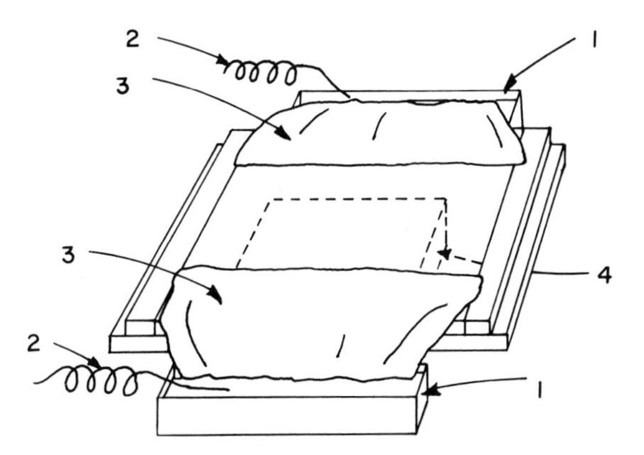

Figure 7. Preparation of gel for electrophoresis. Plastic butter dish covers (1) or similar containers are filled two-thirds full with the electrode (tray) buffer and platinum electrodes (2) are placed in the buffer. An absorbent wick (3), e.g., a cloth towel, is placed in the buffer solution and firmly pressed on to the surface of the gel, covering the exposed surface. The gel frame is elevated about 4 cm by any convenient material (4), placed beneath the glass plate.

3.3 Staining gels

Staining recipes we have found to be effective in a diversity of organisms (see *Tables 3* and *4*) are listed in Appendix 1 (see also refs 4,6,7,14,16,28,29). A most comprehensive guide to enzymatic staining protocols can be found in Manchenko (30). General notes for staining are presented below.

(a) The components in some of these recipes can be reduced substantially after the stain has been found to work in a particular organism.

(b) All recipes are made up in 100 ml of R gel buffer (see *Table 1*) unless otherwise noted. Careful trimming of the gels and use of smaller staining trays permits smaller volumes of buffer (and smaller quantities of stain components) to be used.

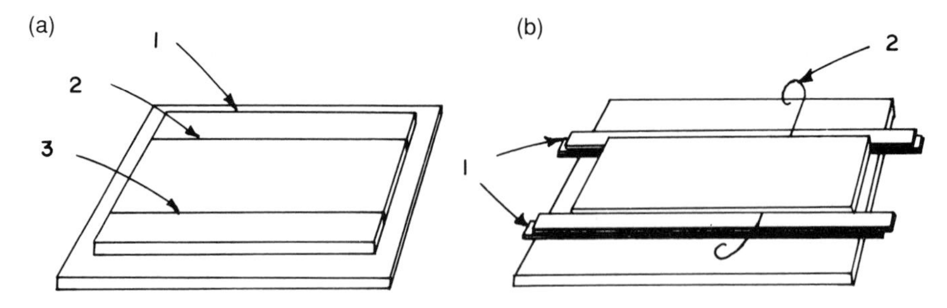

Figure 8. Procedure for slicing a starch gel. (A) When electrophoresis has finished, the Plexiglass strips are removed and those portions of the gel anodal to the buffer boundary or marker dye (1) are discarded; 2, buffer boundary on RSL buffer; 3, origin. (B) Sets of Plexiglass strips of 1.6 mm thickness (1) are placed on each side of the gel and the gel is cut by pushing or pulling a tightly held nylon sewing thread (2) along the surface of the 1.6 mm strips.

(c) For agar overlays, bring to the boil a solution of agar in distilled water (2 g/100 ml). Approximately 10 ml are needed for each stain. Maintain in a water bath at 60°C. Immediately prior to staining, add dyes to stain, mix with an equivalent volume of agar, and pour over gel.

(d) Stains can be mixed several hours before use. MTT and PMS should be added just prior to pouring the stain over the gel.

(e) Powdered enzymes can be put into a few millilitres of water and refrigerated so that it is easier to dispense them volumetrically.

(f) For all NADP stains, add 1 ml of a $MgCl_2$ solution in water (5 g/500 ml water).

(g) A critical enzyme or cofactor should be left out of fluorescent stains until immediately prior to staining.

(h) Tetrazolium dyes can be destained in a 1:4:5 dilution of acetic acid:water:ethanol. Leave on gels for 10 min and rinse in water. MTT gels destain rapidly (< 10 min.). Other staining salts, i.e. Fast Blue BB, Fast Garnet GBC, can be destained with 1:1 glycerine:water (left indefinitely).

(i) NAD can be put in a solution of 2 g NAD/100 ml water. Use 1 ml per stain. NADP can be put in a solution using 1 g NADP/100 ml water. Use 1 ml per stain.

(j) Mix MTT (1 g/100 ml) and PMS (0.3 g/100 ml). Use 1 ml of each.

(k) We use an 8 mm wide standard laboratory spatula (Fisher Scientific, no. 21-401-5) to speed the process of measuring chemicals for stains. The boldface numbers on the left of each recipe in Appendix 1 are increments measured from the end of the spatula as follows:

numbered increment	**1**	**2**	**3**	**4**	**5**
distance from tip (mm)	3	6	10	20	30

(l) After staining the gels can be photographed with Polaroid, 35 mm slide, or video camera. Using tungsten-light-sensitive colour slide film, we photograph any gels that show polymorphisms (colour slides permit much greater resolution for subsequent re-examination).

Note: read all hazardous warnings on bottles or with packing materials that come with chemicals, e.g., peptidase stains contain an amino acid oxidase from snake venom.

3.4 Reading gels

Interpreting banding patterns on gels is a skill which matures with experience. Below are cited some rules and guidelines which can help the researcher.

(a) Examine the gel for phenotypic classes. All individuals should fit clearly into a finite number of discrete classes.

(b) Count the maximum number of bands seen in any individual.

(c) Decide what the quaternary structure is for the enzyme. In a presumptive heterozygous phenotype

  (i) Two bands of equal intensity represent a monomer (e.g., phosphoglucomutase).

  (ii) Three bands in a 1:2:1 ratio represent a dimer (e.g., MDH).

  (iii) Four bands in a 1:3:3:1 ratio represent a trimer (e.g., nucleoside phosphorylase).

  (iv) Five bands in a 1:4:6:4:1 ratio represent a tetramer (e.g., LDH).

(d) Use the smallest possible number of loci and alleles per locus that explain the banding pattern seen.

(e) To determine whether the bands seen are artefacts or real enzyme products, check to see if the staining intensity ratios within the banding patterns are correct for the subunit composition of the enzyme and that all banding patterns fit readily into a discrete number of unambiguous phenotypic classes. Use parents and offspring where possible to confirm Mendelian inheritance.

(f) Determine how many alleles are represented on the gel. Note that heteromeric bands are not representative of separate alleles. They result from the different subunits coming together to form the bands.

(g) Be consistent in naming alleles. Run standard samples between gels to ensure consistency.

(h) Alleles can be pooled into a single class (because of difficulty in scoring) as long as a given allele is always pooled into the same class.

(i) Loci coding for discrete zones of activity can be numbered sequentially from the most cathodal to the most anodal bands. One allele, usually the most common, is designated 100 and all the other alleles are designated by the rate of travel of their homomeric protein products relative to that of allele 100. For gel scoring and subsequent data analysis, the most common allele (100) is designated 1, and successive alleles are given the next highest number, regardless of mobility. We recommend the enzyme, locus, allele nomenclatural standards presented by Shaklee *et al.* (31).

3.5 Data handling and applications

The resultant genotypic data can be analysed like any other kind of co-dominant Mendelian data (Appendix 3). Computer programs are available in many laboratories for treatment of data (see ref. 32 and Appendix 5 of this volume). We use one called 'Genes in Populations' designed by B. May and C. C. Krueger and written in C by W. Eng and E. Paul (`http://animalscience.ucdavis.edu/extension/Gene.htm`) that calculates allele frequencies, F-statistics, Nei–Rogers genetic distances, and clustering algorithms.

As stated earlier, the basic Mendelian nature of the raw data (predictability of transmission from parent to offspring and from one generation to the next) suggests the types of applications which are possible. The types of application made in our laboratory are listed in *Table 5*.

Table 5. Applications of allozyme elcetrophoresis data

Type of application	Explanation
Population analysis	These analyses examine the variability at 20–100 loci and compare the variation within and among populations. Similarity algorithms determine which populations have the most genetic similarity and which the least. These populations can then be associated in a dendrogram illustrating relationships.
Systematics	Similar to population analysis, these analyses determine the systematic relationship of a group of taxa (usually species within a genus or family) based on percent shared alleles at single loci.
Parental analysis	Allozymes permit the identification of putative parents in a number of situations (e.g., genomic manipulation, chimeric tissue typing, multiple mating by sires).
Mixed stock analysis	Allozyme frequency data can be used to partition a mixed population into its components or to assign contributions of parental populations to a second generation mixed population.
Linkage	These studies assign the relationship of allozyme coding loci close to one another along chromosomes.
Isolate identification	A combined genotype over a number of loci can be used to discriminate a finite number of individuals (or clones; e.g., fungi) from one another.
Hybrid identification	Fixed locus differences between species permit unambiguous recognition of interspecies hybrids as heterozygotes.

3.6 Laboratory set-up

The procedures we use were designed to be cost- and labour-effective. For instance, we use the tops of quarter-pound butter dishes as buffer trays to decrease the amount of buffer used, and only an inch of platinum wire is used in the electrode. The gel forms are easily dismantled to aid the cutting of gels and clean-up.

4. Alternative methods

Starch has been the predominant medium for allozyme electrophoresis for the past 30 years because of its low cost, ease of handling, resolution, and non-toxicity. Cellulose acetate (4,33) and polyacrylamide (34) are two other media which have particular advantages in certain situations (see Table 1 in ref. 7). Cellulose acetate provides separation in only 20 min, requires smaller amounts of sample and staining ingredients, and is often easier to learn. On the other hand, only a single enzyme can be stained and fewer individuals can be run on each gel (strip). Investigators who need to examine only a few enzymes on an intermittent basis may find cellulose acetate a reasonable alternative to starch gels. Polyacrylamide gels offer the ability to incorporate ampholites (for separation based on charge, isoelectric focusing) or SDS (for separation based on molecular weight), but they are more difficult to work with and toxic. Neither of these media allow the researcher to gather as much allozyme data per unit effort per unit of time. By this criterion horizontal starch gel electrophoresis of allozymes remains the system of choice for most population studies.

A variety of other techniques are available for separating non-enzymatic proteins (reviewed in ref. 35), each having particular advantages. Isoelectric focusing uses the isoelectric point differences of allozymes to achieve greater resolving power (36). By combining molecular weight sensitive separation in the presence of SDS and isoelectric focusing, sequential (or two-dimensional) gel electrophoresis provides even greater resolving power to reveal almost all amino acid substitutions. However, silent mutations (e.g., changes in the third position of codons or substitution of an amino acids with another of the same charge state) and mutations in non-coding regions are still missed. This technique can be quite useful for examining non-enzymatic proteins (see review of applications of this technique in Appendix 5 of ref. 37). Immunoelectrophoresis in its various forms permits the differentiation and identification of homology of proteins through immunological reactions (38). This procedure is especially useful for determining homologies between taxa where multiple forms (multiple coding loci) of the protein exist.

5. Future of the methodology

Many different biological disciplines (e.g., physiology, development, and plant breeding) are currently driven by an interest in the ability to study DNA variation directly (also note the interest in mapping whole genomes from *Escherichia coli* to humans). The fields of evolutionary and conservation biology are no exception. The ability to examine DNA sequence variation directly puts us at the ultimate level of genetic interest, namely variability in the genetic code itself. Mitochodrial DNA haplotypes (Chapter 3), restriction fragment length polymorphisms (RFLPs, Chapter 5) of nuclear DNA (nDNA), random amplified polymorphic DNA (RAPD, Chapter 6), microsatellites (Chapter 7), single-stranded conformational polymorphisms (SSCPs, Chapter 8), multilocus DNA fingerprinting (Chapter 9), and sequencing (Chapters 6 and 10) are examples of techniques available for examining genomic variation within and among individuals, families, populations, and various taxonomic levels.

Has the availability of these techniques made allozyme analysis obsolete? In choosing an appropriate methodology for any new investigations, we must examine the nature of the question being addressed and determine which types of data will answer the particular question sufficiently and efficiently. Each of the aforementioned techniques has advantages (e.g., more alleles per locus or more loci) and disadvantages (e.g., cost during preliminary project phases or dominant/recessive expression of alleles) relative to allozyme electrophoresis for addressing particular types of questions. Throughout the rest of this book, the reader will be shown the protocols behind the tremendous accomplishments of each of these techniques (see examples throughout refs 39–42). Some of these techniques are so new that we are only beginning to recognize some of their limitations (e.g., the extensive homoplasy of microsatellites). Decisions on choice of technique need to be made on the basis of cost (supplies, equipment, labour, and start-up), data type, and probability of success and not on the basis of newness nor current popularity.

In the future we will see allozyme analysis continue to play an important role in many different biological disciplines, including the analysis of populations, because of the ease and cost of data acquisition, the large number of organisms for which allozyme data exist and the co-dominant nature of the resultant data. Not only are the data simple, co-dominant, Mendelian characters, but these data come from important structural genes which code for the enzymes of such critical pathways as glycolysis and the citric acid cycle.

Acknowledgements

I am indebted to J. Ellen Marsden, Catherine G. Schenck, and Adrian P. Spidle, who have helped me put the preceding information down on paper where it could be used by the many students and investigators who have learned allozyme analysis in my laboratory.

References

1. Lewontin, R.C. and Hubby, J.L. (1966). *Genetics*, **54**, 595.
2. Harris, H. (1966). *Proc. Royal Soc. B*, **164**, 298.
3. Hunter, R. and Markert, C. (1957). *Science*, **125**, 1294.
4. Richardson, B.J., Baverstock, P.R., and Adams, M. (1986). *Allozyme electrophoresis: a handbook for animal systematics and population studies*. Academic Press, Sydney.
5. Ryman, N. and Utter, F. (eds) (1987). *Population genetics and fishery management*. University of Washington Press, Seattle, WA.
6. Pasteur, N., Pasteur, G., Bonhomme, F., Catalan, J., and Britton-Davidian, J. (1988). *Practical isozyme genetics*. Ellis Horwood, Chichester.
7. Murphy, R.W., Sites, J.W., Buth, D.G., and Haufler, C.H. (1996). In *Molecular systematics* (ed. D.M. Hillis, C. Moritz, and B.K. Mable), pp. 51–120. Sinauer Associates, Sunderland, MA.
8. Whitmore, D.H. (ed.) (1990). *Electrophoretic and isoelectric focusing techniques in fisheries management*. CRC Press, Boca Raton, FL.
9. Herman, B. and Hummel, S. (eds) (1994). *Ancient DNA*. Springer-Verlag. New York, NY.
10. Utter, F.M., Allendorf, F.W., and May, B. (1979). *Biochem. Genet.*, **17**, 1049.
11. Ferris, S.D. and Whitt, G.S. (1978). *Biochem. Genet.*, **16,** 811.
12. Buth, D.G. (1990). In *Electrophoretic and isoelectric focusing techniques in fisheries management* (ed. D.H. Whitmore), pp. 1–22. CRC Press, Boca Raton, FL.
13. Utter, F., Aebersold, P. and Winans, G. (1987). In *Population genetics and fishery management* (ed. N. Ryman and F. Utter), pp. 21–46. University of Washington Press, Seattle, WA.
14. Morizot, D.C. and Schmidt, M.E. (1990). In *Electrophoretic and isoelectric focusing techniques in fisheries management* (ed. D.H. Whitmore), pp. 23–80. CRC Press, Boca Raton, FL.
15. Leary, R.F. and Booke, H.E. (1990). In *Methods for fish biology* (ed. C.B. Schreck and P.B. Moyle), pp. 141–170. American Fisheries Society, Bethesda, MD.
16. Aebersold, P.B., Winans, G.A., Teel, D.J., Milner, G.B. and Utter, F.M. (1987). *Manual for starch gel electrophoresis: a method for detection of genetic variation*. NOAA Technical Report NMFS 61, US Department of Commerce, Washington, DC.
17. Kephart, S. (1990). *Am. J. Bot.*, **77**, 693.
18. Dessauer, H.C., Cole, C.J., and Hafner, M.S. (1996). In *Molecular systematics* (ed. D.M. Hillis, C. Moritz, and B.K. Mable), pp. 29–47. Sinauer Associates, Sunderland, MA.
19. Clayton, J.W., Tretiak, D.N., Billeck, B.N., and Ihssen, P. (1975). In *Isozymes IV* (ed. C.L. Markert), pp. 433–448. Academic Press, New York, NY.
20. Cross, T.F., Ward, R.D., and Abreu-Grobois, A. (1979). *Comp. Biochem. Physiol.*, **64B**, 403.
21. Ayala, F.J., Hedgecook, D., Zumwalt, G.S., and Valentine, J.W. (1973). *Evolution*, **27**, 177.

22. Clayton, J.W. and Tretiak, D.N. (1972). *J. Fish. Res. Bd. Can.* **29**, 1169.
23. Cardy, B.J., Stuber, C.W. and Goodman, M.M. (1980). *Techniques for starch gel electrophoresis of enzymes from maize (Zea mays L.).* Department of Statistics Mimeo Series no. 1317, North Carolina State University, Raleigh, NC.
24. Markert, C.L. and Faulhaber, I. (1965). *J. Exp. Zool.*, **159**, 319.
25. Ridgway, G.J., Sherburne, S.W. and Lewis, R.D. (1970). *Trans. Am. Fish. Soc.*, **99**, 147.
26. Selander, R.K., Smith, M.H., Yang, S.Y., Johnson, W.E. and Gentry, J.B. (1971). *Stud. Genet.*, **6**, 49.
27. International Union of Biochemistry and Molecular Biology, Nomenclature Committee (1992). *Enzyme nomenclature 1992.* Academic Press, Orlando, FL.
28. Harris, H. and Hopkinson, D.A. (1976). *Handbook of enzyme electrophoresis in human genetics.* North Holland, Amsterdam.
29. Vallejos, C.E. (1983). In *Isozymes in plant genetics and breeding* (ed. S.D. Tanksley and T.J. Orton), pp. 469–516. Elsevier, Amsterdam.
30. Manchenko, G.P. (1994). *Handbook of detection of enzymes on electrophoretic gels.* CRC Press, Boca Raton, FL.
31. Shaklee, J.B., Allendorf, F.W., Morizot, D.C., and Whitt, G.S. (1990). *Trans. Am. Fish. Soc.*, **119**, 2.
32. May, B.P. and Krueger, C.C. (1990). In *Electrophoretic and isoelectric focusing techniques in fisheries management* (ed. D.H. Whitmore), pp. 157–172. CRC Press, Boca Raton, FL.
33. Hebert, P.D.N. and Beaton, M.J. (1989). *Methodologies for allozyme analysis using cellulose acetate electrophoresis: a practical handbook.* Helena Laboratories, Beaumont, TX.
34. Hames, B.D. (1990). In *Gel electrophoresis of proteins: a practical approach* (ed. B.D. Hames and D. Rickwood), p. 1. IRL Press, Oxford.
35. Hames, B.D. and Rickwood, D. (eds) (1990). *Gel electrophoresis of proteins: a practical approach.* IRL Press, Oxford.
36. Righetti, P.G., Gianazza, E., Gelfi, C., and Chiari, M. (1990). In *Gel electrophoresis of proteins: a practical approach* (ed. B.D. Hames and D. Rickwood), p. 149. IRL Press, Oxford.
37. Rickwood, D., Chambers, J.A.A., and Spragg, S.P. (1990). In *Gel electrophoresis of proteins: a practical approach* (ed. B.D. Hames and D. Rickwood), p. 217. IRL Press, Oxford.
38. Bøg-Hansen, T.C. (1990). In *Gel electrophoresis of proteins: a practical approach* (ed. B.D. Hames and D. Rickwood), p. 273. IRL Press, Oxford.
39. Avise, J.C. (1994). *Molecular markers, natural history and evolution.* Chapman and Hall, New York, NY.
40. Avise, J.C. and Hamrick, J.L. (1996) *Conservation genetics: case histories from nature.* Chapman and Hall, New York, NY.
41. Schierwater, B., Streit, B., Wagner, G.P. and DeSalle, R. (1995) *Molecular ecology and evolution: approaches and applications.* Birkhäuser Verlag, Berlin.
42. Ferraris, J.D. and Palumbi, S.R. (1996) *Molecular zoology.* Wiley-Liss, New York, NY.

2

Total DNA isolation

BROOK G. MILLIGAN

1. Introduction

Population geneticists are relying increasingly on the study of DNA variation within and among populations. However, the feasibility of such studies is often limited by one's ability to isolate DNA from many individuals. The problem of dealing with many samples is particularly acute for the study of plants and ancient or forensic samples, because often intact DNA must be isolated from a complex and usually unknown chemical mixture. In the case of plant samples, difficulties encountered include DNA degradation due to the presence of either native DNases or secondary plant compounds, and the co-isolation of polysaccharide contaminants that inhibit further enzymatic reactions. In the case of ancient or forensic samples, difficulties include contamination with modern DNA and *in situ* degradation under inevitably sub-optimal preservation conditions. The focus of this chapter is on methods that have been developed to overcome the problems associated with total DNA isolation from a variety of samples encountered by population biologists. Particular reference is given to streamlining the procedures to make them applicable to population surveys. The techniques discussed here are useful for studying variation in any of the plant or animal genomes; other references should be consulted for additional descriptions of techniques specifically adapted to isolation of nucleic acids from purified nuclei, mitochondria, or chloroplasts (1–5). In addition, the general principles of cell fractionation have been reviewed elsewhere (6–13).

One of the motivations for collecting these procedures is to encourage more widespread use of genetic information to answer questions in population biology (14). The most broadly applicable approach for obtaining useful genetic information is to isolate total DNA and subsequently to select a specific subset using the polymerase chain reaction (PCR) for study via such methods as gel electrophoresis, sequencing, restriction site analysis, or cloning. Because of the generality of this approach, the greatest emphasis is given to isolation protocols optimized for small tissue samples, such as might subsequently be used in PCR-based genetic analysis. Even in largely PCR-based studies, however, large quantities of DNA are often required from a small subset of the individual samples. For example, DNA from a few exemplar individuals may be needed on each gel as standard allele references to maintain continuity across multiple gels, or additional DNA may be needed for cloning or Southern blots. In such cases it is often most convenient to increase the scale of a working small-scale protocol to accommodate larger quantities of tissue. Any of the protocols given here may simply be scaled up to accommodate more tissue.

Of course, the use of the isolation procedures described here is not limited to PCR-based studies. Once DNA has been successfully isolated, the numerous techniques now available to molecular biologists can be incorporated into a particular study. These include construction of genomic libraries to study restriction fragment variation (see Chapters 4 and 5), PCR amplification of specific segments from the nuclear, chloroplast, or mitochondrial genomes to study either restriction fragment or nucleotide sequence variation (see Chapters 5, 6, and 10), or detection of microsatellites (see Chapter 7) or individual-specific 'fingerprints' (see Chapter 9). Uses differ in their demands for purity, length, and quantity of isolated DNA, therefore techniques are described in this chapter that yield a broad range of DNA quality. Coupled with those described elsewhere in this book, they provide the population biologist with a repertoire applicable to a broad range of problems.

2. Sample preservation and preparation

DNA isolation involves a sequence of several steps. Specimens must be collected from the environment and given any pre-treatment necessary to remove exogenous contamination prior to further isolation. DNA must be released into solution, isolated from other cellular components, and concentrated for use in subsequent analysis. Finally, additional purification may be required to remove compounds that initially co-purify with DNA. Many of the techniques associated with extraction, isolation, and concentration of DNA are broadly applicable across a wide array of sample types. However, one of the most important determinants of DNA quality is the means used initially to collect and preserve the tissue. By necessity, these techniques must be adapted to the specific nature of the sample being investigated.

2.1 Plant samples

Although DNA has been successfully isolated from such non-optimal sources as herbarium sheets and fossils (15,16), the yield and quality of DNA can be greatly affected by the condition of the original tissue and the means of preserving it prior to extraction (17,18). This is particularly true for species that produce large quantities of tannins, phenolics, or other secondary metabolites that interfere with successful extraction of DNA. In general, it is best to harvest the freshest material possible. New growth early in the season often yields the best results. Freshly harvested material should be kept cool and moist (but not submerged), for example, in an ice chest, until further processing can be undertaken. If necessary, subsequent preservation is best done by freezing, although dried specimens may also be useful sources of DNA (15,17). One novel means of rapidly freezing samples is with a freeze spray used in histology preparations (e.g., Cytocool II (Stephens Scientific 8323, VWR 48212-190) or Histo-Freeze 22 (Fisher 12-645-32)). It is also important to keep in mind that once frozen, tissues should remain frozen until thawed in an extraction buffer.

When adequate refrigeration is not possible, the best means of preserving plant tissue for DNA analysis appears to be rapid drying in individual bottles containing silicic acid (silica gel Sigma S7500 and S7625) or anhydrous $CaSO_4$ (Drierite VWR 22890-229 and 22891-028) (19,20). In my laboratory we routinely use 15 ml tubes containing approximately 5 g silica gel (5–10% of which changes from blue to pink upon absorbing water vapour) under a cotton plug to preserve approximately 1 g of leaf tissue; however, experimentation is necessary to avoid saturation of the drying medium. Tissue dried rapidly (<3 days) is an excellent source of DNA for years. Because individual tubes of dry tissue can conveniently be stored at ambient temperature and because subsequent manipulations are easy and yield plenty of DNA, this means of preservation and storage of plant tissue is now routine in my laboratory. It should be noted, however, that in some cases degradation of DNA during drying or subsequent isolation is observed, so this may not be universally useful as a preservation method.

Chemical preservation of plant material generally results in highly degraded DNA (17); however, this may be due to the inability of the extraction procedure to release DNA from protein–DNA complexes, rather than due to degradation of the DNA itself (18). One notable exception is a method of chemical preservation (*Protocol 1*) (21) that has been shown to yield excel-lent DNA from oak leaves full of tannins, phenolics, and other secondary compounds even after storage at ambient temperature for 3–5 weeks in the field and subsequently at –20°C for 3–6 months (R. Spellenberg, personal communication).

Protocol 1. NaCl–CTAB preservation of tissues

Reagents

- Saturated NaCl–CTAB solution: H_2O saturated with NaCl and hexadecyltrimethylammonium bromide (CTAB). Mix NaCl into a sufficient volume of H_2O until a 1 cm layer remains insoluble. Add CTAB slowly with mixing until the solution achieves the viscosity of honey or motor oil; this will require approximately 30–40 g CTAB per litre of NaCl-saturated solution. It may take at least several hours to hydrate the CTAB. The exact concentration of CTAB is not extremely important.

Method

1. If penetration of the leaf tissue by the NaCl–CTAB solution is likely to be a problem, remove the thickest portions of the leaves (e.g., the largest veins and petioles).

Protocol 1. *Continued*

2. Cut the remaining leaf tissue into small (e.g., 1 cm^2) pieces. If the tissue is extremely thick, cut it into smaller pieces to aid penetration of the NaCl–CTAB solution.
3. Immediately immerse the leaf tissue in NaCl–CTAB solution at a ratio of approximately one part leaves to three parts solution.
4. Samples may be stored in the dark at ambient temperature for at least a month and at –20°C indefinitely.

As always care in the initial collection and preservation of tissues may greatly reduce problems encountered later that result from degradation due to mature or senescent material, or due to inadequate preservation (22,23). Further, because plant tissues differ so greatly in their secondary chemistry and anatomical structure, potential preservation techniques and subsequent DNA isolation methods should be tried with material at least taxonomically similar to the target species prior to reliance on them in the field.

2.2 Animal samples

As with plants, care in the initial collection of animal tissue samples greatly reduces the problems encountered during DNA isolation. Many of the techniques and important considerations in collecting and preserving animal tissue in the field are reviewed by Dessauer *et al.* (24). Ideally, samples will be preserved immediately in ultra-cold freezers or liquid nitrogen, preferably after dissection into component tissues to enhance both later usefulness and long-term preservation. Vertebrate tissues providing good sources of DNA include blood (particularly the red blood cells of all vertebrates except mammals and some amphibians, and the buffy coat of mammalian white blood cells), heart, liver, kidney, stomach, intestine, cerebrospinal fluid, and skeletal muscle. Clotting of blood samples may be prevented with 10 mM EDTA or 0.001% heparin. Blood cells should be separated from the plasma prior to freezing, and the DNA may be preserved by diluting samples with sterile 50 mM Tris pH 7.5, 1 mM EDTA, 100 mM NaCl. High molecular weight DNA in animal tissue can be preserved at ambient temperature (for up to two years in laboratory trials) if cut into 0.5 cm square blocks or smaller, and immersed in 5 vol. or more of 20% dimethylsulfoxide (DMSO) solution (in water), saturated with NaCl (25).

Archival tissue samples are also valuable sources of DNA (26). For example, fresh tissue placed in 10% buffered, neutralized formalin for 3–48 h, dehydrated with xylene, and infiltrated with liquid paraffin yields DNA for at least five years; in fact, DNA has been analysed from blocks older than 40 years (26). Not all fixatives are optimal in this application, however; in addition to buffered formalin, alcohol, acetone, and Omnifix work well.

Unfavourable tissues for isolation of DNA include those treated with picric acid or mercuric chloride, or those fixed for long periods in unbuffered or excess formaldehyde.

Cryopreservation or formalin–paraffin fixation is often difficult or impossible under field conditions, so alternative procedures may also be used for sample preservation. One common means of preserving soft tissue, including plant material, is in alcohol (27). Although isopropanol or propanol may be used, ethanol is preferred because it leads to better preservation (18).

Protocol 2. Ethanol preservation of tissues

Equipment and reagents

- 95% ethanol
- Shallow dish
- Scalpel or razor blade

Method

1. Place tissue in a clean but nonsterile shallow dish and cut it into pieces of 2–4 mm in diameter (the smaller the better).
2. Cover the tissue with twice its volume of 95% ethanol. Let the ethanol diffuse into the tissue for 1–2 h.
3. Replace the original ethanol. Soak the tissue in ethanol for at least 2 days.
4. Place the moist tissue, covered by fresh 95% ethanol, in a suitable shipping or storage container.

2.3 Ancient and forensic samples

From the standpoint of DNA extraction, the primary distinction between modern samples and ancient or forensic samples lies in the conditions under which the specimens were initially preserved. Modern samples are typically preserved immediately or shortly after removal of tissue from a living organism, with the intent of maintaining the specimen for study of its nucleic acids. Thus, the preservation methods can be chosen to maximize the stability of the DNA as discussed above. In contrast, ancient or forensic samples are typically encountered some time, perhaps millions of years, after the initial preservation. The preservation conditions are those prevailing during the intervening time, and are always completely beyond the control of the investigator. These conditions present some of the most challenging samples for DNA isolation.

Several problems are particularly acute for ancient or forensic samples. The samples are generally small and may contain degraded DNA, they may be contaminated by exogenous DNA (e.g., of bacterial or human origin), and they may yield inhibitory substances co-purifying with the DNA especially if they were preserved in a sub-optimal chemical or biological environment.

Thus, the basic prerequisites for successful extraction of DNA from ancient or forensic samples are to handle or pre-treat specimens so as to avoid or remove contamination, to use extraction procedures that do not themselves cause additional DNA degradation, and to remove or to render inactive inhibitory compounds. The following are specific points that are important to take into account in order to minimize the possibility of contamination while working with ancient or forensic samples; they are elaborated on in general discussions of the handling of these materials (28–34).

(a) Laboratories involved with sample preparation and pre-PCR work should be physically separated from those involved with PCR and post-PCR analysis.

(b) Laminar flow hoods, sterilized by continuous UV illumination while not in use, should be used for sample preparation.

(c) Benches and equipment should be cleaned regularly with 5–10% sodium hypochlorite.

(d) Water, reagents (except enzymes, *Taq* DNA polymerase, and primers), and tubes used for sample preparation and amplification should be treated with microwave or UV radiation. Filtration of PCR cocktails (except target DNA and *Taq* DNA polymerase) through 30000 or 100000 mol. wt cut-off spin columns will also remove contaminating DNA.

(e) A set of pipettes should be dedicated for pre-PCR work, regularly cleaned with 5–10% sodium hypochlorite, and UV irradiated.

(f) A set of protective clothing (e.g., laboratory coats, face masks, and gloves) should be dedicated to sample preparation and not other laboratory activities. Gloves should be changed regularly to prevent cross-contamination.

(g) Extraction controls (i.e., samples lacking tissue but extracted in parallel with those containing tissue) and amplification controls (i.e., PCRs with water instead of target DNA template) should be used to monitor contamination. Amplification controls containing non-target DNA template are also useful to identify situations in which the plastics are carriers of contamination.

Clearly, these are excellent principles for guiding any DNA work; however, the nature of ancient and forensic samples makes them particularly important.

A further complication arises with ancient and forensic samples in that different specimens may have been preserved under very different chemical and biological conditions. Some may have been continually frozen in glacial ice (35,36), others may have been entombed in amber (37), while still others may have been buried in bogs (38) or archaeological sites (34,39–41) or preserved in museums and herbaria (23,42,43). Because of this diversity of preservation conditions and because of the importance of the initial sample preparation steps, e.g., to remove contamination and to prevent further degradation, distinct procedures are appropriate for different types of samples. Other than to reiterate the basic prerequisites for working with ancient and forensic samples, it is impossible to provide detailed protocols that cover the full range of contingencies. However, excellent summaries of successful sample preparation appropriate to a wide range of preservation conditions have been collected (44) and should be consulted prior to adapting methodologies for a particular sample.

3. Total DNA isolation

Most population studies of DNA are based on isolation of total cellular DNA followed by subsequent selection of subsets of the genome by the PCR, Southern blotting, or cloning. This approach yields the greatest quantity of DNA, but requires the additional steps of selecting genomic subsets compared with the isolation of purified mitochondrial or chloroplast DNA. Ability to isolate most of the DNA present in a sample via a total DNA preparation is particularly important for studies of ancient or forensic samples where much of the native DNA may be degraded. Further, availability of total cellular DNA allows independent study of many distinct subsets of the genome, thereby providing much more information than would be possible from extractions of only organellar DNA. As population genetics increasingly relies on data for multiple independent loci, this feature will be particularly important.

Depending on the ultimate intended use for the DNA sample, total DNA isolations may be readily scaled up or down to accommodate the available tissue sample. For example, because the PCR is so sensitive and requires little initial DNA template, very small-scale DNA isolations are suitable. This greatly expedites population surveys based on many samples. In contrast, techniques based on either cloning or Southern blotting generally require much greater quantities and much more intact DNA. Because most population surveys will use the PCR, this section focuses on a wide variety of small-scale procedures suitable for a wide range of tissue types. However, these same protocols may be scaled up to accommodate the larger tissue samples useful for restriction fragment length polymorphism (RFLP) analysis or other survey methods.

3.1 Overview of isolation procedures

Because the chemical composition of tissue samples derived from the diversity of sources considered here is highly variable and may include both endogenous (e.g., plant secondary compounds) and exogenous (e.g., degradation products in ancient samples) compounds, the array of available isolation techniques is extensive. Further, laboratories often independently modify existing techniques without fully characterizing the effects of each ingredient or step in the procedure, increasing the variety of options. To provide some sense of comparison among major classes of isolation protocols, I have organized them along functional lines according to the strategy adopted for each of the major steps: initial disruption of tissue, solubilization of cellular components following disruption, and purification of the DNA from the complex cellular solution (*Table 1*). Additionally, I have included some information on the type of tissue commonly used for each protocol. It should be noted, however, that some distinctions are traditional rather than necessary; for example, the use of proteinase K to digest cellular proteins has primarily been used for animal tissue whereas the cationic detergent hexadecyltrimethylammonium bromide (CTAB) has been used for solubilizing plant tissue.

These protocols differ in both the time involved and the quality of isolated DNA. Although the time required depends greatly on the number of samples extracted at once, *Table 1* provides some information on the relative length of each procedure. Note that the longer procedures involve incubation steps during which one can perform other tasks, so they do not necessarily require a proportionately longer investment of activity devoted exclusively to the isolation. In my own experience, I have often found that the slight extra time involved in a more complete purification is more than recovered in the ease with which subsequent analysis may be performed.

Many of these protocols provide high quality DNA that is useful for a variety of analytical techniques. However, they may differ greatly in their ability to purify DNA away from secondary contaminants in the cellular solution. This is perhaps the single most difficult aspect of choosing a procedure, because the nature of the contaminants is often unknown; it may even be difficult to anticipate whether or not contaminants will be present. Thus, one may first need to try different protocols to determine which one provides enough purification. The functional summary of each protocol may provide a guide. Finally, it should be noted that the functional components may be combined in different permutations to yield a wide array of novel DNA isolation methods.

Two purification protocols have been left out of *Table 1* because of their specialized nature. The first (*Protocol 15*) describes the purification of DNA using a caesium chloride density gradient. This may prove useful as a final step following some of those listed in *Table 1*, especially if large quantities of extremely pure DNA are needed. The second (*Protocol 17*) describes the isolation of intact chloroplasts from other cellular components. This is useful if chloroplast, rather than total, DNA is needed.

3.2 Buffer components

Extraction buffers for animal DNA, of either modern or ancient origin, typically contain a buffer (almost universally Tris) to maintain pH, EDTA to chelate divalent cations such as calcium and magnesium and thereby inhibit DNases, a sodium or potassium salt to stabilize the nucleic acids in an isotonic

Table 1. Overview of total DNA isolation protocols

Protocol	Tissue disruption	Tissue solubilization	Primary purification	Secondary purification	Tissue type	Time (h)	DNA Quality	Comments	References
3. Boiling	Boiling	H_2O	None	None	Tiny samples (10^2–10^4) cells without cell walls	1	Denatured	Cellular debris may inhibit PCR	59
4. NaOH	Grinding	Alkaline pH	None	None	Alternative to boiling for plant	1	Denatured		60, 61
5. Chelex	Grinding	H_2O	None	None	Any	1	Denatured	Chelate metal ions involved with DNA degradation	62, 65, 67
6. SDS	Grinding	SDS	Isopropanol precipitation	None	Any	2	High	No separation of complex components	68
7. CTAB	Grinding	CTAB	Chloroform	Isopropanol precipitation	Traditionally plant	2–4	High		15, 58, 71
8. CTAB precipitation	Grinding	CTAB	Chloroform	CTAB precipitation	Traditionally plant	2–4	High		72, 74, 75, 77
9. Protein digestion	Grinding, protein digestion	SDS	Phenol:chloroform (2×), chloroform	Ethanol precipitation	Traditionally animal	4–8	High	PCR inhibitors may co-purify	27, 31, 42, 79, 81, 82, 86
10. Protein precipitation	Grinding	SDS	Protein precipitation	Isopropanol precipitation (2×)	Any	4–8	High		57, 90, 91
11.GuSCN	GuSCN	GuSCN	Ethanol precipitation	None	Any, preferred for ancient	2	High		92, 97, 98
13. GuSCN-silica	GuSCN	GuSCN, Triton X-100	Silica binding	None	Any, preferred for ancient	2	High		32, 99, 105, 106, 107
14. PEX	PEX	SDS	Protein precipitation	Ethanol precipitation	Plant	4–8	High		108, 109

Abbreviations: SDS, sodium dodecyl sulfate; CTAB, hexadecyltrimethylammonium bromide; GuSCN, guanidinium isothiocyanate; PEX, potassium ethyl xanthogenate.

medium, and a protease (usually proteinase K) to digest the proteins. An anionic detergent (usually sodium dodecyl sulfate (SDS)) is often included to solubilize cellular membranes and denature proteins. The buffer pH and concentrations of each component vary—for example, much higher than normal concentrations of EDTA (500 mM) are used to demineralize bone tissue during extractions (45)—however, extraction buffers generally maintain the same overall composition. Undoubtedly, this reflects the relative chemical uniformity of animal cells.

Unlike animal cells, plant cells commonly contain numerous secondary compounds, the composition of which varies greatly from species to species. As a result many alternative isolation buffers have been devised to overcome problems encountered in particular species. One notable difference is that the cationic detergent CTAB is commonly used instead of SDS. To make the protocols as easy to follow as possible I have included those that are most generally useful. I have included some important variations, however, to provide ideas for modifications that may improve either yield or quality of DNA. Keep in mind that portions of many of the techniques may easily be combined or rearranged to overcome a specific problem.

The components of the isolation buffers may also be modified in numerous ways. Unfortunately, when faced with contamination, degradation, or low yield, one is usually unsure about the exact cause. Both the pH of the isolation buffer and the specific protective agents or detergents included may need to be optimized for a particular species. The pH should be selected to avoid the optima of the degradative enzymes. For example, most lipolytic enzymes and lipoxygenases have pH optima between 5.0 and 6.0 whereas nuclear DNases have pH optima around 7.0 (46). As a result, most plant DNA extraction buffers have a pH of 8.0, though some are as high as pH 9.0.

Apparently essential components of DNA isolation buffers used for plant material are those aimed at protecting the DNA from degradation by native enzymes or secondary compounds released upon disruption of the cells (9,46). These include a wide array of compounds; it may prove beneficial to determine which are necessary in a particular case. EDTA is usually included to inhibit metal-dependent enzymes by chelating such divalent cations as Mg^{2+} and Ca^{2+}; however, at least one EDTA-stimulated DNase is known from wheat (47). Bovine serum albumin is included to cause surface denaturation of degradative enzymes (9). Reducing or sulfhydryl agents such as β-mercaptoethanol, glutathione, cysteine, dithiothreitol (DTT) and other thiols are commonly included to protect DNA against quinones, disulfides, peroxidases, and polyphenoloxidases (48).

Polyvinylpyrrolidone (PVP) is added to decrease the effect of polyphenols, quinones, and tannins. Less common additives are ascorbic acid as protection against quinones; diethylpyrocarbonate, bentonite, spermine, spermidine, and other polyamines as protection from RNases (49–51); diethyldithiocarbamate to chelate Cu^{2+}; cyanide as protection against heavy metal oxidases; NH_3

infiltration as protection against H^+; ethidium bromide (52); and coconut milk. Aurintricarboxylic acid has also been used to inhibit nucleases in the isolation of plant DNA and RNA (53,54); however, since it protects the nucleic acids from any enzyme action by coating them (55) it must be removed following the isolation. The additional steps required may not be warranted when many samples must be processed.

Finally, DNA may co-isolate with a polysaccharide contaminant. Typically, such contaminants either prevent resuspension of the DNA after a precipitation or prevent digestion with restriction endonucleases. One means of eliminating this problem is to isolate DNA using one of the methods described, followed by separation of the DNA from the contaminant using anion-exchange chromatography. Several commercially available columns may be useful in this application; one has specifically been studied for its ability to remove a variety of plant polysaccharides known to reduce or eliminate restriction endonuclease activity (56). Cheaper and more flexible approaches are either to increase the concentration of NaCl to 2 M prior to the first alcohol precipitation of DNA or to resuspend the sample in a low ionic strength buffer, increase the concentration of NaCl to 2 M, and reprecipitate the DNA (57,58). Polysaccharides remain soluble in high salt concentrations and therefore do not co-precipitate with the DNA. This single step can have a profound impact on the success of a DNA isolation, although I have encountered DNA that cannot be purified chemically from polysaccharides.

3.3 Boiling techniques

The simplest means of DNA isolation is to disrupt cells by boiling them in water (59). This is only useful if the DNA released is intended for PCR reactions, because only a few (10^2–10^4) cells can be used without accumulation of enough cellular debris to inhibit enzymatic manipulations or to actively degrade the DNA. This is, of course, insufficient to perform any manipulations prior to amplifying the quantity of DNA. Nevertheless, if appropriate, this remains the quickest, simplest, and cheapest means of extracting small quantities of DNA.

Protocol 3. Isolation of total DNA by boiling

Equipment and reagents

- Phosphate-buffered saline: 137 mM NaCl, 2.7 mM KCl, 10 mM Na_2HPO_4, 1.8 mM KH_2PO_4, adjusted to pH 7.4 with HCl
- Boiling water bath

Method

1. Starting with a liquid suspension sample of 10^2–10^4 cells, pellet them at 1200–1500 × *g* for 10 min.

Protocol 3. *Continued*

2. Resuspend the cells in 1.5 ml phosphate-buffered saline and pellet as before. If necessary, re-pellet the sample and resuspend the cells again to remove traces of the original suspension buffer.
3. Resuspend the cells gently in 25–50 μl distilled H_2O. Incubate at 95–100 °C for 3–5 min.
4. Centrifuge briefly to collect any condensate.
5. Optional: pellet the cellular debris at 12 000 × *g* for 3 min and transfer the cleared lysate to a new tube.
6. Use the entire cell lysate in a single PCR.

3.4 Alkaline extraction techniques

Boiling plant tissue is less effective at releasing DNA than for animal tissue, because of the cell walls present in plants. However, isolation of DNA at high pH alone is effective at yielding DNA suitable for PCR (60). The following protocol was developed from that observation and is almost as simple as the previous one.

Protocol 4. Alkaline isolation of total DNA

Reagents

- 0.5 M NaOH
- Micropestle (Kontes 749521-1500)
- Storage buffer: 100 mM Tris pH 8.0, 1 mM EDTA pH 8.0

Method

1. Add a small (several milligram) tissue sample to a microcentrifuge tube. For every milligram of tissue, add 10 μl of 0.5 M NaOH.
2. Grind the tissue with a micropestle until no large pieces are left.
3. Transfer 5 μl quickly to a new tube containing 495 μl storage buffer. Mix well.
4. Use 1 μl directly in a PCR reaction.
5. Store the isolated DNA at –20°C.

A variation is to grind the tissue on a solid support (e.g., Hybond N nylon membrane (Amersham)), rather than in a microcentrifuge tube. After a brief wash, pieces of the solid support are used directly in the PCR (61).

3.5 Chelating resin techniques

The previous two protocols are based on the facts that high temperatures (100°C) and high pH (pH > 9) disrupt cell membranes and release DNA into solution. However, not only is DNA denatured under these conditions, but active degradation in low ionic strength solutions is stimulated by metal ions also released from the cells. Chelating resins, notably Chelex 100 (Bio-Rad), have been used to remove metal ions from solution, thereby making feasible the rapid isolation of DNA under these extreme conditions (62).

Protocol 5. Chelex 100 isolation of total DNA

Equipment and reagents

- Chelex (Bio-Rad), 5% (w/v) in sterile, distilled H_2O
- Incubator at 56°C
- Boiling water bath

Method

1. Add tissue sample to 1 ml of sterile distilled H_2O in a microcentrifuge tube. Incubate at 20–25°C for 15–30 min. Mix occasionally by inversion or gentle vortexing.
2. Centrifuge at 12000 × *g* for 2–3 min to pellet cellular debris.
3. Discard all but 20–30 μl of the supernatant.
4. Add 5% Chelex to a final volume of 200 μl.[a]
5. Incubate at 56°C for 15–30 min.
6. Vortex at high speed for 5–10 sec.
7. Incubate at 100°C (e.g., in a boiling water bath) for 8 min.
8. Centrifuge at 12000 × *g* for 2–3 min to pellet the resin beads.
9. Use 20 μl of the supernatant in a typical PCR.
10. Store the sample at either –20°C or 4°C. To re-use the sample in further amplifications, repeat the vortexing and centrifugation steps before removing an aliquot.

[a]When pipetting Chelex solutions, the resin beads must be distributed evenly in the solution. This can be accomplished by gently stirring a small quantity (e.g., 10 ml) of the stock solution in a beaker while removing the necessary quantity with a pipette. A large-bore pipette tip (e.g., a standard 1000 μl pipette tip) should be used.

The primary disadvantage of this procedure is that because of the extreme conditions present during the isolation procedure, the resulting DNA is denatured. As a result it is not suitable for cloning, Southern blots, or restriction fragment analysis. It is, however, perfectly suitable for the PCR, and therefore is both highly effective as a protocol for population surveys and adaptable to a wide range of ancient and forensic tissues (63–67). It should also be noted that because of the high alkalinity (pH 10–11) of the chelating resin, samples prepared with it may degrade much more rapidly than those prepared in other ways (23,67).

3.6 Detergent isolation techniques

The previous methods of isolating DNA are most suitable for cell types that can be readily solubilized in simple solvents. Thus, unless mechanical disruption of the cells precedes these protocols, they may not be optimal for plant samples, mineralized bone, ancient samples, or certain forensic or clinical samples. One means of enhancing the range of suitable tissues is through inclusion of detergents in the isolation buffer. Most commonly SDS is used (68,69), although CTAB is commonly used for isolation of plant DNA. Subjecting samples to microwave radiation (69) or vortexing them in the presence of washed, sterile sand (70) are also means of releasing DNA while simultaneously reducing the possibility of cross-contamination during mechanical disruption of cells.

Protocol 6. SDS isolation of total DNA

Reagents

- Extraction buffer: 200 mM Tris pH 7.5, 25 mM EDTA, 250 mM NaCl, 0.5% (w/v) SDS
- Micropestile (Kontes 749521-1500)
- 100% isopropanol, –20°C
- TE: 10 mM Tris pH 8.0, 1 mM EDTA pH 8.0
- Speed-Vac (Savant) or incubator at 65°C

Method

1. Quickly grind 10 mg tissue sample with a micropestle in a microcentrifuge tube without buffer.
2. Add 400 μl extraction buffer and vortex the sample for 5 sec. Incubate at 20–25°C until other samples are prepared (up to 1 h or more).
3. Centrifuge at 12 000 × *g* for 1 min to pellet cellular debris.
4. Transfer 300 μl of the supernatant to a new tube. Add 300 μl of isopropanol and incubate at 20–25°C for 2 min.
5. Centrifuge at 12 000 × *g* for 5 min.
6. Dry the DNA pellet in a Speed-Vac, in a 65 °C oven, or at 20–25 °C.
7. Dissolve the DNA in 100 μl TE.
8. Use 2.5 μl of the dissolved DNA for a typical PCR reaction.
9. The dissolved DNA may be stored at 4°C for over one year.

One of the most widely used techniques for isolating plant DNA, particularly from small samples, is based on the cationic detergent CTAB. It is used to solubilize the plant membranes, and will form a complex with the DNA. The protocol described here (*Protocol 7*) is largely based on a previous description (71), with modifications and comments added based on my own experience and the reports of others (15,57,58,72–75).

One of the advantages of this method is that extensive preparation of the plant tissue is not required, and it is adaptable to numerous types of tissue, including leaves, roots, seeds, embryos, endosperm, pollen, and suspension cultures (75). Additionally, it can easily accommodate a wide range of sample sizes from milligram quantities of herbarium, mummified, or fossil tissue to many grams of freshly harvested tissue (16,75).

Protocol 7. CTAB isolation of total DNA[a]

Reagents

- 2× CTAB isolation buffer: 100 mM Tris pH 8.0, 1.4 M NaCl, 20 mM EDTA, 2% (w/v) CTAB, 1.0% (w/v) polyvinylpyrrolidone (PVP-360); 0.2% (v/v) β-mercaptoethanol (added immediately prior to use in a fume hood)
- 1× CTAB isolation buffer: same as above, except half the concentration
- Micropestile (Kontes 749521-1500)
- Water bath at 60°C
- Chloroform:isoamyl alchohol (24:1 by vol.)
- 5 M NaCl
- Isopropanol, –20°C
- Wash buffer: 76% (v/v) ethanol, 10 mM ammonium acetate
- Resuspension buffer: 10 mM ammonium acetate, 0.25 mM EDTA pH 8.0

Method

1. Heat the isolation buffer to 60°C.
2. Grind 10–50 mg of fresh or frozen leaf tissue in isolation buffer using a micropestle. For wet tissues (e.g., fresh leaves), use a volume of 2× buffer equal to the volume of the tissue and add additional 1× buffer to increase the volume to 500 μl. For dry tissues, use 500 μl of 1× buffer. The 2× buffer is only to compensate for the water contained within the tissues. A pinch of sterile sand and/or brief vortexing may aid the tissue disruption.
3. Incubate at 60°C for 30–120 min with periodic gentle swirling.
4. Extract once with 500 μl chloroform:isoamyl alcohol. Mix gently but thoroughly. Spin at 12000 × *g* for 30 sec at 20–25°C to separate the phases.
5. Avoiding the interface, pipette the aqueous (top) phase into new tubes.
6. Add 0.5 vol. of 5 M NaCl. Add cold isopropanol to 40%. Mix gently to precipitate nucleic acids. If no precipitate is visible, place at –20°C for 20 min or longer.
7. Spin at 12000 × *g* for 1 min at 20–25°C. If no pellet or precipitate is visible, place on ice for 20 min and spin again. In the worst case, spin for 10 min at 12000 × *g* .
8. Gently pour off as much of the supernatant as possible without losing the nucleic acid pellet. Add 0.5–1.0 ml of wash buffer and swirl gently to wash the pellet. Let the nucleic acids sit in the wash buffer for 15–20 min. Generally, nucleic acids will become much whiter at this step.

Protocol 7. *Continued*

9. Spin at 12 000 × *g* for 1 min at 20–25°C. If this is not sufficient, spin harder and longer as before. Pour off wash buffer and allow the pellet to dry briefly (2–4 min) by inverting the tube on a paper towel. Be careful that the pellet does not slide out!
10. Resuspend DNA in resuspension buffer in small increments (e.g., 10–100 μl) depending on the size of the pellet.

[a]Although this protocol is useful for tissue derived from many sources, in some cases it may be necessary to further purify the DNA particularly if the tissues contain tannins or other secondary compounds. This may be accomplished by following this procedure with either a final phenol extraction or a caesium chloride gradient. Refer to *Protocol 15* for the caesium chloride purification procedure. Alternatively, up to three sequential chloroform extractions may be used at steps **4** and **5**.

Numerous variations on *Protocol 7* exist. Foremost are the alternative compositions of the extraction buffer. For example, many authors (15,71) do not include PVP in the buffer. Inclusion does not seem to impair extraction of DNA, and is important for certain species with high concentrations of secondary compounds, so it seems prudent to include PVP routinely. A second area in which protocols differ is in the choice of reducing agents. β-Mercaptoethanol is almost always included, but may be increased in concentration as high as 25 mM; however, ascorbic acid (0.1%) or DTT (10 mM) may also be included. An additional chelating agent, diethyldithiocarbamate (0.1%), may be included to chelate Cu^{2+} (76). Both ascorbic acid and diethyldithiocarbamate are required for extraction of DNA from cotton (J. F. Wendel, personal communication).

Alternative detergents and pHs are possible as well. For example, SDS has been successfully substituted for CTAB (16), and the pH of the buffer may be adjusted to pH 9.5 rather than pH 8.0 (E. Nevo, personal communication). Finally, 10 mM Tris, 1 mM EDTA, pH 8.0 (TE) may be used as a resuspension buffer. Given the complexity, and often unknown nature, of plant secondary chemistry, one is wise to experiment with some of these variations in buffer composition to identify which yields the best DNA from the target samples.

In my experience, precipitation of DNA from 2× CTAB isolation buffer in the presence of either ethanol or isopropanol often yields a large gelatinous mass of unknown composition. The DNA is usually clearly visible within the mass, but difficult to separate from it. Several solutions to this problem are available. If successful, the simplest solution is either to dilute the aqueous phase obtained from the chloroform extraction or, as in the protocol above, to increase the NaCl concentration prior to alcohol precipitation. Separation may also be accomplished with either anionic exchange chromatography (56), or the caesium chloride procedures discussed in Section 3.11. Finally, precipitation of the DNA in the absence of alcohol (*Protocol 8*) is a useful alternative. CTAB is a cationic detergent that will form a precipitable complex with DNA when the NaCl concentration is below 0.7 M (72,74,75,77). *Protocol 7* can therefore be modified slightly as described in *Protocol 8*.

Protocol 8. CTAB precipitation of total DNA

Reagents

- 1× and 2× CTAB isolation buffers (see *Protocol 7*)
- Chloroform:isoamyl alchohol (24:1 by volume)
- 10% (w/v) CTAB in H_2O
- Ethanol: 100% and 80%, –20°C
- 1× precipitation buffer: 50 mM Tris pH 8.0, 10 mM EDTA, 1% (w/v) CTAB
- Resuspension buffer: 10 mM Tris pH 8.0, 1 mM EDTA pH 8.0, 1 M NaCl
- TE: 10 mM Tris pH 8.0, 1 mM EDTA pH 8.0

Method

1. Follow steps **1–5** from *Protocol 7*.
2. Add 0.1 vol. of 10% CTAB solution and mix.
3. Perform a second chloroform extraction as in steps **4** and **5** of *Protocol 7*.
4. Add an equal volume of precipitation buffer to reduce the concentration of NaCl to 0.35 M. Mix gently and incubate at 20–25°C for 30 min. Note that it is important to measure the sample volume so that the concentration of NaCl is reduced to the proper level.
5. Recover the precipitated DNA by centrifugation at 12000 × *g* at 20–25°C for 10–60 sec.
6. Rehydrate the DNA pellet in 200 μl resuspension buffer.
7. Add 2 vol. of cold 100% ethanol and mix gently to precipitate the nucleic acids. Recover the precipitated DNA by centrifugation at 12000 × *g* at 4°C for 5–15 min.
8. Wash the DNA pellet in 200 μl cold 80% ethanol and centrifuge at 12000 × *g* at 4°C for 5 min.
9. Resuspend the DNA pellet in resuspension buffer in small increments (e.g., 10–100 μl) depending on the size of the pellet.
10. At this point it may be necessary to purify the DNA further in a caesium chloride gradient (*Protocol 15*). This is especially true for those tissues that contain tannins or other secondary compounds, although more than one chloroform extraction or a final phenol extraction may be sufficient.

3.7 Protein digestion techniques

Perhaps the most widely used method of isolating DNA, particularly from animal sources, is based on solubilization of the tissue in a detergent buffer, usually containing SDS, followed by protease digestion of cellular proteins, often with proteinase K, less often with pronase E (18,26,27,78–81).

Protocol 9. Proteinase K isolation of total DNA

Reagents

- Extraction buffer: 50 mM Tris pH 7.5, 1 mM EDTA, 100 mM NaCl, 1% (w/v) SDS
- Proteinase K, 10 mg/ml in H_2O, stored at −20°C
- Chloroform:isoamyl alcohol (24:1 by vol.)
- Speed-Vac (Savant)
- Phenol:chloroform:isoamyl alcohol (25:24:1 by vol.)
- 3 M sodium acetate, pH 7.6
- 100% ethanol, −20°C
- TE: 10 mM Tris pH 8.0, 1 mM EDTA pH 8.0

Method

1. If the sample cannot be disrupted easily in solution, grind it to a fine powder in liquid nitrogen, e.g., using a mortar and pestle.
2. Place 100 mg of tissue into 500 μl extraction buffer. Add 25 μl of proteinase K and mix well. If necessary, grind the tissue. Incubate at 55°C for 2 h with occassional mixing.
3. Add an equal volume of phenol:chloroform:isoamyl alcohol, mix gently but thoroughly, and incubate at 20–25°C for 5 min.
4. Centrifuge at 7000 × *g* for 5 min to separate the phases.
5. Remove the aqueous layer[a] using a wide-bore pipette tip and transfer it to a new tube. Be careful not to collect or disrupt the cellular debris at the interface.
6. Re-extract the aqueous phase with phenol:chloroform:isoamyl alcohol.
7. Add an equal volume of chloroform:isoamyl alcohol to the aqueous phase, mix gently, and incubate at 20–25°C for 5 min. Re-mix once each minute to prevent the phases from separating.
8. Centrifuge at 7000 × *g* for 3 min to separate the phases.
9. Carefully transfer the aqueous phase (upper layer) to a new tube. Again, be careful to avoid the interface.
10. Re-extract the aqueous phase with chloroform:isoamyl alcohol.
11. Add 0.1 vol. (about 45 μl) of sodium acetate and mix. Add 1 ml (2–2.5 vol.) cold ethanol to precipitate the DNA. Incubate on ice for 10–20 min.
12. Centrifuge the DNA at 4°C. If DNA is clearly visible, centrifuge at 7000 × *g* for 20 sec; otherwise, centrifuge at 7000 × *g* for 1–2 min. Excessive force will make resuspension difficult; however, longer times are necessary to pellet DNA in low concentration.
13. Discard the supernatant and dry the DNA pellet either by inverting the tubes on paper towels for 10 min or as long as necessary, or by

centrifuging it in a Speed Vac for 1–2 min. Overdrying will make resuspension difficult.

14. Incubate the DNA in an appropriate volume (e.g., 250 μl, but dependent on the size of the pellet) of TE at 65°C for 30 min to resuspend it.

[a] Normally the aqueous layer is above the organic layer, but the phases may invert in high salt concentrations.

Like the detergent-based protocols, this procedure overcomes some of the limitations of the earlier ones. It is adaptable to a wide range in quantity of tissues, because it is much more effective at separating DNA from other cellular components, especially proteins. As a result, larger quantities of tissue may be handled while still not inhibiting subsequent reactions. At the same time the protocol is not overly complex as to be useless for the analysis of many samples typically encountered in population surveys. Finally, many different types of starting material may be used with this protocol including animal tissues, plant tissues (78), cells fixed in ethanol (82), tissues embedded in paraffin (83–85), blood or clinical swabs (27), microscope slides (26), and various tissues from avian museum specimens (42,86,87), extinct mammals (88), or other ancient samples (31). One disadvantage of this method is that it may not remove compounds, such as fulvic acids, often present in ancient samples that are potent inhibitors of enzymatic reactions (89).

3.8 Protein precipitation techniques

A different strategy for separating DNA from proteins involves selective precipitation of the proteins rather than digestion of them. Thus, a series of protocols is based on the solubilization of cellular components in a detergent such as SDS that complexes with proteins followed by precipitation of the SDS–protein complex (57,90). Like other protocols, this one may be scaled up or down to accommodate smaller or larger samples.

Protocol 10. Isolation of total DNA by protein precipitation

Reagents

- Extraction buffer: 100 mM Tris pH 8.0, 50 mM EDTA pH 8.0, 500 mM NaCl, 2% (w/v) SDS, 1% (w/v) PVP-360, 0.1% (v/v) β-mercaptoethanol (added immediately prior to use in a fume hood)
- 5 M potassium acetate, pH 6.5
- 3 M sodium acetate, pH 7.6
- 100% isopropanol, –20°C
- TE: 10 mM Tris pH 8.0, 1 mM EDTA pH 8.0
- 80% ethanol, –20°C
- Micropestle (Kontes 749521-1500)
- Speed-Vac (Savant)

Method

1. Grind 10–50 mg fresh tissue in 200–600 μl extraction buffer with a micropestle. A pinch of sterile sand and/or brief vortexing may aid tissue disruption.

Protocol 10. *Continued*

2. Incubate tissue extracts at 65°C for 20 min.
3. Add one-third the volume (e.g., 67–200 μl) potassium acetate. Shake vigorously and incubate on ice for 5 min. Most proteins and polysaccharides are removed as a complex with the insoluble potassium dodecyl sulfate precipitate.
4. Spin at 12000 × *g* for 20 min at 4°C.
5. Pipette the supernatant into a clean microcentrifuge tube. Try to avoid as much of the particulate matter as possible. Add 0.5 vol. (e.g., 133–400 μl) isopropanol. Mix and incubate the solution for 1 h at 4°C.
6. Pellet the DNA at 12000 × *g* for 15 min at 4°C. Gently pour off the supernatant and lightly dry the pellets either by inverting the tubes on paper towels for 10 min or as long as necessary, or by centrifuging it in a Speed Vac for 1–2 min.
7. Incubate the DNA in 200–500 μl TE at 65°C for 30 min to resuspend it.
8. Transfer the solution to a microcentrifuge tube and spin for 5 min at 4°C to remove any insoluble debris.
9. Transfer the supernatant to another microcentrifuge tube. Add 0.1 vol. (e.g., 20–50 μl) sodium acetate and two-thirds of the volume (e.g., 133–333 μl) of cold isopropanol. Mix well, incubate at 4°C for 1 h, and pellet the DNA for 10 min in a microcentrifuge at 4°C.
10. Wash the pellet with 200–500 μl cold 80% ethanol for 10 min and centrifuge again for 1 min at 4°C. Dry the pellet for 10 min in a Speed-Vac.
11. Re-dissolve the DNA in TE using small increments (e.g., 10–100 μl) depending on the size of the pellet.

While most commonly used for plants, this protocol should be adaptable to a diversity of animals. One important variation is to include 40% polyethylene glycol 6000 in the protein precipitation step; this removed plant secondary compounds that inhibited subsequent PCRs (91).

3.9 Single-step techniques

The search for simple DNA extraction methods that are particularly appropriate to population studies has yielded two distinct single-step extraction approaches. The first, now commercialized in the form of DNAzol (Molecular Research Center or Life Technologies), involves a single extraction buffer that solubilizes all cellular components yet allows selective precipitation of DNA in the presence of alcohol (92). The main ingredient used in these extractions is guanidinium isothiocyanate (93), a strong denaturant of proteins often used in the isolation of RNA (94–96) and capable of dissolving cytoplasmic and nuclear membranes. Protocols following this approach simply require homogenization of the tissue in the buffer and precipitation of the DNA (97,98). The second approach involves binding DNA to glass particles, washing away remaining cellular components, and eluting the DNA. Neither approach requires organic solvents, and both are simple and fast to perform.

Protocol 11. Guanidinium isothiocyanate isolation of total DNA

Reagents

- Extraction buffer: 6 M guanidinium isothiocyanate, 100 mM sodium acetate, pH 5.5
- 100% ethanol, 20–25°C
- 80% ethanol, –20°C
- TE: 10 mM Tris pH 8.0, 1 mM EDTA pH 8.0; or 8 mM NaOH prepared freshly (within one month of use)
- Tris–HCl or Hepes, free acid

Method

1. Add 10 mg tissue sample to 500 μl of extraction buffer in a microcentrifuge tube. Homogenize the tissue if necessary.
2. Incubate at 20–25°C for 10 min. Longer incubation for over 1 h with mixing may be necessary. Additionally, the procedure may be interrupted at this step by storing the extract at 20–25°C for as long as 18 h or at 4°C for as long as 3 days.
3. Centrifuge at 12 000 × *g* for 10 min at 4°C to pellet the cellular debris.
4. Precipitate the DNA from the supernatant by adding 1 ml 100% ethanol at 20–25°C. Mix by inversion and incubate at 20–25°C for 1–3 min. The DNA should become visible as a fibrous or cloudy precipitate.
5. Collect the DNA by spooling on a pipette tip or by centrifugation at 1000 × *g* for 1–2 min at 4°C.
6. Wash the DNA precipitate twice with 0.5–1.0 ml of 80% ethanol. The procedure can be interrupted during the washes by storing the DNA for 1 week at 20–25°C or for 3 months at 4°C.
7. Remove the ethanol wash and allow the DNA precipitate to dry for 5–15 min at 20–25°C.
8. Dissolve the DNA to a concentration of 0.25 μg/μl in TE or 8 mM NaOH; typically this entails addition of 200 μl solvent. The alkaline solvent may solubilize the DNA faster and more completely.
9. If necessary, centrifuge the sample at 12 000 × *g* for 10 min to remove insoluble material such as polysaccharides.
10. If NaOH was used to dissolve the DNA, adjust the pH of the solution to a desired pH by adding Tris–HCl or Hepes (free acid).

The main variation among related guanidinium-based isolation protocols is the composition of the extraction buffer. Detergents added include the sodium salt of *N*-lauroylsarcosine (0.5–1.0%, w/v), 1.3% (v/v) Triton X-100, and 1% (v/v) Nonidet P-40 (98–100) and one version includes proteinase K (98); these components undoubtedly serve under some conditions to improve the release of DNA into solution. Additionally, a wide range of buffer pH (5.5–7.3 and even unbuffered) has been used, though it would seem that the higher pH buffers would cause less DNA damage. Finally, the initial incubation may be performed at 60°C for 1 h; this may release DNA into solution more effectively.

A second highly streamlined approach to isolating DNA takes advantage of the fact that silica particles selectively bind DNA to their surfaces under high salt conditions (101–104). However, very small particles of silica may inhibit some enzymatic reactions and are difficult to remove via filtration. Size fractionation by sedimentation at unit gravity is required to remove those particles.

Protocol 12. Preparation of silica slurry

Reagents

- Powdered or microcrystalline silica: SiO_2 (e.g., Sigma S-5631) or diatomaceous earth (e.g., Sigma D-5384). Other sources of silica include crushed flint glass from scintillation vials, ground glass-fibre filters (e.g., Whatman GF/A, 10 cm diameter), preparative C_{18} silica beads (55–105 μm diameter, Waters Associates) or Celite. Silica from some of these sources may require acid washing or other pre-treatment.
- Resuspension buffer: 6 M guanidinium isothiocyanate, 100 mM Tris pH 7.0, 20 mM EDTA pH 7.0

Method

1. Suspend 10 g of silica in a full graduated cylinder containing 100 ml of H_2O.
2. Allow the silica particles to settle for 2 h.
3. Remove by suction and discard the supernatant.
4. Increase the volume to 100 ml with H_2O and resuspend the silica by vigorous shaking.
5. Allow the silica particles to settle for 2 h.
6. Remove by suction and discard the supernatant.
7. Increase the volume to 100 ml with H_2O and resuspend the silica by vigorous shaking.
8. Centrifuge the silica at 2000 × *g* for 2 min to collect the silica.
9. Resuspend the silica particles at 100 mg/ml (assuming no significant loss of material during size fractionation) in resuspension buffer.

10. Store the solution in the dark. It is stable for at least 3 months. Silica prepared in this way will generally bind over 1 μg DNA per mg silica.

One report (105) indicates that the size fractionation may be streamlined by compressing the two sedimentation steps outlined above into a single, 3 h sedimentation step. Although acid-washing the silica is not necessary, it may be desirable to do so prior to size fractionation to remove impurities or contaminants. This may be accomplished by suspending the silica in excess 35% nitric acid (i.e., 0.5 × concentrated nitric acid), heating it to 85°C for 30–60 min in a fume hood, allowing it to cool, and rinsing it four times in H_2O. Note that addition of nitric acid to H_2O releases both toxic fumes and a significant amount of energy, enough to heat the solution to approximately 40°C.

Isolation of DNA using size-fractionated silica is quite rapid and simple: DNA is bound to the silica particles, other cellular components are removed by washing, and the DNA is eluted from the silica (99,105,106). It is applicable to a wide range of tissue types (32,63,99,100), and is perhaps the best one for ancient or forensic tissues because it overcomes some of the problems associated with inhibitors co-isolating with DNA (32,89). Like the previous protocol (*Protocol 11*), this one takes advantage of the properties of guanidinium isothiocyanate to denature proteins and similar extraction buffers may be used. Several commercially available DNA extraction kits are available that are based on binding DNA to silica. These include the 'Geneclean' and 'RPM' kits from Bio 101, the 'Prep-A-Gene' kits from Bio-Rad, and some of the 'Wizard' kits from Promega; all use guanidinium isothiocyanate as the chaotropic binding salt except the 'Geneclean' kits which use NaI (107).

Protocol 13. Silica-based isolation of total DNA

Reagents

- Extraction buffer: 6 M guanidinium isothiocyanate, 100 mM Tris pH 7.0, 200 mM EDTA pH 7.0, 1.3% (v/v) Triton X-100
- Size-fractionated silica slurry (*Protocol 12*).
- 80% ethanol, –20°C
- Wash buffer: 6 M guanidinium isothiocyanate, 100 mM Tris pH 7.0
- Acetone
- TE: 10 mM Tris pH 8.0, 1 mM EDTA pH 8.0

Method

1. Add 25 mg tissue sample (or 500 mg of bone powder) to 0.5–1 ml of extraction buffer in a microcentrifuge tube. Incubate at 60°C for one to several hours with sporadic mixing.
2. Centrifuge at 12 000 × *g* for 5 min to remove any cellular debris.
3. Transfer 500 μl of the supernatant to a new tube containing 1 ml silica slurry (100 mg silica). Incubate at 20–25°C for 10 min.
4. Pellet the silica at 12 000 × *g* for 5–10 sec.

Protocol 13. *Continued*

5. Wash the silica pellet twice with 100 μl wash buffer. Pellet the silica as above after each wash.
6. Wash the silica pellet twice with 100 μl 80% ethanol. Pellet the silica as above after each wash.
7. Wash the silica pellet once with 100 μl acetone to aid the removal of residual ethanol. Pellet the silica as above.
8. Dry the silica pellet at 65°C in an oven or heat block.
9. Elute the nucleic acids at 65°C in two aliquots of 65 μl TE.
10. Use 5 μl in a typical 30 μl PCR.
11. Store the eluted DNA at 4°C or –20°C.

Variations on this protocol use filtration to wash the silica particles, and consequently simplify the washing steps (63,100).

3.10 Non-grinding technique

One of the disadvantages of the techniques outlined above is that many, especially those intended for isolation of plant DNA, require maceration of each tissue sample. Not only is this a time-consuming step, especially if many samples are involved, but it greatly increases the opportunity for contamination of one sample by another or by exogenous sources such as from the laboratory environment or the personnel performing the extraction.

Overcoming the need for a grinding step requires identification of an extraction buffer that will simultaneously solubilize the cellular components, inactivate DNases, and be readily removable so that subsequent manipulations will not also be inactivated. For plants the problem is exacerbated because not only must the membranes be solubilized but so must the cell walls. I have found the non-grinding method described in *Protocol 14* to be especially useful in its ability to extract DNA from plant tissue without grinding, in its relative simplicity and speed, and hence in its applicability to population studies. The important component of the extraction buffer is potassium ethyl xanthogenate, a compound used in the textile industry to solubilize cellulose. In the isolation buffer it may also solubilize polysaccharides, degrade or precipitate proteins, and bind metal ions thereby inhibiting DNase activity (108).

Protocol 14. Potassium xanthogenate isolation of total DNA

Reagents

- Extraction buffer: 200 mM potassium ethyl xanthogenate (Fluka 60045), 100 mM Tris pH 8.0, 50 mM EDTA pH 8.0, 300 mM NaCl, 2% (w/v) SDS, 1% (w/v) PVP-360, 0.2% (v/v) β-mercaptoethanol (added immediately prior to use in a fume hood)
- 5 M potassium acetate, pH 6.5
- 5 M NaCl
- Ethanol: 100% and 80%, –20°C
- 3 M sodium acetate, pH 7.6
- TE: 10 mM Tris pH 8.0, 1 mM EDTA pH 8.0

Method

1. Add 150 μl extraction buffer to 10 mg leaf tissue in a microcentrifuge tube. Incubate at 65°C for 30 min.
2. Vortex the sample at medium speed for 10 sec. Incubate at 65°C for 1.5 h. The leaf tissue will become more transparent as the cell walls break down.
3. Transfer fluid (*c.* 150 μl) from cleared leaf tissue to a new tube. Add 50 μl potassium acetate. A large precipitate of proteins should form immediately.
4. Centrifuge at 12 000 × *g* for 20 min at 4°C.
5. Collect the supernatant. Add 100 μl NaCl and mix thoroughly. Add 500 μl cold 100% ethanol and mix well. Incubate at 4°C for 1.5 h.
6. Centrifuge at 12 000 × *g* for 15 min at 4°C.
7. Discard the supernatant and allow the DNA pellet to dry.
8. Re-dissolve the DNA pellet in 100 μl TE.
9. Add 10 μl sodium acetate and mix. Add 200 μl 100% cold ethanol and mix well. Incubate at 4°C for 2–16 h.
10. Centrifuge at 12 000 × *g* for 15 min at 4°C.
11. Discard the supernatant. Wash the DNA pellet with 100 μl cold 80% ethanol; allow the pellet to soak for 10 min.
12. Centrifuge at 12 000 × *g* for 2 min at 4°C.
13. Discard the supernatant. Allow the DNA pellet to dry.
14. Resuspend the DNA in TE using small increments (e.g., 10–100 μl) depending on the size of the pellet.

Although addition of NaCl prior to alcohol precipitation of DNA should increase the separation of DNA from other cellular components, if that is not sufficient a chloroform extraction step may be added to the protocol just prior to the ethanol precipitation in step **5** (109).

3.11 Caesium chloride techniques

The final means of isolating total DNA is based on density gradient separation of cellular components in caesium chloride. This procedure will yield the most highly purified DNA. However, the procedure is expensive and may be time-consuming unless a suitable ultracentrifuge and rotor are available. This is a significant disadvantage when many samples are to be extracted as is typical of population studies. Nevertheless, I have found that with a Beckman TL-100 tabletop ultracentrifuge and a TLN-100 rotor runs of 3–4 h are sufficient and that these procedures are viable alternatives to those described above that require multiple organic extractions and/or precipitations (110–112). Furthermore, in some cases this may be the only means of extracting high quality DNA in the presence of numerous secondary plant compounds (113).

DNA may be purified by banding it in an equilibrium gradient of caesium chloride in the presence of one of several intercalating dyes: ethidium bromide, bisbenzimide, or propidium iodide. Bisbenzimide and propidium iodide are particularly useful for physically separating genomes based on their buoyant density. For example, fungal mitochondrial DNA and algal chloroplast DNA, two genomes with relatively high AT content and low buoyant density, can often be separated from nuclear DNA in a total DNA preparation on a CsCl–bisbenzimide gradient. Higher plant chloroplast genomes are generally similar enough in density to the nuclear genome that they cannot be physically separated in this way. Nevertheless, a caesium chloride gradient with any of these dyes may be used to purify DNA by removing RNA, proteins, and polysaccharides.

Total DNA is isolated as described in steps **1–4** in *Protocol 7*, and PVP is omitted from the isolation buffer. Care should be taken to match the quantity of extraction buffer used, and hence the volume of the aqueous phase following chloroform extraction, to the size of the caesium chloride gradients. I routinely extract 0.5–1.5 g of fresh tissue in 2.0–3.0 ml extraction buffer for use on 3.9 ml caesium chloride gradients. This total DNA may then be purified on a caesium chloride gradient as described in *Protocol 15*.

One variation on this procedure that may be particularly useful for population studies is to eliminate the organic extraction and rely on the gradient to separate the DNA from the remaining cellular components. This streamlines the procedure especially when DNA is extracted from small tissue samples. In this case it is also possible to include the caesium chloride directly in the extraction buffer; use 6.736 g caesium chloride per 10 ml final volume of extraction buffer; the density (ρ) of this solution is 1.4969. Weeks *et al.* (112) describe such a buffer using the detergent *N*-lauroylsarcosine sodium salt; I have successfully used this approach with the extraction buffer listed in *Protocol 7*, omitting PVP. In this case it is best to limit the amount of tissue to 100–500 mg, although larger caesium chloride gradients may adequately handle more tissue.

Other variations in the extraction buffer are common. One variant useful for cotton involves an extraction buffer composed of 0.3 M sucrose, 50 mM Tris pH 8.8, 5 mM $MgCl_2$, followed by addition of PVP-40 to 1% (w/v) and ethidium bromide to 5 g/ml, followed by centrifugation at 2500 $\times$ *g* for 10 min at 4°C and resuspension of the pellet in a lysis buffer composed of 50 mM Tris pH 8.0, 20 mM EDTA, 1% (w/v) *N*-lauroylsarcosine sodium salt (J. Price, personal communication). A second buffer found to yield 32–376 g DNA per gram of tissue fresh weight in numerous plant species consisted of 200 mM Tris pH 8.3, 300 mM LiCl, 10 mM Na_2EDTA, 1% (w/v) deoxycholic acid sodium salt, 1 mM aurintricarboxylic acid, 5 mM thiourea, 1.5% (w/v) lithium dodecyl sulfate, 1% (v/v) Nonidet P-40, 10 mM DTT (113).

Methods for constructing caesium chloride gradients abound and must be adapted as appropriate for the rotors at hand. I have found the combination of a Beckman TL-100 table-top ultracentrifuge and the near-vertical TLN-100 rotor to be useful because of the rapidity with which samples may be processed. Therefore, I include the protocol for constructing gradients for this combination (see *Protocol 15*); the procedure may be easily scaled up or down for other centrifuge/rotor combinations.

Protocol 15. Caesium chloride purification of DNA

Equipment and reagents

- CsCl (ρ = 1.88). At 20–25°C CsCl-saturated H_2O has a density of approximately 1.88.
- Intercalating dye: 10 mg/ml ethidium bromide, 10 mg/ml bisbenzimide (Hoechst 33258),[a] or 2 mg/ml propidium iodide. Because these dyes bind DNA and may be mutagens, they should be handled with care using protective gloves
- 100% ethanol or isopropanol, –20°C
- Isopropanol saturated in H_2O and NaCl. Combine 100 ml isopropanol, 50 ml H_2O, and 50 g NaCl. Shake well and allow the three phases to separate. Isopropanol is the uppermost phase.
- Syringe needles, 18 and 20 gauge
- mineral oil

A. *Caesium chloride–ethidium bromide gradients*

1. To prepare a CsCl gradient with an average density of 1.546, add 1.65 ml CsCl (ρ = 1.88) for every 1.0 ml of resuspended DNA. (For a gradient with average density of ρ = 1.4969, use 1.31 ml CsCl (ρ = 1.88) for each 1.0 ml sample. For plasmid preparations prepare a gradient with average density of ρ = 1.699; use 3.9 ml CsCl (ρ = 1.88) for each 1.0 ml sample.) Add 10 µl ethidium bromide for each 1.0 ml of original sample. Seal the sample in an ultracentrifuge tube. If the sample does not completely fill the tube, add enough mineral oil to do so to prevent tube collapse during centrifugation. Note that the volumes of CsCl (ρ = 1.88) given above take into account the volumes of both the sample and the ethidium bromide to yield the correct final density.
2. Spin the ρ = 1.546 gradient in a Beckman TLN-100 near-vertical ultracentrifuge rotor at 450 000 × *g* for 3–4 h at 20°C. Make sure that the rotor slows with little or no braking after the run so the gradient remains intact. (For plasmid preparations, spin the ρ = 1.699 gradient at 288 000 × *g* for 12 h at 20°C. Centrifugal force must be reduced for high density gradients to prevent sedimentation of CsCl crystals and consequent overstressing of the rotor.) If a bench-top ultracentrifuge is not available, centrifugation of gradients at 100 000–200 000 × *g* in a vertical rotor for 12–16 h at 20°C or 250 000–350 000 × *g* in a fixed-angle rotor for 24 h at 20°C should be adequate. Be sure to check the rotor specifications to make sure it is rated for high density solutions at these forces.
3. The DNA should form a fluorescing band in the central region of the centrifuge tube. Remove the band with an 18 gauge syringe needle; a 21 gauge syringe needle may be used as a vent. With the TLN-100

Protocol 15. *Continued*

tubes it should be possible to remove the band in approximately 200 μl. Adjust the following volumes according to the actual volume recovered.

4. Extract the sample three times (or until the isopropanol phase is no longer pink) with isopropanol saturated in H_2O and NaCl to remove the ethidium bromide. Each time add excess (e.g., 500 μl) isopropanol, mix gently, let stand to allow phase separation, then remove the isopropanol. After the last extraction, spin the sample in a microcentrifuge for 2 min to ensure good phase separation. Transfer the bottom aqueous layer (e.g., 200 μl) to a new microcentrifuge tube.
5. Add twice the volume of the aqueous layer (e.g., 400 μl) of H_2O to dilute the CsCl, and mix gently. Add six times the volume of the aqueous layer (e.g., 1200 μl) of ethanol. The solution should consist of sample: H_2O:ethanol (1:2:6). Alternatively, the DNA may be precipitated in isopropanol. In this case, add a volume of water equivalent to the volume of the aqueous layer (e.g., 200 μl), and add twice the volume of the aqueous layer (e.g., 400 μl) of isopropanol. The solution should consist of sample:H_2O:isopropanol (1:1:2). Precipitate the nucleic acids at 4°C for 30 min to overnight. Do not place at –80°C or the CsCl will precipitate.
6. Spin in microcentrifuge for 10 min to collect nucleic acids.
7. Wash pellet with 80% ethanol for 10 min. Spin in microcentrifuge for 5 min to recollect the nucleic acids.
8. Resuspend pellet in small increments (e.g., 10–100 μl) of 10 mM Tris, 1 mM EDTA, pH 8.0 (TE) depending on the size of the pellet.

B. *Caesium chloride–bisbenzimide gradients*

1. Prepare the CsCl gradient as described above, except add 50 μl bisbenzimide instead of ethidium bromide. Because a larger volume of intercalating dye is needed for bisbenzimide gradients, the volume of CsCl (ρ = 1.88) must be adjusted to obtain the same final density. To prepare ρ = 1.546, ρ = 1.4969, or ρ = 1.699 gradients, use 1.72 ml, 1.36 ml, or 4.06 ml CsCl (ρ = 1.88), respectively, per 1.0 ml of sample.
2. Proceed with the centrifugation and purification as described above for CsCl –ethidium bromide gradients.

C. *Caesium chloride –propidium iodide gradients*

1. Prepare the CsCl gradient as described above, except add 500 μl propidium iodide (2 mg/ml) instead of ethidium bromide. Because a larger volume of intercalating dye is needed for propidium iodide gradients, the volume of CsCl (ρ = 1.88) must be adjusted to obtain the same final density. To prepare ρ = 1.546, ρ = 1.4969, or ρ = 1.699 gradients, use

2.45 ml, 1.95 ml, or 5.79 ml CsCl (ρ = 1.88), respectively, per 1.0 ml of sample.

2. Proceed with the centrifugation and purification as described above for CsCl–ethidium bromide gradients.

[a]The stock bisbenzimide solution should be stored at −20°C in the dark. Never add bisbenzimide to DNA solutions that are not in high-salt solutions.

As a reliable means of separating DNA from caesium chloride following equilibrium gradient centrifugation, dialysis is a useful alternative to precipitation. In some cases less degradation of the DNA is observed following dialysis than following precipitation. Following removal of ethidium bromide from the DNA as discussed in *Protocol 15*, the dialysis described in *Protocol 16* may be substituted for the precipitation steps. In this procedure, care should be taken to handle the dialysis tubing with gloves.

Protocol 16. Dialysis of caesium chloride

Reagents

- Dialysis tubing: molecular weight cut-off 12 000–14 000
- 10 mM EDTA pH 8.0
- 2% (w/v) sodium bicarbonate, 10 mM EDTA pH 8.0
- TE: 10 mM Tris pH 8.0, 1 mM EDTA pH 8.0

A. *Preparation of dialysis tubing*

1. Cut tubing into many pieces approximately 20 cm in length.
2. Boil the tubing in sodium bicarbonate, EDTA for 10 min. Keep the tubing submerged from this point on.
3. Rinse the tubing several times in distilled H_2O.
4. Boil the thoroughly rinsed tubing in EDTA for 10 min.
5. Allow tubing and solution to cool and store it submerged in 10 mM EDTA, pH 8.0 at 4°C.

B. *Dialysis of DNA samples*

1. Rinse dialysis tubing in distilled H_2O before use.
2. Transfer the DNA–caesium chloride solution to a dialysis tube, sealing both ends with clamps.
3. Submerge the tube in a large volume (e.g., 500–1000 ml) of TE and mix slowly for 12–24 h. During this time change the buffer at least twice.
4. Transfer the DNA solution to labelled tubes.
5. If the volume of DNA is too large after dialysis, extract it with an equal volume of 2-butanol, centrifuge at 12 000 × *g* for 20 sec, and remove the upper, butanol phase. This may be repeated until the desired volume is achieved.

3.12 Chloroplast DNA technique

Although any of the total DNA isolations described above provide access to DNA derived from any of the cell organelles, in some instances it is desirable to limit consideration to DNA derived only from the chloroplast. For example, restriction digestion of chloroplast DNA reveals a set of distinguishable restriction fragments, whereas only an indistinguishable smear of fragments is obtained if the much larger nuclear genome is retained along with the chloroplast genome. The distinct chloroplast fragments may be analysed directly from agarose gels, so other selective techniques such as PCR or Southern blots are not required. Likewise, screening a library of clones for chloroplast-derived fragments is much simpler if the library was constructed from chloroplast DNA.

Essentially all methods for isolating DNA derived from a single type of organelle, including chloroplasts as well as nuclei and mitochondria, rely on separating the desired organelles from the remainder of the cellular components. Thus, the general principles of cell fractionation are especially relevant to understanding and optimizing these techniques (6–13). A number of methods have been described that are specifically designed for isolation of nucleic acids from purified nuclei, mitochondria, or chloroplasts (1–4,114). Because of the utility of chloroplast DNA, specifically, in the study of varation within and among populations and species of plants, I focus this section on the isolation of intact chloroplasts. Once a pellet of intact chloroplasts has been obtained, essentially any of the means of isolating total DNA may be followed to extract the chloroplast DNA from the organelles. It is, of course, only the initial separation of cellular components that is critical in yielding chloroplast as opposed to total DNA. For those specifically interested in a broader range of techniques for isolating chloroplast DNA than considered here, the earlier edition of this book (5) should be consulted. The protocol outlined below is based on isolating intact chloroplasts in a high salt medium that lyses nuclei (115).

Protocol 17. Isolation of chloroplasts

Equipment and reagents

- Isolation buffer: 1.25 M NaCl, 50 mM Tris pH 8.0, 5 mM EDTA, 0.1% (w/v) BSA, 0.1% (v/v) β-mercaptoethanol (added immediately prior to use in a fume hood)
- Blender ot tissue homogenizer
- Cheesecloth
- Miracloth (Calbiochem)
- 50 ml centrifuge tubes

Method

1. Add 10 g tissue to 100 ml cold isolation buffer and grind in a cold blender or tissue homogenizer. Filter through four layers of cheesecloth and squeeze out all liquid. If buffer still contains large quantities

of particulate matter, filter through one layer of Miracloth or another four layers of cheesecloth without squeezing.

2. Pellet the chloroplasts in the isolation buffer at 3000 × *g* for 10 min at 4°C. At this step most of the nuclear DNA is removed with the supernatant.
3. Resuspend the chloroplast pellet in 10 ml cold isolation buffer and transfer to 50 ml centrifuge tubes.
4. Pellet the chloroplasts again at 3000 × *g* for 5 min at 4°C.
5. Resuspend the chloroplast pellet and proceed as if isolating total DNA. Both *Protocol 7* and *Protocol 10* are useful. In either case use 10 ml of the appropriate extraction buffer to resuspend the chloroplast pellet.

4. Conclusion

Population biology is rapidly incorporating genetic information; however, development of the full potential offered by genetics is in its infancy (14). One of the main obstacles has been the lack of reliable techniques for efficiently isolating DNA from many samples. For example, most methods used in studies of animals are hopeless in the face of the myriad secondary compounds commonly found in plants, and the sub-optimal preservation conditions associated with ancient or forensic samples present special challenges. In this chapter I have collected a broad range of techniques that have proven useful both in my own laboratory and in those of others for overcoming the problems of isolating DNA from a taxonomically diverse array of species preserved under a variety of conditions. This collection should by no means be regarded as exhaustive; rather I have included what I consider to be representatives of several major approaches to the isolation of total cellular DNA. These are augmented with references to papers that describe them in further detail or that describe variations useful for particular species or sources of samples. In devising my own techniques I have often relied on combining distinct portions of existing methods into a novel combination that overcomes the problem at hand. In this regard it is best to keep in mind what each step in a protocol accomplishes, and to remember that in many cases what appear to be equivalent steps may often be substituted with non-equivalent results. The first stage of a population survey, therefore, should consist of extracting DNA from a few individuals using the protocol that appears most appropriate. If problems are encountered, however, several alternative isolation techniques may be compared. Once the best method has been identified, a broad survey may be undertaken successfully. The protocols included here should enable the researcher to undertake a population study of DNA variation in most groups of plant and animal species and to accommodate specimens preserved under a variety of conditions.

References

1. De Vries, S., Hoge, H., and Bisseling, T. (1988). In *Plant molecular biology manual* (ed. S. B. Gelvin and R. A. Schilperoort), Chapter B6, p. 1. Kluwer Academic, Dordrecht.
2. Hamby, R. K., Sims, L., Issel, L., and Zimmer, E. (1988). *Plant Mol. Biol. Reporter*, **6**(3), 175.
3. Jofuku, K. D. and Goldberg, R. B. (1988). In *Plant molecular biology: a practical approach* (ed. C. H. Shaw), p. 37. IRL Press, Oxford.
4. Skubatz, H. and Bendich, A. (1990). In *Plant molecular biology manual* (ed. S. B. Gelvin, R. A. Schilperoort, and D. P. S. Verma), Update 1, Chapter A11, p. 1. Kluwer Academic, Dordrecht.
5. Milligan, B. G. (1992). In *Molecular genetic analysis of populations* (ed. A. R. Hoelzel), p. 59. IRL Press, Oxford.
6. Alberts, B., Bray, D., Lewis, J., Raff, M., Roberts, K., and Watson, J. D. (1994). *Molecular biology of the cell*, 3rd edn. Garland Publishing, New York, NY.
7. Lodish, H., Baltimore, D., Berk, A., Zipursky, S. L., Matsudaira, P., and Darnell, J. (1995). *Molecular cell biology*, 3rd edn. Scientific American Books, New York, NY.
8. Hall, J. L. and Moore, A. L. (eds) (1983). *Isolation of membranes and organelles from plant cells*. Academic Press, London.
9. Price, C. A. (1983). In *Isolation of membranes and organelles from plant cells* (ed. J. L. Hall and A. L. Moore), p. 1. Academic Press, London.
10. Quail, P. H. (1979). *Annu. Rev. Plant Physiol.*, **30**, 425.
11. Reid, E. (ed.) (1979). *Methodological surveys*, Vol. 9: *Plant organelles*. Ellis Horwood Limited, Chichester.
12. Schuler, M. A. and Zielinski, R. E. (1989). *Methods in plant molecular biology*. Academic Press, San Diego, CA.
13. Shaw, C. H. (ed.) (1988). *Plant molecular biology: a practical approach*. IRL Press, Oxford.
14. Milligan, B. G., Leebens-Mack, J., and Strand, A. E. (1994). *Mol. Ecol.*, **3**, 423.
15. Doyle, J. J. and Dickson, E. E. (1987). *Taxon*, **36**, 715.
16. Golenberg, E. M., Giannasi, D. E., Clegg, M. T., Smiley, C. J., Durbin, M., Henderson, D., and Zurawski, G. (1990). *Nature*, **344**, 656.
17. Pyle, M. M. and Adams, R. P. (1989). *Taxon*, **38**, 576.
18. Flournoy, L. E., Adams, R. P., and Pandy, R. N. (1996). *BioTechniques*, **20**, 657.
19. Liston, A., Rieseberg, L. H., Adams, R. P., and Do, N. (1990). *Ann. Missouri Bot. Garden,* **77**, 859.
20. Chase, M. W. and Hills, H. H. (1991). *Taxon*, **40**, 215.
21. Rogstad, S. H. (1992). *Taxon*, **41**, 701.
22. Nickrent, D. L. (1994). *BioTechniques*, **16**, 470.
23. Taylor, J. W. and Swann, E. C. (1994). In *Ancient DNA. Recovery and analysis of genetic material from paleontological, archaeological, museum, medical, and forensic specimens* (ed. B. Herrmann and S. Hummel), p. 166. Springer-Verlag, New York, NY.
24. Dessauer, H. C., Cole, C. J., and Hafner, M. S. (1996). In *Molecular systematics* (ed. D. M. Hillis, C. Moritz, and B. K. Mable), 2nd edn, p. 00. Sinauer, Sunderland, MA.

25. Amos, W. and Hoelzel, A. R. (1991). *I.W.C. Special Issue*, **13**, 99.
26. Shibata, D. (1994). In *Polymerase chain reaction* (ed. K. B. Mullis, F. Ferre, and R. A. Gibbs), p. 47. Birkhäuser, Boston, MA.
27. Kawasaki, E. S. (1990). In *PCR protocols. A guide to methods and applications* (ed. M. A. Innis, D. H. Gelfand, J. J. Sninsky, and T. J. White), p. 146. Academic Press, San Diego, CA.
28. Pääbo, S., Higuchi, R. G., and Wilson, A. C. (1989). *J. Biol. Chem.*, **264**, 9709.
29. Kwok, S. (1990). In *PCR protocols. A guide to methods and applications* (ed. M. A. Innis, D. H. Gelfand, J. J. Sninsky, and T. J. White), p.142. Academic Press, San Diego, CA.
30. Orrego, C. (1990). In *PCR protocols. A guide to methods and applications* (ed. M. A. Innis, D. H. Gelfand, J. J. Sninsky, and T. J. White), p. 447. Academic Press, San Diego, CA.
31. Pääbo, S. (1990). In *PCR protocols. A guide to methods and applications* (ed. M. A. Innis, D. H. Gelfand, J. J. Sninsky, and T. J. White), p. 159. Academic Press, San Diego, CA.
32. Handt, O., Höss, M., Krings, M., and Pääbo, S. (1994). *Experientia*, **50**, 524.
33. Hummel, S. and Herrmann, B. (1994). In *Ancient DNA. Recovery and analysis of genetic material from paleontological, archaeological, museum, medical, and forensic specimens* (ed. B. Herrmann and S. Hummel), p. 59. Springer-Verlag, New York, NY.
34. Merriwether, D. A., Rothhammer, F., and Ferrell, R. E. (1994). *Experientia*, **50**, 592.
35. Handt, O., Richards, M., Trommsdorff, M., Kilger, C., Simanainen, J., Georgiev, O., Bauer, K., Stone, A., Hedges, R., Schaffner, W., Utermann, G., Sykes, B., and Pääbo, S. (1994). *Science*, **264**, 1775.
36. Rollo, F., Asci, W., Antonini, S., Marota, I., and Ubaldi, M. (1994). *Experientia*, **50**, 576.
37. Poinar, G. O., Jr (1994). *Experientia*, **50**, 536.
38. Hauswirth, W. W., Dickel, C. D., Rowold, D. J., and Hauswirth, M. A. (1994). *Experientia*, **50**, 585.
39. Brown, T. A., Allaby, R. G., Brown, K. A., O'Donoghue, K., and Sallares, R. (1994). *Experientia*, **50**, 571.
40. Hardy, C., Casane, D., Vigne, J. D., Callou, C., Dennebouy, N., Mounolou, J.-C., and Monnerot, M. (1994). *Experientia*, **50**, 564.
41. Hardy, C., Callou, C., Vigne, J.-D., Casane, D., Dennebouy, N., Mounolou, J.-C., and Monnerot, M. (1995). *J. Mol. Evol.*, **40**, 227.
42. Cooper, A. (1994). In *Ancient DNA. Recovery and analysis of genetic material from paleontological, archaeological, museum, medical, and forensic specimens* (ed. B. Herrman and S. Hummel), p. 149. Springer-Verlag, New York, NY.
43. Roy, M. S., Girman, D. J., Taylor, A. C., and Wayne, R. K. (1994). *Experientia*, **50**, 551.
44. Herrmann, B. and Hummel, S. (eds) (1994). *Ancient DNA. Recovery and analysis of genetic material from paleontological, archaeological, museum, medical, and forensic specimens*. Springer-Verlag, New York, NY.
45. Hagelberg, E. (1994). In *Ancient DNA. Recovery and analysis of genetic material from paleontological, archaeological, museum, medical, and forensic specimens* (ed. B. Herrmann and S. Hummel), p. 195. Springer-Verlag, New York, NY.

46. Dunham, V. L. and Bryant, J. A. (1983). In *Isolation of membranes and organelles from plant cells* (ed. J. L. Hall and A. L. Moore), p. 237. Academic Press, London.
47. Jones, M. C. and Boffey, S. A. (1984). *FEBS Lett.*, **174**, 215.
48. Herrmann, R. G., Palta, H. K., and Kowallik, K. V. (1980). *Planta*, **148**, 319.
49. Calie, P. J. and Hughes, K. W. (1987). *Plant Mol. Biol. Reporter*, **4**(4), 206.
50. Shoemaker, R. C., Atherly, A. G., and Palmer, R. G. (1983). *Plant Cell Rep.*, **2**, 98.
51. Shoemaker, R. C., Palmer, R. G., and Atherly, A. G. (1984). *Plant Mol. Biol. Reporter*, **2**(2), 15.
52. Kislev, N. and Rubenstein, I. (1980). *Plant Physiol.*, **66**, 1140.
53. Hallick, R. B., Chelm, B. K., Gray, P. W., and Orozco, E. M., Jr (1977). *Nucleic Acids Res.*, **4**, 3055.
54. Stern, D. B. and Palmer, J. D. (1984). *Proc. Natl Acad. Sci. USA*, **81**, 1946.
55. Nagy, F., Kay, S. A., and Chua, N.-H. (1988). In *Plant molecular biology manual* (ed. S. B. Gelvin and R. A. Schilperoort), p. 1. Kluwer Academic, Dordrecht.
56. Do, N. and Adams, R. P. (1991). *BioTechniques*, **10**, 162.
57. Fang, G., Hammar, S., and Grumet, R. (1992). *BioTechniques*, **13**, 52.
58. Lodhi, M. A., Ye, G.-N., Weeden, N. F., and Reisch, B. I. (1994). *Plant Mol. Biol. Reporter*, **12**, 6.
59. Saiki, R. K. (1990). In *PCR protocols. A guide to methods and applications* (ed. M. A. Innis, D. H. Gelfand, J. J. Sninsky, and T. J. White), p. 13. Academic Press, San Diego, CA.
60. Wang, H., Qi, M., and Cutler, A. J. (1993). *Nucleic Acids Res.*, **21**, 4153.
61. Langridge, U., Schwall, M., and Langridge, P. (1991). *Nucleic Acids Res.*, **19**, 6954.
62. Walsh, P. S., Metzger, D. A., and Higuchi, R. (1991). *BioTechniques*, **10**, 506.
63. Cano, R. J. and Poinar, H. N. (1993). *BioTechniques*, **15**, 432.
64. Cano, R., Poinar, H. N., Pieniazek, N. J., Acra, A., and Poinar, G. O., Jr (1993). *Nature*, **363**, 536.
65. Ellegren, H. (1994). In *Ancient DNA. Recovery and analysis of genetic material from paleontological, archaeological, museum, medical, and forensic specimens* (ed. B. Herrmann and S. Hummel), p. 211. Springer-Verlag, New York, NY.
66. Poinar, G. O., Jr, Poinar, H. N., and Cano, R. J. (1994). In *Ancient DNA. Recovery and analysis of genetic material from paleontological, archaeological, museum, medical, and forensic specimens* (ed. B. Herrmann and S. Hummel), p. 92. Springer-Verlag, New York, NY.
67. Sensabaugh, G. F. (1994). In *Ancient DNA. Recovery and analysis of genetic material from paleontological, archaeological, museum, medical, and forensic specimens* (ed. B. Herrmann and S. Hummel), p. 141. Springer-Verlag, New York, NY.
68. Edwards, K., Johnstone, C., and Thompson, C. (1991). *Nucleic Acids Res.*, **19**, 1349.
69. Goodwin, D. and Lee, S. (1993). *BioTechniques*, **15**, 438.
70. Liou, G. I. and Matragoon, S. (1992). *BioTechniques*, **13**, 719.
71. Doyle, J. J. and Doyle, J. L. (1987). *Phytochem. Bull.*, **19**, 11.
72. Murray, M. G. and Thompson, W. F. (1980). *Nucleic Acids Res.*, **8**, 4321.
73. Saghai-Maroof, M. A., Soliman, K. M., Jorgensen, R. A., and Allard, R. W. (1984). *Proc. Natl Acad. Sci. USA*, **81**, 8014.
74. Rogers, S. O. and Bendich, A. J. (1985). *Plant Mol. Biol.*, **5**, 69.
75. Rogers, S. O. and Bendich, A. J. (1988). In *Plant molecular biology manual* (ed. S. B. Gelvin and R. A. Schilperoort), p. 1. Kluwer Academic, Dordrecht.

76. Stewart, C. N., Jr, and Via, L. E. (1993). *BioTechniques*, **14**, 748.
77. Bellamy, A. R. and Ralph, R. K. (1968). In *Methods in enzymology*, Vol. 12B: *Nucleic acids* (ed. L. Grossman and K. Moldave), p. 156. Academic Press, New York, NY.
78. Appels, R. and Moran, L. B. (1984). In *16th Stadler Genetics Symposium: Gene manipulation in plant improvement* (ed. J. P. Gustafson), p. 529. Plenum Press, New York, NY.
79. Koebner, R. M. D., Appels, R., and Shepherd, K. W. (1986). *Can. J. Genet. Cytol.*, **28**, 658.
80. Clarke, B. C., Moran, L. B., and Appels, R. (1989). *Genome*, **32**, 334.
81. Hillis, D. M., Larson, A., Davis, S. K., and Zimmer, E. A. (1990). In *Molecular systematics*, p. 318. Sinauer, Sunderland, MA.
82. Smith, L. J., Braylan, R. C., Nutkis, J. E., Edmundson, K. B., Downing, J. R., and Wakeland, E. K. (1987). *Anal. Biochem.*, **160**, 135.
83. Wright, D. K. and Manos, M. M. (1990). In *PCR protocols. A guide to methods and applications* (ed. M. A. Innis, D. H. Gelfand, J. J. Sninsky, and T. J. White), p. 153. Academic Press, San Diego,CA.
84. Banerjee, S. K., Makdisi, W. F., Weston, A. P., Mitchell, S. M., and Campbell, D. R. (1995). *BioTechniques*, **18**, 768.
85. Turbett, G. R., Barnett, T. C., Dillon, E. K., and Sellner, L. N. (1996). *BioTechniques*, **20**, 846.
86. Houde, P. and Braun, M. J. (1988). *Auk*, **105**, 773.
87. Houde, P., Sheldon, F. H., and Kreitman, M. (1995). *J. Mol. Evol.*, **40**, 678.
88. Higuchi, R., Bowman, B., Freiberger, M., Ryder, O. A., and Wilson, A. C. (1984). *Nature*, **312**, 282.
89. Tuross, N. (1994). *Experientia*, **50**, 530.
90. Dellaporta, S. L., Wood, J., and Hicks, J. B. (1983). *Plant Mol. Biol. Reporter*, **1**(4), 19.
91. del Castillo Agudo, L., Gavidia, I., Pérez-Bermúdez, P., and Segura, J. (1995). *BioTechniques*, **18**, 766.
92. Chomczynski, P. (1993). *BioTechniques*, **15**, 532.
93. Cox, R. A. (1968). In *Methods in enzymology*, Vol. 12B: *Nucleic acids* (ed. L. Grossman and K. Moldave), p. 120. Academic Press, New York, NY.
94. Chomczynski, P. and Sacchi, N. (1987). *Anal. Biochem.*, **162**, 156.
95. Puissant, C. and Houdebine, L.-M. (1990). *BioTechniques*, **8**, 148.
96. Chomczynski, P. and Mackey, K. (1995). *BioTechniques*, **19**, 942.
97. Bowtell, D. D. L. (1987). *BioTechniques*, **162**, 463.
98. Jeanpierre, M. (1987). *Nucleic Acids Res.*, **15**, 9611.
99. Höss, M. and Pääbo, S. (1993). *Nucleic Acids Res.*, **21**, 3913.
100. Nollau, P., Moser, C., and Wagener, C. (1996). *BioTechniques*, **20**, 784.
101. Vogelstein, B. and Gillespie, D. (1979). *Proc. Natl Acad. Sci. USA*, **76**, 615.
102. Marko, M. A., Chipperfield, R., and Birnboim, H. C. (1982). *Anal. Biochem.*, **121**, 382.
103. Kaur, R., Kumar, R., and Bachhawat, A. K. (1995). *Nucleic Acids Res.*, **23**, 4932.
104. Beld, M., Sol, C., Goudsmit, J., and Boom, R. (1996). *Nucleic Acids Res.*, **24**, 2618.
105. Carter, M. J. and Milton, I. D. (1993). *Nucleic Acids Res.*, **21**, 1044.

106. Boom, R., Sol, C. J. A., Salimans, M. M. M., Jansen, C. L., Wertheim-van Dillen, P. M. E., and van der Noordaa, J. (1990). *J. Clin. Microbiol*, **28**, 495.
107. Boyle, J. S. and Lew, A. M. (1995). *Trends Genet.*, **11**, 8.
108. Jhingan, A. K. (1992). *Methods Mol. Cell. Biol.*, **3**, 15.
109. Ross, I. K. (1995). *BioTechniques*, **18**, 828.
110. Carr, S. M. and Griffith, O. M. (1987). *Biochem. Genet.*, **25**, 385.
111. Iversen, P. L., Mata, J. E., and Hines, R. N. (1987). *BioTechniques*, **5**, 521.
112. Weeks, D. P., Beerman, N., and Griffith, O. M. (1986). *Anal. Biochem.*, **152**, 376.
113. Baker, S. S., Rugh, C. L., and Kamalay, J. C. (1990). *BioTechniques*, **9**, 268.
114. Palmer, J. D. (1986). In *Methods in Enzymology* Vol. 167: *Plant molecular biology* (ed. A. Weissbach and H. Weissbach), p. 167. Academic Press, Orlando, FL.
115. Bookjans, G., Stummann, B. M., and Henningsen, K. W. (1984). *Anal. Biochem.*, **141**, 244.

3

Mitochondrial DNA isolation, separation, and detection of fragments

P. SCOTT WHITE, OWATHA L. TATUM, HÅKAN TEGELSTRÖM, and LLEWELLYN D. DENSMORE III

1. Introduction

The 'revolution' that began in the 1980s with the application of mitochondrial DNA (mtDNA) analyses to address questions of molecular evolutionary and population biology has now largely become the paradigm for such work (1). Studies ranging in scope from purely systematic problems to identification and recognition of species to documentation of genetic changes in populations that have occurred over a very short period of time, are based on both indirect or direct estimates of nucleic acid variation. While progress continues to be made in analysing coding regions of nuclear DNA, the sequences retained in organellar genomes are still the most widely studied by evolutionary biologists. Restriction site or fragment analyses of mtDNA and direct sequencing of specific regions of the mitochondrial genome following amplification by the polymerase chain reaction (PCR) are currently the methods of choice for the majority of population level studies. The reasons for the increasingly widespread application of mtDNA markers in studies of animal populations are many; however, there are also a few noteworthy drawbacks (reviewed in ref. 1). *Figure 1* shows the typical vertebrate gene order of mtDNA, with birds having a single rearrangement involving ND 6 (a subunit of NAD dehydrogenase) and $tRNA^{Glu}$ not found in other vertebrates examined to date. Other minor rearrangements have been reported, and gene order can provide important synapomorphies (shared derived characters) for deciphering some relationships.

Mitochondria are common in many cells (about 1×10^3 organelles per cell in rat liver), but stoichiometrically a much smaller proportion of most cells' nucleotide bases are mitochondrial as compared with nuclear DNA. This latter characteristic directly influences both the isolation and purification, aspects on which we will concentrate in this chapter. Traditionally, mtDNA is

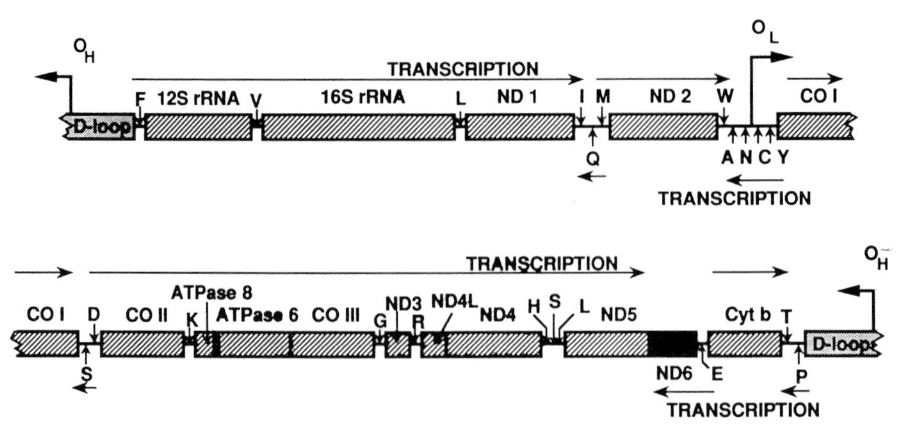

Figure 1. Linearized representation of mammalian mitochondrial genome. O_H and O_L refer to origins of 'heavy' and 'light' strand replication, respectively, with the arrows indicating direction of replication. Small lines and arrows indicate direction of transcription. The standard single letter abbreviations for abbreviations are used to denote various tRNA-coding genes; the regions labelled 12S rRNA and 16S rRNA show the DNA encoding RNA in the small and large mitochondrial ribosomal subunits respectively. Other abbreviations correspond to the following: ND 1–ND 5, genes encoding different subunits of NAD dehydrogenase; CO I–CO III, genes encoding three subunits of cytochrome oxidase; ATPase 6 and 8, genes encoding two subunits of ATPase; and Cyt b, gene encoding cytochrome *b*.

isolated by lysing mitochondria that have been separated from the nuclei and other organelles by differential centrifugation (2). Specific mitochondrial probes have also been extensively used to identify the mitochondrial fraction in Southern transfer and hybridization reactions of total DNA preparations. Today, with a relatively small number of highly conserved, consensus PCR primers the entire mtDNA of most vertebrates and some invertebrates can be 'isolated' using only PCR amplification of very small amounts of total DNA. These amplification reactions are no longer merely being used with previously isolated or cloned mtDNA fragments for the purpose of DNA sequencing. The 'isolation' is actually a specific enhancement of the mitochondrial fraction present in the total DNA sample, with fragments of interest being amplified to microgram or even milligram amounts.

Fragments of mtDNA, size fractionated on agarose or polyacrylamide gels, have traditionally been detected using intercalating dyes or nucleic acid stains or by incorporating radioactively labelled nucleotides, followed by autoradiography. To these standard methods have now been added a large number of other non-radioactive strategies, ranging from the incorporation of biotinylated or fluorescent nucleotides to the use of heterologous chemiluminescent or chromogenic probes. In this chapter we will present protocols for isolating and/or enhancing the concentration of mtDNA and for separating and detecting fragments following digestion with one or more restriction enzymes or PCR amplification.

1.1 Choice of tissue

The quantity and quality of the tissue from which mtDNA is to be extracted determine whether it is more efficient to isolate pure mtDNA or if it is necessary to isolate total cellular DNA, and use a mtDNA probe that has been either purified or PCR amplified. Total DNA can also be used as template for amplification of mitochondrial fragments that can be visualized directly. Heart and pleopod muscle in crustaceans, adductor muscle in bivalve molluscs, and insect flight muscle are all suitable invertebrate tissues for use in isolation of mtDNA. Fresh or frozen organ tissue such as liver, heart, brain, ovary, and kidney are excellent sources of vertebrate mtDNA, and yields are greatly increased if the tissue is first pulverized in liquid nitrogen and then homogenized in a buffer before pelleting the nuclei and cell debris. It is often possible to use skeletal muscle or whole blood, particularly blood obtained from non-mammalian vertebrates, but for routine, non-PCR analyses these may not be dependable sources of mtDNA. Non-mammalian whole blood is a reasonable source of mtDNA in amounts as small as 1.0 ml, since erythrocytes contain not only nuclei, but also mitochondria (see *Figure 2*). Yields of mtDNA from blood are considerably lower than from organ tissues, probably due to the small number of mitochondria per erythrocyte (about 10% of the number in liver).

Techniques for collecting blood, often the primary source of DNA for use in population-level studies, have until recently not considered the effects that various anti-coagulants have on the integrity of the DNA. Blood treatment at collection is discussed and protocols are presented in ref. 11.

2. Methods of isolation

2.1 Differential centrifugation

Differential centrifugation of a cell lysate separates intact mitochondria from intact nuclei and cell debris; an effective procedure is given in *Protocol 1*. Fresh tissues such as unfertilized animal eggs, ovaries, and liver are sometimes available in sufficient quantity to provide high yields of mitochondria, but most tissues still contain enough contaminating nuclear DNA to require further purification. Sucrose step gradients (3) and isopycnic sucrose gradients as in *Protocol 2* (4) can be used to obtain purified mitochondria from a cell homogenate from which DNA can be extracted.

2.2 Ultracentrifugation

Most studies have employed mtDNAs isolated by density gradient ultracentrifugation. This isolation procedure takes advantage of the closed-circular supercoiled structure of mtDNA, and uses intercalating dyes, either ethidium bromide or propidium iodide, to differentiate among the different forms of

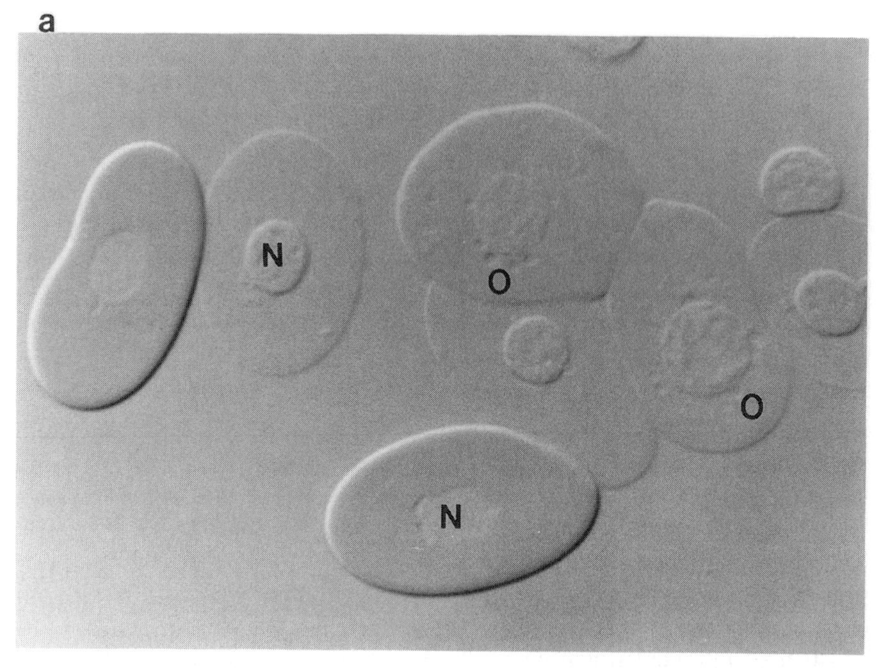

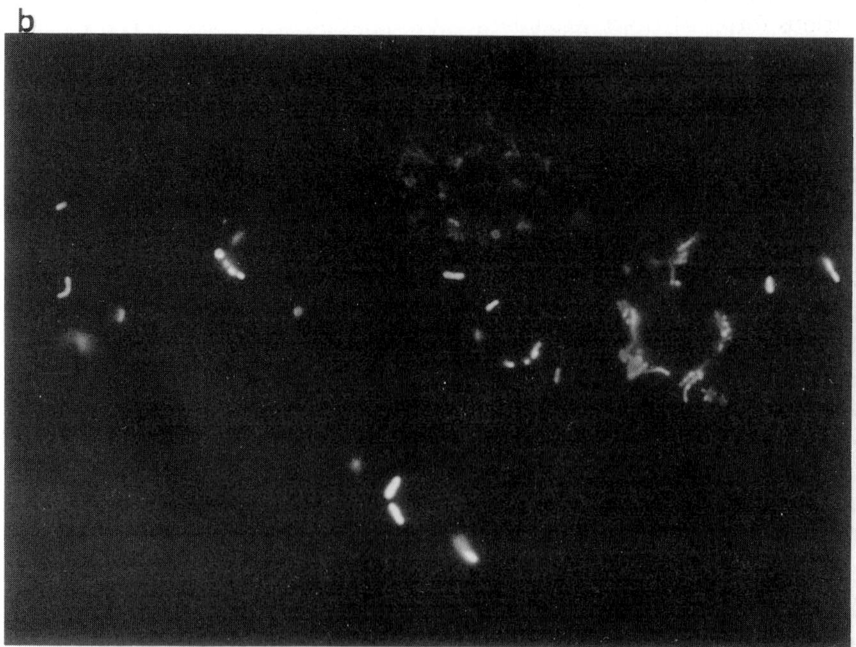

Figure 2. Micrographs of *Alligator mississippiensis* erythrocytes. (a) Light micrograph showing nuclei (N) and organelles (O); (b) $DiOC_6$ fluorescence showing mitochondria inside erythrocytes. Micrographs courtesy of A. Roberts.

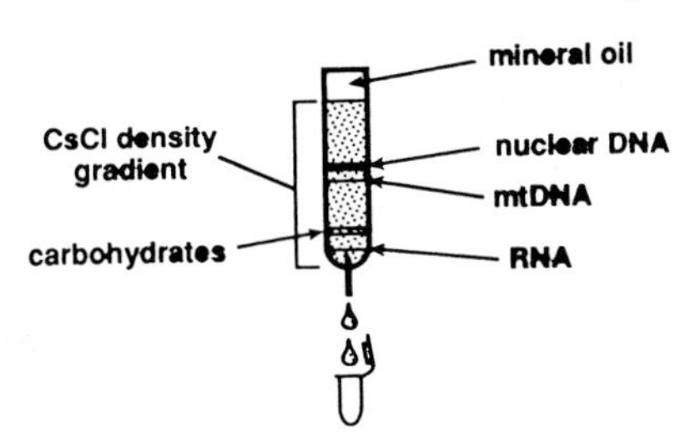

Figure 3. Schematic of CsCl gradient used to isolate mtDNA following 30 h equilibrium ultracentrifugation run. Different nucleic acid and carbohydrate fractions correspond to relative mobilities in the gradient.

DNA. These dyes, when added to a DNA sample, will slide between adjacent base pairs in the double helix causing disruption in helix periodicity. Linear DNA intercalates more ethidium bromide or propidium iodide than the same length of supercoiled DNA due to the unrestrained rotation of the free ends. With less dye intercalated, the density of the supercoiled DNA–dye complex is different from that of the linear complex, and it migrates to a different position in a density gradient. The most commonly used gradient is made with caesium chloride (CsCl) in a solution of Tris–EDTA buffer (TE), and after 20–40 h of ultracentrifugation at $200000 \times g$, will differentiate among the various forms of DNA–propidium iodide complexes. Resolution is sufficient to allow separation of complexes having buoyant densities that differ by less than 0.025 g/ml. *Figure 3* shows the relative positions of the various components in a typical gradient. The mtDNA is separated from the other components by side- or bottom-dripping the gradient and collecting only the mtDNA band.

From higher plants, mtDNA can be isolated by pelleting the mitochondria; any other DNA in solution is degraded with DNase I. The mitochondria are then lysed and the lysate is subjected to density gradient ultracentrifugation to remove contaminating nuclear and chloroplast DNA. Plant mtDNA rearrangements are such frequent events that restriction fragment length polymorphism (RFLP) studies using the plant mitochondrial genome are of limited value unless homology of fragments is determined by probing Southern blots. The expense and labour involved rarely justify their use in estimating phylogenies, but for studying the organization of the rearrangements or for studies involving closely related individuals, RFLP analysis may be useful. We include a protocol for its isolation, with the caveat that for some studies plant mtDNA may not be the ideal molecule.

2.3 Probing total genomic DNA with mtDNA

Often sufficient mtDNA cannot be isolated in pure form from the desired tissue. However, it is usually possible to isolate total cellular DNA and digest it with restriction endonucleases, and subsequently probe for the mitochondrial fragments on a Southern blot. The probe DNA has conventionally been mtDNA isolated using density gradient ultracentrifugation or from fragments that have been cloned. This procedure involves five steps:

(a) isolation of total DNA;
(b) digestion of the DNA sample with appropriate restriction enzyme(s);
(c) electrophoretic separation of restriction fragments;
(d) transfer of fragments on to a nylon support membrane (Southern blotting);
(e) the use of a labelled mtDNA probe isolated from another source to locate the mtDNA fragments in the target DNA bound to the membrane (see Section 6.2.2).

The visualized fragments should be the same size as those determined when pure mtDNA has been end-labelled and electrophoretically separated, except that the smaller fragments produce a weaker signal (in random priming or nick translation, the amount of label incorporated depends upon the length of the fragment, as opposed to end-labelling in which each fragment receives the same amount of label regardless of its length).

This difficulty in obtaining information from smaller fragments makes it desirable to obtain pure mtDNA whenever possible (especially for assessing variation among different individuals in a population), but the quality or quantity of available tissues often precludes this approach. A clear advantage of using probes is that it is less expensive and much simpler to isolate total cellular DNA from a large number of individuals, and probe them with purified mtDNA obtained from a single individual. Also, if probes for nuclear genes or minisatellite sequences are available, as well as PCR-amplified fragments, multiple data sets (ten or more) can be obtained from the same blot, making this approach far more effective than studies employing allozyme screening.

2.4 Amplification of mtDNA

An alternative to the direct isolation of mtDNA is the *in vitro* amplification of portions of the molecule using PCR (further described in Chapter 6 of this volume). Amplification of short pieces of mtDNA has become the basis for sequence comparisons among taxa for systematic and population-level studies. In certain applications, the D-loop (control region) has been amplified and sequenced, with differences in sequences providing the basis for assessing degrees of relatedness or determining genetic variability within or between small populations. PCR-amplified mtDNA products can be used as heterologous probes for Southern blots, increasing the amount of information that can be obtained from a single gel.

By using the appropriate primers, large mtDNA fragments can be amplified, digested with several frequently cutting restriction endonucleases (e.g., those with recognition sequences consisting of four bases), electrophoretically separated, and visualized with either ethidium bromide or silver stain, replacing the need to perform autoradiography. These fragments could be compared among individuals, and this rapid, inexpensive method would allow large numbers of DNA samples to be analysed. Further, restriction maps can be derived from the pieces and together a restriction map of the entire mtDNA can more easily be obtained, as well as the orientation of these restriction sites relative to a gene map. These fragments can also serve as template for direct dideoxy sequencing reactions using these same primers. Recent advances in PCR technology allow the amplification of entire mitochondrial genomes in one or just a very few fragments (see Chapter 6). It is important to verify that a single product is the result, otherwise restriction analyses or sequencing without prior gel purification can give misleading or incomprehensible results. Several smaller PCR products are easier to obtain, and make restriction mapping or other searches for polymorphisms easier to localize. Several sequences for 'universal' primers for this approach are listed in *Table 2*; others are provided elsewhere (Chapters 6 and 17).

Significant advances have also been made that allow a large increase in types of tissues available from which mtDNA can be amplified allowing studies involving smaller organisms (e.g., individual *Drosophila*) as well as many tissues that are unsuitable for other molecular studies. These additional sources include shed skin from reptiles, individual fish scales, formalin-preserved specimens, bone and teeth, dried pelts and hair, feather pulp, invertebrate haemolymph, and many others. The inclusion of museum specimens in these studies allows for comparisons among populations collected at a variety of dates and localities, and greatly increases the value of collections. Preservation and fixation methods must be developed which maintain valuable genetic information that is available from the specimens. Still, difficulties often arise with amplification attempts for many template–primer combinations, and long products are difficult to obtain from some taxa. Using conventional isolation methods, such as those provided here, will allow researchers to clone and sequence mtDNA fragments from a high quality specimen of a desired taxonomic group for the purpose of designing better, more specific PCR primers. For this and the reasons mentioned above, it is clear that mtDNA isolation will remain an important tool for years to come.

3. DNA isolation

Many protocols are available for isolation of animal DNA. We include the standard animal mtDNA protocols, as well as discussing methods that are effective in isolating total DNA from a number of different tissues (5). For studies of fresh or freshly frozen mitochondria-rich tissues such as liver, ovaries or unfertilized eggs, the following protocols may be used to obtain relatively pure mtDNA. Protocols for isolating DNA from plant tissues are also included.

3.1 mtDNA isolation from animal tissues: methods not requiring density gradient ultracentrifugation

For studies where fresh or freshly frozen mitochondria-rich tissues such as liver, ovaries, or unfertilized eggs are available, the following protocols may be used to obtain relatively pure mtDNA without the need for ultracentrifugation.

3.1.1 SDS–phenol method of mtDNA isolation

The following protocol uses differential centrifugation and salt-precipitation, followed by an organic extraction to obtain open-circular mtDNA with yields and purity usually sufficient for visualization without autoradiography. It is a modification of previously reported methods (3,6) that reduces the time and expense involved, making it attractive for large numbers of samples. The preferred tissue for use with this protocol is liver, although pancreas and brain may sometimes provide suitable amounts of mtDNA.

Protocol 1. Isolation of open-circular mtDNA

Equipment and reagents

- TEK buffer: 50 mM Tris, 10 mM EDTA, 1.5% KCl, pH 7.5
- 15% sucrose in TEK
- EST: 100 mM EDTA, 10 mM Tris, 150 mM NaCl, pH 8.0
- 18% SDS solution
- 5 M NaCl solution
- Phenol, buffered with TEK pH 7.5–8.0
- PCI: buffered phenol:chloroform:isoamyl alcohol (25:24:1 by vol.)
- CI: chloroform:isoamyl alcohol (24:1 v/v)
- 100% ethanol (4°C or −20°C)
- 75% ethanol
- RNase A solution (10 mg/ml stock)
- 0.5 mM EDTA pH 8.0
- TE: 10 mM Tris–HCl pH 8.0, 1 mM EDTA
- Glass–Teflon tissue homogenizer (or mortar and pestle and liquid nitrogen)
- Polypropylene tubes
- Pasteur pipettes
- Sorvall HS-4 swing-out rotor (or equivalent)
- Rubber sleeve adaptors for rotor
- Refrigerated centrifuge
- Corex 15 ml centrifuge tubes (or equivalent)
- Sorvall SS-34 fixed angle rotor (or equivalent)

Method

1. Obtain 10 g of fresh or frozen tissue and add 5 vol. of TEK buffer.
2. Homogenize tissue using a single stroke of a motor-driven glass Teflon homogenizer. If one is unavailable, grind the tissue in liquid nitrogen with a mortar and pestle before adding TEK.
3. Transfer the homogenate to two 50 ml disposable polypropylene tubes, and underlayer[a] each with a 15% sucrose–TEK solution (use an approximately equal volume).
4. Centrifuge at 1000 × *g* for 10 min at 4°C, then carefully remove the supernatant.

5. Centrifuge the supernatant at 8000 × *g* for 10 min to pellet mitochondria.
6. Wash the pellet with 30 ml chilled EST.
7. Resuspend each pellet in 2.5 ml EST at room temperature and recombine into a single tube.
8. Add 0.1 vol. of 18% SDS to this suspension to lyse the mitochondrial membranes.
9. Add 0.2 vol. of 5 M NaCl to remove nuclear DNA, SDS, and precipitating proteins, mix well, and centrifuge at 8000 × *g* for 15 min at 4°C.
10. Add an equal volume of water-saturated, buffered phenol, pH 8.0, vortex vigorously and let stand at room temperature for 5 min to remove soluble proteins. Random shearing of DNA will be minimal if high quality re-distilled, buffered phenol is used.
11. Centrifuge the sample at 12 000 × *g* for 10 min. Using a Pasteur pipette, carefully transfer the DNA-containing aqueous (upper) phase to a clean centrifuge tube, being careful to avoid the proteins at the interface.
12. Re-extract with an equal volume of PCI, centrifuge as above, and transfer the aqueous phase to a clean centrifuge tube. Add an equal volume of CI to remove residual phenol, and centrifuge as above.
13. Carefully remove the aqueous phase and transfer it to a clean centrifuge tube. Precipitate the DNA by adding 2 vol. of cold (4°C or −20°C) ethanol, and place in −20°C freezer for 2 h to overnight.
14. Centrifuge at 12 000 × *g* at 4°C for 10 min to pellet the DNA. Carefully decant the supernatant, wash the pellet once with 75% ethanol, and centrifuge again. Decant the 75% ethanol and allow the pellet to dry at 37°C until translucent.
15. Resuspend the DNA in 100 μl TE.[b]
16. Remove RNA by digestion with 6 μl of 10 mg/ml RNase A solution at room temperature followed by dialysis against 0.5 mM EDTA pH 8.0 for at least 12 h, or by adding 100 μl of 10 mg/ml RNase stock to each 100 ml cooled liquid agarose prior to casting a gel.

[a]Underlayer using a soft rubber pipette filler (bulb type) and a Pasteur pipette (for larger volumes use a graduated pipette). Measure 15% sucrose–TEK solution into a suitable container and draw the entire volume into the pipette. Next, touch the upper surface of the sample with the tip of the pipette containing the 15% sucrose–TEK solution, and allow some of the solution to run out of the pipette to remove any air that is in the tip. Carefully lower the pipette tip to the bottom of the tube and begin slowly draining the sample into the bottom. Before all the solution has drained from the pipette, slowly begin lifting the tip off the bottom without stopping the flow of the solution. Once the tip of the pipette has reached the interface, discontinue the flow of 15% sucrose–TEK solution and slowly withdraw the tip. The interface should now be visible due to the differing light refractivity of the layers.
[b]Lower concentrations of TE are sometimes necessary due to inconsistent digestion by some restriction endonucleases. It is therefore desirable to resuspend the DNA in a buffer with lower amounts of Tris salts, such as 0.1 × TE, 0.5 mM EDTA, or water if the sample is to be used as a template for PCR.

3.1.2 Density gradient centrifugation

Some researchers (8) suggest that a continuous (linear) gradient from 1.0 to 2.0 M sucrose provides a good separation of intact mitochondria, and a brief protocol is given here (*Protocol 2*).

Protocol 2. Isolation of mitochondria using isopycnic sucrose gradient centrifugation

Equipment and reagents

- Sucrose solutions: 0.8 M, 1.0 M, 1.3 M, 1.6 M, and 2.0 M sucrose, all containing 1 mM EDTA, 0.1% bovine serum albumin (BSA), and 10 mM Tris–HCl pH 7.5
- TEK buffer: 50 mM Tris, 10 mM EDTA, 1.5% KCl, pH 7.5
- Phenol, buffered with TEK, pH 7.5–8.0
- PCI: buffered phenol:chloroform:isoamyl alcohol (25:24:1 by vol.)
- CI: chloroform:isoamyl alcohol (24:1 v/v)
- 100% ethanol (ice-cold)
- TE: 10 mM Tris–HCl pH 8.0, 1 mM EDTA

Method

1. Homogenize 5–10 g tissue as in steps **1** and **2** of *Protocol 1*, except suspend the homogenate in a solution of 0.8 M sucrose. Prepare all sucrose solutions to the desired molarity, and include 1 mM EDTA, 0.1% BSA, and 10 mM Tris–HCl pH 7.5 to increase the stability of the components.
2. If a gradient-maker is not available, set up the gradient in steps of 1.0 M sucrose underlayered with 1.3 M sucrose, then with 1.6 M sucrose, and finally underlayered with 2.0 M sucrose. Allow these to diffuse overnight at 4°C. Underlayering is described in *Protocol 1*.
3. Carefully add the sample to the top of the gradient, and centrifuge for 2 h at 80 000 × *g* at 5°C.
4. Remove the brown band of intact mitochondria using a Pasteur pipette, and dilute this solution with 2 vol. of TE buffer.
5. Pellet the mitochondria by centrifugation at 20 000 × *g* for 10 min at 5°C. Isolate DNA from the mitochondria as described in *Protocol 1*, steps **6–16**.

Other types of gradient systems are also recommended, such as 25–60% (v/v) Percoll with 0.25 M sucrose centrifuged at 40 000 × *g* for 20 min, or 20–50% (w/v) metrizamide or Nycodenz (Sigma Chemical, see Appendix 6) centrifuged at 80 000 × *g* for 60 min. These provide higher yields of mitochondria than sucrose gradients, though they are more expensive (4).

3.2 Isolation of animal mtDNA using ultracentrifugation

Protocols 3–5 are appropriate for isolating mtDNA from most fresh or frozen tissues, and are among the most aggressive for obtaining mtDNA from problematical tissues. They have been used successfully on frozen tissues, including skeletal muscle, as well as frozen liver obtained from a specimen that was found dead alongside a highway. This frozen liver has served as a source of mtDNA for probing Southern blots of crocodilian DNA fragments, and mtDNA isolated from *Alligator mississippiensis* blood using these protocols has also served as probe. These procedures are relatively expensive and time-consuming, and are generally not used when large numbers of individuals are involved, except when isolating heterologous probe.

Protocol 3. Crude isolation of mitochondria from fresh or frozen animal tissues, including whole blood using differential centrifugation

Equipment and reagents

- STE: 10 mM NaCl, 10 mM Tris–HCl pH 8.0, 25 mM EDTA
- TE: 10 mM Tris–HCl pH 8.0, 1 mM EDTA
- 1.5 M sucrose in TE
- 20% SDS
- CsCl-saturated double-distilled H_2O (add CsCl to double-distilled H_2O until it will no longer go into solution)
- Mortar and pestle
- Liquid nitrogen
- Tissue homogenizer (e.g., Tekmar Tissuemizer)
- Wet ice
- Micropipetter and tips
- 10 ml beakers
- 15 ml screw-capped polypropylene tubes
- 15 ml glass centrifuge tubes (Corex tubes)
- Refrigerated centrifuge
- Sorvall HS-4 or equivalent swinging bucket rotor
- Sorvall SS-34 or equivalent fixed angle rotor
- Rubber adaptor sleeves for rotors

Method

1. Prepare homogenization buffer by combining 1 part 1.5 M sucrose in TE to 5 parts STE. Prepare enough to use 12 ml for each sample, cool to 4°C, and keep homogenization buffer on wet ice.
2. Label the 15 ml polypropylene tubes with sample information and place them on ice.
3. Pre-cool a clean mortar and pestle with liquid nitrogen to prevent breakage upon grinding.
4. Obtain 0.5–1.5 g fresh or frozen tissue, using more skeletal muscle, heart, or kidney than ovary, liver or brain. Immediately place tissue in the pre-cooled mortar, add more liquid nitrogen if necessary, and pulverize the tissue into a fine powder. Alternatively, finely dice the tissue using a clean razor blade or sharp dissecting scissors. One to five millilitres of whole blood from non-mammalian vertebrates can be added directly to the homogenization buffer in the next step, or acid citrate dextrose (ACD solution B, see Section 3.3) can be substituted for homogenization buffer. If the available tissue is mammalian

Protocol 3. *Continued*

blood, add the pellet from 10–20 ml whole blood to homogenization buffer or ACD as described in the next step of this protocol.

5. Add powdered (or diced) tissue, whole blood, or blood cell fraction as described in step **4** above to 5 ml of homogenization buffer in a 10 ml glass beaker, and homogenize further, using two or three short (5 sec) bursts of a high-speed tissue homogenizer. Allow homogenate to settle for a few minutes, then pour some of the liquid homogenate into the appropriate 15 ml polypropylene tube and place it on ice. Add additional homogenization buffer to the solid contents in the glass beaker and homogenize a second time, using two or three short bursts. Continue pouring off portions of the liquid and homogenize further with more homogenization buffer until the final volume is close to 12 ml. By combining the grinding in liquid nitrogen with the homogenization it is possible to improve yields, although either step alone will suffice where the quality or quantity of fresh tissue is not limiting. Over-homogenization will cause a reduction in yield, so a balance must be found for each type of tissue used.
6. Pour homogenate into the appropriately labelled 15 ml polypropylene tube, and bring the volume to 12 ml with cold homogenization buffer or ACD solution B (where appropriate). Spin at 4°C for 5 min at 3000 $\times g$ in a Sorvall HS-4 rotor (or equivalent) using rubber sleeves. The pellet contains nuclei as well as intact cells and cell debris, and can be used as a source of total DNA by entering *Protocol 4*, adding the pellet to 2 ml SDS–urea, and continuing with the procedure.
7. Transfer the supernatant to a 15 ml Corex tube, and centrifuge at 20 000 $\times g$ in SS-34 or equivalent rotor for 20 min at 4°C. Discard the supernatant, and drain excess fluid from the tube by allowing it to remain inverted for 5 min. Resuspend the mitochondria-enriched pellet in 1 ml TE at room temperature.
8. Add 0.125 ml of 20% SDS to each sample to lyse the mitochondrial membranes, vortex the contents, and incubate at room temperature for 10 min.
9. Add 0.187 ml CsCl-saturated double-distilled H_2O to each tube to precipitate the SDS, lipids, and proteins. Allow sample to precipitate at 4°C overnight, or on ice for 2–4 h.
10. Spin the sample in the same tube at 20 000 $\times g$ in SS-34 or equivalent rotor for 20 min at 4°C. Remove the supernatant, which contains mtDNA of all three forms (i.e., supercoiled, nicked-circular, and linear), as well as a substantial amount of nuclear DNA. Further purify mtDNA using density-gradient ultracentrifugation as described in *Protocol 4*.

Because ultracentrifuge swinging bucket rotors usually have only six buckets, and the length of the first gradient run averages about 30 h, the following protocol presents a formidable functional bottleneck, although the use of vertical rotors having eight or more spaces can reduce the gradient run to 4–16 h. Nevertheless, the resulting mtDNA is of suitable quality for use as probe, and yields from each sample are usually sufficient for 25–100 primer-extension-labelled probes (discussed in Chapter 5 of this volume).

Protocol 4. Purification of animal mtDNA using density-gradient ultracentrifugation

Equipment and reagents

- CsCl (dry, technical grade)
- Propidium iodide, 20 mg/ml in double-distilled H_2O
- CsCl–propidium iodide stock solutions, of densities 1.40, 1.55, and 1.70 mg/ml (see *Table 1*)
- TE: 10 mM Tris–HCl pH 8.0, 1 mM EDTA
- Parafilm
- Mineral oil
- Ultracentrifuge
- Sorvall TST-60.4 or equivalent swinging bucket rotor
- Seton 4.4 ml (11 mm diameter × 60 mm) polyallomer ultracentrifuge tubes (or equivalent)
- 1.5 ml microcentrifuge tubes
- Long-wave UV light source
- Bottom-dripping apparatus or syringe with 21 gauge needle
- Stiff wire, approximately 9 cm long (guitar or piano string, or bristles from a wire brush)
- UV-blocking goggles or face shield
- Latex gloves
- Propidium iodide waste container
- Isobutanol in CsCl-saturated double-distilled H_2O
- 1.0 M NaCl in TE
- Dowex mixed-bed ion-exchange beads or equivalent
- Dialysis tubing (6.4 or 10 mm diameter, mol. wt cut-off = 12 000–14 000)[a]
- Dialysis tubing clamps (optional)

Method

1. Place the sample from *Protocol 4*, step **10**, into a Seton ultracentrifuge tube, and add 0.25 ml of propidium iodide. Adjust the density of this solution to 1.40 g/ml[b] by adding CsCl, and manually shaking the tube while holding a small piece of Parafilm over the opening.
2. Underlayer[c] the sample with 2.0 ml of CsCl–propidium iodide solution, density = 1.7 g/ml.
3. Being careful not to disturb the interface, add mineral oil to within 5 mm of the top of the Seton tube.[d] Balance the tubes against each other in pairs to within 0.05 g by taring the heavier of the pair on a top-loading balance. Carefully exchange the heavier tube for the lighter one, and add mineral oil until the scale reads between –0.05 and +0.05 g. Balance the remainder of the tubes in the same manner, place them into the rotor buckets, and secure the caps.
4. Spin at 40 000 r.p.m. (200 000 × *g*) in a Sorvall TST-60.4 or equivalent swinging-bucket rotor for about 30 h under vacuum at 21 °C. Decelerate the rotor using the brake, making certain that the brake is not used

Protocol 4. *Continued*

for the final phase of deceleration from 500 r.p.m. to stop. Some ultracentrifuges can be programmed to allow this slow deceleration, which prevents disruption of the gradient and keeps the DNA bands sharp. Being careful not to disturb the buckets, remove the rotor, then using forceps or a haemostat gently remove the sample tubes from the buckets.

5. Wearing goggles or a face shield, bottom drip[e] the tubes under long-wave UV using a dripping apparatus (*Figure 4*) collecting only the mtDNA band (the lower DNA band; refer to *Figure 3*). If only one DNA band is visible (in the upper half of the tube), collect the region immediately below the visible (nuclear) band, being careful not to include any of the nuclear DNA. Frequently the mtDNA band is not visible, but is present in sufficient amounts to justify the additional effort of further purification. Add the collected fraction (final volume should be between 0.5 and 0.7 ml) to a 1.5 ml microcentrifuge tube.[f]
6. Remove propidium iodide using isobutanol that has been mixed and stored with CsCl-saturated double-distilled H_2O. The isobutanol will form a layer on top of the aqueous layer. Fill the microcentrifuge tube containing the sample with isobutanol and vortex to mix well. Spin down the aqueous phase in a microcentrifuge for a few seconds, and remove the upper isobutanol phase with a Pasteur pipette. Add more isobutanol, and re-extract the sample as many times as needed to remove residual propidium iodide. This may require as many as eight isobutanol extractions, but usually requires between three and six extractions.
7. When the sample is no longer pink, add the entire volume to a 4–6 cm length of prepared dialysis tubing that has been tied at one end. Secure the other end of the tubing with a knot or a tubing clamp, making sure the identity of the sample is indicated in some manner.[g] Tie all knots well to avoid leakage, making sure the tubing is fairly turgid and contains a small (*c.* 0.25 ml) air bubble to allow for an increase in liquid volume.
8. Place samples in a large ($\geq$2 litre) dark bottle containing 1.0 M NaCl in TE and several grams of Dowex MR-12 (Sigma) or equivalent mixed resin beads. Add a magnetic stirrer bar and dialyse samples while stirring at 4°C, changing the dialysis solution after 2–3 h. After the second change, dialyse twice or more against TE.[h] Remove the sample with a Pasteur pipette and transfer to a clearly labelled microcentrifuge tube. Add TE until the final volume is about 0.5 ml and store at –20°C.

[a] Prepared by boiling for 10 min in 2% (w/v) sodium bicarbonate, 25 mM EDTA, then for 10 min in 25 mM EDTA, and stored in the same at 4°C in a closed container.

[b]This usually requires about 0.75 g CsCl. The density of the sample solution is easily determined by placing the sample on a top-loading balance. Remove 1.0 ml of solution, tare (zero) the balance, then add the sample back into the tube. Sample density is the amount that is subsequently displayed on the weight indicator. Sample density can be increased with CsCl or decreased with water or TE.
[c]Measure 2 ml of CsCl–propidium iodide (density = 1.7 g/ml) into a suitable container, draw the entire volume into a Pasteur pipette, and mark the 2 ml line on the pipette (this can now be used to draw 2.0 ml from a beaker containing stock solution for subsequent samples). Underlayer each sample as described in *Protocol 3*.
[d]The mineral oil is necessary to maintain pressure on the sides of the Seton tubes during extremely high gravity spins. If there is too little mineral oil, the tubes will collapse, so there should be 2–5 mm between the top of the mineral oil and the top of the tube. Excess mineral oil on the outside of the tubes may also cause them to collapse, and will dissolve the marker ink used to identify the sample.
[e]Insert a piece of stiff wire into the needle to remove debris that plugs the opening upon puncture of the tube. When forcing the Seton tube on to the needle, avoid forming a seal over the top with your fingers, which may result in a disruption of the gradient. Alternatively, side-drip the tubes by collecting the mtDNA with a syringe and a 21 gauge needle. Puncture the side of the tube slightly below both DNA bands, and carefully draw the lower DNA band into the syringe.
[f]If the mtDNA is to be further purified for cloning, etc., also collect a single drop of the nuclear DNA band for use as a reference for subsequent gradients. For most enzymatic manipulations, including restriction endonuclease digestion, end-labelling, nick translation, or primer extension labelling (random priming), it is usually not necessary to perform the velocitization and final gradient steps. Cloning sometimes requires mtDNA of higher purity, and it is advisable to perform these steps if the sample is to be cloned. Also, if by mistake some nuclear DNA is collected with the mtDNA during the dripping, simply continue with the velocitization (a rapid clean-up step) and final gradient steps as described in *Protocol 5*. Otherwise, continue with clean-up steps **6–8**.
[g]Clamps can be numbered with a marker pen or tubing containing different samples can be distinguished by varying the number of knots and the lengths of tubing remaining on each end after the knots have been tied. Two samples can be placed in a single piece of tubing that has been tied in the middle, and the samples added through each open end.
[h]Dowex beads will help remove any residual propidium iodide, and may be removed after the final NaCl dialysis. If a lower TE concentration is desired, the last change of dialysis solution should be against 0.1 × TE or double-distilled H_2O for PCR. Some restriction endonucleases perform better if higher TE concentrations are avoided, and PCR enzymes are sensitive to the presence of EDTA.

For some lower eukaryotes (e.g., filamentous fungi), bisbenzimide can be substituted for propidium iodide or ethidium bromide. This causes the mtDNA to migrate to a point above the nuclear DNA. The procedure is described in ref. 7.

Table 1. CsCl–propidium iodide stock solutions[a]

	Density		
	1.4 g/ml	**1.56 g/ml**	**1.70 g/ml**
CsCl (g)	106.6	146.6	186.6
10 × TE (ml)	20	20	20
Double-distilled H_2O (ml)	143.4	133.3	123.4
Propidium iodide[b] (ml)	10	10	10

[a]Solutions need to be stored in light-tight bottles at room temperature.
[b]Use a 10 mg/ml stock solution of propidium iodide in water.

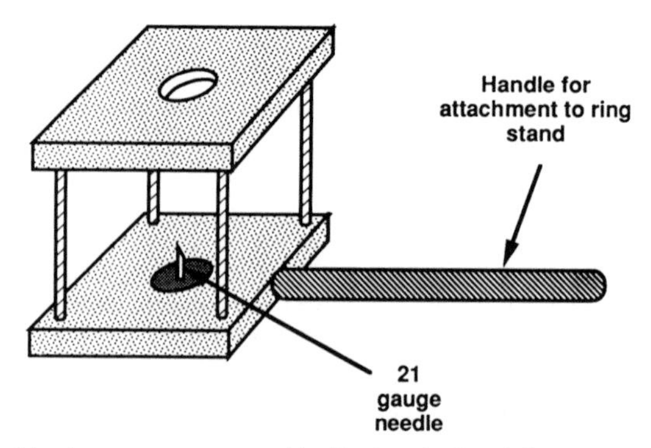

Figure 4. Bottom dripping apparatus used in *Protocols 4* and *5*.

Protocol 5. Further purification of animal mtDNA using velocitization and additional density gradient ultracentrifugation

Equipment and reagents

as in *Protocol 4*

Method

1. Using TE, double the volume of the sample obtained in step **5** of *Protocol 4* above. To another Seton tube, add about 1.5 ml CsCl–propidium iodide, density = 1.4 g/ml. Carefully underlayer this with 0.7 ml CsCl–propidium iodide, density = 1.7 as described in *Protocol 3.*
2. Slowly add the mtDNA sample to the top of the step gradient. Three layers should be visible, but if either or both interfaces are disturbed, the results will not be greatly affected. Balance the tubes with mineral oil as in *Protocol 4,* and place samples in the rotor buckets.
3. Replace the buckets and centrifuge at 45000 r.p.m. (250000 × *g*) for 3.5 h. This step allows for separation of DNA fragments based upon their relative molecular weights, with the smaller linear fragments or nicked circular DNAs migrating more slowly, allowing them to be removed more effectively from the sample.
4. Bottom-drip the lower 1.4 ml of the Seton tube into a 1.5 ml microcentrifuge tube under visible light, and add this fraction to a clean Seton tube. Add 1.0 ml CsCl–propidium iodide, density = 1.55 g/ml, to the sample and fill the remainder of the tube with mineral oil, balancing tubes in pairs.
5. Centrifuge as in *Protocol 4,* step **4**, for about 24 h at 21°C and 40000 r.p.m. (200000 × *g*) under a vacuum. Collect the mtDNA band by

either the bottom-dripping or the side-dripping method, and extract the propidium iodide with isobutanol and dialyse the sample as described in *Protocol 4*, steps **5–8**. Store all samples in clearly labelled microcentrifuge tubes at –20°C.

3.3 Isolation of total DNA from animals

For many studies involving vertebrates, blood is a readily available source of DNA and researchers are able to obtain samples from living specimens. A method for obtaining blood from medium to large reptiles that is both convenient to the worker and safe for the animal is described in ref. 8. Other techniques have been developed for other organisms, including haemolymph of invertebrates, and are reviewed elsewhere (14).

Two procedures for preserving blood are typically used, the first of which involves adding whole blood to a lysis buffer after it has been drawn (9). Blood collected in this manner can be stored at room temperature for up to a year before DNA is isolated, with the resulting DNA being of suitable length for DNA fingerprinting analysis (10) as described in Chapter 5 of this volume. This procedure is very convenient for workers out in the field and provides excellent protection for the DNA in the sample. The second method involves the use of an anti-coagulant to maintain intact cells so that blood components (e.g., plasma proteins) needed for other studies can be isolated. Of the commonly used anti-coagulants, acid citrate dextrose (ACD solution B) is many times more effective than EDTA or heparin for preserving native DNA in a blood sample (11). Both of these procedures allow the samples to be maintained at ambient temperatures for at least several days, making it possible to send them in the mail. If mtDNA is to be isolated in pure form from the sample, it will be necessary to use *Protocols 3–5* so that intact mitochondria can be pelleted.

A standardized method for isolating total cellular DNA from blood or blood fractions, which is readily modified for other tissues is found elsewhere (ref. 5; see also Chapter 2). It involves homogenization of the tissue followed by lysis of intact cells and concurrent disintegration of proteins. These steps are followed by removal of all reagents by dialysis, yielding high molecular weight DNA of extremely high purity. 0.25 g of fresh or frozen tissues and 150–250 µl of non-mammalian whole blood typically yield 1–2 mg DNA with an $A_{260/280}$ ratio between 1.8 and 2.0. As an alternative, we recommend the use of a commercially available kits from Qiagen (QIAamp Blood and Tissue Kits #29106 and 29306), which are not only simple, rapid, and relatively inexpensive, but also can be used to isolate DNA from a wide range of tissues. The extractions do not involve organic solvents (e.g., phenol–chloroform) and can be used for bacteria and viruses, as well as mitochondria-containing organisms such as yeast and insects, and even paraffin-embedded tissues, clinical samples, and crude cell lysates (see Appendix 6).

3.4 Plant DNA isolation

A simple method for isolating DNA from plant mitochondria is included. It involves the isolation of intact mitochondria within a suspension that has been treated with DNase I to remove all other contaminating DNA, followed by CsCl density gradient ultracentrifugation using a vertical rotor. If a vertical rotor is not available, it may be possible to substitute the density gradient as described in *Protocols 4* and *5*. A method that is extremely effective for isolation of total nucleic acids from plants is also included.

3.4.1 Plant mtDNA isolation

In *Protocol 6* we include a standard and effective plant mtDNA isolation method (12, 13).

Protocol 6. DNase I isolation of plant mtDNA

Equipment and reagents

- Isolation buffer: 0.35 M sorbital, 50 mM Tris–HCl pH 8.0, 5 mM EDTA, 0.1% (w/v) BSA, 0.1% (v/v) β-mercaptoethanol
- Wash buffer: 0.35 M sorbital, 50 mM Tris–HCl pH 8.0, 25 mM EDTA
- Lysis buffer: 5% (w/v) sodium sarcosinate, 50 mM Tris–HCl pH 8.0, 25 mM EDTA
- DNase I
- DNase I buffer: 0.35 M sorbital, 50 mM Tris–HCl pH 8.0, 15 mM $MgCl_2$
- TE: 10 mM Tris–HCl pH 8.0, 1 mM EDTA
- Proteinase K (optional): 20 mg/ml frozen stock
- CsCl (dry, technical grade)
- CsCl–ethidium bromide–TE, density = 1.55 mg/ml: 750 mg CsCl, 100 μg ethidium bromide in each ml of TE
- NaCl–water-saturated isopropanol
- 100% ethanol (ice-cold)
- 76% ethanol
- Blender
- Cheesecloth
- Soft paintbrush
- 15 ml glass centrifuge tubes (Corex tubes)
- Refrigerated centrifuge
- Sorvall HS-4 or equivalent swinging bucket rotor
- Sorvall SS-34 or equivalent fixed angle rotor
- Rubber adaptor sleeves for rotors
- Sorvall TV-865 or TV-1665 vertical rotor (or equivalent)
- Ultracentrifuge

Method

1. Cut young green leaves or etiolated shoots into small pieces of approximately 2–10 cm^2 and wash the pieces in tap water, if necessary to remove visible dirt and debris.
2. Place 0.1–1.0 kg of chopped tissue in a 3- to 10-fold excess (w/v) of ice-cold isolation buffer (fresh).
3. Homogenize in a blender using 3–5 short bursts (5 sec or longer if necessary) at high speed. Filter through four layers of cheesecloth, squeezing out excess liquid homogenate.
4. Centrifuge the filtrate at 1000 × *g* for 15 min at 4°C. The pellet can be used to prepare chloroplasts as described in Chapter 2 of this volume or in (11).

5. Centrifuge the supernatant from the previous step at 12 000 × *g* for 20 min at 4°C to pellet the mitochondria. Resuspend this pellet in 100 ml of DNase I buffer per kg of starting material using a soft paint-brush and vigorous swirling.
6. Add 30 mg DNase I (freshly resuspended in DNase I buffer) per kg of starting material, mix gently but well, and incubate for 1 h on ice with occasional mixing.
7. Add three volumes of wash buffer and centrifuge at 12 000 × *g* for 20 min at 4°C. Resuspend the pellet in wash buffer and repeat washing spin twice more. Resuspend the mitochondrial pellet in 1–2 ml wash buffer.
8. Add 0.05 vol. of a 20 mg/ml solution of self-digested (2 h at 37°C) proteinase K and incubate for 2–10 min at room temperature.[a]
9. Add 0.2 vol. of lysis buffer, and slowly invert the tube several times over a period of 15–30 min at room temperature.
10. Bring the final volume of the mitochondrial lysate to about 3 ml with wash buffer, add 3.35 g of CsCl, and dissolve by gentle mixing. If desired, CsCl can be powdered using a mortar and pestle prior to its addition to the sample.
11. Add ethidium bromide to a final concentration of 200 μg/ml and double-distilled H_2O to a final volume of 4.45 ml and a final density of 1.55 g/ml.
12. Centrifuge in a Sorvall TV-865 or TV-1665 vertical rotor (or equivalent) at 50 000–58 000 r.p.m. (220 000–300 000 × *g*) for 4–16 h at 20°C.
13. Using a 1 ml pipette tip with the end cut off, remove any material that has collected on top of the gradient. Then, using a second tip cut at an angle, remove the visible band of DNA in as small a volume as possible, between 0.5 and 1.0 ml.
14. If further purification is necessary, bring the sample volume to 4.45 ml using a stock solution of CsCl–ethidium bromide–TE (density = 1.55 mg/ml). Repeat steps **12** and **13** above.
15. Remove the ethidium bromide and CsCl as described in *Protocol 4*, steps **6–8**, substituting NaCl–water-saturated isopropanol for CsCl–water-saturated isobutanol. The upper layer of the stock solution is to be added to the sample, and after each extraction the upper phase should be discarded.
16. An alternative to dialysis is to ethanol precipitate the sample by tripling its volume with double-distilled H_2O after ethidium bromide removal. Then add 6 vol. of ice-cold 100% ethanol, mix gently, and precipitate for 30 min to overnight at –20°C (not –80°C as the CsCl will precipitate at the lower temperature). Spin at 2000 × *g* or greater for

Protocol 6. *Continued*

10 min, decant the aqueous layer, and wash the pellet with 76% ethanol. Spin again at 2000 × *g* or greater for 2 min, and dry the pellet in air or under a vacuum, avoiding over-drying. Resuspend the mtDNA in 0.1–0.5 ml TE and store at 4°C for immediate use or at –20°C for longer storage times.

[a]This step can be omitted, if not needed; some researchers have found that mtDNA isolated without the proteinase K digestion appears cleaner and is more easily cut with restriction endonucleases (14).

3.4.2 Isolation of total nucleic acids from plants

We include *Protocol 7*, which describes an extremely effective method for isolating nucleic acids of exceptional purity from plants, including even those with recalcitrant tissues. It has been adapted for large numbers of small samples and can be performed in 1.5 ml microcentrifuge tubes.

Protocol 7. Total nucleic acid isolation for plant tissues (or other samples containing high amounts of polysaccharides and/or polyphenols (15))

Equipment and reagents

- Liquid nitrogen
- Extraction buffer: Tris–borate (0.2 M Tris, 0.2 M boric acid, 10 mM Na_2EDTA, pH 7.6), 1/50 volume 25% SDS, 1/50 volume β-mercaptoethanol
- Phenol–chloroform, 1:1 (v/v), buffered with 0.2 M Tris–borate
- Sodium acetate, 0.1 M, pH 4.5
- Ethylene glycol monobutyl ether (2-butoxyethanol, or 2-BE)
- Tris–borate:2-BE, 1:1 (v/v)
- 100% ethanol (*Protocol 5*)
- 70% ethanol, 0.1 M potassium acetate

Method

1. Grind approximately 100 mg tissue in liquid nitrogen in a 1.5 ml microcentrifuge tube, add 500 μl of extraction buffer and 10 μl of β-mercaptoethanol to each sample, and continue grinding.
2. Add 500 μl of phenol–chloroform 1:1 (buffered with 0.2 M Tris–borate) and mix.
3. Spin samples in microcentrifuge at full speed for 10 min.
4. Re-extract with 500 μl of phenol–chloroform.
5. Remove 250 μl of supernatant and add to 750 μl of 0.1 M sodium acetate pH 4.5. Mix well.
6. Precipitate polysaccharides by adding 400 μl of 2-BE to each sample; mix well then place on ice for 30 min.
7. Spin samples in microfuge at full speed at 4°C for 10–15 min.

8. Carefully remove supernatant, and add 650 μl to each of two tubes containing 650 μl each of 2-BE. Mix thoroughly and place on ice for 30 min.
9. Spin in microfuge at full speed at 4°C for 10–15 min.
10. Discard supernatant, and wash pellet with 500 μl of each of the following, spinning for 2–5 min after each: 0.2 M boric acid/Tris, 70% ethanol, and absolute ethanol
11. Vacuum dry pellet.

This protocol can also be used to clean up dirty samples by differential precipitation. Other extraction buffers can be substituted, but yields and purity are superior when borate extraction buffers are used. RNA can be precipitated by adding an equal volume of 6 M LiCl for a final concentration of 3 M LiCl, if necessary. For larger volumes, ratios of 2-BE and diluted aqueous phase of phenol–chloroform extract need to remain the same, as well as the Na^+ concentration, including that contributed by EDTA. We have used this protocol successfully for peanut, banana, and pepper, as well as tobacco.

4. PCR amplification as a method for isolating mtDNA

The popularity of mtDNA as a molecular marker for systematic and population genetic studies, has undoubtedly contributed to the application of new technologies that allow researchers to compare large numbers of mitochondrial characters among a large number of individuals or taxa. Clearly, the most important of these technologies is the polymerase chain reaction (PCR), and recent modifications to the basic method, e.g., long PCR (16), are especially useful. Many PCR applications and protocols are provided elsewhere in this volume (Chapter 6).

Because of its small size and normally conserved gene order and composition, the mitochondrial genome makes an excellent template for producing consensus PCR primers that can be used to amplify DNAs in many groups (17). In addition, some of these consensus primers (*Table 2*) used in the appropriate combinations allow for the amplification of entire mitochondrial genomes from a wide variety of taxonomic groups. This amplified DNA, which has effectively been 'isolated' in high concentrations using PCR, can then be used for restriction endonuclease digestions that allow both site and length polymorphism comparisons, and in addition can provide suitable template for obtaining DNA sequence data (either by cloning in an appropriate vector or by direct sequence analysis). It should be noted that although these kinds of analyses are not being routinely used for plant mtDNAs, they do have applications for chloroplast DNA comparisons (17).

We include a method (*Protocol 8*) that has proven useful for several consensus primers (*Table 2*), allowing amplification of a small number of

Table 2. Consensus primers for the amplification of the vertebrate mtDNA[a]

Primer	Sequence	3′ end in human	Estimated T_m (°C) −5 °C
mt 16sbr H	5′-ccg gtc tga act cag atc acg t-3′	3080	59
mt 16sbr L	5′-acg tga tct gag ttc aga ccg g-3′	3100	59
mt CO I H	5′-ggr tct cct cct ccg gcg gg-3′	6549	65
mt CO I L	5′-ccc gcc gga gga gga gay cc-3′	6569	65
mt CB 2 H	5′-ccc tca gaa tga tat ttg tcc tca-3′	15 175	63
mt CB 3 H	5′-ggc aaa gag aaa rta tca ttc-3′	15 560	51
mt Glu 1 H	5′-caa cga tga ttt ttc asg tca-3′	14 745	51
mt Glu 1 L	5′-tga cat gaa aaa tca tcg ttg-3′	14 724	51
mt CB 3r L	5′-cac atc aaa cca gaa tga tay tt-3′	15 580	55
mt 12S-L	5′-act ggg att aga tac ccc act atg c-3′	1067	69
mt 12S-H	5′-tag tgg ggt atc taa tcc cag ttt g-3′	1042	67

[a]Adapted from Palumbi (17).

(≤4) large fragments that together make up all or almost all of the mitochondrial genome from a wide range of vertebrate taxa.

Protocol 8. PCR amplification of large fragments (>3 kb) of mtDNA as method for isolation[a]

Equipment and reagents

- Master mix[b] (700 μl final volume), stable up to 6 months at 4°C: 100 μl 10× buffer (500 mM Tris–HCl pH 8.3, 20 mM $MgCl_2$, 5% Ficoll 400, 10 mM tartrazine, 2.5 mg/ml BSA)[c], 10 μl 20 mM (each) dNTP mix (e.g., Pharmacia dNTPs), 8 μl *Taq* DNA polymerase, 1 μl *Pfu* DNA polymerase (Stratagene), and 581 μl HPLC-grade water or Milli-Q water
- PCR reaction mix (10 μl final volume); make up only at time of running PCR reaction: 7 μl master mix, 1 μl each of 5 μM forward and reverse primer stocks (*Table 2*), 1 μl template DNA (20–50 ng), diluted in water or 10 mM Tris–HCl pH 8.0 (not in TE)

Method

1. After making the Master Mix, set up the required number of 10 μl reactions into PCR reaction tubes[d] (always use plugged tips for all pipetting)
2. Initiate denaturation reaction with a 'hot start' by placing in pre-heated thermal cycler heat block for an initial 15–30 sec denaturation step (94°C); link this immediately to the appropriate program below.
3. PCR programs:
   (a) Primer pair 1: Glu 1 L/16sbr H: denaturation, 94°C for 15 sec; annealing, 51°C for 15 sec; elongation, 72°C for 3 min; 32 cycles (fragment about 4.4 kb).

(b) Primer pair 2: 16sbr L/CO I H: denaturation, 94°C for 15 sec; annealing, 59°C for 15 sec; elongation, 72°C for 1 min; 32 cycles (fragment about 3.5 kb).

(c) Primer pair 3: CB 3 RL/COI H: denaturation, 94°C for 15 sec; annealing, 55°C for 15 sec; elongation, 72°C for 4 min; 32 cycles (fragment about 8.6 kb).

[a] Conditions optimized for Perkin–Elmer 9600; for other thermal cyclers slight modification may be required.
[b] For fluorescent labelling (see *Section 6.1.3*) include 0.5 μl fluorescently labelled dUTP.
[c] Ficoll and tartrazine are included to allow direct loading of PCR products on to agarose gel. Tartrazine is yellow and migrates slightly ahead of primer dimers (see Idaho Technologies, Inc. WWW page at `http://128.110.195.115/BRC1.html`)
[d] Depending on the thermal cycler being used, samples may need to be covered with a small amount (40–60 μl) of sterile mineral oil after making PCR reaction mix, and the hold times will need to be lengthened.

5. Separation and detection of mitochondrial fragments

Once the mtDNA has been isolated/amplified and digested with either restriction endonucleases or fragments characterized using DNA sequencing techniques, fragment separation and detection become critical.

5.1 Separation of mtDNA fragments

To a certain extent the choice of restriction enzymes with four- or six-base recognition sites will determine the choice of separation medium and detection method (*Table 3*). Separation of DNA fragments is performed in gels of either agarose, which can vary in concentration, or in polyacrylamide gels, which can vary in pore size. Polyacrylamide gels give poor resolution of fragments larger than about 2000 bp and are therefore not suitable for use with fragments generated by restriction enzymes with six-base recognition sites (these enzymes cut mtDNA into only 1–8 fragments with average lengths greater than 2000 bp). Enzymes with four-base recognition sites will cleave the DNA on average every 4^4 (256) bp, while those with six-base recognition sites cleave on average every 4^6 (4096) bp. Thus, agarose gels are always preferred when using enzymes with six-base recognition sites.

There are many designs of apparatus that can be used for electrophoresis of DNA fragments and some of them are easy to make from glass or Perspex. A number of vertical or horizontal slab gel systems are also commercially available. Agarose gels usually are run horizontally submerged in buffer (submarine gels) and smaller gels are cast directly on glass plates of desired size. Owl Scientific (see Appendix 6) manufactures gel boxes that recirculate buffer using bubbles collected at the cathode, allowing for faster, more uniform runs.

Table 3. Comparison of methods for the visualization of mtDNA fragments

Method	**Sensitivity**	**Cost**	**Time consumption**	**Separation medium**	
				Agarose	**Polyacrylamide**
Non-isotopic					
Ethidium bromide	Low	Low	Low	+	(+)
Silver staining	High	Low	Low	–	+
Fluorescence	High	Moderate[a]	Low	+	+
Direct sulfonation	Moderate	High	Moderate	+	–
Hybridization	Moderate	High	Moderate	+	–
Isotopic					
End-labelling	High	High	High	+	+
Nick translation	Moderate	High	High	+	–
Oligolabelling	Moderate	High	High	+	–

[a]Initial cost of detection equipment is significant, however, once in place the cost per reaction is low to moderate.

More detailed descriptions of electrophoresis apparatus, and the theoretical and practical background to electrophoresis can be found in refs 1, 18 and 19.

TAE buffer (which often gives a better resolution for smaller gels run at higher voltages) and TBE buffer (which is sometimes preferred for longer gel runs; 18) are made as stock solutions that are diluted for use. The TAE stock solution (×50) is made up from 850 ml water and 242 g Tris base which is adjusted to pH 7.5–8 with about 57 ml of glacial acetic acid. Ten millilitres of 0.5 M EDTA (adjusted to pH 8.0 with 10 M NaOH) are the added. The buffer is stored cold and diluted 1:50 (stock solution/water) for electrophoresis or gels. This buffer loses its buffering capacity relatively quickly during electrophoresis and should be replaced at intervals or re-circulated when longer gels are needed (capacity approximately 20 h at 100 V/l of buffer).

For some applications such as checking small (<1 kb) PCR products, 0.5 × TAE buffer strengths allow faster electrophoresis with minimal heat build-up and improved resolution. Short runs on 1.5–2% agarose submarine gels (using several combs and run lengths of 4–5 cm) can be electrophoresed at up to 300–700 V for 6–20 min total run time, depending on the gel apparatus, gel thickness, and power supply. Constant current settings of up to 400 mA can be used to avoid overheating. Care should be taken to avoid excessive run times at higher voltages, and power supplies with shut-off timers are preferred.

The 5× TBE buffer stock is made up of 950 ml water, 54 g Tris base, 27.5 g boric acid, and 20 ml 0.5 M EDTA, adjusted to pH 8.0 and diluted 1:5 for electrophoresis or gels. Low molecular weight DNA fragments diffuse more easily and are thus best separated at fairly high voltage while the resolution of larger fragments will be best achieved at lower voltages. For fragments between 500 and 6000 bp, gels with 0.8% agarose are appropriate. For smaller fragments a higher agarose concentration may be necessary (up to 2.5%). The optimal resolution for mtDNA digested with different enzymes will be achieved when you know the number and size of the fragments and can adjust the separation characteristics (agarose concentration and voltage) accordingly.

Larger agarose gels generally give better resolution for Southern blotting or end-labelling. To maximize resolution and reproducibility and to avoid temperature gradients in the gel, a voltage of 1–2 V/cm is usually applied. If you have an electrophoretic apparatus that allows cooling and re-circulation of buffer, slightly higher voltages can be applied. Such tanks are recommended since they improve the chance of obtaining straight separations that are reproducible between runs and facilitate interpretations of banding patterns. More detailed descriptions of how to prepare and run such gels are given in Chapter 5 and in refs 18–23.

5.2 Small polyacrylamide gels for silver staining

Polyacrylamide gels are used for the analysis of smaller DNA fragments produced by digestion with restriction enzymes with four-base recognition sites. A 3.5% gel will give optimal resolution for fragments 10–2000 bp long, while a 12% gel will separate 40–200 bp fragments (18; Appendix 4). Gels of different pore size are made by varying the concentration of acrylamide. Varying the amount of bisacrylamide cross-linker also affects the pore size but mainly affects the brittleness and clarity of the gel (21). Polyacrylamide gels are almost always poured between two glass plates held apart by spacers. Gels may vary in length, depending upon the separation required, and are invariably run in the vertical position. The thickness of gels can be varied from ultra-thin (0.1 mm or less) to the more common thickness of about 1 mm. Precast gels of different kinds, designed to fit many electrophoresis systems (Pharmacia LKB, Hoefer, Bio-Rad, etc.) are available from Palliard Chemical Co. Ltd and other suppliers.

5.3 Longer polyacrylamide gels for end-labelling

The fine resolution of the silver-staining method allows visualization of DNA fragments separated in small gels. Isotopic methods of fragment visualization give coarser bands and, in order to visualize the 25–50 fragments generated after digestion with restriction enzymes with four-baser recognition sites, longer polyacrylamide gels must be used to separate fragments. The pore size of such gels can be varied (2.5–6% or gradient gels) and the gels are prepared using essentially the same method as described above. Depending upon whether wet- or dried-gel autoradiography will be used, the application of Bind-Silane (Sigma Chemical, see Appendix 6) to one glass plate may be omitted. Appropriate guidance for making longer polyacrylamide gels (DNA sequencing gels) can be found in Chapter 6 and refs 18 and 24.

6. Methods for the visualization of mtDNA fragments

The easiest way to detect DNA fragments is by staining with ethidium bromide (25) which can be used with both agarose and polyacrylamide gels, but with somewhat reduced sensitivity in polyacrylamide gels. Ethidium bromide staining is a relatively insensitive technique for visualizing fragments (requiring >20 ng of DNA that is >50 bp in length). Both silver staining and isotopic methods of fragment visualization offer a much higher sensitivity than ethidium bromide staining and are suitable for use with small samples of mtDNA. The two most used isotopic techniques are end-labelling of relatively pure mtDNA and hybridization with mtDNA probes. The easiest techniques (ethidium bromide and silver staining) will be described below in more detail than isotopic methods. There are excellent protocols for isotopic visualization of mtDNA fragments described in refs 1 and 26–28.

6.1 Non-isotopic methods

6.1.1 Ethidium bromide

The planar group of the ethidium bromide molecule intercalates into the stacked regions of the DNA helix and this complex absorbs UV light at 300 nm, loses some energy, and re-emits visible light at 590 nm as fluorescent orange bands. Lamps are available with emission spectra maxima at 254, 302, or 366 nm. Illumination at 254 nm damages the DNA while illumination at 366 nm gives less fluorescence than illumination at 302 or 254 nm. A photographic technique that increases the sensitivity of ethidium bromide staining by a factor of 10 has recently been described (29) as well as the use of other fluorochromes such as 4,6′-diamidino-2-phenylindole (DAPI) (30).

Ethidium bromide is a powerful mutagen and may also be carcinogenic. Avoid using it in gels or electrophoretic tanks. Design the staining process and the handling of gels so that contamination of the laboratory environment is prohibited and minimize the volumes of staining solutions. Collect the staining solutions and the stained gels for proper disposal. Suggested methods for the disposal of ethidium bromide solutions include treatment with potassium permanganate/hydrochloric acid (31), adsorption in activated charcoal and subsequent commercial incineration (32), and the collection of all the ethidium bromide solutions for commercial destruction. When handling dry ethidium bromide, use a mask and gloves for protection and clean the balance and surrounding areas carefully with several wet paper towels. Small particles of ethidium bromide are easily dispersed in the air and great caution is recommended in handling this substance.

6.1.2 Silver staining

Many silver-staining protocols are time-consuming or make use of harmful chemicals. The sensitivities of the protocols are more or less the same (33)

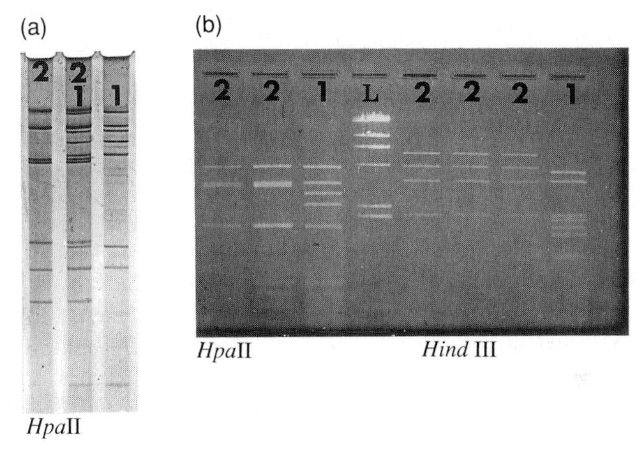

Figure 5. (a) 5% polyacrylamide gel, silver stained using *Protocol 9*. mtDNA fragments are from the yellow-necked mouse (*Apodemus flavicollis*; lane 1) and the wood mouse (*A. sylvaticus*; lane 2). mtDNA was isolated by CsCl ultracentrifugation (22) and about 20 ng of DNA was digested with *Hpa*II and applied to the gel. The lane in the middle contains mixed mtDNA from both species to allow fragment comparison between species (22). (b) Agarose gel (1.2%) stained with ethidium bromide. mtDNA fragments are from the yellow-necked mouse (1) and wood mouse (2). mtDNA was isolated by CsCl ultracentrifugation (22) and about 100 ng of DNA was digested with *Hin*dIII and *Hpa*II and applied to each lane. The lane marked L contains 500 ng of lambda DNA digested with *Hin*dIII to produce size markers (236130, 9416, 6557, 4361, 2322, 2027, and 564 bp).

and therefore we have adopted (34) the protocol of Guillimette and Lewis (35) which is rapid (less than 2 h, which means that the whole process from undigested mtDNA to band analysis can easily be performed in one day), avoids many reactive chemicals, and is inexpensive.

The silver-staining protocol (*Protocol 9*) is designed for polyacrylamide gels and has a sensitivity of about 10–30 pg of DNA, allowing the use of samples as small as 10–40 ng of mtDNA digested with an enzyme with a four-base recognition site (*Figure 5*). There are also commercial silver-staining kits available (e.g. from Bio-Rad) which are more expensive than *Protocol 9*. DNA in agarose gels can be stained with silver but the sensitivity is only as good as for ethidium bromide staining.

Protocol 9. Silver staining of polyacrylamide gels

Reagents

- 0.1% CTAB (Sigma T4762) in double-distilled H_2O
- 0.3% ammonia in double-distilled H_2O
- $AgNO_3$
- 1 M NaOH solution, freshly prepared
- 25% ammonia
- 2% sodium carbonate solution in double-distilled H_2O
- 35% formaldehyde
- Glycerol

Protocol 9. *Continued*

Method

1. After electrophoresis, rinse the gel briefly in distilled water for a few minutes. We use a separate plastic container for each gel and for each of the solutions during staining. Use double-distilled H_2O in all solutions for silver staining.
2. Move the gel to a new container with 0.1% CTAB. Let the gel soak under gentle agitation for 20 min.[a]
3. Move the gel to a new container with 0.3% ammonia (freshly prepared, 1.3 ml/100 ml of a commercial 25% solution) in double-distilled H_2O. Let the gel soak for 15 min under gentle agitation.
4. Prepare the ammoniacal silver solution as follows.
   (a) Dissolve 0.2 g of $AgNO_3$ in 10 ml of double-distilled H_2O and then pour the solution into 125 ml of double-distilled H_2O.
   (b) Prepare a 1 M NaOH solution (1 g of NaOH in 25 ml of double-distilled H_2O). Make a new solution each day.
   (c) While stirring vigorously, add exactly 0.5 ml of 1 M NaOH to the $AgNO_3$ solution which will turn brown and cloudy.
   (d) While stirring, add 0.5–0.6 ml of 25% ammonia to the silver–NaOH solution drop by drop until the solution becomes completely clear and then add two more drops of 25% ammonia.
5. Pour the ammoniacal silver solution into a new plastic container with the gel and leave it under agitation for 20 min.
6. During the ammoniacal silver treatment prepare the appropriate amount of developer as follows.
   (a) Make a 2% sodium carbonate solution (4 g in 200 ml of double-distilled H_2O) by rapid mixing followed by vigorous stirring.
   (b) Add 120 μl of 35% formaldehyde (the usual commercial concentration) to the sodium carbonate solution (final concentration of formaldehyde, 0.02%). **Warning**: This step and the developing should be performed in a fume cupboard. Formaldehyde may cause allergic reactions and some persons are extremely sensitive to it.
7. Give the gel a quick (*c.* 15 sec) rinse in distilled water in a new plastic container and then transfer it to a glass container.[b]
8. Pour the developer on to the gel and agitate so that a 'wave' passes over the gel surface. The time for development will range from 5 to 25 min depending on DNA concentration and the freshness of the solutions.
9. Stop the development by a quick rinse in distilled water and then soak the gel in 1.5% glycerol for 30 min.

10. Dry the gel at room temperature overnight. The gel can be stored for at least 7 years.

[a]The cationic detergent CTAB is believed to increase the sensitivity and linearity of the staining process (11) but is optional in our experience. If the CTAB step is omitted, let the gel soak in distilled water for 20 min.
[b]Silver particles will be released during gel development and, in order to be able to keep the developing container as clean as possible, a glass container is preferable to a plastic one for this step.

6.1.3 Incorporation of fluorescent label during PCR amplification

Chapter 10 includes a large sampling of the new fluorescence-based technologies. As applied to detection of mtDNA fragments amplified by PCR, the minor modification noted in *Protocol 8* allows for incorporation of fluorescent dyes into the final PCR product. These fragments can then be visualized and the sizes very accurately estimated using automated sequencing equipment such as Applied Biosystems models 310, 373, or 377.

6.1.4 Direct sulfonation of DNA fragments on membranes

A technique for staining DNA fragments bound to nylon membranes without using hybridization has recently been reported (29). Direct sulfonation can be used for detecting mtDNA fragments in preparations with low nuclear DNA contamination. Direct sulfonation of mtDNA fragments converts cytidine residues to sulfonate derivatives and sulfonation is followed by the application of mouse antibodies specific to sulfonated DNA. A second antibody–alkaline phosphatase conjugate is then applied and a chromogenic tetrazolium developer will give dark bands against a clear background. This direct immunohistochemical method is said to be able to detect down to 10 pg of DNA—a sensitivity comparable to that of silver staining. This method is somewhat time-consuming since the gels must be transferred either using a vacuum- or standard Southern-blotting apparatus (22) and the staining process requires about 6 h. The resolution is probably not sufficient for mtDNA digested with restriction enzymes with four-base recognition sites but, if only small amounts of pure mtDNA are available, this immunohistochemical method may be worth trying.

6.1.5 Non-isotopic methods for the detection of mtDNA fragments after hybridization

Although there are only a few reports of non-isotopic methods being used for detecting mtDNA fragments after hybridization (26, 36) these methods are worth mentioning in order to indicate their possibilities. Three methods predominate (27; see also Chapter 5):

(a) Incorporation of biotinylated nucleotides into probe DNA by standard methods (random primer method or nick translation, Chapter 5) followed by hybridization to filter-bound DNA. The biotin-labelled DNA is detected by incubation with avidin or streptavidin coupled to an enzyme that catalyses a reaction with a coloured end-product.

(b) Covalent linkage of an enzyme to single-stranded DNA which is used directly as a probe. DNA fragments are detected as in (a).

(c) The probe DNA is modified by insertion of an antigenic sulfone group on to the cytosine moieties and then used for hybridization. Visualization of the probe is carried out by a combined immunoenzymatic reaction where monoclonal antibodies specifically bind to the sulfone groups and the enzyme converts a chromogenic substance to an insoluble visible dye that precipitates where the probe has hybridized.

There are a number of commercially available kits based on these principles: the ECL detection system (Amersham); a digoxigenin-antibody conjugate method (Boehringer Mannheim GmbH); Chemiprobe (FMC Bioproducts); BluGENE and Photobiotin Labeling without nick translation (BRL); Gene Images (United States Biochemical Corporation). It is also possible to use the methods without ready-made kits. The reagents are comparatively inexpensive, whereas some of the commercial kits are extremely expensive. Protocols for preparation of biotinylated DNA probes and for the whole process from preparation of mitochondria and mtDNA to preparation and detection of probes can be found in ref. 26.

The advantages of non-isotopic probes include the following:

- There are none of the hazards associated with radioisotopes and no radioactive waste disposal problems. No licence is required for handling of the substances.
- Non-isotopically labelled probes are stable and can be stored for long periods while ^{32}P has a half-life of about 14 days.
- Results are on film within a few minutes or hours.

The disadvantages of non-isotopic probes are as follows:

- Higher probe concentrations are required, making it necessary to clone the probe DNA. Very low probe concentrations can be used with ^{32}P-labelling.
- ^{32}P-labelled probes can detect lower amounts of a specific DNA sequence in most cases and thus non-isotopic probes are most suitable for lower sensitivity applications.

6.2 Isotopic methods

Isotopic methods are extremely sensitive and allow the detection of small amounts of DNA. The sensitivity of some of the isotopic methods can be increased by using more than one type of labelled nucleotide simultaneously. If very pure mtDNA is available (after isolation by CsCl density centrifugation) the most common method is to end-label the fragments directly after

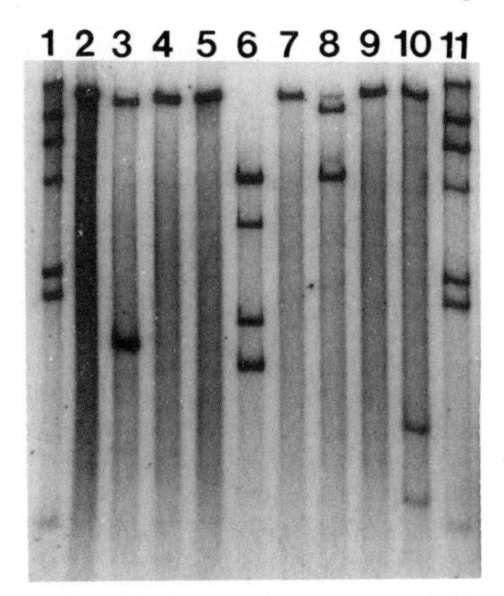

Figure 6. 1.2% agarose gel electrophoresis of restriction enzyme digested mtDNA (lanes 2–10) labelled with [α-^{32}P]dNTPs (described in *Protocol 10*) isolated from frozen *Crocodylus acutus* liver as described in *Protocols 3–5*. Lanes 1 and 11 contain *Hin*dIII-digested lambda DNA as size standard.

digestion and separate the labelled fragments in either agarose or polyacrylamide gels (*Figure 6*). The gel is then autoradiographed wet or dried on to filter paper. The DNA ends produced by the restriction enzyme are blunt or contain 5′ or 3′ single-stranded extensions (26). The large (Klenow) fragment of DNA polymerase I can be used to label recessed 3′ termini by 'filling in' when there is a 5′ extension, and by replacement of the nucleotide present at the 3′-hydroxy terminus when the ends are blunt. Klenow is inefficient at labelling DNA termini with 3′ extensions; instead use bacteriophage T4 DNA polymerase (at 37 °C in the following assay conditions: 33 mM Tris–acetate pH 8.0, 66 mM potassium acetate, 10 mM magnesium acetate, 0.5 mM dithiothreitol (DTT), 100 μg/ml BSA, fraction V; see ref. 5). Bacteriophage T4 polynucleotide kinase can be used to label the 5′ termini of dephosphorylated DNA fragments (for protocols see Chapter 5 and ref. 5). Labelling is usually carried out with ^{32}P or ^{33}P. The alternative isotope, ^{35}S, which gives sharper bands than ^{32}P, has a longer half-life (60 days) but requires much longer exposure times.

6.2.1 Direct labelling of fragments

Slight modifications of the procedure for end-labelling given below, background information concerning different steps in the procedure, and troubleshooting can be found in refs 1, 23, and 26–28. Digested samples (10–40 ng) can be labelled directly without addition of any stop solution or samples can be stored frozen after digestion and labelled later. All restriction enzyme buffer conditions, except when NaCl or KCl are omitted, function well with the DNA polymerase. When a low salt buffer has been used for the digestion of mtDNA, add NaCl to a final concentration of 50 mM.

Protocol 10. End-labelling by end-filling of mtDNA fragments

Equipment and reagents

- 10× labelling buffer: 6 mM KCl, 10 mM Tris–HCl pH 7.5, 50 mM NaCl, 10 mM $MgCl_2$, 1 mM DTT or 7 mM 2-mercaptoethanol (this is a common medium-salt restriction buffer)
- 1 M NaCl in distilled water
- Unlabelled nucleotides (0.1 mM each)
- Labelled nucleotide
- Large (Klenow) fragment of DNA polymerase I
- Sephadex G-50 columns
- *Escherichia coli* tRNA
- Cold 99% ethanol
- 0.3 M sodium acetate

Method

1. Prepare a mixture of the unlabelled nucleotides (0.1 mM of each). The choice of unlabelled nucleotides is dictated by the choice of restriction enzyme, which determines the base sequence in the free 3′ end, and the choice of labelled nucleotide.
2. For 15 samples, prepare the following labelling mix (work behind a protection shield and be aware of procedures for handling radionucleotides).
   - 137 ml sterile distilled water
   - 16 μl of buffer
   - 5 units large (Klenow) fragment of DNA polymerase I
   - 2 μl of unlabelled nucleotide mixture
   - 2 μl of labelled nucleotide (*c.* 20 μCi [α-^{32}P]dNTP)
3. Add 10 μl of labelling mix to each sample and leave at the appropriate temperature for 20–30 min. Usually the reaction is carried out on ice (for enzymes that create a 5′ overhang, such as *Mbo*I), but for enzymes that create blunt ends (like *Rsa*I) or 3′ overhangs (rarely used) labelling is best when carried out at 37°C (or label 3′ overhangs using T4 DNA polymerase; see Section 6.2).
4. Remove unincorporated nucleotides by chromatography through small Sephadex G-50 columns or by ethanol precipitation using yeast or *E. coli* tRNA as a carrier (20 μg/sample in 340 μl of 0.3 M sodium acetate). Add 2.5 volumes of cold 100% ethanol and incubate for 20 min at –70°C. Spin for 10 min in a microfuge. Decant the supernatant, gently rinse the pellet in cold 70% ethanol, and dry the pellet.
5. Resuspend the pellet in an appropriate volume of sterile, distilled water. Then separate the dissolved DNA either in larger agarose gels (concentration depending on the size of the fragments) or in 3.5–5.0% polyacrylamide gels together with a molecular-weight standard.

6. Transfer the gel to a support (sheet of Whatman 3MM paper), covered with Saran wrap, and dry it on a pre-warmed (80°C) automatic gel dryer. Press the dried gel tightly against a piece of X-ray film and expose with or without intensifying screens.[a]

[a]Dried gel autoradiography results in sharper bands because of a reduced scattering angle between the radioactive sample and the film.

The technique of end-labelling mtDNA fragments after digestion has been successfully applied to a range of organisms. The original, much-cited reference is that of Brown (37) who investigated human mtDNA. Among many of the recent reports where the end-labelling technique has been used are investigations on mtDNA variation in populations of fishes (38, 39), fruit flies (39), lizards (41), crocodilians (5), birds (42), fruit bats (43) and deer mice (44). There are two reports on mtDNA variation in fish and birds where ^{32}P has been replaced by ^{35}S in the end-labelling reaction (45, 46).

6.2.2 Hybridization with mtDNA probes

mtDNA fragment variation can be identified by hybridization with mtDNA probes (*Figure 7*). This is an alternative to the more traditional techniques and permits RFLP analysis when only small or impure samples are available.

Southern blots with total cell DNA can also be used to study nuclear genes with appropriate genomic or cDNA probes, providing a rapid means for comparing population structure using different genetic markers. There is a possibility that mtDNA and nuclear DNA may share common sequences that may confuse the interpretation of fragment patterns (47). Such homology is easy to check for by increased stringency washes, comparisons of fragment patterns generated by different probes, or by using a few samples of purified mtDNA. Hybridization with mtDNA probes to reveal mtDNA variation will only be summarized here and the reader is referred to Chapter 5 and references given in the discussion that follows:

(a) Total DNA isolated by proteinase digestion of 0.2–0.5 g of tissue (which will yield 0.5–1.0 mg of DNA), or from isolated blood cells (such as platelets, which lack nuclei and are consequently relatively rich in mtDNA, ref. 48), or directly from blood (100–200 μl mammalian blood or 10 μl from fish-reptiles, amphibians or birds). Sample collection is not dependent on refrigeration because biopsies or blood can be stored for extended periods in salt-based preservatives or alcohol (49, 50). DNA is purified by repeated phenol–chloroform (51; see also Chapter 6) or by salt–chloroform extractions (see Chapter 2).

(b) Restriction enzyme digested DNA is separated in larger agarose gels of the appropriate concentration and Southern blotted (Chapter 5 and refs 18 and 52) to nylon membranes.

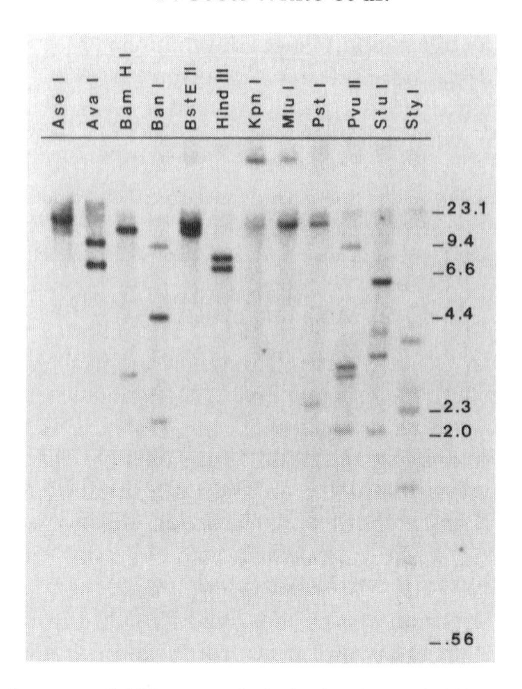

Figure 7. Autoradiogram of *Alligator mississippiensis* total DNA, isolated from whole blood as described in ref. 5, that has been digested with restriction enzymes, subjected to electrophoresis on a 1.2% agarose gel, Southern blotted, and probed with mtDNA isolated from *A. mississippiensis* whole blood as described in *Protocols 3–5*. Numbers at the side indicate the positions of the *Hin*dIII-digested lambda DNA fragments used as size standards (size in kb).

(c) Probes used for hybridization may be either the total, highly purified, mtDNA molecule or smaller fragments. For small-scale investigations enough probe mtDNA can be obtained by isolation using density centrifugation, an extraction procedure (Chapter 2 and ref. 53), or using PCR (17, *Protocol 8*). It is not necessary to use mtDNA from the same species because hybridization is possible between as distantly related species as Dall's porpoise and humpback whale (54), jackal, fox, and dog (55), deep sea scallop and fruit fly (56), or house sparrow and rat (49). If larger quantities of probe mtDNA are needed it is better to construct recombinant plasmids containing the complete mitochondrial genome or smaller fragments of mtDNA which can be used for hybridization (Chapter 4).

(d) If the complete mtDNA molecule is used it is ^{32}P-labelled by nick translation (for the procedure see refs 1, 23, and 27 or Chapter 5). If smaller fragments are used they may be ^{32}P-labelled by nick translation or the random primer method (Chapter 5 and ref. 57).

(e) Hybridization, washing, and autoradiography are performed as described in refs 1, 18, 23, and 58 and Chapter 5.

7. Conclusions

mtDNA isolation can be performed in a variety of ways and the quality of a given tissue is as critical as using the appropriate procedure(s). We have provided several protocols, with hopes that one or more will be of use to the majority of researchers. Amplification of mitochondrial sequences (including the entire genome in some cases) is now capable of replacing standard mtDNA isolation procedures. A wider variety of techniques is also becoming available for separating and detecting mtDNA fragments, increasing the sensitivity, while reducing both costs and health concerns.

Acknowledgements

This work was supported in part by NSF grant BSR-8607420, Texas ARP Award No. 0036440140, the Ford Foundation, Texas Tech University, the US Department of Energy and the Swedish National Research Council.

References

1. Dowling, T.E., Moritz, C., Palmer, J.D., and Rieseberg, L.H. (1996). In *Molecular systematics*, 2nd edn (ed. D. Hillis, C. Moritz, and B. Mable), pp. 249–321. Sinauer, Sunderland, MA.
2. Hillis, D.M., Moritz, C., and Mable, B.K. (eds) (1996). *Molecular systematics*, 2nd edn. Sinauer, Sunderland, MA.
3. Chapman, R.W. and Powers, D.A. (1984). *Maryland Sea Grant Prog. Technical Report* UM-SG-TS-84-05.
4. Rickwood, D., Wilson, M.T., and Darley-Usmar, V.M. (1987). In *Mitochondria: a practical approach* (ed. D. Rickwood, M. Wilson, and V. Darley-Usmar), pp. 1–16. IRL Press, Oxford.
5. Densmore, L.D. and White, P.S. (1991). *Copeia*, 1991, 602.
6. Wada, S., Kobayashi, T., and Numachi, K. (1991). *Rep. Int. Whaling Commission*, Special Issue **13**, 203.
7. Hudspeth, M.E.S, Shumard, D.S., Tatti, K.M., and Grossman, L.I. (1980). *Biochem. Biophys. Acta*, **610**, 221.
8. Gorzula, S., Arocha-Pinango, C.L., and Salazar, C. (1976). *Copeia*, 1976, 838.
9. Dessauer, H.C., Cole, C.J., and Hafner, M.S. (1996). In *Molecular systematics*, 2nd edn (ed. D. Hillis, C. Moritz and B. Mable), pp. 29–51. Sinauer, Sunderland, MA.
10. Longmire, J.L., Lewis, A.K., Brown, N.C., Buckingham, J.M., Clark, L.M., Jones, M.D., Meinke, L.J., Meyne, J., Ratliff, R.L., Ray, F.A.,Wagner, R.P., and Moyzis, R.K. (1988). *Genomics*, **2**, 14.
11. Gustafson, S., Proper, J.A., Bowie, E.J.W., and Sommer, S.S. (1987). *Anal. Biochem.*, **165**, 294.
12. Kolodner, R. and Tewari, K.K. (1972). *Proc. Natl Acad. Sci. USA*, **69**, 1830.
13. Palmer, J.D. (1992). In *Plant genomes: methods for genetic and physical mapping* (ed. T. Osborn and J. Beckman). Kluwer Academic, Dordrecht.

14. Hsu, C.L. and Mullin, B.C. (1988). *Plant Cell Rep.*, **7**, 356.
15. Manning, K. (1991). *Anal. Biochem.*, **195**, 45–50.
16. Barnes, W.M. (1994). *Proc. Natl Acad. Sci. USA*, **91**, 2216.
17. Palumbi, S.R. (1996). In *Molecular systematics*, 2nd edn (ed. D. Hillis, C. Moritz, and B. Mable), pp. 205–249. Sinauer, Sunderland, MA.
18. Maniatis, T., Fritsch, E.F., and Sambrook, J. (1982). *Molecular cloning: a laboratory manual.* Cold Spring Harbor Laboratory Press, Cold Spring Harbor, NY.
19. Dalgleish, R. (1987). In *Gene cloning and analysis. A laboratory guide* (ed. G.I. Boulnois), p. 47. Blackwell Scientific Publications, Oxford.
20. Andrews, A.T. (1981). *Electrophoresis: theory, techniques and biochemical and clinical applications.* Clarendon Press, Oxford.
21. Sealy, P.G. and Southern, E.M. (1982). In *Gel electrophoresis of nucleic acids: a practical approach* (ed. D. Rickwood and B.D. Hames), p. 39. IRL Press, Oxford.
22. Southern, E.M. (1979). In *Methods in enzymology*, Vol. 68 (ed. R. Wu), p. 152. Academic Press, London.
23. Lansman, R.A., Shade, R.O., Shapira, J.F., and Avise, J.C. (1981). *J. Mol. Evol.*, **17**, 214.
24. Davies, R.W. (1982). In *Gel electrophoresis of nucleic acids: a practical approach* (ed. D. Rickwood and B.D. Hames), p. 117. IRL Press, Oxford.
25. Sharp, P.A., Sugden, B., and Sambrook, J. (1973). *Biochemistry*, **12**, 3055.
26. Hauswirth, W.W., Lim, L.O., Dujon, B., and Turner, G. (1987). In *Mitochondria: a practical approach* (ed. V.M. Darley-Usmar, D.Rickwood, and M.T. Wilson), p. 171. IRL Press, Oxford.
27. Arrand, J.E. (1985). In *Nucleic acid hybridization: a practical approach* (ed. B.D. Hames and S.J. Higgins), p. 17. IRL Press, Oxford.
28. Cunningham, M.W., Harris, D.W., and Mundy, C.R. (1990). In *Radioisotopes in biology: a practical approach* (ed. R.J. Slater), p. 137. IRL Press, Oxford.
29. Chapman, R.W. and Brown, B.L. (1989). *Anal. Biochem.*, **177**, 199.
30. Avise, J.C., Lansman, R.A., and Shade, R.O. (1979). *Genetics*, **92**, 279.
31. Quillardet, P. and Hofnung, M. (1988). *Trends Genet.*, **4**, 89.
32. Bensaude, O. (1988). *Trends Genet.*, **4**, 89.
33. Beidler, J.L., Hilliard, P.R., and Rill, R.L. (1982). *Anal. Biochem.*, **126**, 374.
34. Tegelstrom, H. (1986). *Electrophoresis*, **7**, 226.
35. Guillemette, J.G. and Lewis, P.N.L. (1983). *Electrophoresis*, **4**, 92.
36. Afonso, J.M., Volz, A., Hernandez, M., Ruttkay, H., Gonzalez, M., Larruga, J.M., Cabrera, V.M., and Sperlich, D. (1990). *Mol. Biol. Evol.*, **7**, 123.
37. Brown, W.M. (1980). *Proc. Natl Acad. Sci. USA*, **77**, 3605.
38. Ovenden, J.R. and White, R.W.G. (1990). *Genetics*, **124**, 701.
39. Wirgin, I.I., Proenca, R., and Grossfield, J. (1989). *Can. J. Zool.*, **67**, 891.
40. Monnerot, M., Solignac, M., and Wolstenholme, D.R. (1990). *J. Mol. Evol.*, **30**, 500.
41. Densmore, L.D., Wright, J.W., and Brown, W.M. (1985). *Genetics*, **110**, 689.
42. Ovenden, I.R., Mackinlay, A.G., and Crozier, R.H. (1987). *Mol. Biol. Evol.*, **4**, 526.
43. Pumo, D.E., Goldin, E.Z., Elliot, B., Phillips, C.J., and Genoways, H.H. (1988). *Mol. Biol. Evol.*, **5**, 79.
44. Ashley, M. and Wills, C. (1987). *Evolution*, **41**, 854.
45. Avise, J.C., Bowen, B.W., and Lamb, T. (1989). *Mol. Biol. Evol.*, **6**, 258.

46. Zink, R.M. and Avise, J.C. (1990). *Syst. Zool.*, **39**, 148.
47. Plante, Y., Boag, P.T., and White, B.N. (1987). *Can. J. Zool.*, **65**, 175.
48. Giles, R.E., Blanc, H., Cann, H.M., and Wallace, D.C. (1980). *Proc. Natl Acad. Sci. USA*, **77**, 6715.
49. Arctander, P. (1988). *J. Ornithol.*, **129**, 205.
50. Hoelzel, A.R. and Amos, W. (1988). *Nature*, **333**, 305.
51. Tegelstrom, H. and Wyoni, P.-I. (1986). *Electrophoresis*, **7**, 99.
52. Southern, E.M. (1975). *J. Mol. Biol.*, **98**, 503.
53. Gyllensten, U., Leary, R.F., Allendorf, F.W., and Wilson, A.C. (1985). *Genetics*, **111**, 905.
54. Baker, C.S., Palumbi, S.R., Lambertsen, R.H., Weinrich, M.T., Calambokidis, J., and O'Brien, S.J. (1990). *Nature*, **344**, 238.
55. Wayne, R.K., Van Valkenburgh, B., Kat, P.W., Fuller, T.K., Johnson, W.E., and O'Brien, S.J. (1989). *J. Hered.*, **80**, 447.
56. Snyder, M., Fraser, A.R., LaRoche, J., Gartner-Kepkay, K.E., and Zouros, E. (1987). *Proc. Natl Acad. Sci. USA*, **84**, 7595.
57. Feinberg, A.P. and Vogelstein, B. (1983). *Anal. Biochem.*, **132**, 6.
58. Sambrook, J., Fritsch, E.F., and Maniatis, T. (1989). *Molecular cloning: a laboratory manual*, 2nd edn. Cold Spring Harbor Laboratory Press, Cold Spring Harbor, NY.

4

Genomic libraries and the development of species-specific probes

J. P. WARNER

1. Introduction

The development of standard methods for reliable cloning and propagation of DNA fragments of various sizes has transformed our understanding of gene structure and regulation. The ability to generate representative libraries of sequences comprehensively covering the genome regardless of size, complexity, or type of organism has enabled population geneticists to undertake fine resolution analysis of sequence variation. The purpose of this chapter is to describe general protocols for the construction and screening of genomic libraries and for the development of species-specific probes. Although the construction of libraries was traditionally regarded as technically demanding, modern high-quality enzymes, commercially prepared packaging extracts, competent cells, and complete cloning kits have made genomic cloning universally accessible. A high quality representative library of clones available for rapid re-screening whenever new sequences suitable for population analysis are described is ample reward for the time and effort invested in the initial library construction.

2. Construction of genomic libraries

2.1 Strategy

The ideal library would contain a complete set of overlapping clones and all of the sequences in the genome would be equally represented. Individual clones would be sufficiently large to contain complete genes or sequence clusters allowing analysis of the sequences of interest in isolation from the rest of the genome. The clones themselves would be mainly made up of insert with little or no vector sequence to interfere with subsequent analysis. It would have been a simple process to construct this library and the clones would be stored in a permanent stable format for easy distribution and screening. Such ideal libraries are already a reality for some model prokaryotic organisms with small, simply organized genomes. Population geneticists, however, make phylogenetic comparisons and study sequence variation in populations from the simplest prokaryotes to the most complex higher eukaryotes. There is no such thing as a perfect library appropriate for all species or situations and often real constraints such as difficulty obtaining large amounts of high molecular weight DNA are the main consideration in the choice of system used for library construction.

Multiple different cloning vehicles are currently used for the construction of representative libraries from genomic DNA. These range from lambda vectors with a maximum insert length of about 25kb to yeast artificial chromosome (YAC) vectors with up to 2 Mb capacity (1). Cosmid, P1 (2), and bacterial artificial chromosome (BAC) (3) vectors cover the middle range between 40 kb and 300 kb. All of these individual sytems have their merits but in this chapter the discussion will be limited to the use of lambda and cosmid vectors as these are the most universally applicable. Small insert (300–500 bp) library construction using plasmid and M13 vectors is presented in a separate section. This type of library is designed specifically for the isolation and sequencing of small inserts containing polymorphic microsatellite sequences with only a subset of the genome being represented.

2.1.1 Size of library

It is important to establish the size of library required since this will determine the vector used. This size (i.e. the number of recombinants, N) for a probability (P) that a sequence is represented and for the fractional proportion (f) of the genome in a single recombinant is given (4) by:

$$N = \frac{\ln(1 - P)}{\ln(1 - f)} \quad [1]$$

For example, to be 99% sure that a given sequence is found in a library of 40 kb fragments from the human haploid genome (3×10^9 bp)

$$N = \frac{\ln(1 - 0.99)}{\ln(1 - (4 \times 10^4/3 \times 10^9)} = 3.45 \times 10^5 \quad [2]$$

Table 1 shows, for various vectors, the size (N) of library required when the probability of a given sequence being contained in the library is set at 99%. The following average insert sizes have been assumed; plasmid, 5 kb; lambda insertion vector, 10 kb; lambda replacement vector, 17 kb; cosmid vector, 40 kb. The calculations are based on the assumption that a restriction enzyme cuts at random. This is not strictly true for a partial digest. However, the numbers are likely to be a sufficient approximation for most purposes (see Chapter 5 for further discussion).

Plasmid vectors are most suitable for genomic libraries of organisms with small genomes (e.g. *Escherichia coli* or *Saccharomyces cerevisiae*), while for

Table 1. Variation of library size with vector and genome size

Haploid genome size (bp)	Organism	Size (N) of library required for $P = 99\%$ Plasmid	Lambda insertion	Lambda replacement	Cosmid
4×10^6	*Escherichia coli*	3.7×10^3	1.8×10^3	1.1×10^3	4.6×10^2
1.4×10^7	*Saccharomyces cerevisiae*	1.3×10^4	6.4×10^3	3.8×10^3	1.6×10^3
1.1×10^8	*Drosophila melanogaster*	10^5	5×10^4	3.3×10^8	1.3×10^4
3×10^9	*Zea mays*	2.8×10^6	1.4×10^6	8.1×10^5	3.5×10^5

libraries of organisms with larger genomes ($> 5 \times 10^7$ bp) lambda or cosmid vectors are suitable. The latter two vector types are discussed here.

2.1.2 Lambda or cosmid vector?

Lambda vectors are more frequently used for genomic library construction for the following reasons:

- the efficiency of recovery of recombinant DNA molecules is high;
- slightly degraded source DNA (*c.* 100 kb) can be used (cosmids require source DNA of *c.* 200 kb);
- plaques are easier to screen than colonies.

However, there are advantages to cosmid libraries:

- clones contain larger fragments so that representative libraries are smaller and complete genes are more likely to be cloned;
- restriction mapping of recombinant DNA yields more information since vector sequences make up a smaller proportion of the clone.

There is a bewildering variety of vectors for genomic library construction and the following general design features common to many modern vectors should be considered in making an informed choice. The presence of promoters for phage-specific DNA-dependent RNA polymerases (T7, T3, or SP64) flanking the cloning site allows riboprobes specific to either extremity of the insert to be produced *in vitro*. This is especially important for chromosome walking and for rapid restriction mapping using partial digestion techniques. Similarly, unique restriction sites on either side of the cloning site enable rapid restriction mapping and flanking 'rare-cutter' restriction sites (e.g *Not*I) allow the majority of inserts to be excised as a single fragment for further mapping and cloning.

2.1.3 Lambda vectors

The bacteriophage lambda chromosome is a 48.6 kb linear DNA molecule. Nearly 40% of its genome is not essential for propagation. Moreover, infectious particles can be prepared with DNA molecules in the range 37–52 kb inserted. Cloning strategies make use of these and other features of lambda biology (see ref. 5). Most strategies involve *in vitro* construction of recombinant DNA molecules followed by 'packaging' of these into infectious, viable phage particles. This packaging requires the presence in the recombinant molecules of the lambda determinant cohesive termini (*cos*) and of certain lambda proteins. Packaging of DNA is comparatively efficient: 1 μg of DNA can give 10^9 plaques.

There are two principal types of lambda vector:

(a) *Insertion vectors* have a single restriction site for the insertion of source DNA and are generally used for cloning completely digested genomic DNA. The smaller insert size tolerated by these vectors (< 10 kb) means that libraries constructed using these vectors will not be fully representative. For this reason, several libraries should be constructed using different restriction enzymes so that more complete representation can be obtained. Despite being slightly unfashionable lambda insertion libraries remain the easiest to make for the beginner. We recommend the vector NM1149. This vector is lysogenic (in the host NM514, but not in NB78) but the insertion of source DNA at the cloning site renders the recombinant non-lysogenic (in both NM514 and NB78). NM1149 requires the host *recA* recombination function which can lead to instability of some recombinants.

(b) *Replacement vectors* contain a pair of restriction sites flanking a non-essential segment of lambda DNA, called a 'stuffer fragment'. This stuffer is removed and source DNA is then inserted in its place. Use of size-fractionated, partially digested DNA results in a representative library. Replacement vectors have a cloning capacity of 9–23 kb. We recommend three vectors: EMBL3, lambda DASH II (Stratagene), or the EMBL3 derivative lambda GEM-12 (Promega). We strongly recommend use of commercial preparations of vector arms for EMBL3/4 and lambda GEM-12 (hosts NM538, NM539, and LE392) and lambda DASH II (hosts XL1-Blue MRA and XL1 MRA(P2)) digested with *Bam*HI (allowing insertion of *Mbo*I or *Sau*3A partial digests) and a second enzyme to prevent stuffer re-ligation. The bacterial strain KW251 is recommended for sequences that are difficult to clone or for methylated DNA sequences for the EMBL3-based vectors. Lambda DASH II and lambda GEM-12 have improved multiple cloning sites and flanking RNA polymerase promoters that facilitate restriction mapping, walking, and sub-cloning. An alternative strategy using smaller amounts of genomic DNA (10 μg) with the Lambda GEM-12 vector involves the use of partially filled-in *Xho* I sites in conjunction with partially filled-in *Sau*3A I or *Mbo* I digested genomic DNA (Promega corporation genomic cloning manual).

In general, if source DNA is limiting and genome size is average to large then lambda replacement vectors are the method of choice for producing representative genomic libraries. The lambda insertion vectors offer very few advantages over plasmid cloning vectors especially now that high efficiency electroshock transformation methods for *E. coli* routinely give $1–5 \times 10^9$

transformants per microgram of test plasmid DNA. Electroshock methods place plasmid cloning efficiency into the same range as the best lambda packaging extracts (see below).

2.1.4 Cosmid vectors

Cosmids are cloning vectors 5–8 kb in length that bear the *cos* region of bacteriophage lambda. This region contains *cis*-acting elements that determine the ability of source DNA in the size range 37–52 kb to be packaged into lambda heads to form infectious particles. The elements must be present at both ends of the DNA. Recombinant molecules are thus *c*. 40 kb structures wherein source DNA is flanked at both ends by cosmid sequences. Cosmids have a large cloning capacity and transfection is efficient. Choice of cosmid vector in library construction is determined by:

(a) The presence of an appropriate cloning site.
(b) The presence of a suitable selective marker: kanamycin and neomycin are more stable than ampicillin, although their use requires longer pre-expression times.
(c) The source of the cosmid replication origin: with vectors derived from pBR322 based on the pMB1 replicon, slow-growing recombinants may be under-represented or lost. Further, transcription from adventitious promoters in the source DNA can interfere with the replication origin. The Lawrist family of vectors contain the lambda origin of replication which partially overcomes these problems. Additionally, in some members of this family, cloning sites are protected by transcription terminators.
(d) The presence of transcriptional promoters flanking the cloning site: these permit generation of riboprobes from sequences at one end or another of the source DNA.

We would recommend two vectors with two *cos* elements, SuperCos 1 and Lawrist4. The cloning sites in these vectors are flanked by promoters for riboprobe synthesis and both give high transfection efficiencies and grow well in the *recA*$^-$ hosts NM554 or ED8767.

(a) The vector pSuperCos 1 (Stratagene) is 7.6 kb long and contains an ampicillin resistance gene. The *Bam*HI cloning site is flanked by *Not*I sites allowing excision of most inserts as a single restriction fragment. The cloning site is flanked by the T7 and T3 bacteriophage promoters.
(b) The vector Lawrist4 (6) carries the kanamycin resistance gene and has *Bam*HI and *Hin*dIII cloning sites. The cloning site is flanked by the SP6 and T7 polymerase promoters.

For library construction we recommend using the following procedure. A restriction site between the two *cos* sites is cleaved and dephosphorylated. The linearized vector is then cleaved at the cloning site with enzyme B. Source

DNA, digested with a compatible enzyme, is added to the resulting two vector arms. This mixture is then ligated prior to packaging (*Figure 1*).

2.2 Cloning procedure for lambda insertion vectors

2.2.1 Preparation of vector and source DNAs

Genomic DNA in the size range 50–100 kb is adequate. *Protocol 1* is suitable and yields DNA in excess of 200 kb which is digested to completion with *Bam*HI or *Hin*dIII and dephosphorylated (*Protocol 2*). Alternatively, rapid extraction of high quality DNA of *c.* 50 kb average size starting with blood samples from many species can be achieved using the Nucleon extraction kits from Scotlab. This DNA is suitable for cloning in plasmid and lambda insertion vectors. Lambda DNA is prepared as in *Protocol 4*. Lambda arms are prepared for ligation as described *in Protocol 5*.

Protocol 1. Preparation of high molecular weight DNA

Equipment and reagents

- Liquid nitrogen
- 1 × SSC: 0.15 M NaCl, 0.05 M trisodium citrate
- TES: 100 mM Tris–HCl pH 7.5, 100 mM EDTA, 1% sarcosyl
- TE: 10 mM Tris–HCl pH 7.5, 1 mM EDTA
- Proteinase K
- Phenol saturated with TE
- Phenol–chloroform (1:1 v/v) saturated with TE
- 3 M sodium acetate, pH 6.0
- Chloroform
- Absolute ethanol
- 70% ethanol
- TKO 100 mini-fluorimeter (Hoefer Scientific)
- 0.3% agarose gel
- Polymerized lambda DNA ladder (50 kb steps to 1 Mb), T4 DNA (160 kb), and high molecular weight standards (6–50 kb)

A. *Tissue specimens*

1. Cool a pestle and mortar with liquid nitrogen. Pour liquid nitrogen into the mortar and add the sample[a] (0.5–1 g). Before the liquid nitrogen has evaporated, grind the sample to a fine powder with the pestle.
2. Using a cold spatula or paint brush, transfer the sample to a Wheaton 15 ml homogenizer containing 5 ml of ice-cold 1 × SSC. Homogenize and decant into a 50 ml Sorvall tube. Continue with Part B, step **2**.

B. *Cultured cells, sphaeroplasts, or protoplasts*

1. Resuspend cells gently but completely in 5 ml of 1 × SSC.
2. Add 5 ml TES. Mix carefully with a glass rod for 2–3 min until cells lyse. Add proteinase K to 100 μg/ml and incubate at 55°C for 2–3 h.
3. Add an equal volume of phenol saturated with TE. Mix by gentle inversion. Extract by slow rolling at 4°C for 15 min. Centrifuge at 6000 × *g* for 20 min at 4°C. Carefully remove the aqueous (top) layer to a fresh tube, using a wide-mouthed pipette and avoiding protein at the phase interface.

4. Extract with an equal volume of phenol–chloroform, as in step **3**.
5. Add 0.1 vol. of 3 M sodium acetate and extract with an equal volume of chloroform as in step **3**.
6. Precipitate DNA by adding 2 vol. of cold (–20°C) absolute ethanol. Mix by gentle inversion. Spool out the DNA on a glass rod. Wash the spool three times by immersion in 70% ethanol, air-dry briefly, and disperse into 1.0 ml of TE. Leave at 4°C for DNA to re-dissolve. If the DNA has not re-dissolved after 48 h, mix gently with a disposable pipette tip cut to have a 3 mm diameter end. The solution should be very viscous. It contains some RNA but this does not interfere with subsequent steps.
7. Estimate the DNA concentration using a TKO 100 mini-fluorimeter. The fluor, Hoechst 33258, binds double-stranded DNA strongly but RNA poorly.
8. Check DNA quality by electrophoresis through a 0.3% agarose gel (Sections 6.9–6.19 in ref. 7) using polymerized lambda DNA ladder (50 kb steps to 1 Mb), T4 DNA (160 kb), and high molecular weight standards (6–50 kb) to assess the size. Source DNA for cosmid or lambda replacement vector cloning must have an average molecular weight of 200 kb or greater and the method outlined generally gives the highest quality product.

[a] Use fresh living cells for the highest molecular weight source DNA. Otherwise use tissue cut or minced into small pieces with surgical scissors, frozen in liquid nitrogen, and stored at –70°C. For cultured mammalian cells start with 10^8–10^9 cells. For very tough plant tissue or yeast cells start with sphaeroplasts or protoplasts equivalent to 1 g of tissue.

Protocol 2. Complete digestion of genomic DNA

Equipment and reagents

- Restriction enzyme with manufacturer's buffer
- 1% agarose gel
- Reagents for phenol–chloroform extraction and ethanol precipitation (*Protocol 3*)
- Calf intestinal alkaline phosphatase (CIP) (Boehringer, molecular biology grade) with manufacturer's buffer
- TK100 mini-fluorimeter (Hoefer Scientific Instruments)

Method

1. Incubate source DNA (30 μg) for 1 h with an excess (4 U/μg) of a suitable restriction enzyme. Remove a 1 μg aliquot and check on a 1% agarose gel for complete digestion (Sections 6.9–6.19 in ref. 7). Add further enzyme and incubate for another 2 h if required.
2. Extract and ethanol precipitate the DNA (*Protocol 3*).

Protocol 2. *Continued*

3. Re-dissolve DNA in 40 μl of CIP buffer. Add 0.1 U of CIP and incubate for 30 min at 37 °C.
4. Extract and precipitate DNA again. Re-dissolve at 1 μg/μl. Use a fluorimeter to estimate concentration.

Protocol 3. Extraction and precipitation of DNA

Reagents

- Phenol saturated with TE (10 mM Tris–HCl pH 8.0, 1 mM EDTA)
- Chloroform
- 3 M sodium acetate, pH 6.0
- Absolute ethanol
- 70% ethanol

Method

1. It is often necessary to remove all traces of restriction or modifying enzymes from DNA. Extract the DNA solution with an equal volume of phenol saturated with TE. Mix by gentle shaking for 2–5 min.
2. Centrifuge the mixture for 5 min at room temperature at full speed in a microcentrifuge. Transfer the aqueous (upper) layer into a fresh tube, avoiding the protein interface.
3. Extract the solution with an equal volume of fresh chloroform. Mix by gentle shaking for 2–5 min.
4. Centrifuge at full speed in a microcentrifuge at room temperature for 3 min. Transfer the aqueous (upper) layer to a fresh tube, avoiding the protein interface and the chloroform.
5. Add 0.1 vol. of 3 M sodium acetate, pH 6.0. Mix by gentle inversion. Add 2 vol. of cold (–20 °C) absolute ethanol. Mix by gentle inversion. Incubate at –70 °C for 20 min or at –20 °C for 1–16 h.
6. Centrifuge for 10 min at full speed in a microcentrifuge. Carefully remove the supernatant without disturbing the pellet (which may be invisible).
7. Add 100 μl of cold (–20 °C) 70% ethanol. Mix to dissolve salts by gently flicking the bottom of the tube.
8. Centrifuge for 10 min and remove the supernatant without disturbing the pellet.
9. Centrifuge for 5 sec to bring any ethanol off the sides of the tube and then remove this ethanol without disturbing the pellet.
10. Air-dry the DNA for 5 min and resuspend in an appropriate volume of TE.

Protocol 4. Preparation of lambda and cosmid DNA

A. *Lambda DNA*

Reagents

- 0.7% agarose
- LB broth (1 % NaCl, 0.5% Difco Yeast Extract, 1% Difco Bacto-tryptone in distilled water) containing 1.5% agarose
- SM: 5.8 g NaCl, 2 g $MgSO_4 \cdot 7H_2O$, 50 ml 1 M Tris–HCl pH 7.5, 5 ml 2% gelatin, made up to 1 litre with deionized water, sterilized, and stored in 50 ml aliquots
- Chloroform
- RNase A
- Pancreatic DNase
- Polyethylene glycol (PEG; mol. wt 8000)
- NaCl
- 10% sodium dodecyl sulfate (SDS)
- 0.5 M EDTA pH 8.0
- TE: 10 mM Tris–HCl pH 8.0; 1 mM EDTA
- Phenol equilibrated with TE
- Phenol–chloroform (1:1 v/v) equilibrated with TE
- 7.5 M ammonium acetate
- 95% ethanol
- 75% ethanol

Method

1. To grow the phage:
   (a) Mix *c.* 10^6 purified phage with 1 ml of plating bacteria (see *Protocols 8, 9,* and *23*) to achieve confluent lysis (the number of phage will vary with phage strain). Incubate at 37 °C for 20 min.
   (b) Add and mix 30 ml of 0.7% agarose at 42 °C.
   (c) Pour 10 ml of the mixture on to three freshly prepared 140 mm Petri dishes, each containing 100 ml LB with 1.5% agarose. Incubate at 37 °C until lysis is detected (about 10 h).
2. To harvest the phage:
   (a) Scrape agarose top layer into a 50 ml polythene (Falcon) tube. Add 20 ml SM and 300 μl chloroform. Shake at 37 °C for 20 min and then centrifuge for 20 min at 2500 × *g* in a bench-top centrifuge.
   (b) Remove supernatant, add 5 ml SM to the pellet, and shake at 37 °C for 20 min.
   (c) Centrifuge at 2500 × *g* for 20 min and pool supernatants.
   (d) Centrifuge at 2500 × *g* for 20 min to remove all traces of agarose.
3. To remove bacterial DNA and RNA, add RNase (25 μg/ml) and pancreatic DNase (10 μg/ml) to the supernatant and incubate for 1 h at 37 °C.
4. To precipitate phage particles:
   (a) Add PEG to 20% and NaCl to 1 M. Store for 1–16 h at 4 °C.
   (b) Centrifuge at 4 °C in 30 ml Sorvall tubes at 13 000 r.p.m. for 15 min (Sorvall RC-5B). Pour off the supernatant and let the tube drain.
   (c) Re-dissolve the phage pellet in 3 ml SM (e.g. in a 13 ml Sarstedt tube).

Protocol 4. *Continued*

5. To extract phage DNA:
   (a) Add SDS to 0.1% and EDTA to 5 mM. Incubate for 15 min at 68°C. Allow to cool.
   (b) Extract twice with phenol, twice with 1:1 phenol–chloroform, and once with chloroform.
   (c) To the aqueous phase add ammonium acetate to 3.75 M and 2 vol. 95% ethanol. Chill to –70°C for 20 min and then centrifuge at 10 000 r.p.m. for 10 min.
   (d) Pour away the supernatant and wash the pellet in 75% ethanol. Let the pellet dry and then re-dissolve in 1 ml TE.
6. Residual proteins may be removed by adding 500 μl 3.75 M ammonium acetate, mixing, incubating on ice for 20 min, and centrifuging at 4°C for 20 min at 10 000 r.p.m. Transfer supernatant (avoiding the translucent precipitate) to a clean tube. Precipitate the DNA by adding two volumes of 95% ethanol and resuspend in TE as described in steps 5c and d.

B. *Cosmid DNA*[a]

Reagents

- LB broth (see *Protocol 4A*)
- SET: 15% sucrose; 25 mM Tris–HCl pH 7.5, 10 mM EDTA
- TE (see *Protocol 4A*)
- 0.2% NaOH, 1% SDS
- 3 M sodium acetate, pH 4.6
- 10 mg/ml RNase A in 0.05 M NaCl heat-treated at 95°C for 5 min
- Phenol:chloroform (1:1 v/v) equilibrated with TE
- Chloroform
- Isopropanol
- PEG (see *Protocol 4A*)
- 4 M NaCl
- 70% ethanol

Method

1. Inoculate 250 ml LB broth containing antibiotic (e.g. ampicillin at 100 ng/ml) in a 1 litre conical flask with the clone of interest and incubate in a shaker at 37°C overnight.
2. Transfer culture to centrifuge tube and centrifuge for 5 min at 6000 r.p.m. Resuspend the pellet in 6 ml SET and transfer to a 50 ml polypropylene centrifuge tube.
3. Lyse the cells by adding 12 ml 0.2 M NaOH, 1% SDS. Mix gently to avoid shearing bacterial DNA. Incubate on ice for 10 min.
4. Add 7.5 ml cold 3 M sodium acetate, pH 4.6, mix gently, and incubate on ice for 10 min. This causes proteins and bacterial DNA to form a clotted precipitate. Centrifuge at 4°C for 15 min at 15 000 r.p.m.
5. Decant supernatant into a 50 ml polythene tube and add 5 μl 10 mg/ml RNase A; incubate for 20 min at 37°C.
6. Extract the solution with 30 ml of 1:1 (v/v) phenol–chloroform, followed by 30 ml of chloroform.

7. Transfer the aqueous phase to a 50 ml polypropylene tube and add an equal volume of isopropanol. Incubate for 10 min on ice and centrifuge for 15 min at 15 000 r.p.m. at 4°C.
8. Re-dissolve the pellet in 1.6 ml of water, transfer to a 15 ml polypropylene tube, and add 2 ml of 13% PEG and 0.4 ml of 4 M NaCl. Mix. Incubate on ice for 60 min and then centrifuge for 10 min at 10 000 r.p.m. Wash the DNA pellet with 70% ethanol, air-dry, and dissolve pellet in an appropriate volume (e.g. 500 μl) of TE.

[a] This is a rapid, simplified protocol (8,9) suitable for plasmids and cosmids.

Protocol 5. Digestion of vector arms for lambda insertion vector

Reagents

- Restriction enzyme and manufacturer's buffer
- 0.5% agarose gel
- Solutions for *Protocols 3* and *6–10*

Method

1. Digest 30 μg of vector DNA for 1 h with 3 U/μg of the appropriate high-quality restriction enzyme. Remove a small aliquot, heat to 65°C for 10 min (to disrupt the cohesive termini), and fractionate on a 0.5% agarose gel to test for complete digestion.
2. As in *Protocol 3*, extract and ethanol-precipitate. Re-dissolve DNA in TE at 1 μg/μl.
3. Check the quality of the digest by packaging and titrating 1 μg of digested vector and 1 μg of undigested vector (see *Protocols 7–10*). The titre of the latter should exceed that of the former by $>10^4$-fold.
4. Ligate (as in *Protocol 6*) 1 μg of digested vector, package, and titre. This should restore the titre to >40% of that of undigested vector.
5. Before setting up ligation reactions, incubate vector arm DNA for 1 h at 42°C. This allows annealing of the cohesive termini, increasing packaging efficiency.

2.2.2 Ligation of source DNA with lambda vector arms

A test should first be performed to establish the optimum amount (30,60,120,240 or 480 ng) of source DNA to ligate (*Protocol 6*) with 1 μg of lambda arms. Package (*Protocol 7*) and titre (*Protocols 8–10*) an aliquot from each test ligation. Using the optimum ratio of DNA to arms, ligate and package 1–3 μg of genomic DNA. The resulting titre should be 10^6–10^7 plaques/μg donor DNA. For a discussion of ligation theory, see ref. 7.

Protocol 6. General method for the ligation of DNA

Reagents

- T4 DNA ligase
- 10 × T4 DNA ligase buffer: 0.675 M Tris–HCl pH 7.5, 0.1 M $MgCl_2$, 0.1 M dithiothreitol (DTT), 0.15 M spermidine, 10 mM ATP, pH 7.0[a] (store at –70°C)

Method

1. For present purposes, carry out ligation reactions in as small a volume as possible. Commercial packaging extracts use 1–5 μl of ligated DNA. Typically we use:
   - 1 μl lambda/cosmid vector arm DNA (1 μg or 0.5 μg of each arm)
   - 1–2 μl source DNA (30 ng–1 μg)
   - 0.5 μl 10 × buffer
   - 0.2 μl T4 DNA ligase (1 U)[b]
   - H_2O to 5 μl

   For self-ligation the volume need not be minimal: use 10 μl.
2. Incubate for 4–16 h at 16°C.
3. For test reactions, package 1 μl. For library construction, use as much DNA as possible (in the optimized ratio) in a 5 μl ligation reaction. Package all of the ligated material.

[a] Adjust pH of ATP using dilute NaOH. Check pH by spotting 2 μl on to pH paper.
[b] The proportion of ligase should be no greater than 5% because of the inhibitory effects of its storage buffer.

Protocol 7. Packaging of lambda and cosmid DNA

Reagents

- Commercial packaging extracts[a]
- SM (*Protocol 4A*)
- Chloroform

Method

1. Follow manufacturer's instructions to package the ligation reaction volume with the following provisos.
   - Do not be tempted to package more as the titre drops.
   - When testing gradient fractions for lambda replacement or cosmid vectors one tube of packaging mixture can be split and used to package small aliquots (e.g. 1 μl) from several ligations.
2. Increase the volume of the packaged phage to 500 μl with SM.

3. Add 40 μl of chloroform and mix by gentle inversion.
4. Centrifuge briefly at low speed in a microcentrifuge to pellet cell debris.
5. Store at 4°C.[b]

[a]We routinely obtain 10^6–10^7 recombinant clones per microgram of source DNA for lambda vectors using Gigapack Gold (Stratagene) and 10^6/μg for cosmid vectors using Gigapack XL (Stratagene). Other commercial extracts give similar yields. If the cost of commercial extracts is prohibitive, they can be prepared readily (Sections 2.95–2.107 in ref. 7). Strains BHB2688 and BHB2690 give extracts suitable for packaging both cosmid and lambda recombinants. Our extracts compare very favourably with commercial ones. We use our extracts to test gradient fractions and optimize ligation ratios.
[b]Packaged material can be stored for up to 6 months at 4°C without loss of infectivity.

Protocol 8. Preparation of plating bacteria

Reagents

- LB broth (*Protocol 4A*) supplemented with 10 mM $MgSO_4$, 0.2% maltose
- 10 mM $MgSO_4$

Method

1. Grow the required *E. coli* strain overnight in LB broth supplemented with $MgSO_4$ and maltose.
2. Centrifuge the cells at room temperature for 10 min at 4000 × *g*. Pour off supernatant and drain well.
3. Resuspend the pellet in 0.5 vol. of 10 mM $MgSO_4$.
4. For the highest plating efficiencies, use plating bacteria immediately; however, they can be stored for a week at 4°C if necessary.

Protocol 9. Plating bacteriophage lambda

Reagents

- 1.5% LB agarose: 1% NaCl, 0.5% Difco Yeast Extract, 1% Difco Bacto-tryptone, 1.5% agarose in distilled water
- SM (*Protocol 4A*)
- 0.7% top agarose

Method

1. Pour melted 1.5% LB agarose into a Petri dish. Allow to set. Reduce moisture content by leaving the dish inverted with the lid open for 2 h in a 37°C incubator. Store sealed at 4°C. Warm for 30 min at room temperature prior to use.

Protocol 9. *Continued*

2. Take 50 μl of the bacteriophage stock diluted with SM into a sterile universal tube.
3. Add 0.3 ml plating bacteria (*Protocol 8*). Mix by gentle swirling. Incubate at 37 °C for 20 min.
4. To the tube add 0.7% agarose that has been melted and cooled to 45 °C. Use 3 ml for an 80 mm diameter Petri dish, 6.5 ml for a 150 mm diameter dish. Mix rapidly by gentle swirling, avoiding bubbles. Pour this on to the centre of the agar in the dish and swirl on a flat surface: the mixture forms an even layer. Let the Petri dish stand for 5 min to allow the top agarose to harden.
5. Incubate at 37 °C. After 7 h, plaques can be detected and after 10–12 h they grow to 1 mm diameter. This size is appropriate for replica plating. Count the number of plaques.

Protocol 10. Titration of packaged cosmid and lambda DNA

Reagents

- SM (*Protocol 4A*)
- Solutions for *Protocols 9* and *23*

Method

1. Perform 10-fold serial dilutions in SM of phage particles, starting with 2 μl of the 500 μl stock from *Protocol 4*.
2. For packaged lambda vector, remove 50 μl of each dilution and plate out as described in *Protocol 9*. Packaged lambda has a titre in the range 2×10^3–2×10^4 plaque-forming units (p.f.u.) per μl.
3. For packaged cosmids, use 2 μl of each dilution and plate out as described in *Protocol 23*. From a packaging reaction with 1 μg genomic DNA ligated, packaged, and resuspended in 500 μl, approximately 2×10^3 colonies/μl will be obtained.

2.2.3 Plating and screening the library

If the titre is satisfactory, plate out $2N$ phage for $P = 0.99$ (*Equation 1*). This will be a small fraction of the library. Make replica membranes (*Protocols 8* and *15*) and screen these by hybridization (*Protocol 19*). Positive clones are purified (*Protocols 22* and *23*) and DNA can be prepared (*Protocol 4*).

The remainder of the library should be amplified by the plate lysis protocol (Sections 2.61–2.81 in ref. 7) and can then be stored at 4°C for several years. An amplified library tends to be less representative, requiring the plating of larger numbers of phage to find a particular clone.

2.3 Cloning procedure for lambda replacement vectors

2.3.1 Preparation of genomic and vector DNA

With replacement vectors, the source DNA must be of very high quality (*Protocol 1*) since cloning involves use of large digestion products. Always mix by gentle inversion and use wide-mouthed pipettes or tips cut to give a 3 mm diameter at the end. The protocol involves a minimum number of steps; for instance, it does not include RNA removal. Genomic DNA and vector arms DNA are prepared by *Protocols 1, 4*, and *11* respectively. The process is summarized in *Figure 1*.

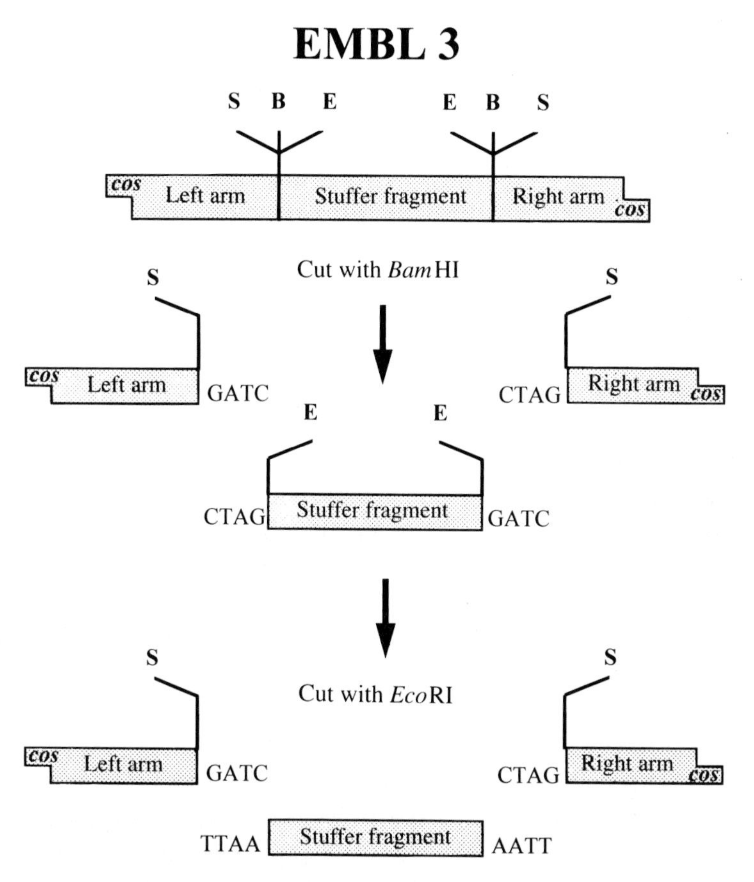

Figure 1. Preparation of *Bam*HI-cut vector arms from the lambda insertion vector EMBL3. (note: not to scale). The stuffer fragment in EMBL3 is flanked by polylinkers in inverse orientation. Unique sites are shown (S, *Sal*I; B, *Bam*HI; E, *Eco*RI (there is also a *Sal*I site in the stuffer fragment). To prepare *Bam*HI-cut vector arms, EMBL3 is cleaved with *Bam*HI generating a single-stranded GATC projection. This allows ligation of fragments generated by partial digestion with the restriction enzyme *Mbo*I. Cleavage with *Eco*RI enzymically isolates the stuffer fragment so it no longer participates in any ligation steps.

Protocol 11. Preparation of EMBL vector arms for cloning

Reagents

- *Bam*HI and manufacturer's buffer
- *Eco*RI and manufacturer's buffer
- Solutions for *Protocols 3* and *5*

Method

1. Incubate 25–50 μg of EMBL3 DNA with high quality *Bam*HI at 3 U/μg DNA in a volume of 200 μl for 1 h following manufacturer's instructions. Transfer the tube to ice.
2. Test a 0.5 μg aliquot for complete digestion (*Protocol 5*). If digestion is incomplete add *Bam*HI at 2 U/μg DNA and incubate for 30 min more.
3. Extract and precipitate DNA (*Protocol 3*). Re-dissolve the DNA in 190 μl *of Eco*RI buffer.
4. Digest the DNA with high-quality *Eco*RI at 4 U/μg DNA for 4 h.
5. Extract twice using phenol–chloroform (1:1 v/v) and once using chloroform and then precipitate DNA (*Protocol 3*). Re-dissolve the arms at 1 μg/μl.
6. Before setting up ligation reactions, incubate vector arm DNA for 1 h at 42°C. This allows annealing of the cohesive termini, increasing packaging efficiency.

2.3.2 Partial digestion and fractionation of DNA

This is summarized in *Figure 2*. It involves the following steps:

(a) Using *Protocol 12*, genomic DNA is partially digested (ref. 8, p. 22) with *Mbo*I to the range 15–25 kb). The resulting digested DNA is derived from the pooling of several digests using varying concentrations of enzyme. This is to allow equivalent coverage within the library of products from regions where the *Mbo*I recognition sequence is under and over represented.

(b) Size selection is achieved either using physical fractionation techniques or using the size selectivity inherent in packaging. The first route requires more genomic DNA (at least 200 μg). The second route is appropriate for small haploid genomes or where DNA is limiting but produces smaller libraries as treatment of the insert material with phosphatase (Sections 3.38 and 3.39 of reference 7) to prevent self ligation, reduces ligation efficiency.

(c) Physical fractionation is achieved by gel electrophoresis (10) or on a sucrose (10–40%) velocity gradient (*Protocol 13*).

2.3.3 Ligation with vector arms

Initial test ligations are recommended (*Protocol 14*) using gradient fraction aliquots with 0.5 μg vector DNA. The fraction with the desired size of source DNA should give the maximum number of plaques after packaging and titration (*Protocols 7–10*). The remaining DNA from this fraction should yield 10^6–10^7 recombinants/μg DNA.

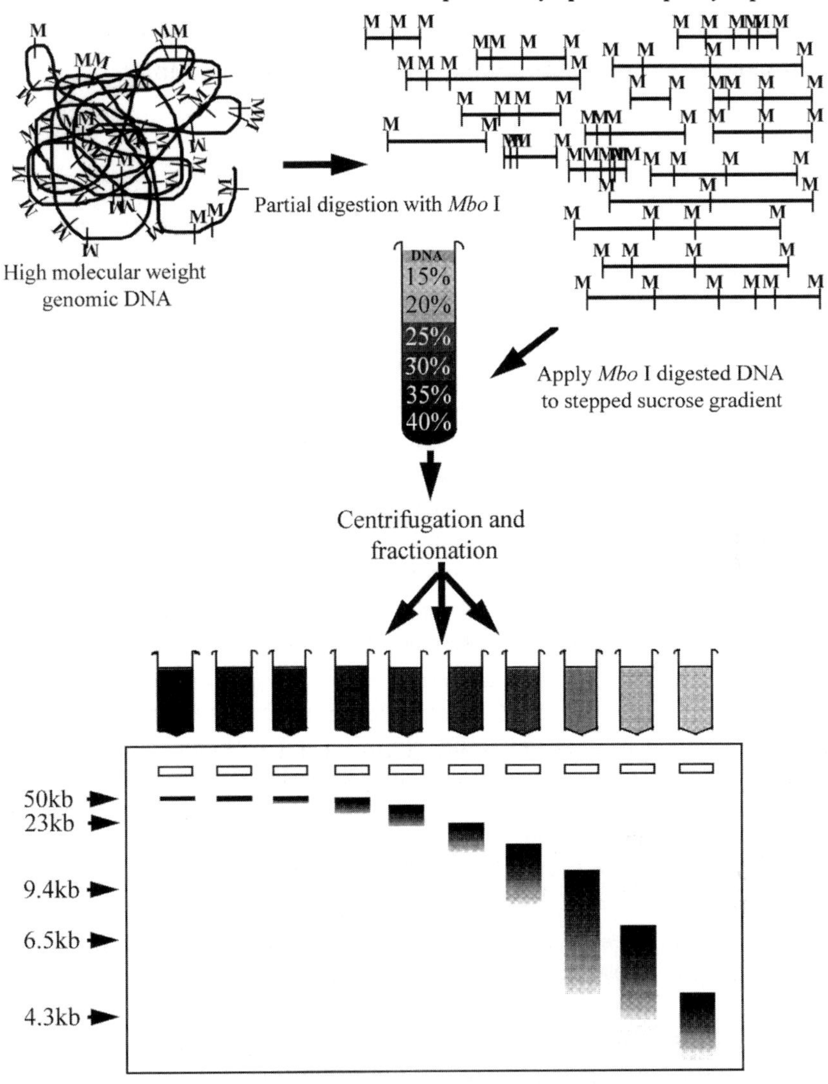

Figure 2. Illustration of partial digestion and velocity gradient fractionation of genomic DNA. High molecular weight genomic DNA is partially digested with *Mbo*I (sites shown as M) to the required size range (15–25 kb for lambda insertion cloning, 35–45 kb for cosmid cloning). Pooled partially digested DNA is applied to a 15–40% sucrose velocity gradient and, following centrifugation, fractions are collected. Fractions are viewed by the electrophoresis of aliquots on a 0.3% agarose gel stained with ethidium bromide.

Protocol 12. Partial digestion of genomic DNA with *Mbo*I

Reagents

- Source DNA (*Protocol 1*)
- *Mbo*I and manufacturer's buffer
- 1% agarose gel
- 0.3% agarose gel
- Ethidium bromide
- 1 kb ladder and a high molecular weight ladder (BRL)
- Reagents for phenol extraction and ethanol precipitation (*Protocol 3*)
- TE: 10 mM Tris–HCl pH 8.0, 1 mM EDTA

Method

1. Working at room temperature, add 20 μl of 10 × *Mbo*I buffer to 30 μg of source DNA (*Protocol 1*) and add water to 200 μl.
2. Place 40 μl into tube 1 and 20 μl into each of tubes 2–9.
3. Add 4 U of *Mbo*I to tube 1 and mix carefully. Using a clean pipette tip, remove 20 μl into tube 2. Now, similarly remove 20 μl from tube 2 to tube 3. Repeat, but add nothing to tube 9.
4. Incubate all tubes at 37 °C for 1 h and at 65 °C for 15 min.
5. Fractionate 5 μl from each tube on a 1% agarose gel. Digestion should be virtually complete in tube 1 and increasingly partial towards tube 8.[a]
6. Now fractionate 15 μl from each tube on a 0.3% agarose gel (containing ethidium bromide at 50 ng/ml) and photograph the gel. Estimate the size range of digestion products using a 1 kb ladder and a high molecular weight ladder (BRL). Identify the track that has most of its DNA within 5 kb of the required size. The corresponding *Mbo*I concentration is not the optimum (11): half that concentration is the optimum. But also digest two DNA aliquots, one with double and the other with half the optimum amount of *Mbo*I.
7. Scale up the three partial digestion conditions carefully: pre-warm all solutions to allow for the larger volumes; do not change buffer, enzyme, or DNA concentrations. Use as much DNA as possible: a 38 ml gradient can fractionate 200–300 μg DNA.
8. Check the digestions by fractionation of aliquots on a 0.3% gel.
9. Pool the three partial digests.[b] Extract and precipitate the DNA (*Protocol 3*). Leave the DNA to precipitate overnight at –20 °C and collect the DNA by centrifugation. Resuspend the DNA gently in TE at approximately 200 μg/ml. Dot not vortex the DNA. If difficulty is experienced re-dissolving the DNA, pipette the DNA up and down with a wide-mouthed pipette.

[a] If RNA contamination of genomic DNA is excessive, add RNase A (10 mg/ml in 0.05 M NaCl, heat-treated at 95 °C for 10 min) to the partial digestion reactions to 0.1 μg/μl.
[b] The products of different partial conditions are used in subsequent steps since the rate of digestion is context-dependent.

Protocol 13. Size fractionation of partially digested genomic DNA

Equipment and reagents

- 40, 35, 30, 25, 20, and 15% sucrose in 20 mM Tris–HCl pH 7.4, 20 mM EDTA
- Syringe with a long needle
- Pump for fraction collection
- 0.3% agarose gel
- 3 M sodium acetate, pH 6.0
- Molecular weight markers made up in *c.* 30% sucrose
- Isopropanol
- 70% ethanol
- TE: 10 mM Tris–HCl pH 7.5, 1 mM EDTA
- TKO 100 mini-fluorimeter (Hoefer)

Method

1. We use a Beckman SW 27 rotor and 38 ml gradients. Using a syringe with a long needle, introduce 5.4 ml of each sucrose solution (starting with the least dense) into a tube. Interfaces at each step can be seen if the tube is held up to the light. We set up the gradient in the morning and allow it to diffuse at 4°C for 10–16 h during the day.
2. Carefully layer up to 300 μg DNA in up to 1 ml on to the gradient. Centrifuge at 26000 r.p.m. (average radius, 84000 × *g*) for 16 h at 10°C.
3. Pierce the bottom of the tube with a needle attached to tubing. Collect 0.6–0.9 ml fractions, using a pump. Test on a 0.3% agarose gel the DNA size range in 25 μl aliquots of each fraction. Use molecular weight markers made up in *c.* 30% sucrose since this affects DNA mobility.
4. Identify six fractions that span the correct size range for cloning. Add 0.1 vol. of 3 M sodium acetate, pH 6.0 and 0.6 vol. of isopropanol to each fraction, mix thoroughly, and leave at –20°C overnight. Collect the precipitated DNA by centrifugation and wash the pellets twice with 70% ethanol. Gently resuspend the DNA in about 10 μl of TE. A 10 min incubation at 65^°C may help. Estimate DNA concentrations in a fluorimeter using a 1 μl aliquot. 300 μg of partially digested DNA should yield 5–12 μg of DNA in each fraction.

Protocol 14. Ligation of EMBL3 vector arms with source DNA

Reagents

- Solutions as in *Protocols 6–10*

Method

1. Test-ligate (*Protocol 6*) 0.5 μg of source DNA from gradient fractions (*Protocol 13*) that include the optimum insert length (15–25 kb) to 0.5 μg vector DNA. Package and titrate (*Protocols 7–10*) using Escherichia

Protocol 14. *Continued*

coli strain NM538 or NM539. Maximum efficiency should be seen in fractions of the desired size. Efficiency falls suddenly with fractions larger than the maximum size acceptable to the vector.

2. Titrate source DNA against vector DNA using source:vector molar ratios of 0.25:1, 0.5:1, 1:1, and 2:1, with a constant amount of vector DNA.
3. Identify the most efficient ratio (maximum efficiency should be seen with a source DNA:vector DNA molar ratio of 1) and ligate and package the donor DNA.

2.3.4 Plating out the lambda library

This is carried out at 50 000 plaques per 150 mm diameter Petri dish. Make replica membranes with a nylon membrane (e.g. Amersham Hybond-C extra) since the nylon backing withstands repeated probings better than nitrocellulose.

Protocol 15. Replica plating a lambda library

Equipment and reagents

- Plating bacteria (*Protocol 8*)
- 1.5% LB agarose plates (*Protocol 9*)
- Top agarose
- Hybond-C extra membrane (Amersham)
- Soft lead pencil
- 18 G needle containing waterproof ink
- Blunt-ended forceps
- Blotting paper
- Denaturing solution: 0.5 M NaOH, 1.5 M NaCl
- Neutralizing solution: 1.5 M NaCl, 0.5 M Tris–HCl pH 7.4
- 2 × SSC: 0.3 M NaCl, 0.1 M trisodium citrate
- Oven at 80°C

Method

1. Mix aliquots of packaging reaction or amplified phage stock containing 50 000 p.f.u. in 50 μl or less with 0.3 ml of plating bacteria (*Protocol 8*).
2. Plate as described in *Protocol 9* on to 150 mm diameter plates.
3. When the plaques reach a diameter of approximately 1 mm, chill the dishes at 4°C for at least 1 h to allow the top agarose to harden.
4. Number Hybond-C extra membranes with a soft lead pencil.
5. At room temperature place a dry circular membrane neatly on to the plaques. Avoid air bubbles. Do not let the membrane slip. Mark the membrane by stabbing through it with an 18 G needle containing waterproof ink.
6. After 30–60 sec, use blunt-ended forceps to peel off the membrane with a straight even motion of the forceps away from yourself.

7. Repeat steps **4–6** to make a second replica but leave the membrane on the plaques for 1–2 min.
8. Place the membranes plaque side up on to blotting paper saturated with denaturing solution for 5 min. Transfer the membrane to blotting paper saturated with neutralizing solution for 5 min. Rinse the membranes with 2 × SSC and place DNA side up to dry for 30–60 min.
9. Fix the DNA to the membranes by baking for 2 h at 80 °C.

2.4 Cloning procedure for cosmids

2.4.1 Preparation, digestion, and fractionation of DNA

Follow the protocols outlined in Sections 2.3.1 and 2.3.2, but note that cosmid library construction calls for the use of fractions in the 35–45 kb size range.

2.4.2 Ligation of sized DNA to cosmid vector arms

Prepare cosmid DNA (*Protocol 4*) and purify it further on a CsCl gradient (ref. 7; Chapter 3). Arms can be obtained from this vector as described in *Protocol 16*, where we describe the preparation of cosmid vector SuperCos I arms for ligation with partially *Mbo*I-digested, size-sorted source DNA (4). See also *Figures 3* and *4*.

Test ligate 0.3 μg aliquots of gradient-fractionated DNA from *Protocol 13* (including the 35–45 kb fraction) with 0.3 μg vector arms. The fraction with the desired size of DNA should give the maximum number of colonies after packaging and titration (*Protocols 7–10*). The remaining DNA from this fraction should yield 10^6 recombinants/μg DNA. Note that cosmid vectors vary in size and hence may give best results with different fractions. Using more than one vector can thus yield substantially more recombinants.

Protocol 16. Cosmid vector arms

Reagents

- SuperCos I DNA
- *Xba*I and manufacturer's buffer
- 1.0% agarose gels
- *Bam*HI and manufacturer's buffer
- CIP (Boehringer molecular biology grade)
- TE: 10 mM Tris–HCl pH 7.5, 1 mM EDTA
- Solutions as for *Protocols 3, 6,* and *12–14*

Method

1. Completely digest 20 μg SuperCos I DNA with *Xba*I using 5 U/μg at 37 °C for 1 h following manufacturer's instructions. Check the digestion is complete by fractionating on a 1.0% agarose gel. Complete digestion yields a single sharp band of size 7.6 kb. Uncut DNA will migrate faster than digested DNA and appear to be made up of more than one species as small amounts of nicked supercoiled DNA, linearized DNA,

Protocol 16. *Continued*

and concatemers will be seen as well as supercoiled DNA in most vector preparations. Remove a small aliquot for a test ligation (step **3**).

2. Add 2 U/μg vector CIP to the digested vector DNA. Incubate for 30 min at 37 °C. Extract and precipitate DNA (*Protocol 3*).
3. Use aliquots from steps **1** and **2** in overnight ligation reactions (*Protocol 6*). Test the products by fractionation on a 1% agarose gel. Ligation should occur in aliquots from step **1**, but not from step **2**.
4. Completely digest the DNAs from step **2** with the minimum quantity of high-quality *Bam*HI. Check the completeness of digestion using a 1% agarose gel.
5. Extract and precipitate the DNA (*Protocol 3*). Re-dissolve the arms at 1 μg/μl in TE.
6. Test for arms self-ligation by ligating 1 μg of arm DNA (*Protocol 6*) and fractionating the ligation products on a 1% agarose gel. Self-ligation should yield one product at about 8 kb.
7. Test ligate 0.3 μg aliquots of gradient-fractionated DNA for fractions in the size range 35–45 kb (*Protocols 12* and *13*) with 0.3 μg vector arms, then package and titrate (*Protocols 7–10*). Select the fraction giving the maximum titre (see *Protocol 14*) and ligate 1.0 μg size-fractionated partially digested source DNA with 1.0 μg of vector arms in a total volume of 5 μl, package, and titrate. Either ligate and package the remainder of this source DNA gradient fraction or stop when you have packaged at least 5*N* cosmids for $P = 99\%$ (*Equation 1*).

Protocol 17. Adsorption and plating of packaged cosmids

Equipment and reagents

- SM (*Protocol 4*)
- 200 mm × 200 mm Hybond C extra membranes (Amersham)
- LB broth (*Protocol 4*) containing 1.5% agar
- Solutions for *Protocol 8*

Method

1. Combine 10 μl of freshly prepared plating baceria (*Protocol 8*) for every 1 μl of packaged cosmids in SM buffer. Incubate at room temperature for 15 min.
2. Dilute with 4 vol. of pre-warmed (37 °C) LB broth (no antibiotic) and incubate at 37 °C for 1 h.
3. Spread a volume containing 1.25×10^5 colonies on to a Hybond C extra membrane on a 200 mm × 200 mm Petri dish containing 500 ml

1.5% LB agar and the appropriate antibiotic. Spreading is easier if plates have been aged for a few days at room temperature. For titration tests, spread different volumes on to 80 mm diameter 1.5% LB agar Petri dishes containing the appropriate antibiotic.

4. Incubate the plates at 37°C for 12 h.

2.4.3 Plating out the cosmid library

A master copy of the library is made by plating out adsorbed packaged cosmids on to a nitrocellulose membrane. From the master copy we recommend making a viable replica (on Amersham Hybond C extra) and two screening

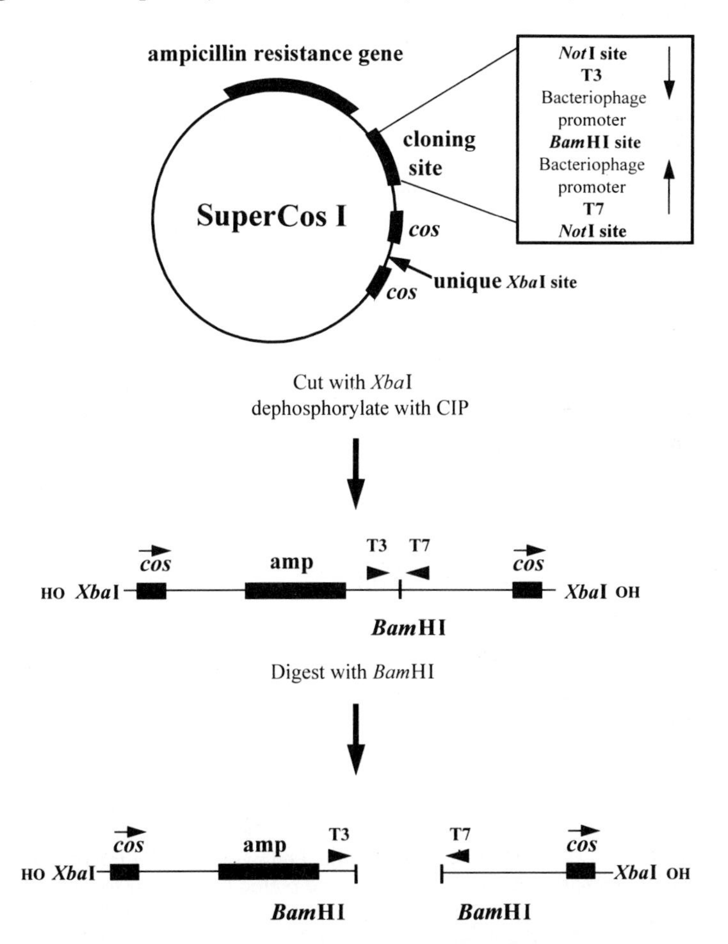

Figure 3. Preparation of *Bam*HI-cut vector arms from the cosmid vector SuperCos I (note: not to scale). SuperCos I vector is cleaved at the unique *Xba*I site, dephosphorylated, and subsequently digested with *Bam*HI to give two fragments or arms, each of which contains a *cos* site.

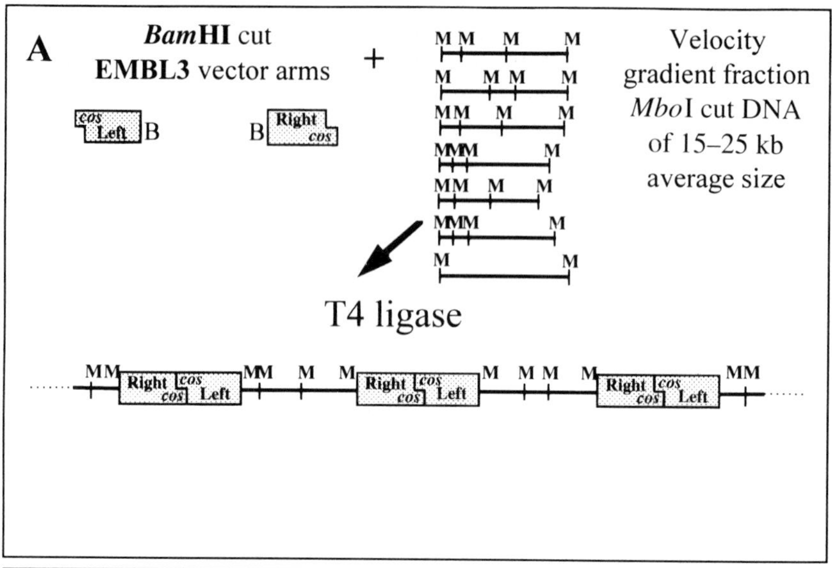

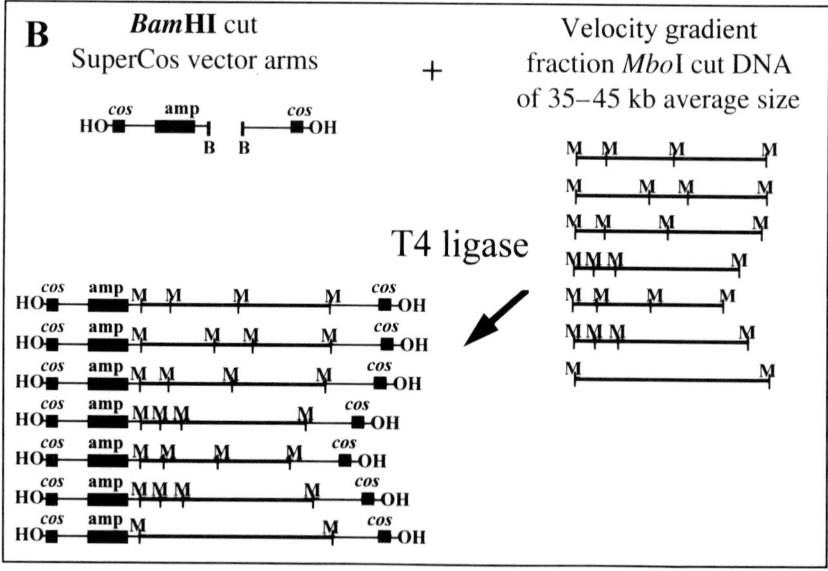

Figure 4. Illustration of the ligation of vector arms for EMBL3 and SuperCos I with *Mbo*I-cut size-fractionated insert DNA. (A) *Bam*HI-cut EMBL3 vector arms are ligated with gradient-fractionated DNA in the right size range. Following ligation the concatemers are packaged *in vitro*, adsorbed on to *E. coli* cells, and plated in top agar to yield plaques. (B) *Bam*HI-cut SuperCos I arms are ligated with gradient-fractionated DNA of the right size. Following ligation recombinants are packaged *in vitro*, adsorbed on to *E. coli* host cells, and spread on L agar plates containing ampicillin to yield colonies.

replicas (on Amersham Hybond N). Cosmid libraries can be amplified (12), but we do not recommend this as it affects the representation of clones. It is better to plate out the primary library, store it frozen (at –70°C), thaw it when needed, and re-freeze it as described in ref. 12.

Protocol 18. Replica-plating cosmid libraries

Equipment and reagents

- Hybond-N membranes (Amersham), 200 mm × 200 mm
- Hybond-C extra membranes (Amersham), 200 mm × 200 mm
- LB broth (*Protocol 4*) containing 1.5% agar
- Antibiotic
- Whatman 3MM paper
- Waterproof ink
- 10% SDS
- 0.5 M NaOH, 1.5 M NaCl
- 0.5 M Tris–HCl pH 7.5, 1.5 M NaCl
- 20 × SSC: 3 M NaCl, 1 M trisodium citrate

Method

1. Working in a hood, wet a Hybond-C extra membrane (for viable replica) or Hybond-N membrane (for screening replica) by placing it on a 1.5% LB agar with antibiotic (30 μg/ml kanamycin for Lawrist4) in a Petri dish.
2. Place Whatman 3MM paper on a glass plate. Place the master membrane on the paper, colony side up. Gently lower the wetted membrane on to the master membrane. Do not let the membranes slip. Place Whatman 3MM paper on the wetted membrane. Place a glass plate on the 3MM paper. Press the 'sandwich' together firmly. Rotate the plates through 90° in a horizontal plane. Press the sandwich together firmly and evenly. Remove the top plate and the top 3MM paper.
3. Pierce the wetted membrane and the master membrane in several places using a needle and waterproof ink for orientation.
4. Starting from one of the corners closest to yourself, peel away the wetted membrane. Place it and the master membrane on unused agar plates. Incubate the master membrane on its plate for 1 h at 37 °C.
5. Repeat steps **1–4** if additional replicas are to be made.
6. Incubate replica membranes until the colonies are approximately 0.5 mm in diameter.
7. Place the replica membrane on to blotting paper saturated with 10% SDS for 3 min.
8. Transfer on to 0.5 M NaOH, 1.5 M NaCl for 5 min.
9. Transfer on to 0.5 M Tris–HCl pH 7.5, 1.5 M NaCl for 15 min.

Protocol 18. *Continued*

10. Air-dry the membrane and bake at 80°C for 2 h. Rinse membranes in 5 × SSC before pre-hybridization.

11. Store the master copy as described in ref. 12

3. Development of polymorphic probes

3.1 Strategy

A species-specific probe may be obtained from a genomic library either by screening the library with a heterologous probe of choice to find the homologue or simply by picking a clone at random. In both cases, an appropriate polymorphism associated with that species-specific probe must then be identified.

While heterologous probes may be used directly to study polymorphisms in another species (Chapter 5), the isolation of their homologues in the species under investigation becomes necessary as sequence divergence increases from the heterologous species.

The genes of the major histocompatibility complex (MHC) locus and the ribosomal DNA (rDNA) gene clusters are suitable regions from which to derive homologues. Clones from loci encoding an essential cellular function shared by all living cells (e.g. alcohol dehydrogenase) may also be useful.

The MHC locus contains a cluster of genes of the same family that specify cell surface proteins involved in immunity. Three classes (I, II, and III) of MHC gene have been described. Many MHC genes (particularly class I) have been shown by restriction fragment length polymorphism (RFLP) and gene number analysis to be highly polymorphic in most species (13,14). Moreover, the availability of many cloned sequences from many species makes them attractive as heterologous probes for screening genomic libraries. However, note that MHC genes will, clearly, only be of use among species where an immune system has been developed. Further, the extreme polymorphism of MHC class I genes is believed to reflect their role in an adaptive strategy for dealing with rapidly evolving infectious agents in natural populations. As this may vary considerably, the value of probes at the MHC locus for studying populations is affected.

An advantage of the rDNA genes (28S, 18S, and 5S) is their almost universal presence in the genomes of plants and animals. The organization of the 18 and 28S rDNA genes as tandem repeat clusters (*Figure 5*) has been studied in a wide variety of organisms.

The tandem repeat unit can be divided into a number of segments which have been defined in the transcribed region of rDNA genes of many organisms. The repeat transcription unit (15) is composed of a leader promoter region known as the external transcribed spacer (ETS), an 18S rRNA coding region, an internal non-coding transcribed spacer region (ITS), a 28S rRNA coding region, and an intergenic non-transcribed spacer segment (NTS). In a

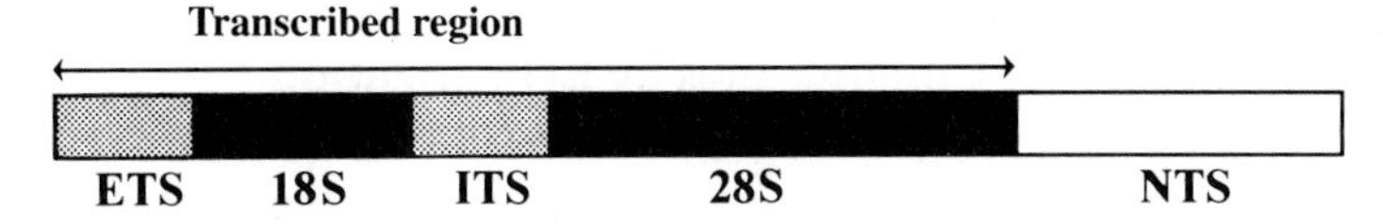

Figure 5. rDNA transcription unit. The transcribed region is indicated by the arrow. ETS, external transcribed spacer; ITS, internal non-coding transcribed spacer region; NTS non-transcribed spacer; 18S and 28S indicate the regions encoding 18S and 28S rRNA respectively.

wide range of organisms, a high degree of polymorphism has been found associated with rDNA genes almost entirely confined to the non-coding segments of the repeat unit. Restriction enzyme site variation and spacer repeat length variation due to deletion or amplification of repeat sequences in spacer regions have been observed. For example, in rodent species intra-individual variation; intra- and inter-specific site variation confined to the NTS region, and length variation in both transcribed and non-transcribed spacer regions have been described.

Using a randomly picked clone from the library avoids the need for screening. However, some characterization of this anonymous probe may be needed if investigating, for example, time-dependent changes: the probe must, at least, be shown to give rise to a hybridization signal with Southern-blotted DNA (see Chapter 5) from all the individuals under investigation. Note that an anonymous probe derives from a locus with unknown constraints on sequence variation.

In this chapter we describe how to find polymorphisms arising from:

- the presence or absence of a restriction enzyme site
- any base pair change in a short sequence of DNA
- microsatellite length variation.

Other DNA polymorphisms are dealt with elsewhere (Chapters 6 and 8). The type of DNA polymorphism that an investigator may wish to work with will be determined by:

- the amount of DNA that can be readily obtained from an organism
- the number of DNA samples under investigation
- the number of probes with which it is intended to study a population
- the need to use highly informative probes (e.g. in pedigrees) or less informative probes (e.g. population studies).

In general, RFLPs will tend to be favoured where there is ample DNA (at least 100 μg/individual plus a pool with milligram quantities), where sample numbers are large, where the number of probes is large, and where less polymorphism may be desirable. The other two types of polymorphism we shall discuss involve use of the polymerase chain reaction (PCR; see Chapter 6) which can be used to examine very small quantities of DNA.

3.1.1 Screening library membranes by hybridization

Both replica membranes are screened by hybridization (*Protocol 19*) (a true positive will give signal at the same position on both replicas) and positive clones are purified (*Protocols 22* and *23*). DNA is then prepared (*Protocol 4*) for further analysis. Avoid stripping the probe off the screening replicas after use allow radioactive decay to deplete signals.

Isolation of a species-specific probe with a heterologous probe may require using reduced stringencies for both hybridization and washing. The melting temperature (T_m) of a hybrid longer than 100 bp decreases by 1–1.5°C with every 1% decrease in homology (16) and T_m is given (16) by

$$T_m = 81.5°\text{C} - 16.6(\log_{10}[\text{Na}^+]) + 0.41(\%\text{G} + \text{C}) - 0.63(\%\ \text{formamide}) - (600/l)$$

where l is the length of the hybrid in base pairs.

We advise empirically establishing reduced stringencies for hybridization and washing. This is best achieved on Southern blot strips of restriction-digested source DNA. In practice, it is convenient to substitute water for formamide during hybridization (*Protocol 19*, step **2**) and to lower the 65°C washing temperature (*Protocol 19*, step **4**).

Hybridization may require pre-treatment of the probe to improve the signal-to-noise ratio (*Protocol 20*). Repeated sequences are found dispersed throughout eukaryotic genomes. A genomic DNA fragment is likely to contain such sequences and a probe that does will hybridize to many clones (and yield a smear on a Southern blot). If a probe hybridizes along much of the length of a track on a Southern blot, obscuring signal due to unique sequence, this 'noise' can be eliminated or substantially reduced by pre-reassociation: hybridizing probe with excess unlabelled source DNA ('driver' DNA) prior to use of the probe in hybridization (18).

It may be convenient to avoid any need for probe pre-reassociation. This is achieved by identifying a fragment of the source DNA in the probe which reveals the polymorphism but which lacks substantial repetitive sequence (*Protocol 21*).

Protocol 19. Screening library membranes by hybridization

Equipment and reagents

- 50 × Denhardt's solution: 1% w/v Ficoll (Type 400 Pharmacia), 1% w/v bovine serum albumin (Fraction V Sigma), 1% w/v polyvinyl pyrrolidone (Sigma) dissolved in distilled water and filtered
- 20 × SSC: 3 M NaCl, 1 M trisodium citrate
- 0.5 M sodium phosphate, pH 6.5
- 10% SDS
- 10 mg/ml salmon sperm DNA (sonicated and denatured)
- Formamide (deionized)
- 50% dextran sulfate
- 0.5 M sodium phosphate, pH 7.5; 7% SDS
- Pre-hybridization buffer: 5 × Denhardt's solution, 5 × SSC, 50 mM sodium phosphate pH 6.5, 0.2% SDS, 100 μg/ml salmon sperm DNA, 50% formamide
- Autoradiography film (Fuji RX)
- Intensifying screens (Dupont)

A. *Screening the membranes*

1. Incubate the membranes at 42°C for 4–16 h in 10 ml of pre-hybridization buffer.
2. Remove membranes to 10 ml fresh pre-hybridization buffer to which has been added 10% dextran sulfate and 2–5 $\times 10^6$ c.p.m. of denatured ^{32}P-labelled DNA or RNA probe per ml of solution and hybridize at 42°C for 1–2 days (see Chapter 5).
3. Wash membranes at room temperature for 20 min in 2 × SSC, 0.5% SDS.
4. Wash membranes at 65°C for 20 min at increasing stringency:
   (a) 2 × SSC, 0.5% SDS;
   (b) 1 × SSC, 0.5% SDS;
   (c) 0.5 × SSC, 0.5% SDS;
   (d) 0.1 × SSC, 0.5% SDS.

   At each step measure background counts on the membrane and stop washing when this has fallen sufficiently. Remove excess moisture from membranes using blotting paper and expose overnight at –70°C to autoradiogaphy film using intensifying screens.
5. Pick and purify (*Protocols 22* and *23*) clones that give a positive hybridization signal on both replica membranes.

B. *Detecting clones containing microsatellites*

1. Prehybridize duplicate membranes for 15 min at 65°C in 10 ml of 0.5 M sodium phosphate buffer pH 7.5, 7% SDS.
2. Add 100 ng of end-labelled microsatellite oligonucleotide (e.g. $(TG)_{15}$ for cytosine–adenosine microsatellites) (*Protocol 26*).
3. Hybridize for 4 h at 65°C.
4. Rinse membranes in 1 × SSC, 0.5% SDS at room temperature for 5 min.
5. Wash the membrane in 0.5 × SSC, 0.5% SDS at 55°C for 15 min and expose membrane to X-ray film.[a]

[a] Do not strip the probe off the screening replicas after use, since boiling or alkali will shorten the life of the membranes. Allow radioactive decay itself to deplete signal strength prior to re-use of the membranes.

Protocol 20. Probe pre-reassociation

Equipment and reagents

- Source DNA, 10 mg/ml in TE (10 mM Tris–HCl pH 7.5, 1 mM EDTA)
- 5 ml syringe with 21 G needle
- 1.5% agarose gel
- 5 × SSC: 3 M NaCl, 1 M trisodium citrate
- Boiling water bath
- 68°C incubator

Protocol 20. *Continued*

Method

1. Prepare a stock of 'driver' DNA by shearing 5 ml of 10 mg/ml source DNA in TE. Shear the DNA by forcing the solution in a 5 ml syringe about 20 times through a 21 G needle until viscosity is much reduced. Check average DNA fragment size of 3–500 bp on a 1.5% agarose gel.
2. Combine 1 mg of this 'driver' DNA and 10–100 ng probe in 100 μl 5 × SSC in a 1.5 ml tube.[a]
3. Denature the mixture for 10 min at 100°C and plunge into ice water for 1 min.
4. Incubate the mixture for 10 min at 68°C so that only rapidly re-annealing sequences become double-stranded.
5. Use this preparation as a probe in the normal way (*Protocol 19*).

[a] The values given are appropriate for genomic DNA with the complexity and size (3×10^9 bp) of the human genome. For less complex genomes, reduce the driver DNA concentration and/or the time for pre-reassociation.

Protocol 21. Preparation of a 'clean' probe

Equipment and reagents

- *Hae*III, *Pst*I, *Taq*I, and manufacturer's buffers
- 1 kb ladder (BRL)
- 1% agarose gels
- Ethidium bromide
- Solutions for *Protocol 3*
- Microcentrifuge tubes with caps removed
- Glass wool
- Equipment and reagents for Southern blotting and probing
- Scalpel
- Sheared source DNA (*Protocol 24*), labelled as described in Chapter 5
- Vector DNA, labelled

Method

1. Digest to completion 1 μg DNA aliquots of the clone with a range of restriction enzymes (e.g. *Hae*III, *Pst*I, and *Taq*I).
2. Fractionate the digestion products on a 1% agarose gel, including a track with molecular weight standards (a 1 kb ladder is suitable). Visualize the fragments after ethidium bromide staining. Photograph the gel.
3. Southern blot (Chapter 5) the gel twice: blot with the first membrane for 1 h, then blot overnight with a second membrane.
4. Probe one blot with labelled sheared source DNA and the other with labelled vector DNA.
5. By comparing the two autoradiograms and the photograph of the stained gel it will be possible to identify bands that contain unique sequence DNA without vector.

6. Now digest about 10 μg clone DNA with the appropriate enzyme and fractionate the products on a 1% agarose gel, along with a 1 kb ladder.
7. After ethidium bromide staining, use a new scalpel to slice out of the gel a band considered to contain unique sequence DNA.
8. Place the slice in a 0.5 ml tube (cap removed) pierced with a needle and containing a small wad of glass wool at the bottom of the tube. Place this in a 1.5 ml tube (cap removed). Centrifuge for 15 sec at 3000 r.p.m. in a microcentrifuge.
9. Transfer the eluate to a fresh 1.5 ml tube. Extract and precipitate (*Protocol 3*) the aqueous eluate. Resuspend the DNA pellet at about 10 ng/μl.
10. Confirm that this fragment is 'clean' by using it as a probe to a Southern blot strip (Chapter 5).

Protocol 22. Cosmid colony purification

Equipment and reagents

- LB broth (*Protocol 4*)
- LB broth containing 15% glycerol and antibiotic
- Solutions for *Protocols 18* and *19*
- Sterile flat-ended toothpick
- Nitrocellulose membranes (80 mm diameter)

Method

1. Positive signals identified by hybridization of replica membranes cannot usually be matched to a single colony. Therefore identify an area of about 5 mm^2 corresponding to the location of the signal and scrape this area of the master membrane using a sterile flat-ended toothpick. Suspend the bacteria in 150 μl of LB broth containing the appropriate selective antibiotic and 15% glycerol. Store this sample at –70°C.
2. Take dilutions in LB broth of the sample and plate on to nitrocellulose membranes (80 mm diameter) on an LB agar plate containing the appropriate antibiotic. Aim for 100–500 colonies per membrane. Incubate overnight at 37°C.
3. Make a replica of this membrane, and prepare it for hybridization as in *Protocol 18*.
4. Hybridize this replica as in *Protocol 19* and identify single positive colonies for further analysis.

Protocol 23. Lambda clone purification

Equipment and reagents

- SM (*Protocol 4*)
- Chloroform
- Solutions for *Protocols 15* and *19*
- Sterile toothpick

Method

1. Because of the high density of plaques associated with libraries, it is only possible to identify an area of the agarose associated with a positive plaque. To recover an area containing the desired plaque, punch a disk out of the top agarose using the wide end of a flamed Pasteur pipette. Re-suspend this disk in 1 ml of SM buffer containing a drop of chloroform. Store the sample at 4°C.
2. Plate out 50 μl of 10^2 and 10^3 dilutions of the sample to obtain 50–200 plaques per 80 mm diameter plate.
3. Replica plate and screen these plaques (*Protocols 15* and *19*).
4. Now it should be possible to pick single positive plaques using a sterile toothpick. Suspend the phage in 1 ml SM containing a drop of chloroform.
5. Prepare DNA for further analysis (*Protocol 4*).

3.2 Identification of RFLPs

Preliminary identification of an RFLP is made by probing a Southern blot of digested source DNAs from a number of individuals and identifying altered numbers or sizes of bands on an autoradiogram. Confirmation is made by ensuring that the new band(s) is not an artefact of partial digestion and by demonstrating Mendelian inheritance of alleles in pedigrees (*Protocol 24* and see *Protocols 19–21*).

Protocol 24. Identification of an RFLP

Equipment and reagents

- *Hin*dIII, *Eco*RI, *Bgl*II, *Bam*HI, *Msp*I, *Taq*I, and manufacturer's buffers
- 0.8% agarose gel
- Ethidium bromide
- Solutions for *Protocols 19–21*

Method

1. Digest 8 μg of genomic DNA from each individual[a] with 16 units of each of a number of restriction enzymes. Cheap 'six-cutters' (e.g. *Hin*dIII, *Eco*RI, *Bgl*II, and *Bam*HI) are suitable and MspI[b] and *Taq*I are useful since these 'four-cutters' recognize sequences containing a CpG

dimer. This is a mutational hot spot (19) and hence more likely to reveal an RFLP.

2. Test for complete digestion by fractionating 1 μg of each digest on a 0.8% agarose gel and visualizing the tracks after ethidium bromide staining. All DNAs digested with the same enzyme should give a similar smear along the length of the track.
3. Prepare Southern blots of all the digests (Chapter 5), ensuring that all DNAs digested with the same enzyme are in adjacent tracks.
4. Probe the blots (*Protocols 19–21*) and from the autoradiograms identify enzymes revealing putative polymorphisms.
5. Test for Mendelian inheritance of the identified alleles by probing Southern blots of appropriate enzyme digests of DNA from informative pedigrees. The relative likelihood of Mendelian inheritance is the ratio of the likelihood that the observed sets of progeny would be obtained from Mendelian inheritance (given the genotypes of the parents) to the likelihood that the observed sets of progeny would be obtained if the variant restriction fragments were distributed two at a time at the frequencies observed among unrelated individuals (for details, see ref. 19).

[a] To detect a frequent RFLP (e.g. for pedigree studies), a Southern blot of DNA (Chapter 5) from five diploid individuals is adequate since this is quite likely to detect a variant present on only 10% of chromosomes. To detect an infrequent RFLP (e.g. for population studies) more individuals will need to be examined.

[b] *MspI* is best used at room temperature for complete digestion.

3.3 Identification of single nucleotide substitutions

The mobility of a single-stranded DNA molecule in a native gel is thought to depend critically on the secondary structure the molecule adopts. This structure, and hence the molecule's mobility, seems to change significantly with just a single base substitution. Hence if two short DNA duplexes varying at one position (i.e. two alleles) are denatured, snap-chilled, and gel-fractionated, each denatured duplex may well give rise to two bands. This is the basis of single-strand conformation polymorphisms (SSCPs) (20–22; Chapter 8). Since nucleotide substitutions occur every few hundred base pairs in, for example, *Homo sapiens* (23,24), examination of virtually any short sequence in a panel of individuals could well reveal a polymorphism. SSCP is thus an attractive alternative to examining sequences at a locus.

4. Construction of small insert genomic libraries and the identification of microsatellites

A microsatellite is the term coined (25) for a DNA sequence consisting of a reiterated simple sequence motif (see Chapter 7). A commonly investigated motif is dA–dC. Alleles show considerable variation in the number of iterations when more than about 20 nucleotides are present in a microsatellite (26). Purine–pyrimidine microsatellites are abundant in primates, rodents, invertebrates, and yeast, but not in prokaryotes (27). They are distributed randomly every 50 000–100,000 bp in the human genome (25). We present here a series of protocols aimed to allow small libraries of genomic clones to be constructed with a view to allowing rapid screening for these abundant and useful markers. This type of library is designed to facilitate the sequencing of positive clones to yield flanking sequence allowing PCR assays to be developed. The important feature of this type of library is that it is fairly simple to construct and handle and need not be fully representative. The small insert size is chosen because the complete DNA sequence of inserts containing the microsatellite repeat can usually be obtained from one sequencing reaction or at most two using primers flanking the cloning site in plasmid clones. The use of the filamentous phage M13 as an alternative to plasmid cloning is presented as many experimenters prefer the sequencing quality associated with single-stranded DNA templates.

4.1 Cloning procedure for plasmid vectors

4.1.1 Preparation of genomic DNA

As with lambda insertion vectors, source DNA in the size range 50–100 kb is adequate. *Protocol 1* is suitable and genomic DNA is digested and size fractionated as described in *Protocol 25*, which is a modified version of a procedure described by Pulido and Duyk (28). The source DNA can be fragmented either by complete digestion to produce sticky ends (Section A) or blunt ends (Section B), or by sonication (Section C). These are in order of difficulty with B and C requiring more expertise. The blunt-ended DNA is either ligated directly or linkered (Section F) prior to cloning and the sonicated DNA is repaired (Section E) and linkered prior to cloning. Low melting point (LMP) agarose gels are used for size fractionation (Section D) for all three fragmentation methods to yield insert material in the 250–500 bp size range. The base composition of the source DNA determines restriction site distribution so the representation of sequences in such a library is better for the mixed blunt end approach and best if sonicated DNA is used.

Protocol 25. Preparation of size fractionated insert material for small insert (250–500 bp) library construction

A. *Complete digestion of genomic DNA to yield sticky ends*

Equipment and reagents

- High molecular weight DNA (*Protocol 1*)
- Restriction enzyme[a] with manufacturer's buffer
- 1% agarose gel
- Reagents for phenol extraction and ethanol precipitation (*Protocol 3*)
- TK100 mini-fluorimeter (Hoefer Scientific Instruments)

Method

1. Incubate source DNA (100 μg) for 1 h with excess (4 U/μg) of a suitable restriction enzyme. Remove a 1 μg aliquot and check on a 1% agarose gel for complete digestion (Sections 6.9–6.19 in ref. 7). Add further enzyme and incubate for another 2 h if required.
2. Extract and ethanol-precipitate the DNA (*Protocol 3*) or heat inactivate the enzyme if appropriate by incubating at 65°C for 15 min.
3. Use a fluorimeter to estimate concentration.
4. Proceed to Section D of this protocol for size fractionation.

B. *Complete digestion of genomic DNA to yield blunt ends*

Equipment and reagents

- High molecular weight DNA (*Protocol 1*)
- 1% agarose gel
- Restriction enzyme(s)[b] with manufacturer's buffer

Method

1. Incubate source DNA (4 × 25 μg) for 1 h with excess (4 U/μg) of *Alu*I, *Hae*III, *Ssp*I, and *Eco*RV. Remove a 1 μg aliquot and check on a 1% agarose gel for complete digestion (Sections 6.9–6.19 in ref. 7). Add further enzyme and incubate for two more hours if required.
2. Pool the four digests and extract and ethanol-precipitate the DNA (*Protocol 3*).
3. Use a TK100 mini-fluorimeter (Hoefer Scientific Instruments) to estimate concentration.
4. Proceed to Section D of this protocol for size fractionation.

C. *Sonication of genomic DNA to yield blunt ends*

Equipment and reagents

- High molecular weight DNA (*Protocol 1*)
- 123bp ladder (Gibco-BRL)
- 1.4% agarose/TBE gel
- Size marker: 123 bp ladder (Gibco-BRL)
- Cup horn sonicator (calibration of a conventional sonicator equipped with a microtip probe is unreliable and not recommended)

Method

1. Place 50 μl of TE containing 400 μg/ml of genomic DNA (use DNA from an easily renewable source!) in a sterile microcentrifuge tube.
2. Fill the cup horn of the sonicator with ice and clamp your tube directly above the probe.
3. Sonicate at maximum power for 20 sec. Collect the DNA by centrifugation for 30 sec in a microcentrifuge at 12^000 × g. Leave the DNA on ice for at least 1 min. Remove 3 μl of the DNA to a fresh tube.

Protocol 25. *Continued*

4. Repeat step **3** until all of the DNA has been transferred to fresh tubes.
5. Analyse the size of the sonicated DNA by electrophoresis on a 1.4% agarose/TBE gel. Sonicated DNA runs as a smear on agarose gels. Use a 123 bp ladder (Gibco-BRL) as a size marker. Select the sonication conditions producing a good yield of fragments in the 300 bp–1 kb size range. It is better to under sonicate as repair of DNA that has been excessively sonicated is inefficient resulting in poor cloning efficiencies.
6. Process between 100 and 200 μg of source genomic DNA using exactly the conditions derived above. Do not attempt to scale up the reaction volume but repeat 5–10 times until the required amount of DNA has been processed and pool the samples.
7. Size-fractionate sonicated DNA as described in Section D of this protocol.

D. *Size fractionation of DNA by LMP agarose gel electrophoresis and agarase treatment*

Equipment and reagents

- Agarase (Boehringer)
- Agarase buffer
- Sodium acetate, 3 M, pH 5.5
- Ethanol
- Glycogen (Boehringer)
- 120 ml (11 cm × 14 cm × 2.5 cm) 0.8% LMP agarose/TBE/ethidium bromide made with LMP agarose (Ultrapure, Gibco-BRL)
- Ethidium bromide
- TBE
- 123 bp ladder size marker
- Long wavelength (366 nm) UV transiluminator
- Scalpel
- Sterile 50 ml Falcon tube
- 65°C incubator/water bath
- 45°C water bath
- Sterile 30 ml high-speed centrifuge tubes

Method

1. Fractionate the restriction digested/sonicated DNA on a 120 ml (11 cm × 14 cm × 2.5 cm) 0.8% LMP agarose/TBE/ethidium bromide gel. Use a 20-well, 1.5 mm thick comb. For convenience create one long well by taping together the 17 wells on the right-hand side of the comb. Separate the 123 bp ladder size marker by at least two lanes from the large well containing the genomic DNA. Run 100 μg DNA per gel and run the gel for 3–5 h at a low voltage (60 V).
2. Visualize under **long wavelength** (366 nm) UV light. Cut out the region between the 246 and 492 bp bands of the marker with a sterile scalpel blade. Trim any excess agarose.
3. Weigh an empty sterile Petri dish and then weigh the gel slice in the dish.
4. Chop the gel slice into pieces, transfer into a sterile 50 ml Falcon tube, and incubate at 65°C for 20 min to melt the LMP agarose. Add 1/25th

volume of the supplied agarase buffer assuming that the gel slice has a density of 1 g/ml. Place in a water bath at 45°C and allow to cool for 10 min before adding the agarase. Add 2 U/100μg of gel slice and incubate at 45°C for 1 h.

5. Transfer the agarase-treated material to a sterile high speed centrifuge tube, add 0.1 vol. of 3 M sodium acetate pH 5.5, and incubate on ice for 15 min.
6. Remove the precipitated oligosaccharides by centrifugation at 12000 × *g* for 15 min. Carefully transfer the supernatant to sterile 30 ml high speed centrifuge tube(s) taking care not to disturb the pellet, add 2 μl of glycogen and 3 vol. of absolute ethanol, and leave at –20°C for 1 h.
7. Collect the precipitated DNA by centrifugation at 12000 × *g* for 15 min. Remove the supernatant and wash once with 5 ml of ice cold 70% ethanol. Re-centrifuge at 12000 × *g* for 15 min.
8. Remove the supernatant and allow to air dry at room temperature for 30 min to remove any ethanol.
9. Resuspend by vigorous vortexing for 2 min in 450 μl of distilled H_2O. If the DNA was prepared by sonication proceed to Section E to repair the ends. Alternatively, concentrate by ethanol precipitation (*Protocol 3*, steps **5–10**) and resuspend the DNA in 40 μl of TE. Estimate the DNA concentration using a fluorimeter. For blunt-ended DNA either proceed to Section F to add linkers or use blunt or sticky ended insert DNA for ligation into appropriately cut M13/plasmid vector DNA.

E. *Repair of DNA ends*

Equipment and reagents

- 100 mM ATP
- 10 × T4 polynucleotide kinase buffer as supplied by manufacturer
- 0.5 mg/ml BSA (molecular biology grade, Boehringer)
- T4 polynucleotide kinase
- Mixture of all four dNTPS at 2 mM each
- 10 × T4 DNA polymerase buffer as supplied by the manufacturer
- T4 DNA polymerase

Method

1. Treat the DNA with T4 polynucleotide kinase by mixing (700 μl per tube):
   - 100 μg (approximately) of DNA in 450 μl of distilled H_2O
   - 42 μl 100 mM ATP
   - 70 μl 10 × T4 polynucleotide kinase buffer
   - 66 μl 10 U/μl T4 polynucleotide kinase
2. Incubate at 37°C for 1 h.
3. Phenol–chloroform extract and ethanol precipitate as described in *Protocol 3.*

Protocol 25. *Continued*

4. Resuspend DNA in 383 μl dH_2O.
5. Treat with T4 DNA polymerase by mixing (660 μl per tube):
   - 65 μg (approximately) of DNA in 383 μl dH_2O
   - 83 μl dNTP mix
   - 66 μl 10 × T4 DNA polymerase buffer
   - 62 μl 3U/μl T4 DNA polymerase
6. Incubate for 30 min at 37 °C.
7. Phenol–chloroform extract and ethanol precipitate as described in *Protocol 3*.
8. Resuspend DNA in 20 μl distilled H_2O. Store DNA at –20 °C or proceed to Section F to ligate linkers.

F. *Linkering of size fractionated repaired sonicated or blunt-ended genomic DNA*

Equipment and reagents

- 5′-phosphorylated *Eco*RI linker sequence, p d(CCGGAATTCCGG), from Stratagene (p=phospho).
- Blunt end ligation buffer: 660 mM Tris–HCl pH 7.6, 50 mM $MgCl_2$, 50 mM DTT
- TK100 mini-fluorimeter (Hoefer Scientific Instruments)
- 5 mM ATP made by diluting 100 mM stock (Pharmacia Biotech) stored in small aliquots at –20 °C
- 10 × *Eco*RI buffer supplied by manufacturer
- *Eco*RI
- T4 DNA ligase

Method

1. Set up the following ligation reaction:
   - 10 μl insert DNA (approximately 30 μg)
   - 18 000 pmol linkers
   - 6 μl 10 × blunt ligation buffer
   - 6 μl 5 mM ATP
   - 6 μl 5 U/μl T4 DNA ligase
   - sterile dH_2O to 60 μl
2. Incubate overnight at room temperature.
3. Remove unattached linker molecules using the size fractionation and agarase treatment method detailed in Section D but select DNA migrating between the 369 and 492 bp bands of the marker and only create one long well taping together six wells as less DNA is being purified. Following the second ethanol precipitation resuspend in 88 μl of distilled H_2O.
4. Add 10 μl of 10 × *Eco*RI buffer and 2 μl of *Eco*RI at 20 U/μl and incubate for 1 h at 37 °C.

5. Extract and ethanol precipitate as described in *Protocol 3*. Resuspend in 10 μl of distilled H_2O and use fluorimeter to estimate concentration. This DNA can be used for ligation into *Eco*RI-cut M13/plasmid vector DNA.

[a]For this type of library construction restriction enzymes with 4 bp recognition sites produce fragments of the required size range. Chose an enzyme producing ends compatible with an enzyme in the multiple cloning site e.g. *Mbo*I into *Bam*HI.
[b]For mammalian DNAs a combination of *Alu*I, *Hae*III, *Ssp*I, and *Eco*RV produce blunt-ended fragments suitable for the attachment of linkers

4.1.2 Preparation of plasmid vector DNA

We recommend the use of a plasmid vector with a multiple cloning site and flanking sequencing primers such as the vector pGEM-3Z/4Z (Promega) or pUC18/19 (Pharmacia) in conjunction with *E. coli* strains JM109 or JM105 respectively. Alternatively the phagemid vectors pGEM-11Zf(+/–) (Promega) or pBluescript II SK (+/–) (Stratagene) could be used in conjunction with *E. coli* strains JM109 and XL1-Blue MRF′ respectively. The rescue of single-stranded DNA templates for sequencing is beyond the scope of this chapter but is well documented in the companies protocol sheets and in Pulido and Duyk (28). Plasmid DNA is purchased from the supplier or prepared as described in *Protocol 4* and purified by caesium chloride density gradient centrifugation (7). Vector DNA is cut and dephosphorylated (*Protocol 5*, steps **1** and **2**, followed by *Protocol 16*, step **2**) with the appropriate enzyme to ligate the insert material (e.g. *Bam*HI for *Mbo*I-cut size-fractionated insert, *Hin*cII for blunt ends or *Eco*RI for linkered insert). The DNA is resuspended following quantification at 200 ng/μl.

4.1.3 Ligation of size fractionated source DNA with plasmid vector DNA

Ligate 200 ng of dephosphorylated vector DNA and 200 ng of insert material (*Protocol 6*). Following ligation, 1 μl aliquots of ligated DNA are used to transform competent *E. coli* of the appropriate strain. A choice of a traditional transformation protocol (*Protocol 26*), essentially as described by Chung *et al.* (29) and an electroshock transformation protocol (*Protocol 27*), essentially as described by Hanahan *et al.* (30) is offered. Slightly higher transformation efficiencies are routinely achieved by the electroshock method but care must be taken not to use more than 1 μl of ligation mix as high salt can cause arcing in the cuvettes. Ligations can be desalted prior to transformation by precipitation or by microdialysis (30). Controls should include cut vector DNA and re-ligated vector which should yield very few ($< 10^3$/μg) blue or no colonies. Transformation efficiencies between 10^3 and 10^5 cfu/μg of insert should be achieved with at least a 20:1 white:blue colony ratio. Colonies can be replica-plated to produce filters for screening as described for cosmids following transformation as described in *Protocol 17* steps **3** and **4** and *Protocol 18*.

Protocol 26. Transformation of *E. coli*

Equipment and reagents

- LB broth (*Protocol 4*)
- Fresh *E. coli* colonies on a plate of LB agar (LB broth with 1.5% agar poured at 50°C into plastic Petri dishes)
- Sterile 50 ml Falcon tube
- 3 litre conical flask
- Sterile plastic 1 ml Pasteur pipette
- Dedicated 500 ml polypropylene centrifuge bottles with caps
- LB agar plates containing the appropriate antibiotic
- Filter sterilized TSS medium: LB broth containing 10% (w/v) polyethylene glycol (either PEG 3350 or 8000), 5% (v/v) DMSO and 50 mM $MgCl_2$; adjust to pH 6.5 with HCl (transformation efficiency is pH dependent)
- X-Gal (5-bromo-4-chloro-3-indolyl-β-D-galactoside), 2% in dimethylformamide; store at –20°C
- IPTG (isopropyl-β-D-thiogalactopyranoside), 100 mM (24.8 mg/ml) in sterile water; store at –20°C

A. *Preparation and freezing of competent cells*

1. Pick a single 2–3 mm colony from a freshly streaked LB agar plate and mix in 10 ml of LB by vortexing in a sterile 50 ml Falcon tube. Incubate overnight at 37°C and 200 r.p.m.
2. The following day use 2 ml of this culture to inoculate 400 ml of LB in a 3 litre conical flask.
3. Incubate at 37°C at 200 r.p.m. until an OD_{600} of 0.3–0.4 units is reached (this corresponds to 1–3 $\times$ 10^8 cells/ml).
4. Decant the cell suspension into two chilled sterile 500 ml centrifuge pots.
5. Pellet the cells by centrifugation at 2500 $\times$ *g* for 15 min at 4°C. Carefully discard the supernatant, minimizing the disturbance of the pellet.
6. Resuspend cells are resuspended in a total of 40 ml chilled TSS using a sterile plastic 1 ml Pasteur pipette. These cells are competent and can be transformed immediately. Alternatively, freeze cells for up to 6 months (this causes only a slight (10%) drop in titre).
7. Store the cell suspension in 500 μl aliquots in cryo tubes. Rapid freeze the tubes by partial submersion in a dry-ice/ethanol bath and store at –80°C.

B. *Transformation of competent cells with plasmid DNA*

1. Pipette aliquots of cells (0.1 ml) into cold polypropylene Eppendorf tubes. Add 1 μl of DNA and gently mix the cell/DNA suspension by flicking the base of the tube. If frozen cells are used allowed them to thaw on ice.
2. Incubate the cell/DNA suspension at 4°C for 60 min.

3. Add 0.9 ml of TSS (or LB broth) plus 20 mM glucose and incubate cells at 37 °C with shaking for 1 h to allow expression of the antibiotic resistance gene.
4. Spread varying fractions (10 and 100 μl) on LB agar plates containing the appropriate antibiotic, X-Gal[a], and IPTG[a] to allow blue/white selection and incubate inverted at 37 °C for 12–14 h.

C. *Transfection of competent cells with M13 DNA*

1. Perform steps **1** and **2** are as in Part B above.
2. Following incubation on ice, varying fractions can be mixed with plating bacteria and plated in top agar as described in *Protocol 28*, step **3** and subsequent.

[a] Incorporate 100 μl each of X-Gal and IPTG into 100 ml of agar and antibiotic at about 50°C. Avoid leaving plates under UV irradiation, as this destroys X Gal.

Protocol 27. Electroshock transformation of *E. coli*

Equipment and reagents

- Mg^{2+}-free SOB: 2% Bacto-tryptone, 0.5% Bacto Yeast Extract, 10 mM NaCl, 2.5 mM KCl
- Fresh *E. coli* colonies on a plate of Mg^{2+}-free SOB agar (Mg^{2+}-free SOB with 1.5% Bacto-agar autoclaved and poured at 50°C into plastic Petri dishes)
- Dedicated 500 ml polypropylene centrifuge bottles with caps (see below for washing instructions)
- 10% re-distilled glycerol (Gibco-BRL), at 4°C
- SOC recovery medium: 98 ml of Mg^{2+}-free SOB plus 1 ml of 2 M $MgCl_2$ + $MgSO_4$ and 1 ml of 2 M glucose
- SOB agar plates containing the appropriate antibiotic (produced by adding antibiotic and 1 ml of 2 M ($MgCl_2$ + $MgSO_4$) to 99 ml of Mg^{2+}-free SOB autoclaved with 1.6% agar at 50°C and poured into plastic Petri dishes)
- Gibco-BRL 0.15cm cuvettes or Bio-Rad 0.2cm cuvettes.

A. *Preparation and freezing of electrocompetent cells*[a]

1. Pick a single 2–3 mm colony from a freshly streaked Mg^{2+}-free SOB agar plate and mix in 1 ml of Mg^{2+}-free SOB medium by vortexing in a sterile plastic bijou bottle.
2. Transfer to a 500 ml conical flask containing 50ml of Mg^{2+}-free SOB and incubate overnight at 37 °C and with shaking (200 r.p.m.).
3. The following day use 7.5 ml of this culture to inoculate 750 ml of Mg^{2+}-free SOB in a 3 litre conical flask.
4. Incubate at 37°C with shaking (150 r.p.m.) until an OD_{550} of 0.75–1.0 units is reached (this corresponds to 2–6 × 10^8 cells/ml).
5. Decant the cell suspension into two chilled 500 ml centrifuge pots.
6. Pellet the cells by centrifugation at 2500 × *g* for 15 min at 4°C.

Protocol 27. *Continued*

Carefully discard the supernatant, minimizing the disturbance of the pellet.

7. Resuspend the pellet in 1 vol. of ice-cold 10% glycerol by vigorous agitation and vortexing. Take care to keep the cells and centrifuge tubes cool on ice from this step onwards.
8. Repeat steps **6** and **7**.
9. Recentrifuge as in step **6** and decant the supernatant. Leave to stand for 30 sec and carefully remove the last of the supernatant with a sterile plastic 1 ml pipette while attempting to minimize loss of cells.
10. Leave to stand for 30 sec and resuspend in the residual liquid by gentle aspiration up and down with a 1 ml pipette.
11. Measure the volume of the resuspended cells and measure the OD_{550} of a 200-fold dilution of 5 μl of the suspension relative to Mg^{2+}-free SOB.
12. Estimate the absorbance of the concentrated cell suspension and adjust the concentration to approximately 200 OD units/ml. If the concentration of the cell suspension is lower than this repeat steps **9–12**.
13. Store the cell suspension in 120 μl aliquots in cryo tubes. Rapidly freeze the tubes by partial submersion in a dry-ice/ethanol bath and store at –80°C

B. *Electroshock transformation*

1. Thaw electrocompetent cells on ice and aliquot into sterile chilled 0.5 ml Eppendorf tubes (use 20 μl of cells for Gibco-BRL 0.15 cm cuvettes or 40 μl for Bio-Rad 0.2 cm cuvettes).
2. Add 1 μl of DNA solution.
3. Mix the cells and DNA by gently aspirating up and down using a pipette tip and transfer into chilled electroporation cuvettes.
4. Electroshock the cells at 12.5 kV/cm, 25 μF, and 200 Ω or, if your apparatus can achieve it, 17.7 kV/cm, 2 μF, and 4000 Ω.
5. Remove the electroshocked cells immediately from the chamber and add 1 ml of pre-warmed (42°C) SOC recovery medium to the cuvette. Mix rapidly by inversion. Transfer to a universal tube and incubate for 1 h at 37°C and 200 r.p.m.

C1. *Plating out plasmid transformants following electroshock transformation*

1. Spread varying fractions (1, 10, and 100 μl) on to SOB agar plates containing the appropriate antibiotic and X-Gal and IPTG to allow blue/white selection.

2. Incubate inverted at 37 °C for 12–14 h.
3. Store the remaining cells at 4°C; they can be plated the next day with only a slight drop in titre.

C2. *Plating out M13 transformants following electroshock transformation*

1. Following electroshock transformation with M13, plate out transformants on to a lawn of cells as described in *Protocol* 28. The 1 h incubation in SOC recovery medium is not necessary and can be omitted or reduced to a 10 min recovery step. Any remaining cells can be stored at 4°C and appropriate fractions plated the next day with a slight drop of titre.

[a] All solutions used in this protocol are made up in ultra-pure water. All glassware used to make up or keep solutions used in this protocol must have been extensively rinsed to remove all traces of detergent as this dramatically reduces transformation efficiency. Two or three rinses with tap water and two with ultra-pure or double-distilled water usually suffices. Wherever possible the use of pre-packed sterile plasticware is recommended.

4.2 Cloning procedure for M13 vectors

4.2.1 Preparation of genomic DNA

The preparation of genomic DNA is identical to that described in Section 4.1.1 for plasmid cloning.

4.2.2 Preparation of M13 vector DNA

We recommend the use of an M13 vector with a multiple cloning site and flanking sequencing primers such as the vectors M13BM20/21 RF (Boehringer) in conjunction with *E. coli* strains JM109 or JM105. M13 RF DNA can be purchased directly from the supplier or prepared as described in Section 4.31 of ref. 7 and purified by caesium chloride density gradient centrifugation. Cut vector DNA with the appropriate enzyme to ligate the insert material (e.g. *Bam*HI for *Mbo*I-cut size-fractionated insert, *Eco*RV for blunt ends or *Eco*RI for linkered insert) and dephosphorylate it (*Protocol 5* steps **1** and **2** followed by *Protocol 16* step **2**). Following quantification, resuspend the DNA at 200 ng/μl.

4.2.3 Ligation of size fractionated source DNA with M13 vector DNA

Ligation and transformation steps are identical to those described for plasmid DNA in Section 4.1.3 except that the transformed *E. coli* are plated in top agar onto a lawn of plating bacteria. These steps are described at the end of *Protocols 26* and *27* and *Protocol 28*. Transformation efficiencies should be similar to those described for plasmid DNA. Replica plating following transformation to produce filters for screening is essentially as described for bacteriophage lambda in *Protocol 15*.

Protocol 28. Transfection of M13 transformed *E. coli*

Reagents

- 2 × TY broth: 1.6% Bacto-tryptone, 1.0% Bacto Yeast Extract, 0.5% NaCl
- Bottom agar: 1.0% Bacto-tryptone, 0.8% NaCl, 1.2% Bacto Agar
- Top agar: 1.0% Bacto-tryptone, 0.8% NaCl, 0.8% agar
- 2% X-Gal in dimethylformamide
- 100 mM IPTG

Method

1. Pick a single 2–3 mm colony from a freshly streaked plate and mix in 1 ml of 2 × TY broth by vortexing in a sterile plastic bijou bottle. Use 50 μl of this suspension to inoculate 10 ml of 2 × TY broth and incubate overnight at 200 r.p.m. at 37 °C.
2. Inoculate 50 ml of 2 × TY medium in a sterile 400 ml conical flask with 1 ml of overnight culture and shake at 200 r.p.m. at 37 °C for 3 h to produce the plating cells, taking care not to exceed an OD_{600} of 1.0. Chill cells on ice until required; they can be stored at 4 °C for up to a week.
3. In plastic 5 ml screw-capped tubes make up the following mixture. To 200 μl of plating cells add 40 μl of 2% X-gal and 40 μl of 100 mM IPTG.
4. Add varying fractions (1,10 and 100 μl) from each transformation to individual plating cell/X-Gal/IPTG mixtures.
5. Add to the tubes top agar melted and cooled to 42 °C. Use 3 ml for an 80 mm diameter Petri dish or 6.5 ml for a 150 mm diameter dish. Mix rapidly by gentle swirling, avoiding formation of bubbles. Pour this on to the centre of the bottom agar in the dish and swirl on a flat surface: the mixture forms an even layer. Let the Petri dish stand for 5 min to allow the top agarose to harden.
6. Incubate at 37 °C. After 4 h, plaques can be detected and after 8–12 h they grow to 2 mm diameter. Any blue (non-recombinant) plaques should be visible at this stage. The intensity of the blue colour increases with storage at 4 °C for a couple of hours. At this point plaque lifts can be taken for radioactive screening as described in *Protocol 15* starting at step **3**.
7. Pick individual positive plaques using a sterile blunt-ended wooden toothpick and leave the toothpick left for 1–2 h at room temparature in 1 ml of 2 × TY medium. Keep these phage suspensions are kept at 4 °C; they are viable indefinitely. Alternatively, pick positive plaques and prepare single-stranded DNA template as described in *Protocol 29*.

Protocol 29. Small-scale preparation of single-stranded M13 DNA for sequencing

Equipment and reagents

- 2 × TY broth (*Protocol 28*)
- Plating bacteria (steps **1** and **2**, *Protocol 28*)
- Sterile flat-ended wooden toothpicks
- 20% PEG 8000 in 2.5 M NaCl
- TE: 10 mM Tris–HCl, 1 mM EDTA pH 8.0
- Phenol equilibrated with Tris–HCl pH 8.0
- Chloroform–isoamyl alcohol, 24:1 (v/v)
- Absolute ethanol–3 M sodium acetate (pH 5.2), 25:1 (v/v)

Method

1. Add 50 μl of plating bacteria to 2 ml of 2 × TY medium in a sterile 50 ml culture tube. Using a blunt-ended sterile wooden toothpick pick the plaque of interest and inoculate the 2 ml culture or inoculate with one tenth (200 μl) of a plaque suspension.
2. Incubate the culture at 37 °C for 4–5 h (toothpick) or 5–6 h (suspension) at 250 r.p.m. in an orbital shaker.
3. Transfer approximately1.5 ml of the culture to a sterile 1.5 ml microfuge tube and centrifuge at 12 000 × *g* for 5 min.
4. Transfer the supernatant (1.2 ml) to a fresh microcentrifuge tube taking great care not to disturb the pellet.
5. To the supernatant add 200 μl of PEG/NaCl, mix by inversion, and leave to stand for 20 min at room temperature.
6. Harvest the precipitated phage particles by centrifugation at 12 000 × *g* for 10 min in a microcentrifuge.
7. Carefully remove the supernatant, taking care not to disturb the small precipitate at the bottom of the tube. Re-centrifuge the tube for 1 min to allow the last of the PEG/NaCl to be removed from the sides of the tube.
8. Resuspend the pellet in 100 μl of TE by vigorous vortexing for 1 min.
9. Extract and ethanol precipitate as described in *Protocol 3* . After the 70% ethanol wash take care not to aspirate and discard the pellet as it is only loosely attached to the base of the tube.
10. Air dry the tube for 10 min at room temperature to allow any residual ethanol to evaporate. Resuspend the single-stranded DNA in 15 μl of TE; the average yield of DNA should be approximately 10–15 μg.

4.3 Screening of libraries for microsatellite sequences

Given their ease of detection in a library (*Protocol 19*), microsatellite repeats are a readily available source of DNA polymorphisms. Their flanking sequences required for PCR primer design are also readily obtained (*Protocol 30*) for plasmid (31) as well as M13, lambda, and cosmid clones (unpublished observation). Use *Protocol 29* to make small amounts of single-stranded DNA from positive M13 clones for sequencing. Use *Protocol 5* to prepare double-stranded DNA from positive plasmid or cosmid clones. For further discussion on screening microsatellite libraries see Chapter 7.

Protocol 30. Rapid determination of sequences flanking microsatellites

Reagents

- Proteinase K (Boehringer, molecular biology grade)
- 75 mM NaCl, 24 mM EDTA pH 8.0
- TE: 10 mM Tris–HCl pH 7.5, 1 mM EDTA
- Commercial sequencing kit (Chapter 6)
- Phenol equilibrated with TE
- Six primers of the form $(5'\text{-dG–dT-}3')_7\text{dX}$ (where X = A, C, or T) and $(5'\text{-dT–dG-}3')_7\text{Y}$ (where Y = A, C, or G)
- Phenol–chloroform (1:1 v/v) equilibrated with TE

Method

1. Purify DNA from the clone of interest by digesting 30 μg DNA (*Protocol 4*) with 15 μg proteinase K overnight at 55°C in 75 mM NaCl, 24 mM EDTA pH 8.0.
2. Extract the DNA twice with phenol and four times with phenol–chloroform. Ethanol precipitate (*Protocol 3*). Re-dissolve the DNA at 1 μg/μl in TE.
3. Using a commercial sequencing kit (Chapter 6), follow the manufacturer's instructions, except that all annealing and enzyme steps must be at 37°C or above.
4. Carry out six sequencing reactions each with a different primer of the form $(5'\text{-dG–dT-}3')_7\text{dX}$ (where X = A, C, or T) and $(5'\text{-dT–dG-}3')_7\text{Y}$ (where Y = A, C, or G). (For lambda clones, the template:primer ratio is 1 μg:100 ng.) Just one of the six oligonucleotides will prime base incorporation from the 3′ end of the C–A strand of the microsatellite and therefore yield readable sequence.
5. Using these sequence data, design the first PCR primer.
6. Use this primer in another sequencing reaction to determine the sequence on the other flank of the microsatellite.

References

1. Riley, J. H., Ogilvie, D., and Anand, R. (1992). In *Techniques for the analysis of complex genomes* (ed. R. Anand) . Academic Press, New York, NY.

2. Sternberg, N. (1995). In *DNA cloning 3: a practical approach* (ed. D. M. Glover), p. 81. IRL Press, Oxford.
3. Kim, U. J., Shizuya, H., de Jong, P. J., Birren, B., and Simon , M. I. (1992). *Nucleic Acids Res.*, **20**, 1083
4. Clarke, L. and Carbon, J. (1976). *Cell*, **9**, 91.
5. Kaiser, K. and Murray, N. E. (1995). In *DNA cloning 1: a practical approach* (ed. D. M. Glover), p. 37. IRL Press, Oxford.
6. Cross, S. H. and Little, P. R. F. (1986). *Gene*, **49**, 9.
7. Sambrook, J., Fritsch, E. F., and Maniatis, T. (1989). In *Molecular cloning: a laboratory manual*, 2nd edn. Cold Spring Harbor Laboratory Press, Cold Spring Harbor, NY.
8. Krieg, P. and Melton, D. A. (1984). *Nucleic Acids Res.*, **12**, 7057.
9. Melton, D. A., Krieg, P. A., Rebagliati, M. R., Maniatis, T., Zinn, K., and Green, M. R. (1984). *Nucleic Acids Res.*, **12**, 7035–56.
10. Coulson, A. and Sulston, J. (1988). In *Genome analysis: a practical approach* (ed. K. E. Davies), p. 21. IRL Press, Oxford.
11. Seed, B., Parker, R. C., and Davidson, N. (1982). *Gene*, **19**, 201.
12. Ivens, A. and Little, P. (1995). *DNA cloning 3: a practical approach* (ed. D. M. Glover), p 1. IRL Press, Oxford.
13. Hughes, A. L. and Nei, M. (1990). *Mol. Biol. Evol.*, **7**, 491.
14. Yuhki, N. and O'Brien, S. J. (1990). *Proc. Natl Acad. Sci. USA*, **87**, 836.
15. Allard, M. W. and Honeycutt, R. L. (1991). *Mol. Biol. Evol.*, **8,** 71.
16. Bonner, T. I., Brenner, D. J., Neufeld, B. R. and Britten, R. J. (1973). *J. Mol. Biol.*, **81**, 123.
17. Bolton, E. T. and McCarthy, B. J. (1962). *Proc. Natl Acad. Sci. USA*, **48**, 1390.
18. Sealey, P. G., Whittaker, P. A., and Southern, E. M. (1985*). Nucleic Acids Res.*, **13**, 1905.
19. Barker, D., Schafer, M., and White, R. (1984). *Cell*, **36**, 131.
20. Orita, M., Iwahana, H., Kanazawa, H., Hayashi, K., and Sekiya, T. (1989). *Proc. Natl Acad. Sci. USA* **86**, 2766.
21. Myers, R. M., Lumelsky, N., Lerman, L. S., and Maniatis, T. (1985). *Nature*, **313**, 495.
22. Cawthon, R. M., Weiss, R., Xu, G., Viskochil, D., Culver, M., Stevens, J., Robertson, M., Dunn, D., Gesteland, R., O'Connell, P., and White, R. (1990), *Cell*, **62**, 193.
23. Jeffreys, A. J. (1985). *Cell*, **18**, 1.
24. Cooper, D. N., Smith, B. A., Cooke, H. J., Niemann, S., and Schidtke, J. (1985). *Human Genet.*, **69**, 201.
25. Litt, M. and Luty, J. A. (1989). *Am. J. Human Genet.*, **44**, 397.
26. Weber, J. L. (1990). *Genomics*, **7**, 524.
27. Sarkar, G., Paynton, C., and Sommer, S. S. (1991). *Nucleic Acids Res.*, **19**, 631.
28. Pulido, J. C. and Duyk, G. M. (1994). In *Current protocols in human genetics* (ed. N. C. Dracopoli, J. L. Haines , B. R. Korf, D.T. Moir, C.C. Morton, C.E. Seidman, J.G. Seidman, and D. R. Smith), Vol 1. Current Protocols, New York, NY.
29. Chung, C. T., Miemela, S. L., and Miller, R. H. (1989). *Proc. Natl Acad. Sci. USA*, **86,** 2172–5.
30. Hanahan, D., Jessee, J., and Bloom, F.R. (1995). In *DNA cloning 1: a practical approach,* (ed. D. M. Glover), pp. 1–35. IRL Press, Oxford.
31. Yuille, M. A. R., Goudie, D. R., Affara, N. A., and Ferguson-Smith, M. A. (1991). *Nucleic Acids Res.*, **19**, 1950.

5

RFLP analysis using heterologous probes

CHARLES F. AQUADRO, WILLIAM A. NOON, DAVID J. BEGUN, and BRYAN N. DANFORTH

1. Introduction

DNA sequence variation is ideally assayed by direct determination of nucleotide sequences representing independent copies of a gene region or other homologous segment of the genome sampled from a natural population. We will refer to such multiple copies of single regions as 'alleles'. An estimate of the level and distribution of sequence polymorphism segregating in a population requires multiple alleles to be sampled and sequenced. The sequences must also be determined with great accuracy, since a 1% error rate in determination of the sequence would generate an apparent level of nucleotide polymorphism higher than that actually seen for many species. The large sampling variances of most of the descriptive statistics of variation require the investigator to sample perhaps 10 or more alleles in order to obtain statistically reasonable estimates. The development of the polymerase chain reaction (PCR) and rapid sequencing methodologies has made such efforts practical for many genes and organisms.

An alternative approach is to use clones or PCR-amplified fragments of gene regions of interest as molecular probes of restriction endonuclease digested nuclear, mitochondrial, or chloroplast DNA from the study species or close relatives. The variants detected are in the form of restriction fragment length polymorphisms (RFLPs). While estimation of sequence variation is possible from fragment data alone, sequence length variation in the region studied can bias estimates of nucleotide substitution. Comparison of maps of the relative order of restriction sites is thus preferable when length variation is a possibility. Genomic DNA is isolated, cut with restriction enzymes, size-fractionated on gels, transferred (Southern blotted) to a filter, and probed with clones from the genomic region of interest. Restriction maps are constructed using single and double digests. Comparison of maps for a sample of gene isolates provides estimates of the type, level, and distribution of DNA sequence polymorphism among the sample of alleles. Many of the initial data obtained for nuclear DNA variation in *Drosophila* were obtained using such restriction map surveys (1,2). The method still has application in the rapid survey of average sequence variation over large regions of the genome or for the scoring of specific polymorphisms, and for the analysis of 'minisatellite' variation (in multilocus DNA 'fingerprinting'; e.g., see refs 3 and 4 and Chapter 9). However, direct DNA sequencing of PCR products has become the method of choice for most studies of DNA sequence variation (Chapter 6) because of the large number of sites surveyed and the additional information provided about the location and nature of the variants (e.g., coding, non-coding, replacement, silent). Additionally, very sensitive indirect methods are now available to complement direct sequencing studies. These methods include enzyme mismatch cleavage (EMC) reviewed in Babon *et al.* (5) and Youil *et al.* (6), as well as single-strand conformation polymorphism (SSCP) and denaturing gradient gel electrophoresis (DGGE) methods which are discussed in detail in Chapter 8.

Many RFLP studies have used restriction endonucleases with six-base recognition sequences (so-called 'six-cutters'). These enzymes cut on average once every 4096 bp, yielding DNA fragments of a size easily resolved by standard agarose gel electrophoresis. In addition, phage clones of genomic regions that are generally used as molecular probes contain 10–20 kb of sequence and often completely span the gene of interest (see Chapter 4). The number of cleavage sites for any one enzyme in the probed region is usually in the range of four to eight, and these sites can readily be mapped relative to one another by comparisons with DNA cut with different enzymes singly and in combination with other enzymes. The use of four to 10 endonucleases with different recognition sequences allows the 'sampling' of sequence variation across the region probed.

The 'six-cutter' approach outlined has the advantage of requiring little in the way of sophisticated molecular tools or techniques in order to screen a large number of individuals rapidly for DNA sequence variation. Filters can be re-probed with clones or fragments representing adjacent or unlinked regions of interest. All sequence length variation over approximately 50–100 bp in length is detectable, and the presence or absence of sites scored across individuals allows the assessment of the extent of linkage disequilibrium among variants along the chromosome. However, only a small proportion of the individual nucleotide sites in a region is actually assayed. Many surveys have used four to six different restriction endonucleases, giving a density of roughly one or two cleavage sites per kilobase studied. Taking into account that one screens for changes at the six nucleotides at each recognition site cleaved, as well as approximately six additional potential changes from sites one nucleotide off from being a restriction site, then it can be estimated that many of the six-cutter surveys are assaying only 12–24 nucleotides per kilobase. Variances on estimates derived from this small proportion of nucleotides are expected to be very large. Nonetheless, more detailed studies of these regions have revealed that the original 'six-cutter' estimates are remarkably accurate as to the *average* level of nucleotide substitution variation across a region. Studies of variation in the level of sequence polymorphism on a fine scale, such as between different regions of a gene, however, require more sophisticated approaches such as analysis with four-base recognition sequence restriction enzymes (so called 'four-cutters') or by direct sequencing.

In this chapter, we provide a summary of the procedures for the analysis of DNA variation using both 'six-cutter' and 'four-cutter' restriction endonucleases. We discuss protocols for the isolation of DNA, restriction enzyme digestion, electrophoresis, DNA transfer to nylon membranes, preparation of radiolabelled and chemiluminescent probes and hybridization to membrane-bound DNA, and interpretation of autoradiograms and data. Methods for the estimation of DNA sequence polymorphism and divergence from the RFLP data obtained are reviewed in Appendices 3 and 5. While our primary focus has been on nuclear genes, the approaches and methods generally apply to the analysis with heterologous probes of variation in mitochondrial and chloroplast DNA, and genomic and plasmid DNA in microbes.

2. DNA isolation

Several methods for isolation of high molecular weight genomic DNA necessary for RFLP analysis are available and have been reviewed in Chapter 2. The method of choice is determined by the amount of starting material, the desired amount and purity of DNA isolated, and the nature of the material from which the DNA is to be extracted. CsCl gradient centrifugation methods for isolating genomic DNA result in very pure DNA with virtually no RNA contamination. However, these methods require a gram or more of starting material, are expensive and time-consuming, and require access to an ultracentrifuge and rotor. 'Miniprep' methods isolate total cellular nucleic acids, and thus are useful for both the analysis of nuclear genes and organelle genes of a large number of samples or from a small amount of starting material. These preps are 'dirtier' and can sometimes result in DNAs that are difficult to cut. The use of spermidine in the digestion of these miniprep DNAs is often recommended as a solution to this problem (see Section 4.1.1). DNA isolated by most miniprep protocols contains significant RNA contamination which can interfere with DNA digestion and with hybridization of the Southern blots. RNA can be removed either by digestion of the entire sample with DNase-free RNase after preparation, or at the same time as restriction endonuclease digestion.

3. DNA concentration determination

Before further analysis, it is important to determine the concentration and condition of the DNA that you have isolated. This can be done in three ways:

by examining UV absorbance with a spectrophotometer, by fluorimetry, or by comparison with DNA standards on agarose gels.

Using a spectrophotometer, the concentration of DNA in a sample can be determined given that a 50 μg/ml solution of double-stranded DNA has an absorbance of 1.0 at 260 nm (see *Protocol 1*). Single-stranded DNA and RNA in concentrations of 40 μg/ml have absorbances of 1.0 at 260 nm. The purity of a DNA preparation can be judged by examining the ratio of absorbances at 260 nm and 280 nm. Pure DNA and RNA have 260 nm:280 nm absorbance ratios of 1.8 and 2.0, respectively; protein or phenol contamination will lower the absorbance ratios and result in reduced accuracy of DNA quantification (and poor restriction digestion).

DNA prepared by miniprep protocols often contains a large amount of RNA and pigments that can cause spuriously high estimation of DNA concentration on a spectrophotometer. For this reason 0.8% agarose gels are useful for assessing both the quantity and the quality of the genomic DNA (is it high molecular weight, or is there substantial shearing or degradation?) and the amount of RNA present (see Chapter 9). The use of a low percentage (0.4%) agarose gel allows separation of nuclear and mitochondrial DNA for vertebrates and most invertebrates, potentially allowing the assessment of the proportion of nuclear versus mitochondrial DNA.

Protocol 1. Spectrophotometric determination of DNA concentration

Equipment and reagents

- UV spectrophotometer
- Quartz cuvettes
- TE: 10 mM Tris–HCl pH 7.5, 1 mM EDTA

Method

1. Turn on the spectrophotometer's UV lamp and, after it has warmed up, set the wavelength at 260 nm.
2. Determine the volume of DNA to use by estimating its concentration. Most genomic DNA preparations yield DNA at a concentration of about 100 μg/ml while plasmid DNA preparations will often result in a DNA concentration of 1 mg/ml. An absorbance (optical density or OD) of 0.05–0.1 is high enough for an accurate assessment. Thus the sample volume generally ranges from 5 to 50 μl.
3. While the machine is warming up, prepare test samples by taking the appropriate volume of each sample and adding TE to a final volume of 500 μl in a microcentrifuge tube or microtitre plate well. Uniform mixing of the sample is essential.
4. Fill one cuvette (the 'zero' or 'blank' cuvette) with 500 μl TE and place

in the spectrophotometer at 260 nm. 'Zero' the machine or note the reading. (The same 'zero' setting should be sufficient for both 260 and 280 nm readings, but check.)

5. Fill the other cuvette (the 'test' cuvette) with 500 μl TE and take readings at both 260 and 280 nm. Any differences between the cuvettes need to be noted so adjustments to the sample readings can be made.
6. Rinse out the test cuvette with sterile water and tamp out on a paper wipe. Fill test cuvette with 500 μl from the test sample. Take absorbance readings at 260 nm, then again at 280 nm.
7. Rinse out cuvette with sterile water, tamp dry, and repeat step **6** for each sample. It is wise to check the 'zero' setting occasionally with the 'blank' cuvette.
8. Taking into account the initial dilution, the original sample concentration is calculated as:

$$\text{DNA concentration } (\mu\text{g/ml}) = \left(\frac{500\ \mu\text{l}}{x\ \mu\text{l}}\right)\left(\frac{50\ \mu\text{g}}{\text{ml}}\right) OD_{260}$$

where x is the volume of DNA used.

4. Restriction endonuclease digestion of DNA

Digestion of DNA with restriction endonucleases that cleave at different recognition sites lies at the heart of RFLP analysis (see Appendix 2). Complete digestion is essential to obtaining interpretable RFLP patterns. Several factors affect endonuclease activity, including the pH, concentration, and type of ions in the buffer, and the temperature of the reaction. Many common six-cutter restriction enzymes have a fairly broad tolerance to buffer conditions. We recommend the use of the enzyme reaction buffer systems and conditions provided by most restriction endonuclease suppliers. The basic reaction protocol is given in *Protocol 2,* with additional comments specific to 'four-cutter' analysis of nuclear genes given in Section 9.

4.1 'Six-cutter' restriction enzyme digestion of DNA

4.1.1 Single enzyme digests

The basic recipe for digestion with a single enzyme is given in *Protocol 2.* For any one digestion reaction, we aim for a final reaction volume of 20 μl because the entire 20 μl reaction plus 4 μl of 6 × gel loading dye conveniently fits in the 1 mm × 5 mm gel-comb wells that we routinely use for 'six-cutter' gels. Reactions can be carried out in individual microcentrifuge tubes. When performing a large number of reactions, the use of microtitre plates is convenient.

4.1.2 Double digests: digestion with two different restriction enzymes

Scoring of small length variants and mapping of restriction sites often require the digestion of a DNA sample with two different restriction enzymes. Often such 'double digests' can be carried out simultaneously if the salt, buffer, and temperature requirements of the two enzymes are similar. Here the basic recipe is simply modified by the addition of an aliquot (1 μl) of each enzyme to the same reaction, but being careful not to let the glycerol concentration exceed 5%. With enzymes having substantially different reaction condition optima, digestions should be done sequentially. Sometimes it is possible simply to digest with an enzyme requiring a low salt concentration first, and then to add more salt. Otherwise, the DNA can be ethanol-precipitated after the first digestion, resuspended in TE, and digested with the second enzyme. In practice, we have rarely found it necessary to do separate reactions as most enzymes have a fairly broad tolerance of conditions under which they will work sufficiently well for complete digestion.

Protocol 2. Digestion of DNA with a single enzyme

Equipment and reagents

- 10 × enzyme reaction buffer
- 50 mM spermidine
- Restriction enzyme
- 6 × Ficoll loading dye (see *Protocol 4*)
- 37 °C heating block or water bath

Method

1. Make up to a final reaction volume of 20 μl the following reagents, adding them in the given order.[a]
   - *x* μl H_2O[b]
   - 2 μl 10 × enzyme reaction buffer
   - 1 μl of 50 mM spermidine (only if DNA preparations are relatively crude and difficult to cut)[c]
   - *y* μl sample DNA[d,e]
   - 1 μl restriction enzyme (*c.* 2–5 units of enzyme per microgram of genomic DNA)[f]

   Mix by flicking the tube repeatedly with your finger (vortexing will result in inactivation of the restriction enzyme) and then centrifuge for a few seconds to bring the reaction mixture to the bottom of the tube.
2. Incubate at 37 °C (or the temperature appropriate for the enzyme, see Appendix 2) in a heating block or water bath for 2–12 h. When digesting 1–2 μg of DNA, 2 h is usually sufficient; however, larger amounts of genomic DNA may take longer.[g]

3. When the digestion is finished, add 4 μl of 6 × Ficoll loading dye to the 20 μl reaction mixture. This loading dye contains EDTA, which will arrest the enzyme action by chelating Mg^{2+}, thus preventing further nuclease activity.

[a] It is important to add the reagents in the order given in step **1** since spermidine will precipitate DNA when added directly to it, and high salt concentrations will inactivate the enzyme.
[b] The amount of water will vary depending on the concentration of the sample DNA.
[c] Do not store reactions on ice or at 4°C before or after digestion if you have included spermidine in your reaction since these conditions may cause the precipitation of the DNA.
[d] For detection of single-copy nuclear genes, the amount of DNA to be added depends on the genome size of the organism. For organisms with a genome size on the order of 10^8 bp (like *Drosophila*) 1.0 μg of genomic DNA is sufficient. Use proportionately more or less depending on comparative genome size. For total genomic DNA from minipreps, 10–100 ng DNA is usually sufficient to detect mitochondrial DNA from insects and vertebrates.
[e] If digesting several samples with the same enzyme, it is useful to make (on ice) a mixture of reaction buffer, H_2O, spermidine if used(but see note *c*), and restriction enzyme, and then to aliquot out the appropriate volume of the mix, before adding sample DNA. Similarly, if you are cutting the same DNA with several different enzymes that can use the same buffer, make a mixture with reaction buffer, H_2O, spermidine if used (but see note *c*), and DNA and, then aliquot to separate tubes and add appropriate enzymes. This is a useful approach for mapping gels containing single and double digests of a single DNA sample.
[f] Most restriction enzymes come in a 50% glycerol solution to prevent denaturation of the enzymes when stored at –20°C. Be sure that the glycerol concentration does not exceed 5% in the final reaction volume; excess glycerol can lead to non-specific cutting or 'star' activity (cutting at additional sites) by many endonucleases.
[g] With CsCl-purified DNA, digestions can be left overnight. Long incubations with DNA from crude preparations may result in degradation of your sample due to contaminating nucleases.

4.2 Concentration of DNA after restriction digestion reaction

Analysis of RFLP variation in single-copy sequences of mammals and birds often requires large amounts of genomic DNA to be run in each lane of a gel (often 8–10 μg). Digestion of this quantity of genomic DNA may require a digestion volume larger than can be loaded in a single well in an agarose gel due to the concentration of the DNA sample. Hence, digestion must be followed by concentration of the DNA. We find that elimination of restriction endonucleases by phenol extraction prior to concentration will often improve the resolution of bands. Digest the genomic DNA by scaling up the 20 μl procedure in *Protocol 2*. For example, if you need to use 50 μl of DNA then make the final reaction volume 100 μl. Use *Protocol 3* to clean up and concentrate the DNA fragments prior to electrophoresis.

Protocol 3. Phenol extraction and ethanol precipitation of restriction digested DNA

Equipment and reagents

- TE (*Protocol 1*)
- Phenol–chloroform–isoamyl alcohol (25:24:1 by vol.)
- Chloroform–isoamyl alcohol (24:1 v/v)
- 7.5 M ammonium acetate
- Cold (–20°C) absolute ethanol
- Cold 70% ethanol
- Vacuum centrifuge

Protocol 3. *Continued*

Method

A. *Removal of protein*

1. Add TE to bring the reaction volume to 400 μl.
2. Add 400 μl phenol–chloroform–isoamyl alcohol (25:24:1) to tube, cap the tube, and vortex to mix well.
3. Spin in a microcentrifuge for 10 min to separate the phases.
4. Remove the aqueous (upper) phase to another tube avoiding any material at the interface.
5. Add 400 μl of chloroform–isoamyl alcohol (24:1 v/v) to the new tube and vortex to mix.
6. Spin in microcentrifuge for 5 min.
7. Remove the aqueous (upper) phase to a new tube, noting the approximate volume.

B. *Precipitation of digested DNA*

1. Add 0.5 vol. of 7.5 M ammonium acetate and mix.
2. Add 2.5 vol. of cold (–20°C) absolute ethanol.
3. Mix well then place on ice or at 4°C for at least 2 h.
4. Spin in a microcentrifuge at 4°C for 30 min. Pipette off the supernatant.
5. Rinse the tube and pellet with 1 ml cold 70% ethanol; spin for 20 min.
6. Pipette off ethanol being careful to avoid the pellet even if you have to leave a few microlitres of ethanol behind.
7. Dry the pellet in a vacuum centrifuge for 20 min.
8. Resuspend pellet in 20 μl TE and store at 4°C.

5. Agarose gel electrophoresis

The size range of fragments that result from 'six-cutter' digestion of most DNAs makes agarose the ideal matrix for size fractionation of the resulting DNA fragments by electrophoresis. Two buffers are commonly used for agarose electrophoresis of DNA fragments. *Table 1* gives some of the considerations governing the choice of the appropriate buffer. Gels of 0.8% agarose are useful for the broadest range of fragment sizes encountered in routine 'six-cutter' analysis, though gel concentrations down to 0.4% are useful for distinguishing among fragments up to 30 kb, and concentrations of 1.5% are useful to increase resolution of small fragments. Fragments of less than approximately 300 bp in length are generally not detectable with standard agarose gels, with fragments often of less than 600 bp not readily detected due both to diffusion and to reduced intensity of hybridization on Southern blots.

Table 1. Comparison of TAE and TBE buffers

	1 × TAE	**1 × TBE**
Fragment size	1–15 kb (0.8% agarose)	2–5 kb (0.8% agarose) 300–2000 bp (1.5% agarose)
Buffer circulation	Essential	Not necessary
Electrophoresis conditions	Run slowly overnight (40–50 V)	Can be run in *c.* 4 h (≤150 V) or at 40–50 V overnight
Application	Best for normal Southerns	Best for sizing fragments of <10 kb

Protocol 4. Agarose gel electrophoresis

Equipment and reagents

- Large submarine horizontal gel rigs including 20 cm × 25 cm gel trays (e.g., Life Technologies Model H4)
- Laboratory tape
- 500 ml Erlenmeyer flask
- Agarose
- 50 × TAE gel and electrode buffer: 242 g Tris base, 57.1 ml glacial acetic acid, 18.6 g Na_2EDTA; distilled H_2O to 1 litre (final pH will be about 8.0)
- Plastic film
- 10 × TBE gel and electrode buffer: 108 g Tris base, 55 g boric acid, 7.44 g Na_2EDTA, distilled H_2O to 1 litre (final pH will be about 8.3)
- Ficoll loading dye for agarose gel electrophoresis (6× stock): 0.25% (w/v) Xylene Cyanol, 0.25% (w/v) Bromophenol Blue, 15 % (w/v) Ficoll (type 400; mol. wt 400 000) in distilled water, 60 mM EDTA pH 8.0
- Size standard DNA

Method

1. For a 20 cm × 25 cm gel bed, add the appropriate amount of agarose to 250 ml of the appropriate electrophoresis buffer (1 × TAE or TBE) in a 500 ml Erlenmeyer flask. For example, to make 250 ml of a 0.8% agarose gel, combine 2 g of agarose with 250 ml 1 × electrophoresis buffer. Loosely cover the top of the flask with plastic film to prevent excessive evaporation, and heat in a microwave with occasional swirling until no agarose particles can be seen (it is important to check the final volume and add water if necessary, being careful to reheat to ensure the solution is well mixed).
2. While the agarose is cooling, seal the ends of the gel bed with laboratory tape and place it on a level bench. If necessary, the bed can be levelled with paper towels. When the agarose has cooled to about 50°C (when the flask is cool enough to hold it comfortably in your bare hand) swirl to mix, pour the gel, insert the appropriate comb, and remove any bubbles.
3. When the gel has completely cooled and solidified, remove the tape and place in the gel box with the buffer. Remember that the DNA will migrate from the origin towards the positive (red) electrode. The final level of buffer should be *c.* 5 mm above the gel. Gently remove the comb.

Protocol 4. *Continued*

4. Prepare the DNA samples for loading. The samples should all be in about the same concentration of buffer. Add 0.2 vol. of 6 × Ficoll loading dye to each sample and mix well. Spin tubes briefly in a microcentrifuge if necessary to collect the sample in the bottom of the tubes. When loading the gel be sure not to poke the bottom or score the sides of the well. Using a 20 μl adjustable micropipette, slowly add the sample to allow it to sink to the bottom of the wells. Load standards of appropriate size and concentration. It is important that the size standard DNA be loaded in a buffer of approximately the same salt concentration as the sample DNAs since variation in salt concentration will affect DNA mobility. *Hin*dIII-digested lambda DNA is a convenient, general-purpose size standard covering a range of sizes from 500 bp to 23 kb. For later UV visualization use 0.25 μg; only 50 ng is necessary if the standard is to be blotted and probed.
5. Connect the power supply and turn on. Again the DNA moves towards the positive electrode. Check the current reading to confirm a complete circuit. If using TAE buffer for runs of more than a few hours, the buffer should be circulated with a pump between the anode and cathode buffer trays to prevent ion depletion of the buffer which results in poor fragment resolution. Wait until the dyes have moved about 1 cm into the gel before turning on the pump, and adjust the flow rate so as not to wash the gel off the gel bed.
6. After electrophoresis has finished, stain in 0.5 ng/ml ethidium bromide with agitation for 20 min (25 μl of a 10 mg/ml ethidium bromide stock solution per 500 ml).[a] **Caution**: ethidium bromide is a powerful mutagen and carcinogen. Avoid contact with skin or other surfaces.
7. Pour off the ethidium bromide stain solution by holding the gel with a gloved hand and pouring the solution into ethidium bromide liquid waste container. Rinse off gel with distilled H_2O and agitate in water for 10–15 min to leach out ethidium bromide not bound to DNA.
8. Pour off the water and slide the gel on to a piece of plastic film. Place the gel on a UV transilluminator, examine the gel to confirm complete digestion of all DNA samples, and photograph. Some gel beds are constructed with UV-transparent plastic, allowing the gel to be placed directly on the transilluminator without removing it from the bed. However, the resulting photograph is often of poor quality.

[a] Rather than staining the DNA in the gel at the end of electrophoresis, some workers prefer to stain the DNA during the electrophoretic run by adding 10 μl of a 10 mg/ml ethidium bromide solution (see Appendix 4) to the electrode buffer (to the anodal tank, since ethidium bromide will migrate towards the cathode through the gel). This is not recommended as the binding of ethidium bromide slightly alters the mobility of DNA and can cause shifts in band mobilities in adjacent lanes if DNA concentrations differ.

5.1 DNA transfer to nylon membranes

DNA fragments are generally transferred from the electrophoresis gel matrix and bound to a membrane for subsequent hybridization to the probe of choice. While hybridization to DNA in dried agarose gels is possible, subsequent re-probing is not possible at present and we find this a serious limitation. Transfer from agarose gels typically used for 'six-cutter' genomic Southern blots is most easily accomplished by capillary transfer, although vacuum and electrophoretic blotting procedures are available. The small size of fragments that result from the use of most 'four-cutter' restriction endonucleases requires the use of polyacrylamide gels for good resolution. The denser gel matrix of acrylamide requires that the transfer be made by electroblotting (but see Chapter 7).

There are a variety of commercially available membranes suitable for DNA transfer. We prefer charged nylon membranes for capillary blots of 'six-cutter' Southern blots, and have found Zetabind nylon membrane—sold by AMF-Cuño in 100 foot rolls (20 cm wide; the width of our gels)—to be a sturdy membrane, capable of tens of rounds of probing, washing, and re-probing, with very low background. Similar membranes include Zeta-Probe (Bio-Rad) and Hybond-N+ (Amersham). We advise you to ask your sales representatives for samples of their products and do your own side-by-side comparison, choosing the membrane that best suits your needs.

For all protocols, the quality of the resulting membrane is dependent upon careful handling of the gel, membrane, and filter paper that comes in contact with the membrane. Gloves should be worn at all time when handling the above components; fingerprints or smudges of dirt or oils on the membrane will result in blotches of high background.

Two capillary transfer protocols are discussed below that are applicable for the analysis of organelle and nuclear DNA using 'six-cutters'. Electrophoretic transfer from polyacrylamide gels for 'four-cutter' blots is discussed in Section 9.

5.2 Blotting 'six-cutter' genomic fragments to nylon membranes

The ammonium acetate transfer method (*Protocol 5A*; ref. 7) is a robust protocol that is useful for virtually all applications. An alternative protocol (*Protocol 5B*) modified from Reed and Mann (8) provides a more rapid transfer and often yields sharper bands and increased signal. If the size of DNA fragments are to be large (>20 kb) then the efficiency of both protocols can be increased by breaking the DNA into smaller fragments. Treatment with a strong acid followed by a base will depurinate the DNA and then cleave the phosphate backbone. One needs to be careful not to over-treat the gel because, if the DNA is broken into very small fragments, it will not hybridize well to the probe.

Protocol 5. DNA transfer from agarose gels

Equipment and reagents

- Whatman 3MM paper (18 in. × 24in.); cut to 8 in. × 15.5 in. for bottom layer with wicks on ends and to 8 in. × 9 in. for top layer[a]
- Zetabind nylon membrane (AMF-Cuño), or similar positively charged nylon membrane
- Plastic box made from 1/4 in. thick plastic: outer dimensions of 11 in. × 10 in. × 2 in.
- Transfer bridge made from 1/4 in. thick plastic: top dimensions 9.5 in. × 8.5 in.; leg dimensions 1.75 × 9.5 in.
- Glass rod
- Razor blade
- 1 M ammonium acetate/0.04 M NaOH
- 0.4 M NaOH
- 0.25 M HCl
- 1.5 M NaCl, 0.5 M NaOH
- Plastic wrap
- Filter paper
- Paper towels
- 20 cm × 20 cm Perspex sheet
- 20 × standard saline citrate (SSC): 3 M NaCl, 300 mM sodium citrate (adjusted to pH 7.0)
- Heat lamp
- Oven at 80°C

A. *Ammonium acetate transfer method*

1. Set up the box, bridge, and the first three layers of filter paper as in *Figure 1*. Wet the filter paper with 1 M ammonium acetate/0.04 M NaOH. Remove bubbles from the underside of the paper with a glass rod. Leave several millilitres of liquid on top of the paper.
2. After electrophoresis, cut through the centre of the wells with a razor blade and discard the gel behind the lanes. If you wish to perform an acid depurination (optional), go to step **3**; otherwise proceed to step **4**.
3. This is the acid depurination step. Place the gel (on bed) in a tray containing *c.* 250 ml of 0.25 M HCl and agitate the gel by hand (wear gloves) until it rides up upon a layer of acid. Put it on a shaker with mild agitation at room temperature just until the Bromophenol Blue dye in the gel (from the loading buffer) turns yellow (*c.* 3–5 min). Do not over-treat. Pour off the acid and rinse the gel in distilled H_2O.
4. Soak the gel (loosened from but still in the gel bed) in a tray containing 1.5 M NaCl, 0.5 M NaOH twice for 15 min each.
5. Soak gel in 1 M ammonium acetate/0.04 M NaOH twice for 15 min each.
6. Carefully flip the gel over so the smooth bottom of the gel is facing up and slide it on to filter paper wicks on the transfer bridge (an old piece of 8 in. × 10 in. X-ray film is useful as a support for these gel manipulations).
7. Use a glass rod or 5 ml pipette to squeeze out bubbles, taking care not to stretch or distort the gel.
8. Have a nylon membrane cut (7.75 in. × 8 in.) and wetted in distilled H_2O.

9. Place the filter on the gel with the label side downwards. Be sure to put it on as straight as possible (moving the membrane around on the gel may create 'shadow' bands).
10. Repeat step **7**.
11. Place plastic wrap over the top and cut out section just above filter.
12. Add filter paper wetted with transfer solution (step **1**), squeeze out air bubbles, and then add opened paper towels. Pile on two stacks of closed towels, cover with a 20 cm × 20 cm Perspex sheet, and weigh down with a 1 litre bottle of water to ensure good contact between the gel, membrane, and wicking paper.
13. Let the transfer proceed. This will take 12–18 h (as little as 4 h with acid depurinaton (step **3**)). Add more ammonium acetate/0.04 M NaOH solution as needed (you will need about 2 litres per transfer including that used in soaking the gel (step **5**)).
14. Remove wet towels and filter paper from transfer set-up and dispose of in plastic bag.
15. Lift nylon membrane off gel (wearing gloves) and let it soak in a tray containing 200 ml 2 × SSC for several minutes to neutralize the transfer solution and to remove any agarose particles stuck to the filter.
16. Blot excess moisture from the membrane, place it on a clean piece of filter paper, and dry under a heat lamp for 30 min.
17. Place dry filter in a folded scrap of filter paper to protect it, and bake dry in an oven for 2 h at 80°C. Carefully remove filter packet (the filter can be brittle at this point). The filter can be stored dry for several months or used immediately.

B. *Alkaline transfer method*

1. Perform step **1** in *Protocol 5A*, but using 0.4 M NaOH as the transfer solution instead of ammonium acetate/NaOH.
2. Perform step **2** of *Protocol 5A*. If acid depurination is desired, perform the (optional) step **3** of *Protocol 5A*.
3. Perform steps **6–12** of *Protocol 5A*.
4. Let the transfer proceed. For the alkaline transfer method with acid depurination, this will take 4–12 h and use 300–500 ml of 0.4 M NaOH. Without acid depurination, allow DNA to transfer for 12–18 h.
5. Perform steps **14–17** of *Protocol 5A*.

[a] Dimensions given are appropriate for 8 in. × 8 in. gels.

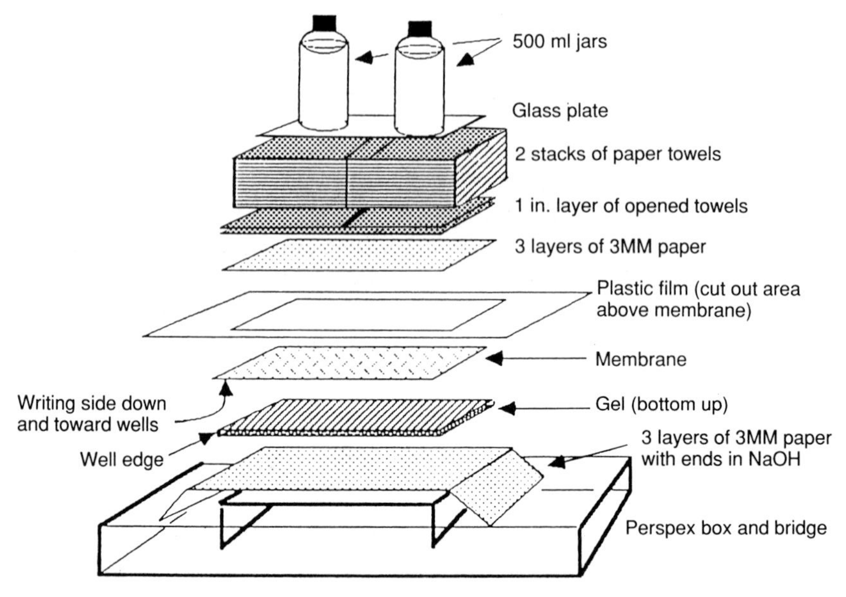

Figure 1. Southern blotting .

6. Probes

6.1 Choice of probes

The 10–20 kb phage clones comprising chromosomal 'walks' or isolated in genomic library screenings are ideal for 'six-cutter' surveys in *Drosophila*. The presence of repetitive sequences in many such clones from mammals, for example, require the use of smaller, unique sequence clones. Plasmid subclones of 2–4 kb are ideal as probes for 'four-cutter' blots. The use of clones from one species to probe homologous regions in related taxa becomes increasingly difficult as sequence similarity decreases below 90–95%, in part due to sequence rearrangements and lower hybridization signal. The analysis of more distantly related species often necessitates the isolation of species-specific clones for use as probes or to sequence directly (see Chapter 4). While this is a relatively straightforward process now, particularly with the widespread commercial availability of the necessary reagents and biologicals, differences in gene and genome organization can mean that substantial characterization of the new species clones is necessary before one can proceed with the population genetics work. The α-amylase locus in *Drosophila pseudoobscura* is an instructive example. While this locus is duplicated in *Drosophila melanogaster*, previous allozyme work suggested it was a simple, single-gene system in *D. pseudoobscura*. However, cloning and characterization of this locus (9,10) revealed a multigene family with one highly expressed gene, one weakly expressed gene, and a pseudogene. These three genes are located adjacent to each other and have extremely similar DNA sequences, making it difficult to interpret genomic Southern blots until the gene family organization was resolved.

Studies of other genes (e.g., *Adh* and *Xdh* in *D. pseudoobscura*; refs 11 and 12) have not encountered these problems, but we raise the preceding example as a cautionary illustration of the type of problem that may occur. The extent to which similar problems arise with other species must be determined by trial and error. Of course, for a clear understanding of the reasons for the distribution of sequence variation seen in any species, a careful characterization of the molecular and gene organization of the region of interest in that species is essential.

6.2 Radiolabelling of probes for hybridization to Southern blots

Detection of the DNA fragments corresponding to the specific gene or region of interest on a Southern blot relies on hybridization of the complementary, membrane bound sequences to a single-stranded DNA or RNA probe molecule appropriately tagged with either radioactive material or some other readily detected material. The DNA to be labelled is often a phage or plasmid clone of the gene or region of interest. Occasionally, small restriction fragments from a clone or a PCR-amplified segment are useful as probes to assist in the mapping of sites or variants, or because some portion of the vector or cloned sequence hybridizes to repeated sequences in the genome, making interpretation difficult. A rapid method for isolating small restriction fragments from low melting point (LMP) agarose for direct use in the random-priming labelling reaction is given as *Protocol 6*.

Labelling of probe with radioactivity is usually accomplished by incorporating ^{32}P-labelled deoxynucleotide triphosphate ([^{32}P]dNTP) into a newly synthesized DNA either by a random-priming reaction (13) or by nick translation (14). The random-priming reaction is simple and robust and leads to generally high specific activity acceptable for most applications; it is presented in *Protocol 7*. Nick translation yields lower specific activity probes but is less dependent on probe DNA purity and requires fewer specialty reagents (see *Protocol 8*). [^{35}S]dNTPs are typically not used because of the weaker signal and the need to dry the hybridized filter for autoradiography, which prevents stripping off the probe for re-use of the filter. High specific activity RNA probes can also be produced using cloning vectors containing the promoters for phage RNA polymerases (15).

For those working in an environment where radiolabelled probes are unavailable or not desired, several non-radioactive methods of detection are available, including those using precipitated dyes or chemiluminescent compounds. Protocols for non-radioactive probings are described in Section 8.

Protocol 6. Isolation of fragments in low melting point (LMP) agarose[a] for labelling

Equipment and reagents

- Low melting point (LMP) agarose
- TAE (*Protocol 4*)
- Gel apparatus
- Restriction enzyme(s) and buffer
- Loading buffer (*Protocol 4*)
- Size standard DNA
- Ethidium bromide
- UV transilluminator
- Razor blade
- 95°C heating block or boiling water bath

Method

1. Pour a 1% LMP agarose gel in TAE (*Protocol 4*; remember that LMP melts at a 65°C compared with 95°C for regular agarose).
2. Digest DNA with appropriate restriction enzyme(s). Load samples and appropriate size standard. Run at about 40 V for 2–3 h or until there is sufficient separation.
3. Stain gel with ethidium bromide, and visualize DNA fragments on a UV transilluminator.
4. Cut the appropriate bands out of the gel with a razor blade and transfer to a 1.5 ml microcentrifuge tube.
5. Add sterile water in the proportion of 3 ml per 1 g of gel.
6. Heat in a 95°C heating block or boiling water bath for 10 min, then cool to 37°C.
7. Use an aliquot corresponding to approximately 100 ng for the random priming reaction without further purification, or store frozen at –20°C (if stored frozen, repeat step **6** before use).

[a] An alternative is to use regular agarose and gel-purify fragments with powdered glass or similar procedure (e.g., GeneClean, from Bio 101 Inc., see Chapter 2)

Protocol 7. Probe-labelling with ^{32}P: random priming of DNA

Equipment and reagents[a]

- Probe DNA
- Solution A: 1 ml 1.25 M Tris–HCl pH 8.0, 0.125 M $MgCl_2$, 18 μl β-mercaptoethanol, 5 μl each of 0.1 M dATP, dGTP, and dTTP
- Solution B: 2 M Hepes titrated to pH 6.6 with 5 M NaOH
- Solution C: random hexadeoxynucleotides resuspended in TE to a concentration of 90 OD units/ml
- Bovine serum albumin (BSA), 10 mg/ml in distilled H_2O
- 5 × oligolabelling buffer made by adding solutions A, B, and C in a ratio of 100:250:150
- Klenow fragment of *Escherichia coli* DNA polymerase I (2 units/μl in 50% glycerol)
- [α-^{32}P]dCTP, 3000 Ci/mmol; aqueous solution, 10 mCi/ml
- 2 × stop buffer: 500 μl 0.2 M EDTA; 450 μl sterile water, 2.5 mg Bromophenol Blue; 25 mg Blue Dextran, 10 μl 10% sodium dodecyl sulfate (SDS)

- 3 M sodium acetate, pH 5.0
- Absolute ethanol
- Cold 70% ethanol
- TE
- 95°C heating block or boiling water bath
- 37°C heating block

Method

1. In a 1.5 ml microcentrifuge tube, add probe DNA and sterile double-distilled water to a final volume of 32 μl (25–100 ng probe; linear DNA is reported to label better than closed-circular or supercoiled DNA).
2. Denature the DNA by heating the tube for 10 min at 95°C (in a heating block or boiling water bath) and subsequently cooling on ice for at least 1 min. Spin briefly in a microcentrifuge to collect condensate.
3. Add 10 μl of 5 × oligolabelling buffer and 2 μl BSA.
4. Behind appropriate shielding, add 5 μl (50 μCi) of [α-^{32}P]dCTP.
5. Add 2 μl of Klenow, cap the tube, and mix by finger-flicking (final volume of reaction is 50 μl).
6. Incubate the tube in a 37°C heating block for 1–2 h (behind appropriate shielding).
7. After incubating for 1–2 h, check the radioactive incorporation (*Protocol 9*). If incorporation is >65%, then proceed to the next step. Otherwise incubate for longer or start again after cleaning up template DNA.
8. The labelled DNA now needs to be separated from the unincorporated nucleotides. This can be done by spin column chromatography (step **9**) or by ethanol precipitation (step **10**). In our hands spin columns give better separation and thus lower background on autoradiograms.
9. Add 50 μl of 2 × stop buffer and clean up probe by spin column chromatography (*Protocol 10*). Go to step **11**.
10. Add 5 μl 3 M sodium acetate pH 5.0. Mix then add 125 μl absolute ethanol. Incubate on ice for at least an hour. Spin in a microcentrifuge for 10 min. Draw off the supernatant then rinse the DNA pellet with 1 ml cold 70% ethanol. Re-spin for 5 min and remove the supernatant. Resuspend the DNA pellet in 50 μl TE.
11. Boil the probe for 10 min. Cool on ice for 5 min. The labelled probe is now ready for hybridization.

[a] Note that commercial kits are reasonably cost-effective and reliable (e.g. Boehringer Mannheim Random Primed DNA Labelling kit, catalogue no. 1 004 760).

Protocol 8. Probe-labelling with ^{32}P: nick translation

Equipment and reagents

- Probe DNA
- 10 × nick translation buffer: 0.2 mM each dATP, dGTP, and dTTP, 500 mM Tris–HCl pH 7.8, 50 mM $MgCl_2$, 100 mM β-mercaptoethanol, 100 μg/ml BSA
- 2 × stop buffer (*Protocol 7*)
- DNA polymerase I/DNase I mix for nick translation (e.g. Life Technologies catalogue no. 18162-016)
- [α-^{32}P]dCTP, 3000 Ci/mmol; aqueous solution, 10 mCi/ml
- 15°C heating block

Method

1. In a 1.5 ml microcentrifuge tube, add probe DNA (use approximately 1 μg), 5 μl 10 × nick translation buffer, and sterile double distilled H_2O to a final volume of 40 μl.
2. Behind appropriate shielding, add 5 μl (50 μCi) of [α-^{32}P]dCTP and 5 μl of DNA polymerase I/DNase I mix. Cap and mix by finger-flicking (final volume of reaction is 50 μl).
3. Incubate the tube in a 15°C heating block for 60 min (behind appropriate shielding).
4. After incubating, check the radioactive incorporation (*Protocol 9*).
5. Continue with *Protocol 7*, step **8** .

Protocol 9. Monitoring ^{32}P-incorporation

Reagents

- 25 mg/ml stock of calf thymus carrier DNA, or *E. coli* or yeast tRNA
- 5% trichloroacetic acid (TCA)

Method

1. To a microcentrifuge tube containing 96 μl of distilled H_2O, add 4 μl of a 25 mg/ml tRNA stock (put on ice and have ready).
2. Use a 200 μl pipette tip and *just touch it* into the reaction mixture to be assayed. Place the reaction mixture on the pipette tip into the microcentrifuge tube. Use a hand monitor to verify that radioactivity has been transferred (10 000 c.p.m. is sufficient).
3. Add 300 μl of 5% TCA. First mix (vortex) and then spin 2 min in a microcentrifuge. Transfer the supernatant into another microcentrifuge tube and save.
4. Rinse the pellet with 300 μl of 5% TCA. Rotate the tube once or twice (do not vortex). Spin for 2 min. Pool the second supernatant with the first. Add 700 μl of distilled H_2O to the pellet.[a]
5. Use either a hand monitor or a scintillation counter (the latter is more

accurate) to measure the radioactivity in the tube with the pellet and the tube with the supernatant .

6. Calculate percentage incorporation as follows:

$$\frac{\text{c.p.m. in pellet}}{(\text{c.p.m. in pellet}) + (\text{c.p.m. in supernatant})} \times 100\% = \%\ \text{incorporation.}$$

[a] Step **4** can be eliminated with only a modest decrease in accuracy.

Protocol 10. Spin column protocol for separating radiolabelled probe DNA from unincorporated nucleotides after random priming

Equipment and reagents

- 10 mm × 75 mm glass culture tube
- Glass wool
- Long Pasteur pipette
- Suspension of Sephadex G-50 beads in TE
- TE
- 2 × stop buffer (*Protocol 7)*

Method

1. Insert a 1 ml plastic pipette tip (for adjustable micropipette) into a 10 mm × 75 mm glass culture tube (see *Figure 2*).
2. Put a small piece of glass wool into the bottom of tip, using a long Pasteur pipette to push it into place.
3. Fill the plugged tip with a suspension of Sephadex G-50 beads in TE.
4. Spin at full speed (5125 × *g*) in a clinical centrifuge for 3 min, timed from the switch being on to the switch being off. The tube and tip fit into adaptors for the clinical centrifuge. It is important to spin the columns consistently for the same time each time it is done.
5. Pour off TE from tube. The Sephadex will look very dry.
6. Fill the tip up again with Sephadex in TE, spin again as before, and discard surplus TE from tube. Ideally, you want 1 ml of packed Sephadex in the tip.
7. Add 200 μl of TE and spin for 3 min. Repeat three times to rinse and equilibrate.
8. Add labelling reaction (that has had an equal volume of 2 × stop buffer (*Protocol 7)* added) to the top of the Sephadex and spin 3 min as before.[a] The unincorporated nucleotide will be in the tip with the Sephadex. The probe DNA will be in the bottom of the glass tube.[b]

[a] These spin columns work well for a range of 50–200 μl of reaction volume.
[b] The light blue dye (Dextran Blue) marks the labelled probe DNA and should pass through the column and collect in the tube, while the dark blue dye (Bromophenol Blue) marks the unincorporated nucleotides and should remain in the tip in the Sephadex.

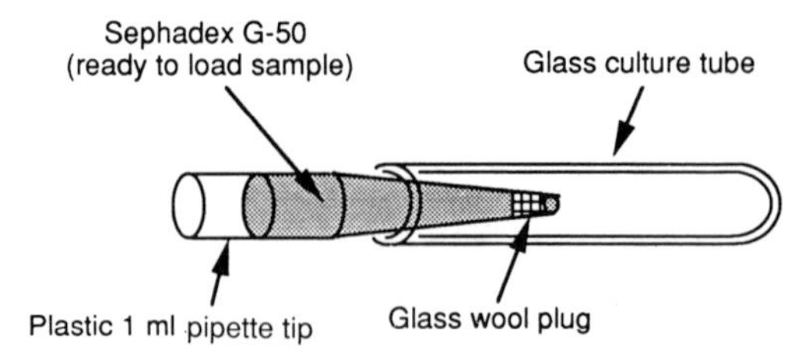

Figure 2. Spin column for purifying probes.

7. Hybridization of radiolabelled probes to genomic Southern blots

Labelled probes are denatured and hybridized to their complementary, single-stranded sequences which are bound to the Southern blot membrane. The stringency level determines the amount of mismatch tolerated between the probe and the target sequence while remaining bound. The primary factors determining stringency are salt concentration and temperature, though formamide is included in some protocols to reduce the temperatures needed for a given level of stringency (15).

There are a number of protocols for Southern blot hybridization. We have provided one that in our hands has proven to be remarkably versatile, simple, and robust. An important component is the use of SDS to block non-specific binding of probe DNA directly to the membrane which can be a major source of 'background' problems in the resulting autoradiographs. The pre-wash step is particularly important in reducing non-specific labelling. Pre-heating all washes and solutions before adding them to the membranes ensures that the appropriate temperatures are achieved.

The choice of container for hybridization depends on what equipment is available and the number of filters to be probed at one time. Heat-sealable plastic bags are convenient for probing only a few filters at a time and require a minimum of special equipment. Plastic boxes are often used to hybridize a large number of filters at a time but require a large volume of hybridization solution (and therefore labelled probe). We have found that doing the hybridization in either plastic tubes (i.d. *c.* 2.5 cm) for three or four filters or larger plastic canisters (i.d. *c.* 10 cm) for up to 10 filters is extremely convenient and effective. The acrylic plastic is sufficiently thick to shield the user from the radioactive probe and the addition or removal of the hybridization fluid can be done neatly. The tubes and canisters need to be rotated during the hybridization to distribute the small volume of hybridization fluid evenly over the filters. We use a tissue culture roller apparatus placed in an incubator. The following protocol is designed for these conditions. Commercially available hybridization ovens with rotators and screw-capped bottles are convenient and available from several vendors (e.g., Hybaid and Biometra).

Protocol 11. SDS/phosphate hybridization

A. *Pre-washing the filter*

Reagents

- 1 × SSC
- Pre-wash solution: 0.1 × SSC, 0.5% SDS (the SDS should be added last)
- Incubator at 65°C
- Hybridization tubes or heat-sealable bags
- 1.0 M NaP_i pH 7.2[a]
- 0.2 M EDTA, pH 8.0
- 20% SDS
- BSA (crystalline, Sigma A-2153, Fraction V)
- Hybridization oven at 60°C

Method

1. If the membrane is dry, wet it by floating it on 1 × SSC in a tray.
2. Pre-heat the pre-wash solution to 65°C in a microwave, pour it over membranes in a clean tray, cover, and incubate with shaking at 65°C for 1 h (make sure that filters are not stuck together). 300 ml of pre-wash solution is sufficient for up to eight filters.
3. Lift the filters from the liquid, allow to drain briefly, and put into hybridization tubes or heat-sealable bags (when putting two filters in one bag, ensure that the DNA side is facing outwards). Drain off or squeeze out all remaining pre-wash by rolling a pipette along the bag.

B. *Pre-hybridization of the filter*

1. While the filters are in pre-wash, prepare the pre-hybridization/hybridization solution. The amount required depends on the number of filters to be labelled in the tube or bag. The following recipe makes 10 ml of solution, which is sufficient for a standard-size (20 cm × 25 cm) filter in a heat-sealable bag (scale up the recipe if using hybridization tubes: 30 ml is sufficient for five filters in a hybridization tube). Make up the following solution, adding the SDS last:
   - 5.25 ml 1.0 M NaP_i pH 7.2
   - 50 μl 0.2 M EDTA pH 8.0
   - 0.1 g BSA
   - 3.5 ml 20% SDS

   Make up to 10 ml with sterile H_2O. Heat to 60°C with mixing to dissolve BSA and to prepare for filter.
2. For hybridization in a canister, pour in the pre-hybridization/hybridization solution and replace lid. Try to roll out large bubbles in between the filters.
3. Place canister in incubator on rollers. Incubator should be set at 60°C and rollers turned on to medium pace (5–10 r.p.m.). Pre-hybridize for 6–8 h or overnight.

Protocol 11. *Continued*

C. *Hybridization of filter to radiolabelled probe*

1. Take boiled, radiolabelled probe from either labelling method (*Protocol 9* or *10*) and add probe to pre-hybridization/hybridization solution in canister (do *not* pour out pre-hybridization/hybridization solution).
2. Replace lid on canister and return it to the incubator roller. Hybridize for 24–48 h at 60°C.

D. *Washes*

Reagents

- Wash solution: 40 mM NaP_i,[a] 1 mM EDTA, 1% SDS (add the SDS last)
- 2 × SSC

Method

1. To wash off the excess pre-hybridization/hybridization solution, put the filters in wash solution (enough to allow filters to float) and shake gently at room temperature for 5 min. Pour off wash into radioactive liquid waste. Repeat wash.
2. Perform a second wash to remove probe weakly annealed to regions of low sequence similarity.
   (a) For DNA probes that are $\leq$5% divergent from the DNA samples, pour on wash solution heated to 55°C in a microwave and shake at 55°C for 30 min. Pour off wash into radioactive liquid waste. Go to *Protocol 11E*.
   (b) For DNA probes that are $>$5% divergent from the DNA samples, use a less stringent wash: wash the filters at a lower temperature (30–35°C) and repeat the wash at higher temperatures until only single-copy bands appear on an autoradiogram. A crude assay for this is to use a hand monitor to indicate when the signal drops to less than *c.* 150 c.p.m.

E. *Putting on film*

1. Move filters to 2 × SSC at room temperature for a few minutes (can be longer) to wash off second wash. Wrap very carefully in cling-film and squeeze out all extra fluid with a glass rod or pipette.
2. Place wrapped filter in autoradiography cassette with the DNA side upwards. In a dark room, add autoradiography film (and intensifying screens if needed). With cardboard film holders it is necessary to sandwich them individually between two Perspex sheets held with four binder's clamps.
3. Place in a plastic bag (to protect from frost and condensation when thawed) and put in –70°C freezer. Freezing of the filter is recom-

mended even if screens are not used because it helps prevent the filter from drying out—if the filter dries out it cannot be stripped and re-probed.

F. *Stripping probe off filters for re-probing*

Reagents

- Neutralizing solution: 0.1 × SSC, 0.5% SDS, 0.2 M Tris–HCl pH 7.5 (add the SDS last)
- 0.4 M NaOH

Method

1. To wash off the probe, place filter(s) (up to eight) in 300 ml of 0.4 M NaOH heated to 42°C. Cover and put on shaker in 42°C incubator for 30 min. Pour off NaOH.
2. Pour 300 ml of neutralizing solution at room temperature onto the filter. Cover and shake at room temperature for 20–30 min. Pour off neutralizing solution and pre-wash membrane as in *Protocol 11A*.
3. After the pre-wash, the filters can be stored indefinitely in 0.5 × TBE (*Protocol 4)* at 4°C.

[a] For 1 litre of 1 M NaP_i, add to distilled H_2O 134 g $Na_2HPO_4 \cdot 7H_2O$ and *c.* 4 ml H_3PO_4 to bring to pH 7.2. Make up to a volume of 1 litre with distilled H_2O.

8. Non-radioactive methods for RFLP analysis

Non-radioactive methods for nucleic acid detection are now widely available and offer safe and rapid techniques for doing RFLP analysis. Nucleic acids may be labelled by a number of different methods, including labelling with biotin, dinitrophenyl, digoxigenin, and fluorochromes. A review of the benefits and drawbacks of the different approaches is beyond the scope of this chapter, but we refer the reader to Hopman *et al.* (16) and Hughes *et al.* (17) for detailed reviews. For the applications described herein, we selected digoxigenin (DIG)-based labelling methods for several reasons:

(a) They give results (i.e., a pattern of bands on X-ray film) comparable to those obtained with radioisotopes. Hence, it should be possible for someone with experience using radioisotopes to switch easily between the two methods.
(b) DIG-labelled probes may be stripped off membranes, allowing re-probing.
(c) DIG-based labelling systems offer high levels of sensitivity with low levels of background.

While there are numerous commercial sources of DIG-based labelling and detection kits, we have had extensive experience with only one system: the

Boehringer Mannheim Genius System (3,4,18). The protocols given below are modified from the Boehringer Mannheim *Genius System Users' Guide to Filter Hybridization* (version 3.0), and are similar to those for other DIG-based kits. Our reference to the Genius System should not be taken as an exclusive endorsement of that system.

All protocols described above, from DNA extraction through agarose gel electrophoresis and Southern transfer (*Protocol 5*), can be used with non-radioactive methods without modification. For Southern transfer, we recommend using the ammonium acetate transfer method (*Protocol 5A*) rather than the alkaline transfer method (*Protocol 5B*). Ammonium acetate transfers give significantly lower background when used with DIG-labelled probes in our experience.

8.1 Non-radioactive probe labelling

As with radioactive methods, probe labelling is an extremely important step in using chemiluminescent detection systems. We have had the best success using the PCR labelling protocol (*Protocol 12*). This protocol allows labelling of highly specific regions of DNA incorporated into a plasmid or phage vector, and is more sensitive than random primed labelling (19,20). The insert should be small enough for succcessful PCR amplification (*c.* 500–2000 bp in length). Random primed labelling and 3′ tailing are described in *Protocols 13* and *14*, respectively. End-labelling and tailing are effective methods for labelling short oligonucleotide probes. End-labelling involves attaching a single DIG-dUTP to the 3′ end of the oligonucleotide, while tailing involves adding several DIG-dUTPs to the 3′ end. We have successfully used oligonucleotides labelled by 3′ tailing to screen plasmid libraries for simple repeat sequences. *Protocol 15* provides a method for estimating the yield of your labelled probe. Once labelled, DIG-labelled probes may be stored indefinitely at –80°C (a considerable advantage over radiolabelled probes).

8.1.1 DIG-labelling protocols

Protocol 12. Probe labelling with PCR

Equipment and reagents

- 10 × PCR buffer: 100 mM Tris–HCl, 500 mM KCl, pH 8.3
- 25 mM $MgCl_2$
- Unlabelled dNTP mixture: 1 mM each of dATP, dCTP, dGTP, and dTTP
- DIG-labelled dNTP mix: 1 mM dATP, 1 mM dCTP, 1 mM dGTP, 0.65 mM dTTP, 0.35 mM DIG-dUTP; pH 6.5 (e.g., Boehringer Mannheim Digoxigenin DNA labelling mixture; catalogue no. 1277 0650)
- Template DNA
- Forward primer (10 μM)
- Reverse primer (10 μM)
- *Taq* DNA polymerase (5 units/μl)
- Sterile H_2O
- Mineral oil
- Thermal cycler

Method

1. Mix the following reaction in a 0.5 ml microcentrifuge tube, adding ingredients in the order listed (total volume 100 μl)
   - 59 μl sterile H_2O
   - 10 μl 10 × PCR buffer
   - 10 μl 25 mM $MgCl_2$ (giving 2.5 mM $MgCl_2$ in final reaction)
   - 10 μl dNTP mix[a]
   - 1 μl template DNA (*c.* 1 ng)
   - 5 μl forward primer
   - 5 μl reverse primer
   - 0.4 μl *Taq* DNA polymerase (2 units)
2. Add 40 μl mineral oil to each tube and place the tubes in a thermal cycler set to the following conditions: 94°C, 45 sec; 55°C, 45 sec; 72°C, 60 sec with an initial 3 min denaturation (94°C) and a total of 30 cycles. (These are 'generic' conditions which should be optimized for each template–primer combination.)
3. Following completion of the reaction, transfer the reaction mix from each tube to a new, sterile tube by gently pipetting the aqueous phase from beneath the mineral oil overlay.
4. Store at –80°C until use.[b]

[a] It is best to run this reaction first with an unlabelled dNTP mixture. Once you have selected optimal thermal cycler conditions, run the reaction again with the DIG-labelled dNTP mix.
[b] When you run the product from these reactions out on an agarose minigel you will see that the labelled product migrates more slowly than its unlabelled homologue. This is a useful assay of the labelling reaction and results from the large digoxigenin molecule that is attached to the probe. Use a size ladder to confirm that you have amplified the region of interest.

Protocol 13. Random primed labelling

Equipment and reagents

- DNA template[a]
- 10 × hexanucleotide mix: 62.5 A_{260} units/ml random hexanucleotides, 500 mM Tris–HCl, 100 mM $MgCl_2$, 1 mM dithioerythritol, 2 mg/ml BSA, pH 7.2 (e.g., Boehringer Mannheim catalogue no. 1277 081)
- Boiling water bath
- 10 × dNTP mix: 1 mM dATP, 1 mM dCTP, 1 mM dGTP, 0.65 mM dTTP, 0.35 mM DIG-dUTP; pH 6.5 (e.g., Boehringer Mannheim Digoxigenin DNA labelling mixture; catalogue no. 1277 0650)
- DNA polymerase I (Klenow enzyme, large fragment, labelling grade from *E. coli*, 2 units/μl)

Method

1. Heat-denature the DNA template by boiling for 15 min.
2. Mix the following reaction in a 1.5 ml microcentrifuge tube, adding ingredients in the order listed (total volume 100 μl)
   - 0.5–1 μg DNA template

Protocol 13. *Continued*

- 10 μl 10 × hexanucleotide mix
- 10 μl 10 × dNTP mix
- H_2O to make volume up to 95 μl
- 5 μl DNA polymerase I

3. Incubate reaction overnight at 37 °C
4. Store labelled probe at –80 °C until use.[b]

[a]The Genius manual recommends linearizing the plasmid or phage vector with a restriction enzyme digest. We have not found this step to be essential and, since it may lead to lower probe yields, we do not recommend it. It may be necessary to clean the DNA with a phenol extraction as described in *Protocol 3*. If so, resuspend the DNA in 20 μl and start the procedure below with that volume. Reaction volumes may range from 20 to 300 μl.
[b]While Boehringer Mannheim originally recommended ethanol precipitating the probe to remove unincorporated nucleotides and enzyme, they now recommend using the probe directly from the reaction tube. We have found that this gives good results, as unincorporated DIG-dUTP does not chemiluminesce, and hence does not lead to increased background in the way that unincorporated ^{32}P does. The labelling reaction should result in a 3- to 4-fold amplification of the DNA.

Protocol 14. Probe labelling by 3′ tailing

Reagents

- 5 × reaction buffer: 1 M potassium cacodylate, 125 mM Tris–HCl, 1.25 mg/ml BSA; pH 6.6
- $CoCl_2$ solution (25 mM)
- Oligonucleotide to be labelled (100 pmol)
- DIG-dUTP: 1 mM digoxigenin-11-ddUTP in distilled water (Boehringer Mannheim catalogue no. 1363-905)
- dATP, 10 mM in Tris buffer pH 7.5)
- Terminal transferase (50 units/μl)

Method

1. Mix the following reaction in a 1.5 ml microcentrifuge tube, adding ingredients in the order listed (total volume 20 μl)[a]
   - 4 μl 5 × reaction buffer
   - 4 μl $CoCl_2$ solution
   - 1 μl DIG-dUTP
   - 100 pmol oligonucleotide
   - 1 μl dATP
   - 1 μl terminal transferase
   - H_2O to 20 μl
2. Incubate the reaction at 37 °C for 15 min.
3. Place reaction tube on ice.
4. Store labelled probe at –80 °C until use.

[a]The Genius manual recommends using reaction volumes of less than 40 μl . Larger volumes will result in inefficient labelling.

Protocol 15. Estimating yield of DIG-labelled probe using dot blots

Equipment and reagents[a]

- NBT solution: 75 mg/ml Nitroblue Tetrazolium salt in 70% (v/v) dimethylformamide
- 50 mg/ml 5-bromo-4-chloro-3-indolyl phosphate (X-phosphate), toluidinium salt in 100% dimethylformamide
- Labelled control DNA: DIG-labelled pBR328 DNA, 5 μg/ml (e.g., Boehringer Mannheim catalogue no. 1585 738)
- Genius buffer 1: in the original User's Guide, this was composed of 150 mM NaCl, 100 mM Tris–HCl pH 7.5, while in version 3 of the User's Guide, Boehringer Mannheim recommends using a maleic acid buffer (100 mM maleic acid, 150 mM NaCl; pH 7.5). Using the maleic acid buffer allows the same buffer to be used with both DNA and RNA.
- Genius buffer 2: 2% blocking reagent (Boehringer Mannheim catalogue no. 1096 176, which is provided with the Genius 1 kit) dissolved in Genius buffer 1 (you may need to heat the solution to dissolve the blocking agent; **do not boil it**)
- Genius buffer 3: 100 mM Tris–HCl pH 9.5, 100 mM NaCl, 50 mM $MgCl_2$
- DNA dilution buffer: 50 μg/ml herring sperm DNA in 10 mM Tris–HCl, 1 mM EDTA, pH 8.0)
- Alkaline phosphatase-conjugated anti-DIG antibody (e.g., Boehringer Mannheim catalogue no. 1093 274)
- Small piece of nylon transfer membrane (e.g., Boehringer Mannheim catalogue no. 1209 299)

Method

1. Make serial dilutions of your labelled DNA and the control labelled DNA:
   - A: 2 μl DNA + 8 μl DNA dilution buffer (1:5 dilution)
   - B: 2 μl dilution A + 18 μl DNA dilution buffer (1:50 dilution)
   - C: 2 μl dilution B + 18 μl DNA dilution buffer (1:500 dilution)
   - D: 2 μl dilution C + 18 μl DNA dilution buffer (1:5000 dilution)
   - E: 2 μl dilution D + 18 μl DNA dilution buffer 1:50 000 dilution)

   Spot 1 μl of each serial dilution of control DNA and labelled DNA on the membrane and bake the membrane at 80°C for 30 min.
2. While the membrane is baking, mix up the following:
   - Colour substrate solution (store in the dark): 45 μl NBT, 35 μl X-phosphate, 10 ml Genius buffer 3
   - Genius buffer 2: 1 g blocking reagent (Boehringer Mannheim catalogue no. 1096 176), 50 ml Genius buffer 1
   - Antibody solution: 5 μl anti-DIG alkaline phosphatase, 25 ml Genius buffer 2
3. After baking, wet the membrane in Genius buffer 1 for 1 min.
4. Incubate in 25 ml Genius buffer 2 for 5 min.
5. Incubate in 25 ml antibody solution for 5 min.
6. Switch to new tray and wash membrane twice in about 50 ml Genius buffer 1 for 5 min per wash.
7. Incubate membrane in about 50 ml Genius buffer 3 for 2 min.

Protocol 15. *Continued*

8. Pour off Genius buffer 3, add colour substrate solution, and let sit in dark for 30–60 min, with occasional agitation.
9. Compare the intensity of the spots in the control labelled DNA and your labelling reaction to estimate the concentration of your probe. If, for example, the 10^{-3} dilution of your DNA matches the intensity of the 10 pg labelled control DNA spot, then the amount of your DNA on the filter is calculated as 10 pg/μl × 10^3, and your total yield would be 500 000 pg (= 10 000 pg/ml × 50 μl; 1 μl of your DNA was dissolved in 50 μl TE buffer).

[a] All the ingredients for estimating yield come with the Boehringer Mannheim Genius 1 kit (catalogue no. 1093 657).

8.2 Hybridization using DIG-labelled probes

Hybridization can be performed essentially as with radiolabelled probes (see *Protocol 11*). We have had good success using glass hybridization bottles in a Hybaid hybridization oven. Hybridization temperatures need to be determined empirically for each probe. Hybridization temperatures are typically lower for DIG-labelled probes than for the same probe labelled with ^{32}P. We have had good success with temperatures ranging from 55 °C to 65 °C, depending on the probe and template combination.

Boehringer Mannheim recommends using a probe concentration of 5–20 ng/ml. For PCR-labelled probes we usually end up with 0.5–5 μg labelled DNA. For a probe concentration of 20 ng/ml you need to add 300 ng labelled DNA/15 ml hybridization solution, or roughly 10 μl of labelled probe. In practice, 10 μl probe/15 ml hybridization solution works well. Do not add too much probe as high probe concentrations will give increased background.

As with radiolabelled probes, DIG-labelled probes should be denatured by boiling for 5 min prior to use. After denaturing the probe, spin the tube briefly and immediately place on ice.

Add the probe to pre-hybridization solution and heat to 60 °C. We recommend mixing the probe with heated hybridization solution and *replacing* the pre-hybridization solution with hybridization solution containing the labelled probe (contrary to recommendation in *Protocol 11* for radiolabelled probes).

One advantage of DIG-labelled probes is that they can be re-used. Hybridization solutions should be stored frozen between uses. Because of the high SDS concentration in the hybridization solution (*Protocol 11*), it will look milky when cold but will clear up when warm. Place the hybridization solution in a clean Falcon tube and denature the probe by placing the tube in boiling water for 10 min. Add the hybridization solution to the Hybaid bottle when it cools to 60 °C.

Washing protocols are the same as those given in *Protocol 11D*. We typically wash membranes twice in 2 × wash (0.3 M NaCl, 3 mM sodium citrate, 0.1% SDS) at room temperature, once in 2 × wash at 55°C, and once in 0.5 × wash (0.075 M NaCl, 0.75 mM sodium citrate, 0.1% SDS) at room temperature (15 min per wash), or with more stringent conditions.

Methods for detection of DIG-labelled probes are given in *Protocol 16*.

Background can be a real problem with DIG-labelled probes. The following precautions should help to reduce background:

- wear powderless gloves when working with membranes
- filter Genius buffers 1 and 3 before use
- block for longer than is normally recommended in the manual (2–3 h should be good)
- be careful not to add to much probe or antibody when mixing up solutions
- do not let the membranes dry out during washes or detection steps

Protocol 16. Detection of DIG-labelled probes

Reagents

- Genius buffers 1–3 (see *Protocol 15*)[a]
- Anti-DIG alkaline phosphatase (e.g., Boehringer Mannheim catalogue no. 1093 274)
- Lumi-Phos 530 (e.g., Boehringer Mannheim catalogue no. 1413 155)

Method

1. Transfer the membrane from the last wash solution to a new tray with 500 ml Genius buffer 1 and let sit at room temperature for 1 min.
2. Transfer the membrane to a new tray with 200 ml Genius buffer 2, cover tray with cling film, and gently shake it for 2–3 h at room temperature. While the Boehringer Mannheim manual recommends blocking for as little as 30 min, we have found that long blocking significantly reduces background. Save 100 ml of Genius buffer 2 for the antibody solution. During incubation, prepare antibody solution by adding 10 μl an alkaline phosphatase conjugated anti-digoxigenin antibody to 100 ml Genius buffer 2 (= 1:10000 dilution).
3. Transfer the membrane to a new tray and add 100 ml of antibody solution. Cover the tray with cling film and shake gently for 30 min at room temperature.
4. Transfer membranes to new tray with 500 ml Genius buffer 1 and gently agitate for 15 min. Repeat.
5. Transfer membrane to a new tray with 500 ml of Genius buffer 3. Let sit for 2 min.
6. Pour off Genius buffer 3 and add 20 ml of a 1:10 dilution of Lumi-Phos 530 in Genius buffer 3.[b] Move the tray so that the Lumi-Phos mixture

Protocol 16. *Continued*

sloshes over the membrane for about 30 sec. Do not let the membrane become dry during this stage or you will have grainy background problems.

7. Place membrane into a plastic page protector (heavy gauge acetate sheet protectors work well; e.g., Joshua Meier catalogue no. 06198) and seal all edges with masking tape. It is essential that the membrane is not allowed to dry out. Distribute the Lumi-Phos by gently wiping the surface of the plastic with a paper towel.
8. Let the membrane sit for 10–12 h in the dark until the Lumi-Phos activity reaches a maximum, and expose film in a dark-room for about 15–60 min, or as needed. We have found that with PCR-labelled probes and efficient Southern transfers, the exposures are routinely between 20 and 40 min.

[a] Genius buffers 1 and 3 should be filtered before use through a 0.45 μm filter.
[b] Although the Boehringer Mannheim manual says not to dilute Lumi-Phos, we have had better results with dilutions because they reduce background. While the membrane is in the last Genius buffer 1 wash (step **4**), add 2 ml Lumiphos to 18 ml Genius buffer 3 and place in the dark. Use sterile technique when removing Lumi-Phos from the bottle since it is prone to contamination. The dilution can be re-used for up to a week but if you plan to save it wrap the container in aluminium foil and store at 4°C.

8.3 Probe stripping

A number of probe stripping protocols are available. Alkaline probe stripping solutions (e.g., *Protocol 11F*) can give high backgrounds when used with DIG detection systems unless precautions are taken to neutralize membranes afterwards (and even that does not always solve the problem). A simpler procedure is to avoid exposing membranes to high pH and simply use a 0.1% SDS solution (*Protocol 17*).

Protocol 17. Probe stripping with 0.1% SDS

Reagents

- 0.1% (w/v) SDS solution (5 ml of 20% (w/v) SDS in 995 ml H_2O) heated to 95°C
- 2 × SSC
- 0.5 × TBE
- Metal tray
- Ziplock bags

Method

1. Wash membranes once in distilled water at room temperature.
2. Transfer membranes to a metal tray and pour 500 ml of 0.1% SDS heated to 95°C over them.

3. Gently agitate tray until solution cools to 60°C, and repeat.
4. Wash membranes once in 2 × SSC.
5. Store membranes in 2 × SSC or 0.5 × TBE in Ziplock (Dow, Inc) bags at 4°C.

9. High resolution 'four-cutter' restriction site analysis

For many questions, a high density of restriction sites is necessary. One solution to the low number of sites screened in a typical 'six-cutter' survey has been to use endonucleases that cleave more often. For example, 'four-cutter' enzymes are expected to cleave on average once every 256 nucleotides. However 'four-cutter' sites cannot readily be mapped with the traditional restriction mapping strategy of single and double digests because of the large number of cleavage sites and because many of the fragments generated are below the size range well resolved on agarose gels. Knowledge of the complete DNA sequence of the *Adh* gene region allowed Kreitman and Aguadé (21) to predict the location of fragments separated and sized on high-resolution gels typically used for DNA sequencing. Coupling this approach with electroblotting and UV cross-linking the DNA to nylon filters has led to a very rapid and general method for surveying sequence variation (see ref. 22). The use of 2–4 kb probes leads to the detection of a manageable number of fragments. Larger regions are best studied by successive probings of the filters with adjacent or slightly overlapping probes. Thus, the virtually impossible task of mapping the many 'four-cutter' sites by traditional single and double digest method is avoided. The location of site gains and losses can also generally be determined based on the sequence.

Kreitman and Aguadé (21) estimated that the use of 10 different 'four-cutter' restriction endonucleases allows the screening of roughly 19% of the nucleotides in the *Adh* region of *D. melanogaster.* Given that some 50–60 chromosomes can be screened per gel for each enzyme, and that the filters can be probed repeatedly with different probes, the large population samples desired are obtainable. In addition, the high density of sites increases the power to detect local variation in levels of polymorphism and linkage disequilibrium. An example autoradiograph is shown in *Figure 3*. A limitation of the 'four-cutter' approach is that small insertions and deletions between species makes the assignment of homologous restriction sites difficult. Thus, inferring percent sequence divergence from gains and losses of restriction fragments is usually not feasible. At this level, DNA sequence data from from the taxa of interest provide better estimates of sequence divergence.

The following discussion and protocols are adapted from Kreitman and Aguadé (21) and Church and Gilbert (23).

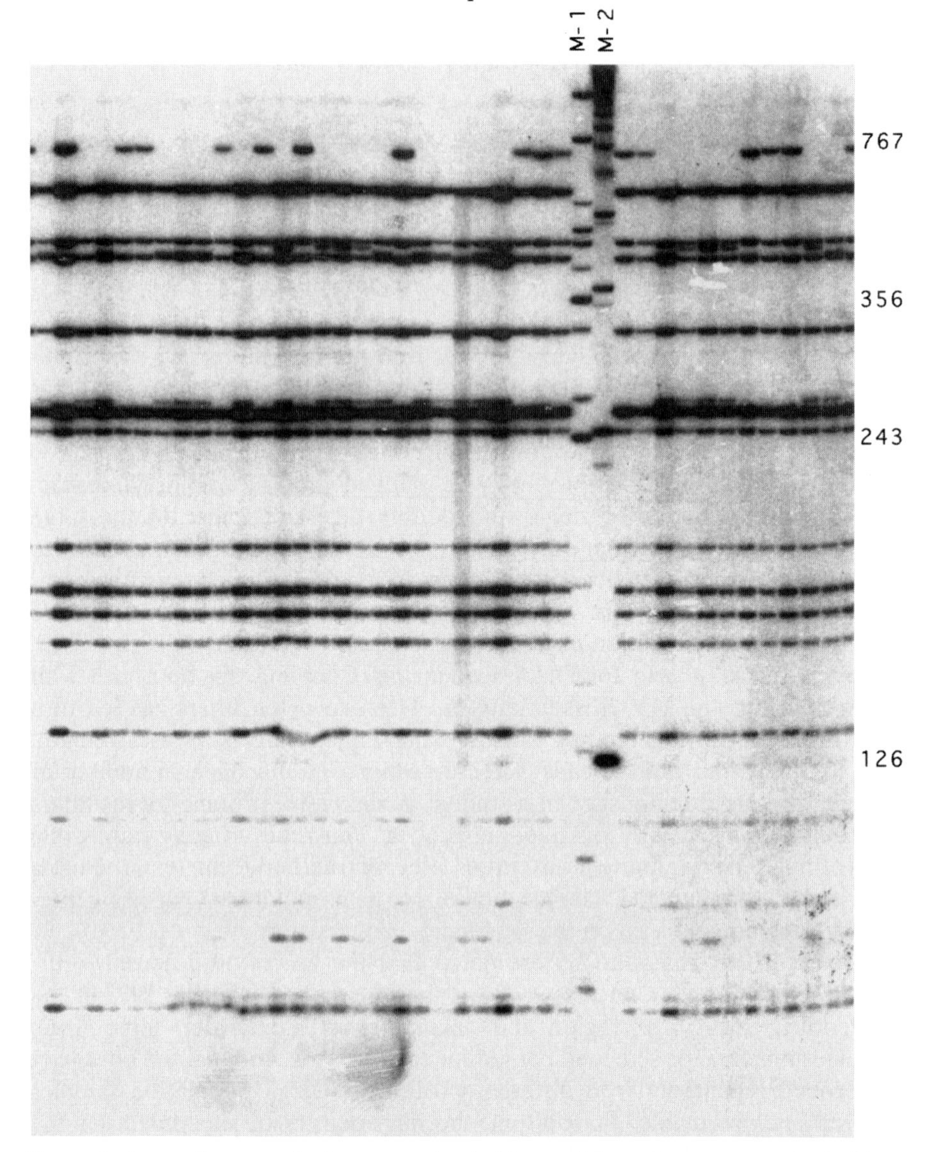

Figure 3. Autoradiogram of a 'four-cutter' gel of *Drosophila melanogaster* probed with a 4.6 kb *Eco*RI fragment of the nuclear gene encoding xanthine dehydrogenase cloned in pBS. Each lane contains genomic DNA (cut with *Taq*I) for an independent homozygous line representing chromosomes sampled from a natural population. Size markers are pBS (Stratagene) digested separately with *Hae*III or with *Hin*fI (M-1 and M-2, respectively). Sizes (in base pairs) are given on the right.

9.1 DNA samples and digestion conditions

The quality of genomic DNA used for 'four-cutter' blots is important. Not only are 'four-cutter' restriction endonucleases often more sensitive to contamination than the common 'six-cutters', but precipitating the DNA and resuspending it in a small volume (required for electrophoresis through sequencing gels) concentrates any contaminants which may reduce the resolution in the gel. Several workers have successfully used DNA isolated using minipreps similar to *Protocol 5* in Chapter 2. However in our hands DNA isolated using a CsCl gradient method (24) has given consistently the best results (see Chapter 2, Section 3.11).

Three micrograms of genomic *Drosophila* DNA is sufficient to allow detection of fragments of single-copy nuclear sequence down to about 60 bases in a 3–4 day exposure. More DNA can be used, but more than about 10 μg will probably overload the lane in the gel and reduce the resolution. Using more DNA will require a proportional increase in the restriction enzyme reaction volumes.

The amount of restriction enzyme used should be determined for each enzyme and the particular method of DNA preparation. We have found that 4 units per μg genomic DNA results in complete digestion in 4 h, but is not so much in excess as to result in 'star' activity. For a more complete discussion of restriction enzyme digestions, see section 4.

After digestion remove 1 μl (50–100 ng) to run on an agarose gel to check for complete digestion of all of the samples (indicated by a smear on the gel when stained with ethidium bromide). If one or more of the samples do not cut completely, you can ethanol precipitate, resuspend, and re-digest the DNA. However, if incomplete or no digestion persists, you should check either the genomic DNA isolation protocol or the restriction enzyme digestion conditions.

After digestion the DNAs need to be ethanol-precipitated (use *Protocol 3*; phenol extraction is unnecessary). Resuspend the dried DNA pellet in 3 μl formamide loading dye (95% formamide, 20 mM EDTA, 0.05% (w/v) Bromophenol Blue, 0.05% (w/v) Xylene Cyanol). The digested DNAs can be stored frozen or at 4 °C at this stage if necessary.

9.2 Polyacrylamide gel electrophoresis for 'four-cutter' blots

Digested DNAs are size-fractionated on a denaturing 5% polyacrylamide sequencing gel (19:1 ratio of acrylamide to bis-acrylamide, 7 M urea, 50 mM TBE (*Protocol 4*) gel and running buffer) to get the best resolution. See Chapters 6 and 10 for a discussion on preparing, loading, and running sequencing gels. We run 30 cm × 40 cm gels with combs that produce distinct wells (not sharks-tooth) to facilitate scoring the variation. To improve the band spacing, we either make the gel with wedge spacers (0.4–1.2 mm) or run the gel using a salt gradient. The salt gradient is made by adding 0.5 vol. of 3 M sodium acetate to the bottom running buffer after the samples have been loaded. The gradient forms during electrophoresis.

Consult the manufacturer's instructions for running conditions specific to your gel rig. The following running conditions are specific to the Life Technologies model S2 sequencing rigs that we use. The gel must be run at 55 °C to ensure that the DNA remains denatured. Pre-electrophorese at 1200 V for 30 min. Boil the DNA samples for 10 min; then put them on ice. If necessary, spin down the samples to ensure that the whole reaction volume is at the bottom of the tube. Use a micropipette with tips that will fit between the glass plates of the gel so that you can load the DNA sample directly to the bottom of the wells. Electrophorese at 1200 V until the Bromophenol Blue dye is at the bottom of the gel (approximately 2 h 40 min, the Bromophenol Blue co-migrates with *c.* 40 bp fragments).

9.2.1 DNA size standards

It is important to run size standards spanning the observable sizes. Life Technologies's 123 bp ladder is a good start but the bands are too far apart for a really good estimation of sizes. A better method is to use the vector in which your probe was cloned as a size standard. Our probes are often cloned in pBS (Stratagene) so we find a combination of restriction enzymes that would give a good range of DNA fragment sizes and load one lane of the gel with those fragments in the same molar concentration as the single-copy genomic DNA loaded in the other lanes. Thus labelling the plasmid containing the DNA probe also labels the size standard DNA. Adding the right amount of size standard DNA ensures that the band intensity will be consistent. We have had success with the following standard: 0.04 ng pBS cut with *Hae*III, 0.04 ng pBS cut with *Hin*f I, and 3 μg sonicated salmon sperm DNA. This is ethanol-precipitated and resuspended in 3 μl formamide loading dye.

9.2.2 Electroblotting from sequencing gels

The DNA fragments need to be electroblotted from the polyacrylamide gel to an uncharged nylon filter. We have used GeneScreen (NEN Products) with good success at retaining small fragments and reprobing. The use of charged nylon membranes may result in excessive background hybridization in autoradiography. Most commercial electroblotters are too small to handle the large gel format that we use (30 cm × 40 cm). Church and Gilbert (23) describe one type which we use (see *Figure 4*); a commercially available electroblotter for large gels can be obtained from Hoefer Scientific Instruments (GeneSweep Sequencing Gel Transfer Unit TE 90). The electroblotting is carried out at 60 V, 4 A (for 10 cm between electrodes) using 50 mM TBE (*Protocol 4*) as the buffer. This generates a lot of heat so the electroblotting is done at 4 °C using pre-cooled buffer. At the end of the run the buffer tempera-

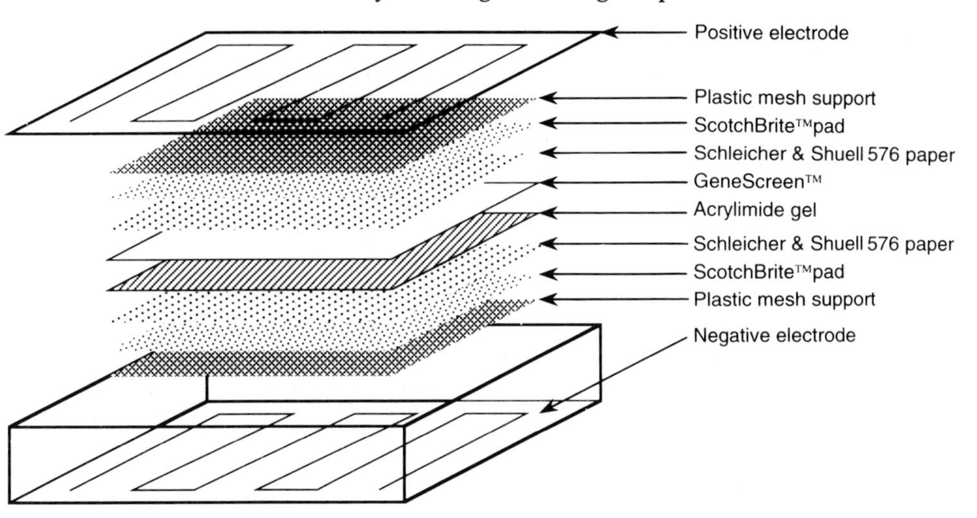

Figure 4. Apparatus for electroblotting DNA from large gels on to nylon membranes.

ture is about 20°C. Under these conditions we have not found re-circulation of the buffer to be necessary.

Protocol 18. Electroblotting from polyacrylamide sequencing gels

Equipment and reagents

- Electroblotting apparatus and power supply
- ScotchBrite pads (from 3M) or other support provided with your electroblot apparatus
- 50 mM TBE (*Protocol 4*)
- Uncharged nylon transfer membrane (e.g., GeneScreen from NEN Products)
- Filter paper to transfer and support gel (e.g., Schleicher and Schuell 576 or GL002 paper cut to be just larger than the gel)

Method

1. Fill the electroblotting apparatus just to the level of the lower ScotchBrite pad (see *Figure 4*), pressing out any air pockets.
2. Separate the glass plates of the gel, making sure that the gel adheres to only one plate. Dampen a sheet of Schleicher and Schuell 576 paper (cut to be just larger than the gel) in buffer and lay it smoothly on the gel. Using a 10 ml pipette, roll out any air bubbles and excess buffer. The gel will now stick to the filter paper. Starting at one end, roll back the filter/gel and lift off of the glass plate.
3. Lay the filter on top of the ScotchBrite pad, gel side up. Wet the nylon filter in buffer and layer it on the gel again using a pipette to remove any air bubbles. Wet another sheet of filter paper and layer it on in a

Protocol 18. *Continued*

similar manner. Another gel can be put on top of this one by repeating steps **1** and **2**. Put the other ScotchBrite pad on top and fill with buffer. It is important to ensure that there are no air bubbles between any of the layers.

4. Electroblot for 45 min (60 min if there are two gels). Remember that the DNA migrates towards the positive electrode.
5. After electroblotting, disassemble pads and filter paper, carefully remove nylon filter, and place it, DNA side down, on a large piece of plastic wrap. Fold over wrap to cover the filter and prevent it from drying out. Immediately proceed to UV cross-linking.

9.3 UV cross-linking of DNA to nylon membrane

The DNA needs to be bound to the nylon filter so that it cannot be removed during the hybridization and washing steps. This is done by using short wavelength UV light to cross-link the DNA covalently to the nylon. Most commercial cross-linkers are too small to handle large (30 cm × 40 cm) filters, although filters can be cross-linked a section at a time with these smaller units. A few commercially available large units are available (e.g., Stratalinker UV Crosslinker Model 2400 from Stratagene). Alternatively, we use a simple apparatus made by mounting six 15 W germicidal bulbs, along with the necessary transformers and hardware, side by side in a shallow metal box, and supported 30 cm above the bench-top. The time required for cross-linking needs to be determined empirically. Too little cross-linking will lead to the DNA fragments being washed off. Too much cross-linking will reduce the hybridization signal presumably by reducing the amount of DNA available for hybridization.

The filter is wrapped in plastic wrap to keep it from drying out. It is then irradiated, DNA side up, for approximately 3 min using the apparatus described. Following irradiation, filters are pre-hybridized and hybridized using a variant of *Protocol 11*. The major change from *Protocol 11* is that pre-hybridization and hybridization must be carried out in 2.5 cm × 40 cm polycarbonate tubes which rotate on a tissue culture roller apparatus in an oven due to the size of the filters. Large heat-sealed bags can be used but are very inconvenient.

Below, we present methods for RFLP detection using radioactively labelled probes. We do not have any experience using non-radioactive detection methods for 'four-cutter' blots, but it is likely that appropriate modifications could be made using the methods given in Section 8.

Protocol 19. 'Four-cutter' blot hybridization and autoradiography

Equipment and reagents

- 2.5 cm i.d. polycarbonate tubes
- Pre-hybridization/hybridization solution (*Protocol 11*) pre-warmed to 60 °C
- Parafilm
- 60 °C incubator
- Boiling water bath
- Probe prepared as in *Protocol 7* or *8*
- Wash solution (*Protocol 11*)
- Incubator at 57 °C
- 50 mM TBE
- Plastic wrap

Method

1. Roll up the filters, DNA side inside, and insert into 2.5 cm i.d. polycarbonate tubes (inserted so that when incubating, the filters unroll against the tube and 'scoop up' hybridization solution as they rotate). Up to three filters can be easily loaded per tube.
2. Drain off the 50 mM TBE and add 20 ml pre-hybridization/hybridization solution (*Protocol 11*) pre-warmed to 60 °C. Insert rubber stopper, wrap with Parafilm, and incubate at 60 °C rolling at *c.* 5 r.p.m. for >5 h.
3. Prepare probe using *Protocol 7* or *8*. (It is important to remove unincorporated nucleotides to reduce background, even with 90–95% incorporation).
4. Prepare 40 ml of pre-hybridization/hybridization solution (*Protocol 11*) per tube, heat to 60 °C, and add 20 ml to the tube.
5. Pre-hybridize filter at 60 °C for at least 5 min, rolling it in a 60 °C incubator.
6. Boil probe for 10 min and then cool on ice; add to remaining 20 ml of pre-hybridization/hybridization solution, and mix well (keep heated to 60 °C). This is the hybridization mix.
7. Pour the pre-hybridization/hybridization solution from the tube and add the hybridization mix. Incubate at 60 °C for 24 h.
8. The washing is done in two stages: the first step occurs in the tube and removes excess hybridization fluid, and the second step is carried out in a large tray for stringency
   (a) Heat 40 ml wash solution (*Protocol 11*) to 57 °C and pour into tube. Place tube in roller apparatus in the 60 °C incubator for 15 min. Pour wash into radioactive waste. Repeat twice.
   (b) For the next three washes unroll the filters into *c.* 1 litre wash solution at 57 °C, cover with a Perspex plate, and shake in an incubator at 57 °C for 30–45 min.

   After each wash, check the wash solution to see if it is so 'hot' as to require disposal in radioactive waste. Up to six filters can be washed

Protocol 19. *Continued*

at one time if the volume of wash solution is sufficient to cover the filters.

9. Rinse the filters in 50 mM TBE and wrap them in plastic wrap for exposure to film. Exposure for 24 h without screens is usually sufficient to get good resolution of the larger bands and 6 days with screens for bands down to 60 bp.
10. Probe can be stripped off filters by pouring 50 mM TBE at 95°C over the filters, placed DNA side up in a plastic tray, followed by shaking at room temperature until the buffer has cooled to *c.* 35°C. Repeat once. Store the filters in the tubes or covered trays in 50 mM TBE at 4°C until future use.

10. Methods and strategies for mapping and scoring restriction site map variation

Statistical methods for estimating sequence variation and divergence exist for both RFLP fragment length data as well as for restriction site map data. While fragment comparisons are simple, their utility is limited to low levels of sequence divergence and these estimates are seriously compromised by insersion/deletion polymorphism. Thus, we strongly recommend the construction and use of restriction site maps.

Though conceptually straightforward, the task of RFLP mapping can, in practice, be quite difficult and time-consuming. If the sample of interest is highly polymorphic and linkage disequilibrium is minimal, then most alleles can have a unique map (i.e. haplotype). In this section we try to familiarize the reader with general strategies and some special techniques for dealing with 'real world' restriction mapping problems.

Keep in mind the following two pieces of advice. First, the construction of a map is a fluid process; the working map should be considered a working hypothesis to be modified as new information accumulates. On several occasions we have observed that a seemingly 'impossible' mapping problem disappears when a dearly held, yet false, assumption regarding the location of a restriction site is discarded. Second, start with the simplest haplotypes available; as one becomes more familiar with the basic map, it becomes easier to determine the haplotypes of other individuals in the population. A good starting point for any mapping experiment is the phage or plasmid clone which will be used to probe genomic Southern blots. This is recommended even if the ultimate goal is to use a clone from one species to generate a map in a related species for which clones are unavailable. Most of the following discussion refers to mapping strategies for genomic Southern blots, though the same principles are generally applicable to most mapping situations (i.e. mapping a clone, PCR fragment, or purified mtDNA). Most of the strategies will apply to both 'six-cutter' and 'four-cutter' interpretation (exceptions are noted in the text). Single and double digests of cloned DNA, PCR fragments, mitochondrial, or chloroplast DNA can be visualized on ethidium bromide stained agarose gels rather than by Southern blotting.

10.1 Constructing 'six-cutter' restriction site maps

10.1.1 Setting up mapping gels

Mapping gels and blots are made by running *all* combinations of single and double digests; the gel should be loaded such that two single digests flank the double digest. For example, the first five lanes of a gel might be *Eco*RI, *Eco*RI/*Hin*dIII, *Hin*dIII, *Hin*dIII/*Sal*I, *Sal*I. We routinely use a '1 kb' size standard (Life Technologies) and 0.8% agarose gels; however, the concentration of agarose and the size standard must be chosen based upon the sizes of fragments to be mapped. Smaller fragments will stain/hybridize less strongly than large fragments. Thus, if one loads the appropriate amount of DNA to resolve fragments in the 2–15 kb range it may be difficult to detect 200–300 bp fragments. Alternatively, if one wants to map several small fragments, the gel may be overloaded and unresolvable for several larger fragments.

10.1.2 Mapping with single and double digests

In *Figure 5* we illustrate a typical autoradiograph with single and double digests of genomic DNA; an *Eco*RI digest results in three fragments of 2, 5, and 8 kb, a *Hin*dIII digestion results in two fragments of 6 and 9 kb, and an *Eco*RI/*Hin*dIII double digest has fragments of sizes 2, 4, and 5 kb. The goal is to determine the orders of the fragments for each enzyme and the relative positions of the sites. Starting with *Eco*RI there are three possibile orders of the three fragments. Either the 2, 5, or 8 kb fragment could be in the middle (*Figure 6*). Next, consider the *Hin*dIII fragments. To construct the map one examines various combinations of hypothetical *Eco*RI and *Hin*dIII maps and then compares the hypothetical double digest pattern with that actually observed. For example, if maps C and E in *Figure 6* are the true maps, then the *Eco*RI/*Hin*dIII double digest should give fragments of 5, 2, 2, and 6 kb. This is not the pattern seen in the actual double digest. Similarly, maps C and D should give fragments of 5, 1, 1, and 8 kb. Since *Eco*RI map C does not fit with either possible *Hin*dIII map we know that the true *Eco*RI map is either A or B. By continuing this process one eventually reaches the conclusion that the only combination of *Eco*RI and *Hin*dIII maps that results in the observed double digest pattern is A and E. Note that the greater intensity of the 4 kb double digest product in *Figure 5* indicates a doublet. Doublets result when two different fragments of similar size co-migrate in a single lane. They are usually more intense and diffuse than single bands of the equivalent size. In general, it is easiest to begin a map with restriction enzymes that cut infrequently in the

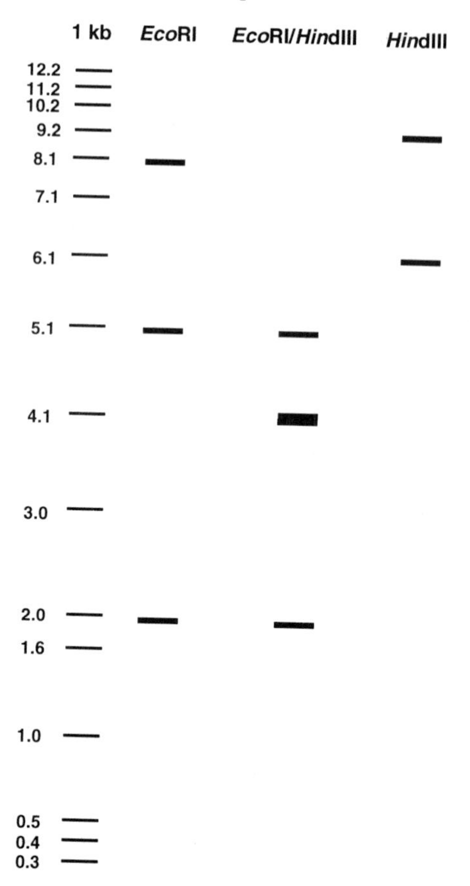

Figure 5. Illustration of typical single and double digests next to a 1 kb size standard.

region of interest; it may be necessary to consider several combinations of enzymes until a suitable pair is found. Once the map for one or two enzymes is determined, the process continues. At each newly arrived at hypothetical map it is important to be sure that the map for each previous enzyme is compatible with the new information. Unlike the example in *Figure 5*, the sizes of the maps often differ across enzymes because of variation in the sizes of flanking fragments for each enzyme.

10.1.3 Flanking fragments

When probing genomic Southern blots one often observes two flanking restriction fragments, one on each side of the probed region. These fragments are only probed by a small part of the clone. Which of these fragments is on the left and right can sometimes be determined with band intensities. An example is shown in *Figure 7*. There are two flanking fragments of 8 kb and

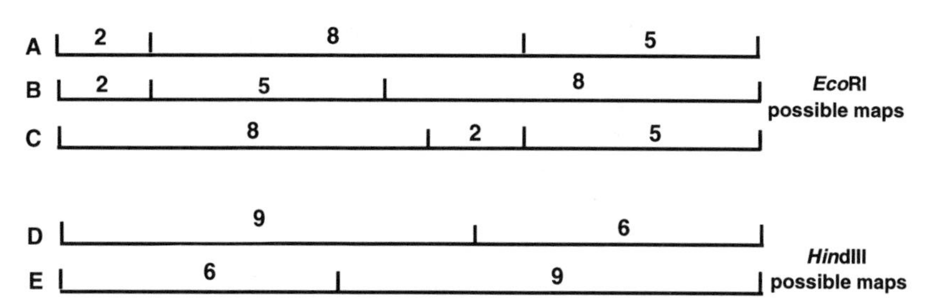

Figure 6. Illustration of the possible restriction maps from the single digests shown in *Figure 5*. Information from the double digest in *Figure 5* reveals that the correct maps are A and E.

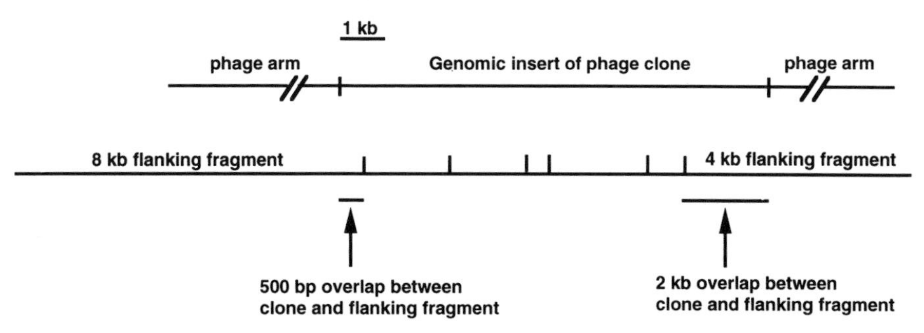

Figure 7. Illustration of how band intensities of flanking fragments can be used in mapping.

4 kb, respectively. All else being equal, an 8 kb fragment should be roughly twice the intensity of the 4 kb fragment. However, for flanking fragments, the intensity can be determined by how many base pairs overlap the clone. Note that if the 8 kb fragment were on the left (as shown in *Figure 7*) it would be probed by about 0.5 kb of cloned DNA, while the 4 kb fragment would be probed by about 2 kb of DNA. If this were the case then the 8 kb fragment would be less intense than the 4 kb fragment, the reverse of what one would expect. This method is not always reliable. For example, very large fragments (>20 kb) can be less intense than smaller fragments because they transfer to membranes less efficiently rather than because they are probed by fewer base pairs of the clone. An always reliable alternative is to probe Southern blots with a leftmost or rightmost fragment from the clone. It is important to remember that, when gel-purifying these fragments, they will often be attached to one of the phage arms. It is also important to hybridize to filters with DNA cut with a second appropriate enzyme in addition to the one in question. This will tell you the side of the clone in which the genomic fragment is located. Furthermore, in the case of cloned DNA, the sizes of the leftmost and rightmost fragments are defined by the restriction enzyme used in constructing the library.

10.1.4 Mapping with partial digests

Sometimes one encounters a number of small fragments consecutively, the order of which can be virtually impossible to determine by combinations of single and double digests. A very effective alternative is to map these sites using partial digests (e.g., ref. 25). The approach is outlined in *Figure 8*. In *Figure 8a* we see a 5 kb *Hin*dIII fragment followed by a large number of small *Hin*dIII fragments of unknown order. Notice that in *Figure 8b* the *Bam*HI fragment encompasses the entire region in question. The first step in completing this map is to do a complete *Bam*HI digest of genomic DNA. The next step is to do a partial *Hin*dIII digest of the same sample. A fairly reliable method of generating partial digests is to add a small amount of enzyme to a sample and then remove aliquots at specific time intervals such as 0.5, 1.0, 1.5 min, etc. The enzyme in these aliquots can be inactivated by adding an EDTA solution or by heating (for most enzymes). The time points must be determined empirically for each DNA–enzyme combination. The products of these partial digests are run on an agarose gel and Southern blotted. Now, note in *Figure 8c* that there is a 2 kb *Sal*I fragment internal to the 5 kb *Hin*dIII fragment. This fragment is gel-purified from the clone and hybridized to the Southern blot. If the partials were performed correctly the order of restriction sites can be read directly from the autoradiograph. For example, let us say the 2 kb *Sal*I fragment hybridizes to fragments of 5.0, 5.7, 6.1, 7.0, 7.7, 8.7, and 9.7 kb. That means the order of restriction fragments from left to right is 5.0, 0.7, 0.4, 0.9, 0.7, 1.0, 1.0. This method may seem laborious, but it can be more efficient than pondering over 'standard' single and double digests and it considerably reduces levels of mental anguish when mapping small fragments.

10.2 Scoring RFLP gels

10.2.1 Setting up survey gels

A restriction map survey of population variation is accomplished by making a series of genomic Southern blots. Gels should be arranged by population and restriction enzyme. For example, a gel may consist of DNA from all the individuals in the population digested with *Eco*RI. It is convenient to load the size standard closer to one edge of the gel (as opposed to the middle) to allow easier orientation of the resulting autoradiograph. A set of mapping gels

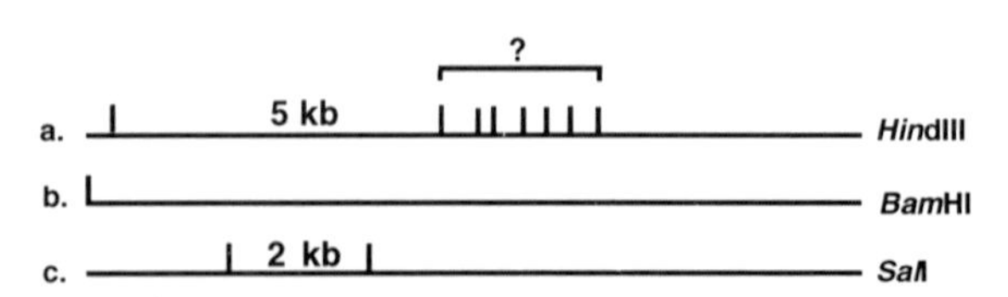

Figure 8. Illustration of mapping small restriction fragments by partial digests. See text for explanation.

should also be made for one or two individuals in the population. The lanes should be set up as described for mapping a clone. Why does one need mapping gels if a clone has already been mapped? First, haplotypes of laboratory stocks from which clones are often isolated can sometimes be quite different from those common in natural populations. Second, when probing Southern blots of one species with clones from another, mapping gels are indispensible if sequence divergence between species is greater than 2–3%. We generally expose our filters for 1–2 days without intensifying screens to obtain accurate size estimates of large fragments and to identify doublets. If necessary, we then expose the filters for a week or more with two intensifying screens per filter in order to see fragments down to about 400 bp.

10.2.2 Gains and losses of restriction sites

Inferring gains and losses of sites is often quite straightforward. For example, a 5 kb fragment present in most individuals may, in some indivduals, be replaced by two fragments of 4 and 1 kb. Not quite as simple is determining if the order of the two fragments is 4–1 or 1–4. The most straightforward way to resolve this ambiguity is to probe the filter with a gel-purified fragment from your clone. For example, in *Figure 9* we show the same map as in *Figure 6* except that the 8 kb *Eco*RI fragment is cut into two fragments of 2 and 6 kb. If the order of the fragments from left to right is 2–6 then the gel-purified 9 kb *Hin*dIII fragment will bind to the 6 kb fragment but not to the 2 kb fragment. Alternatively, if the order is 6–2 the 9 kb *Hin*dIII fragment will bind to both the 6 and 2 kb fragments. An alternative to probing with gel-purified fragments is to make a mapping gel using DNA from the individual with the unknown haplotype. As will be explained in Section 10.2.4, size variation can sometimes be used to determine the order of restriction fragments.

10.2.3 Flanking polymorphism

In addition to fixed flanking sites, one often observes variation resulting from polymorphic sites located outside of the genomic region spanned by the clone. These polymorphisms can be identified because a change in the size of one fragment occurs without any changes in the remaining fragments in the gel lane. For example, imagine a flanking 10 kb fragment which is probed only by 2 kb of the clone. If there was a polymorphism resulting from a gain of site

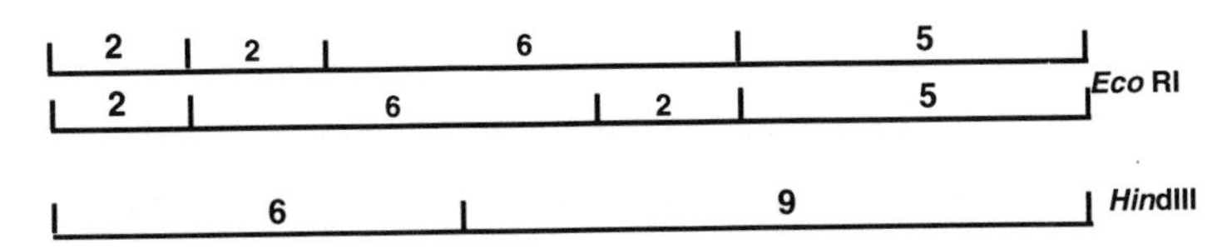

Figure 9. Two possible *Eco*RI restriction maps and a *Hin*dIII map. Which of the two *Eco*RI maps is correct can be determined by hybridization of the 9 kb *Hin*dIII fragment to an *Eco*RI genomic Southern blot.

yielding two fragments of 6 and 4 kb, then your clone would hybridize to either the 6 or 4 kb fragment (depending on the location of the polymorphism), but not both (as it would if the clone spanned the entire 10 kb fragment).

10.2.4 Size variation

Up to this point we have discussed only scoring gains and losses of restriction sites. Another important class of polymorphism may be sequence length variation. Size variation can be identified as such because a set of overlapping fragments from different enzymes will exhibit concordant size increases or decreases. For example, if one individual in a sample had a 5 kb transposable element insert in the surveyed region, one fragment for each enzyme might be 5 kb larger than the most common fragment in the sample. In *Figure 10* we show such a case; only the fragment with the insert is indicated on the maps. The location of the insert can be determined by examining which two insert-containing fragments overlap the smallest number of base pairs; the insert then must be in the region of overlap. In *Figure 10* this is the region of overlap between the *Eco*RI and *Sal*I fragments. In general, one can detect insertion/deletion variation down to about 100 bp in a routine 'six-cutter' survey. The fact that large insertions may have internal restriction sites can sometimes make it appear that information from one restriction enzyme is inconsistent with your working hypothesis about the nature of the insert. If an insert has two internal restriction sites for a given enzyme, the insert DNA released by the enzyme will probably not be probed by the clone. It is also important to remember that a flanking insert variant can look like a flanking restriction site polymorphism for any given enzyme since in both cases one observes a fragment increase in size without the loss of a smaller fragment. Again, examination of maps from all the enzymes in the survey should allow one to resolve the ambiguity in most cases. Size variation can be useful for determining if a flanking fragment is on the left or right side of the map. If one can determine by other means that an insert is in a flanking fragment on the left, then other flanking fragments on the left should have the insert.

Another class of size variation is small tandem repeats. For example, one

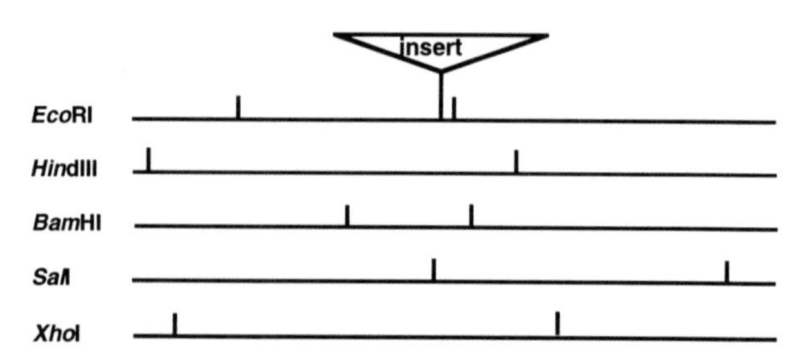

Figure 10. Illustration of how a size variant can be localized. In this example the insert is located in the small region of overlap between the *Eco*RI and *Sal*I fragments.

may observe a fragment assume a range of sizes in increments 100 or 200 bp, concordant across many enzymes. Suffice it to say that if there is a high level of both restriction site variation and complex size variation in the same region, mapping can become quite difficult (this is especially true if the size variation itself has a restriction site for one or several of the enzymes in your survey). In these cases, the best strategy is to employ the probing with gel-purified fragments and the partial digest techniques. In some cases, the variation is such that it cannot be interpreted with any reasonable effort.

10.2.5 Example autoradiograph

Figure 11 shows a schematic of an autoradiogram of six different individual samples cut with two different restriction enzymes. The differences in the band pattern are representative of some of the polymorphisms mentioned above. The first sample is the one with a known map as shown in the first line of *Figure 12*. The 7 and 6 kb bands would be fainter than expected for their length since they only partially hybridize to the probe DNA. By comparing the band patterns of the other samples to the first, the type and position of each polymorphism can usually be deduced.

In the *Bam*HI digest of sample 2, the 0.5 and 4.0 kb bands have been lost and a 4.5 kb band gained. Since there is no obvious change to the *Eco*RI digest, the *Bam*HI site between the 0.5 and the 4.0 kb fragments has probably been lost in this sample as indicated in line 2 of *Figure 12*.

In sample 3 the *Bam*HI fragments appear normal but the *Eco*RI digest shows the loss of the 5.0 kb band and the gain of bands at 4.0 and 1.0 kb. This would indicate the gain of an *Eco*RI site 1 kb from one end of the 5.0 kb *Eco*RI fragment. In this case we are able to determine that the new site is on the left end of the 5.0 kb fragment since the 1.0 kb generated fragment hybridizes to the probe DNA.

In sample 4 the 3.0 kb *Bam*HI band moves to 4.5 kb with no other bands moving. This would indicate the insertion of 1.5 kb of DNA into this fragment which should affect the size of either the 3.0 or the 5.0 kb bands in the *Eco*RI digest. In this case the 3.0 kb *Eco*RI band increases in size to 3.5 kb and a new band appears at 1.0 kb. The combination of these band patterns can be interpreted to be the map shown in line 4 of *Figure 12* where a piece of DNA 1.5 kb in length is inserted somewhere between the *Bam*HI and the *Eco*RI sites indicated. This DNA has an *Eco*RI site in it.

Sample 5 an example of an insert containing multiple restriction sites. Looking at the *Bam*HI digest, the 4.0 kb band increases in size to 5.0 kb and a new, faint, band appears at 2.0 kb. This could be interpreted to be an insert of 3 kb containing one *Bam*HI site. However, looking at the *Eco*RI digest, the 2.5 kb band increases in size to 3.5 kb and a new, faint, band appears at 4.5 kb. This would indicate an insert of 5 kb with one *Eco*RI site in it. Reconciling both of these observations would indicate that the insert contains at least one *Eco*RI site and two *Bam*HI sites and is at least 5 kb long. The only way of

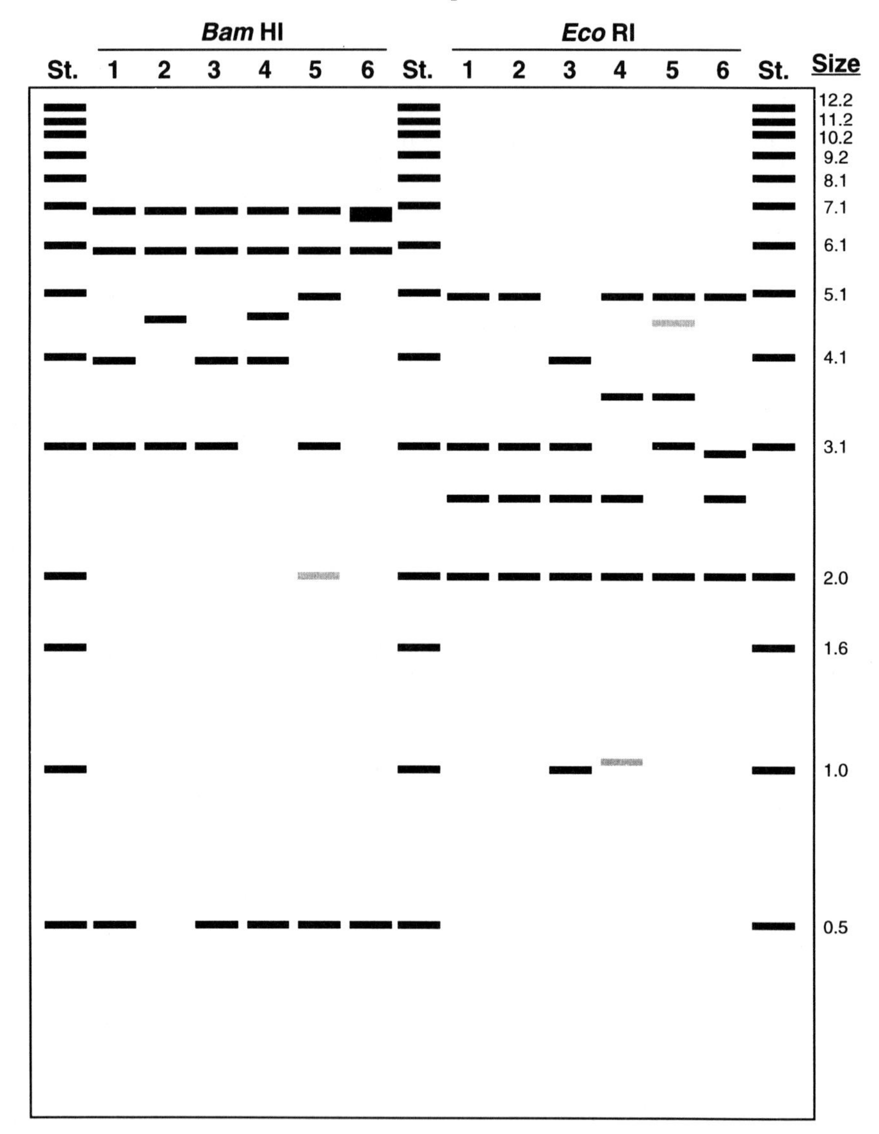

Figure 11. Schematic of DNA samples from six individuals with two different restriction enzymes. See text for interpretation.

determining its exact length is to find an enzyme that does not cut the insert.

In the *Bam*HI digest of sample 6 the 4.0 and 3.0 kb bands are lost with the gain of a band at 6.9 kb. This could indicate simply the loss of the *Bam*HI site between these two fragments (the difference between 6.9 and 7.0 kb could just be measurement error). However, a close examination of the bands in the *Eco*RI digest shows a 0.1 kb decrease in the 3.0 kb band. This information,

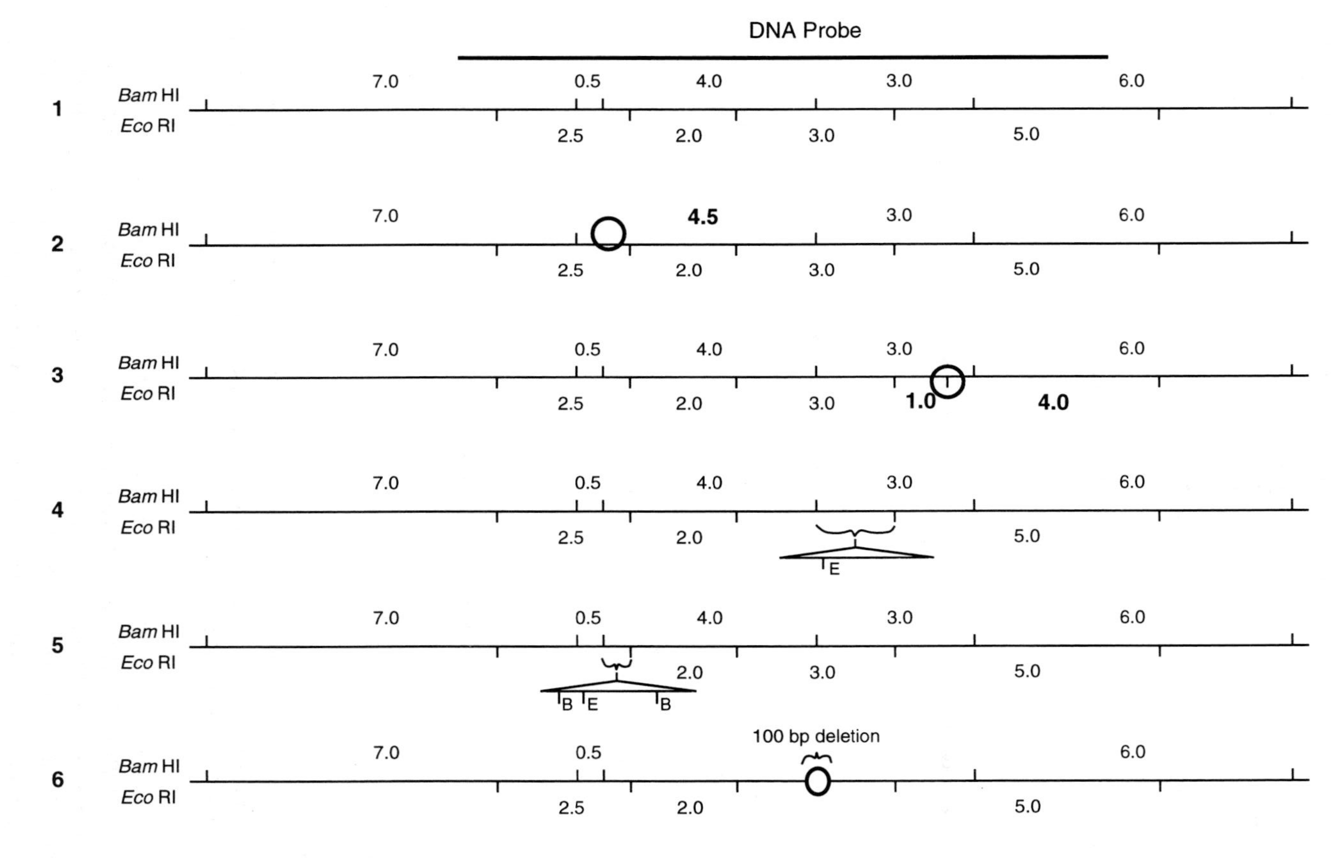

Figure 12. Two enzyme maps for the six individuals presented in *Figure 11*. See text for interpretation.

when taken together, indicates that there is probably a 100 bp deletion spanning the *Bam*HI site as indicated in line 6 of *Figure 12*.

These cases are fairly straightforward since there is only one change in each sample. In a real data set many of these polymorphisms would be in high frequency so there could be several changes between any particular sample and the mapped sample. One other complication is that the mapped sample may not be the most common sample.

10.2.6 Special considerations for scoring 'four-cutter' variation

'Four-cutter' restriction fragments are generally too small to map by combinations of single and double digests. Instead, a predicted restriction map is inferred from DNA sequence data from the region of interest. In terms of scoring the autoradiograph, the strategies outlined above will usually suffice. For example, size variation and gains and losses of sites are distinguished in exactly the same way in 'six-cutter' and 'four-cutter' surveys. The location of a new site (not predicted from the sequence) can be determined by probing with small fragments as described earlier; alternatively, the sequence on which the map is based can be scanned by a computer for 'one-off' sites. For example, assume a predicted 1 kb *Msp*I fragment is cleaved into two fragments of 0.4 and 0.6 kb. A computer search of the 1 kb *Msp*I fragment may reveal that several 4 bp sequences within this 1 kb fragment *would* be an *Msp*I site if there were a different nucleotide in one position. If a potential 'one-off' site was located about 0.4 kb from the left end of the fragment, then it could reasonably be inferred that the new site resulted from a change in that site. Note that sometimes there are a large number of one-off sites such that the precise location of a new site cannot be determined in this way.

10.2.7 Problems resulting from within-line heterozygosity

All the above discussions assume that the individuals from which DNA is isolated are homozygous for each nucleotide in the region examined. In organisms such as *D. melanogaster* this can be accomplished by means of appropriate crosses with marked balancer chromosomes. However, for most taxa, this is not an alternative. The problem when dealing with highly variable diploid organisms is that there may be two divergent haplotypes in each individual (except for sex-linked genes in the heterogametic sex). If one can isolate sufficient DNA from one individual to perform digests with all restriction enzymes in one's survey there is a reasonable chance of resolving the two haplotypes. These are scored as two separate chromosomes. On the other hand, sometimes it is necessary to isolate DNA from several hundred individuals to complete a survey. In these cases there may be several haplotypes in each lane of the gel. Mapping becomes virtually impossible in these situations. One alternative for some species is to inbreed lines for several generations in the laboratory. This can reduce the amount of variation within each line to the point where mapping becomes feasible. Note that this is 'permissible' from a population genetics perspective since variation between lines is unaffected by inbreeding. Variation in band intensities should reflect frequencies of multiple haplotypes within a line. If there are two haplotypes in a line but one occurs at very high frequency, this common haplotype can be scored as the haplotype for the line.

Acknowledgement

Our work is supported by grants from the National Institutes of Health and the National Science Foundation.

References

1. Aquadro, C.F. (1992). *Trends Genet.*, **8**, 355.
2. Aquadro, C.F. (1993). In *Molecular approaches to fundamental and applied entomology* (ed. J. Oakeshott and M.J. Whitten), p. 222. Springer-Verlag, New York, NY.
3. Danforth, B.N., Barretto-Ko, P., and Neff, J.F. (1996). *Evolution*, **50**, 276.
4. Freeman-Gallant, C.R. (1996). *Proc. Royal Soc., Lond. B*, **263**, 57.
5. Babon, J., Youil, R., and Cotton, R.G.H. (1995). *Nucleic Acids Res.*, **23**, 5082.
6. Youil, R., Kemper, B.W., and Cotton, R.G.H. (1995). *Proc. Natl Acad. Sci. USA*, **92**, 87.
7. Westneat, D.F., Noon, W.A., Reeve, H.K., and Aquadro, C.F. (1988). *Nucleic Acids Res.*, **16**, 4161.
8. Reed, K.C. and Mann, D.A. (1985). *Nucleic Acids Res.*, **13**, 7207.
9. Aquadro, C.F., Weaver, A.L., Schaeffer, S.W., and Anderson, W.W. (1991). *Proc. Natl Acad. Sci. USA*, **88**, 305.
10. Brown, C.J., Aquadro, C.F., and Anderson, W.W. (1990). *Genetics*, **126**, 131.
11. Schaeffer, S.W. and Aquadro, C.F. (1987). *Genetics*, **117**, 61.
12. Riley, M.A., Hallas, M.E., and Lewontin, R.C. (1989). *Genetics*, **123**, 359.
13. Feinberg, A.P. and Vogelstein, B. (1984). *Anal. Biochem.*, **137**, 266.
14. Rigby, P.W.J., Dieckmann, M., Rhodes, C., and Berg, P. (1977). *J. Mol. Biol.*, **113**, 237.
15. Sambrook, J., Fritsch, E.F., and Maniatis, T. (1989). *Molecular cloning: a laboratory manual*, 2nd edn. Cold Spring Harbor Laboratory Press, Cold Spring Harbor, NY.
16. Hopman, A.H.N., Speel, E.J.M., Voorter, C.E.M., and Ramaekers, F.C.S. (1995). In *Non-isotopic methods in molecular biology: a practical approach* (ed. E.R. Levy and C.S. Herrington), p. 1. IRL Press, Oxford.
17. Hughes, J.R., Evans, M.F., and Levy, E.R. (1995). In *Non-isotopic methods in molecular biology: a practical approach* (ed. E.R. Levy and C.S. Herrington), p. 145. IRL Press, Oxford.
18. Danforth, B.N., Freeman, C.R., Bogdanowicz, S., and May, B. (1994). *Biochemica* (technical publication of Boehringer Mannheim Corporation), **11**, 7.
19. Emanuel, J.R. (1991). *Nucleic Acids Res.*, **19**, 2790.
20. Lanzillo, J.J. (1990). *BioTechniques*, **8**, 621.

21. Kreitman, M. and Aguadé, M. (1986). *Proc. Natl Acad. Sci. USA*, **83**, 3562.
22. Kreitman, M. (1987). In *Oxford surveys in evolutionary biology* (ed. P. Harvey and L. Partridge), Vol. 4, p. 38. Oxford University Press.
23. Church, G. and Gilbert, W. (1984). *Proc. Natl Acad. Sci. USA* **81**, 1991.
24. Bingham, P.M., Levis, R., and Rubin, G.R. (1981). *Cell*, **25**, 693.
25. Boyce, T.M., Zwick, M.E., and Aquadro, C.F. (1989). *Genetics*, **123**, 825.

6

PCR protocols and population analysis by direct DNA sequencing and PCR-based DNA fingerprinting

A. R. HOELZEL and A. GREEN

1. Introduction

The polymerase chain reaction (PCR) has become a mainstay of molecular ecology and population genetic research in the brief time since its invention (1). By this technique, defined segments of DNA can be amplified to microgram quantities from as little as a single template molecule. Although the procedure is in some ways deceptively simple, and the reaction can entail complex biochemical interactions, it is in most applications a fast, relatively inexpensive and easy way to generate ample material for further analyses. For this reason its application is now widespread and facilitating an ever greater diversity of research. From early applications to previously intractable or difficult procedures, such as the pre-natal diagnosis of sickle-cell anaemia, PCR has now been applied to a great variety of problems in molecular biology, including cDNA cloning (2) and mutagenesis (3). It is a natural for population genetic studies where large numbers of individuals need be screened at as fine a resolution as is practical and feasible. The combination of PCR amplification and DNA sequencing has been especially powerful for these studies.

The principle behind PCR is illustrated in *Figure 1*. Short 'oligonucleotide' sequences, usually 15–35 bp long, are designed, homologous to flanking regions either side of the sequences to be amplified. The primers are added in great excess to the template DNA, in the presence of buffer, DNA polymerase, and free nucleotides. The template DNA is denatured at 95°C and then cooled to allow the oligonucleotides (primers) to anneal (usually to 40–65°C); then the temperature is adjusted to the optimal temperature for the DNA polymerase for an extension phase (usually 72°C), and the cycle is repeated (25–40 times). In the initial cycle, copies are made of the target sequence and, thereafter, more and more copies are made from other copies. The expansion is roughly exponential.

In early applications, three water baths would be set up at the appropriate

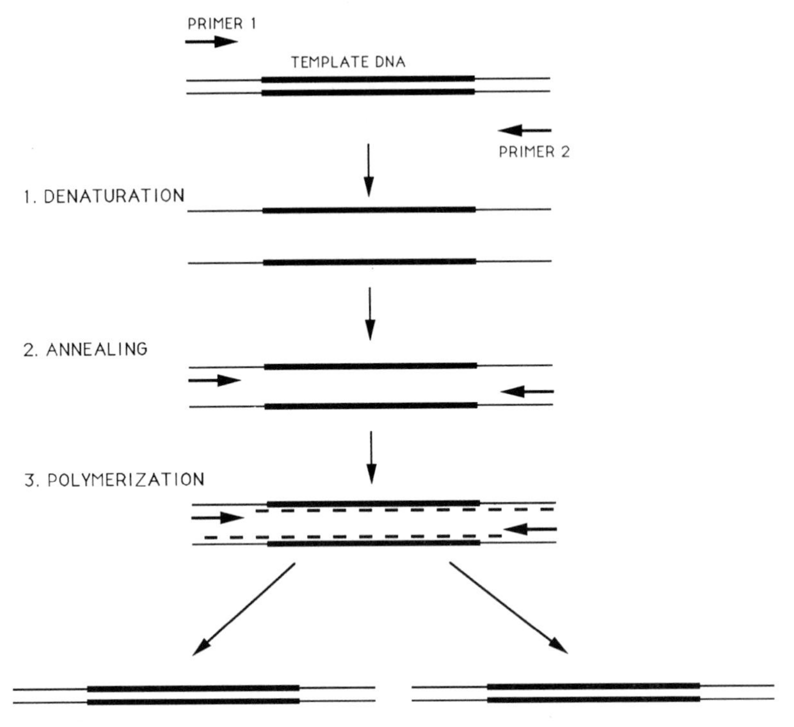

Figure 1. The polymerase chain reaction.

temperatures and the tubes transferred manually between baths. The enzyme used was the Klenow fragment of DNA polymerase I, which is denatured by high temperature, and therefore had to be replenished at the start of each cycle. The use of a thermostable DNA polymerase (such as *Taq* DNA polymerase, isolated from the eubacterial micro-organism *Thermus aquaticus*) instead of the Klenow fragment was a revolutionary development and allowed just one aliquot to be added at the start of the procedure. Another improvement has been the design of 'intelligent' heating blocks that control the thermal cycle, eliminating the need to transfer tubes manually. These are discussed in more detail in Section 2.

This chapter will focus on basic PCR amplification protocols and on those protocols that facilitate the preparation of template for DNA sequencing. In addition, two analytical methods for screening variation will be reviewed: the direct DNA sequencing of PCR product, and PCR-based DNA fingerprinting methods such as RAPD (random amplified polymorphic DNA; ref. 4) PCR, and AFLP (amplified fragment length polymorphism) analysis. Several other important analytical methods are the subjects of separate chapters: microsatellite analysis (Chapter 7), single-strand conformational polymorphism (SSCP) analysis (Chapter 8), and denaturing gradient methods (Chapter 8).

2. PCR machines

Since the initial application of *Taq* DNA polymerase to PCR there has been a proliferation of machines suited to the procedure. The basic design problem is straightforward: to heat and cool the sample tubes quickly and evenly across all wells. Manufacturers have approached the problem in a variety of different ways, and while some early problems (5) have now been largely overcome, there are still some important distinctions between different designs.

One useful and now widespread innovation is the heated lid. The heated cover provides a constant temperature across the tops of all the tubes (usually at about 65°C) and thereby limits evaporation in the tubes, eliminating the need for an oil overlay. Another useful feature is interchangeable blocks, permitting the use of different sized tubes as well as *in situ* PCR on slides.

2.1 Heating and cooling

Many of the early designs for heating and cooling the sample block had some limitations, especially relating to temperature consistency across the block. For example, a machine that was heated with a halogen bulb at the centre of a metal plate (which is indented to form tube wells) and cooled using a fan to blow air across the plate heated tubes most quickly at the centre of the plate, and cooled most quickly near the fan. Designs based entirely on heating and cooling water or air were consistent between samples, but the rate of heating and cooling was limited by thermal inertia. One early design that is still in common use involves the injection of heating and cooling fluids into the corner of a metal block. The result is that the corner where the hot fluid is injected heats faster than the opposite corner and, similarly, that corner falls below the target temperature during the cooling phase (see ref. 5).

The best consistency across the block is achieved by systems that are heated and cooled with a thermo-electric (Peltier) heat pump, and many machines now employ this method. The Peltier pump is based on the principle that thermocouples connected together and arranged back to back will create a temperature differential when a current is applied across them. One side heats up while the other cools down. Which half of the system is hotter than the other depends on the direction of the current. Extreme temperature differentials can be created when the Peltier systems are stacked. The Peltier pumps used in PCR machines are capable of cycling between 4°C and over 100°C. Early versions of thermo-electric pumps were stressed by repeated heating, and not very reliable, but significant advances have been made since then.

2.2 Controlling the cycle

Most machines control the temperature cycle with a thermocouple (or several thermocouples) either connected to the block or set in a tube which can be placed in the block beside the samples. In some designs the timing of the cycle is controlled by a remote thermocouple so that the timing of each phase will not begin until the required temperature is reached within the sample tubes. This design gives a high level of correspondence between the programmed cycle and the temperature cycle in the tube. When the thermocouples are connected to the block their accuracy is increased by using several and averaging across them. However, there will always be some time lag between the block reaching a given temperature and the contents of the tube reaching that temperature; this lag will depend on the resistance of the tube and reaction medium to heat transfer. Therefore, provided that the medium in which the thermocouple is embedded closely approximates the thermal conductivity of the reaction medium, and assuming that the thermocouple is accurate, controlling the cycle with a remote thermocouple in a reaction tube will provide the greatest correspondence between programmed and actual temperatures.

2.3 Software

Most programs either allow an entire cycling profile to be entered into a single file or require that each step be stored in a separate file and later linked. The manufacturer usually lists the number of file spaces available. Therefore, the number of complete profiles which can be entered and stored will depend on which method of programming is employed, For example, if 100 file spaces were available, then this would allow only 30 profiles that required three steps (e.g. initial denaturing, cycling, final extension phase) assuming that each file space allowed only one type of profile (such as constant temperature versus cycling). Some programs allow control over the ramp time between constant temperatures or varying the program over the course of the cycle. Ramp time can be an important factor contributing to increased yield for some templates (see Section 2.4). Changing the profile over the course of the cycle can also improve yield. For example, Yap and McGee (6) found that gradually decreasing the denaturing temperature from 94 to 80°C increased the yield for amplifications of short segments. This is probably due to reduced degradation of the DNA polymerase and template DNA (heat and cations degrade DNA). The small PCR-generated DNA fragments would not require as high denaturing temperatures as the initial genomic template.

2.4 Effect of profile on yield and specificity of amplified product

The rate at which the temperature changes during the transition between increasing or decreasing and stabilizing at a constant temperature (the last few seconds of the approach to a constant temperature) can have an effect on the specificity of the amplification. This is especially true for the transition to the extension phase where a higher proportion of non-specific amplification products are generated when the transition is gradual. This effect may be related to the dynamics of non-specific annealing, as imperfect hybrids may be more likely to persist when the temperature rise is gradual. It can be alleviated to some extent by raising the annealing temperature.

Alteration of the temperature ramp rate (i.e. the duration from one constant temperature phase to the next) can also affect the efficiency of the amplification. Specifically, a slow ramp rate from the denaturation to the annealing phase can increase yield, presumably by facilitating annealing. In a similar way, slow cooling can improve the annealing step for sequencing (7).

3. *Taq* DNA polymerase

In the early 1980s, DNA polymerase activity was detected in several forms of thermophilic bacteria, such as *Bacillus stearothermophilus*, and numerous archaebacteria. Of all these enzymes the most widely characterized DNA polymerase is that from *Thermus aquaticus* (*Taq*), which is a thermophilic bacterium capable of growing at 70–75°C. The purified protein has a molecular weight of 94 kDa, and has an optimum polymerization temperature of 70–80°C. It loses its activity (but is not denatured) at temperatures above 90°C, and its activity is restored on return to lower temperatures. It has a reported half-life of 40 min at 95°C. *Taq* DNA polymerase has very high processivity, and can achieve an activity of up to 150 bases per second per molecule at an optimum extension temperature (8). At lower temperatures, the enzyme activity is reduced to 2 bases per second per molecule. Purified *Taq* DNA polymerase is the most widely available enzyme for PCR, and is the most widely cited. The gene for *Taq* DNA polymerase has been cloned, and the recombinant form is now widely available (9). The recombinant enzyme is reported to have the advantage of minimal batch-to-batch variation, and therefore more consistency than the native form.

More recently, a wide variety of other thermostable DNA polymerases have become available, in both purified and recombinant forms. A summary of most of these is shown in *Table 1*. Not all of the data for each enzyme are available, and many of the data about these enzymes are provided by the manufacturers, as independent evidence is difficult to obtain. These DNA polymerase can be compared with *Taq* DNA polymerase on the basis of half-life, processivity, PCR error rate, proof-reading ability, exonuclease activity, reliability, and cost. An enzyme that has proof-reading ability will have a considerably lower error rate than one without such an ability. Several of the enzymes are considerably more thermostable than *Taq* DNA polymerase. *Taq* DNA polymerase tends to leave a 3′ overhang on the PCR product, which many need to be end-filled prior to cloning. Alternatively, a complementary vector can be used (see below). Several of the newer enzymes produce blunt ends (e.g. Vent, Deep Vent, and *Pwo*), allowing direct cloning into a blunt-ended vector. These enzymes can also be used for end-filling a PCR product produced by *Taq* DNA polymerase prior to cloning. *Tth* DNA

Table 1. Thermostable DNA polymerases

Polymerase	Origin	Species	3′–5′ (error rate per 10^{-5} bp)	5′-3′ exonuclease[b]	Half-life at 95°C	Processivity (bp/sec)[c]	Ends produced
Taq	native	*Thermus aquaticus*	no (2–21)	yes	40 min	150	3′ overhang
Taq	recombinant	*Thermus aquaticus*	no (2–21)	yes	40 min	150	3′ overhang
Taq (exo-)	recombinant	*Thermus aquaticus*	no (5)	no	80 min	40	3′ overhang
Vent	recombinant	*Thermococcus litoralis*	yes (2–5)	no	6–7 h	150	95% blunt
Vent (exo)	recombinant	*Thermococcus litoralis*	no (19)	no	6–7 h	150	70% blunt
Deep Vent	recombinant	*Pyrococcus* sp.	yes	no	23 h		95% blunt
Deep Vent (exo)	recombinant	*Pyrococcus* sp.	no	no	23 h		
$9°N_m$	recombinant	*Thermococcus* sp.	yes[d]	no	6–7 h		
Pwo	recombinant	*Pyrococcus woesei*	yes	no	2 h at 100°C		blunt
Tli	recombinant	*Thermococcus litoralis*	yes	no			
Tth	recombinant	*Thermus thermophilus*	no	yes	2 h	120	
Pfu	recombinant	*Pyrococcus furiosus*	yes (0.16)	no	14 h	30	
Tfl	recombinant	*Thermus flavus*	yes (1.03)	no			

[a]Native and recombinant *Taq* can be acquired from Perkin Elmer, Stratagene, Boehringer Mannheim, Gibco BRL, etc, *Taq* (exo-) from Promega, Vent, Vent (exo-), Deep Vent, Deep Vent (exo-) and $9°N_m$ from New England Biolabs, *Pwo* from Boehringer Mannheim, *Tli* from Promega, *Tth* from Pharmacia, Boehringer Mannheim and Perkin Elmer, *Pfu* from Stratagene and *Tfl* from Promega.
[b]3′–5′ exonuclease proof-reading activity.
[c]At optimum temperature.
[d]1–5% of the wildtype 3′–5′ exonuclease activity.

polymerase has reverse transcriptase as well as DNA polymerase activity, and can be used to generate a PCR product directly from an RNA template.

It is likely that further variant thermostable DNA polymerase with differing properties will become available over the next few years to fill the increasing demand for PCR-derived tests. Several companies are now producing mixtures of DNA polymerases for particular applications, particularly 'long PCR' (see below), and these mixtures are claimed to be more effective. The ideal polymerase depends on the particular application needed. If fidelity of copying from a small amount of template is important, then an enzyme with proof-reading ability should be used. If a PCR product needs to be cloned into a blunt-ended vector, then the type of ends produced will be important. For PCR of a 2 kb or smaller product from a reasonable amount of DNA template, where fidelity is not critical, then *Taq* DNA polymerase is still a sensible and cost-effective choice of enzyme.

4. Sample preparation

4.1 DNA extraction

There are fewer restrictions on the quality of DNA used for PCR than for procedures requiring digestion with restriction enzymes. It is sufficient in some cases to lyse tissue cells and use this cellular material and lysis buffer as the PCR reaction medium. However, contaminating proteins and RNA can in some cases inhibit the reaction. *Protocol 1* is a simple cell lysis procedure (after ref. 10). This is useful when only very small samples are available, such as single plucked hairs. *Protocol 2* is a general-purpose procedure for animal tissue. It produces both nuclear and organelle DNA of sufficiently high quality for PCR in most cases. An OH treatment can be used to clean up crude preparations and in some cases to improve the efficiency of PCR (E. Notarianni, personal communication). This involves simply mixing the DNA solution with equal volumes of 0.1 M NaOH and 0.1 M KH_2PO_4, and incubating at room temperature for 5 min. A portion of this mixture is then used in the PCR reaction (in a volume of less than 5 μl). See Chapter 2 for a much more detailed discussion of DNA extraction methods.

Protocol 1. Cell lysis preparation for PCR

Equipment and reagents

- 10 × PCR/lysis buffer: 500 mM KCl, 100 mM Tris–HCl pH 8.3, 25 mM $MgCl_2$, 1 mg/ml gelatin, 5% (v/v) Nonidet P-40 (NP-40), 5% (v/v) Tween-20; store at −20°C
- 3 mg/ml proteinase K in H_2O; store at −20°C
- Fine-tipped pair of forceps
- Water baths at 65°C and 95°C
- dNTPs, primers, and *Taq* DNA polymerase (see Section 6)

Method

1. Mix 5 μl 10 × PCR/lysis buffer, 1 μl proteinase K solution, and 37 μl

Protocol 1. *Continued*

H_2O in a microcentrifuge (for a 50 μl total PCR volume, scale up or down as appropriate).

2. Immerse a single hair root, or similar-sized tissue sample and use a fine-tipped pair of forceps to cut up the sample. Use the side of the forceps to grind pieces against the side of the tube.
3. Incubate at 65 °C for 2 h or at 37 °C overnight.
4. Incubate at 95 °C for 10 min to inactivate the proteinase.
5. Add dNTPs, primers, *Taq* DNA polymerase, and proceed with PCR (see Section 6), or store at –20 °C.

Protocol 2. Whole-cell DNA extraction

Equipment and reagents

- Phenol (re-distil if required, melt at 68 °C, add 8-hydroxyquinoline to a final concentration of 0.1% (w/v), and equilibrate to pH 7.5–8.0 with 1 M Tris–HCl followed by 100 mM Tris–HCl pH 8.0); store in the dark under buffer at 4 °C
- Digestion buffer: 100 mM NaCl, 10 mM Tris–HCl pH 8.0, 20 mM EDTA, 2% (w/v) sodium dodecyl sulfate (SDS); store at room temperature
- 10 mg/ml proteinase K in H_2O; store at –20 °C
- Phenol:chloroform:isoamyl alcohol (25:24:1 by vol.)
- Chloroform:isoamyl alcohol (24:1 v/v)
- 5 M LiCl
- 70% ethanol
- TE: 10 mM Tris–HCl pH 7.6, 1 mM EDTA pH 8.0; autoclave and store at room temperature
- Mortar and pestle
- 65° incubator

Method

1. Grind tissue under liquid nitrogen in mortar and pestle.[a] Quickly transfer powder into a microcentrifuge tube on ice (do not fill tube(s) more than one-third full).
2. Add enough digestion buffer to bring the volume to 500 μl, and add 30 μl proteinase K solution.
3. Incubate at 65 °C for 2 h with occasional mixing or at 37 °C overnight.
4. Add an equal volume (500 μl) of phenol and shake moderately for 1 min.
5. Spin in a microcentrifuge at 13 000 × *g* for 1 min and transfer the upper (aqueous) phase to a new microcentrifuge tube. Be careful to avoid the interface.
6. Repeat steps **4** and **5** using a 25:24:1 mixture of phenol:chloroform: isoamyl alcohol.
7. Repeat steps **4** and **5** using a 24:1 mixture of chloroform:isoamyl alcohol.

8. Mix aqueous phase with an equal volume of 5 M LiCl and incubate at −20°C for 20 min.
9. Spin for 20 min at 13000 × *g* in a microcentrifuge and carefully remove supernatant. Do not attempt to recover the last 20 μl.
10. Add 2 vol. of 100% ethanol to the supernatant, gently invert tube several times, and leave at room temperature for 20 min if there is a visible precipitate; if not, incubate at −20°C for 1 h.
11. Spin in a microcentrifuge at 13000 × *g* for 10–15 min and carefully pour off supernatant, or remove and discard with a sterile pipette. Add 1 ml of 70% ethanol and spin for 5 min. Discard supernatant as before.
12. Dry pellet under vacuum (or on bench) and dissolve in a suitable volume of TE[b] to a final concentration of about 100–200 ng/μl. Put at 65°C for 10 min and store at 4°C for short periods or −20°C long term.

[a] For whole blood, mix with an equal volume of digestion solution and incubate as for tissue.
[b] It is best to store DNA in TE. However, EDTA can interfere with PCR by binding to magnesium ions. If the volume in which DNA is introduced to the reaction is kept low (1–2 μl out of a 50 μl reaction volume), this is not a problem. Alternatively, store DNA to be used in PCR reactions in a 5:1 dilution of TE.

5. Designing primers

Oligonucleotide primers are chosen to match known DNA sequences, and amplify DNA of a predicted size. Their specificity is critical to the success of the procedure. Certain fundamental concepts apply to the design of primers for specific DNA sequences (11–13). As a rough guide, primers should be 17–24 bases in length, with 40–60% GC content and matching annealing temperatures. Homology with the template should be 100% at the 3′ end of the primer (where polymerization is initiated), while the 5′ end can be completely non-homologous. Annealing between pairs of primers should be avoided, as this can give rise to false PCR products (primer-dimers; see ref. 14), and consumes excess primer which is necessary for re-annealing. Similarly, the predicted secondary structure of each primer should not allow significant folding upon itself, which would make it unavailable for annealing to a template. The primers should not falsely anneal at other sites on the template; this is especially important for the 3′ end of the primer.

There are no universal rules for calculating annealing or melting temperatures for PCR primers. The annealing temperatures are dictated by primer length, relative GC content, the order of nucleotides in the sequence, and the secondary structure formed by the primers. In general, the higher the annealing temperature, the greater the specificity. The simplest rule for determining annealing temperature is that of Suggs *et al.* (15) where melting temperature

Table 2. Approximate annealing temperatures for oligos of varying lengths and composition[a]

Primer length (base)	Annealing temperature (°C) for GC content (bases)												
	6	7	8	9	10	11	12	13	14	15	16	17	18
16	39	41	43	45	57	49	51	53	55	57	59		
17	43	45	47	49	51	53	55	57	59	61	63	65	
18	45	47	49	51	53	55	57	59	61	63	65	67	69
19	47	49	51	53	55	57	59	61	63	65	67	69	71
20	49	51	53	55	57	59	61	63	65	67	69	71	73
21	51	53	55	57	59	61	63	65	67	69	71	73	75
22	53	55	57	59	61	63	65	67	69	71	73	75	77
23	55	57	59	61	63	65	67	69	71	73	75	77	79
24	57	59	61	63	65	67	69	71	73	75	77	79	81
25	59	61	63	65	67	69	71	73	75	77	79	81	83
26	61	63	65	67	69	71	73	75	77	79	81	83	85
27	63	65	67	69	71	73	75	77	79	81	83	85	87
28	65	67	69	71	73	75	77	79	81	83	85	87	89
29	67	69	71	73	75	77	79	81	83	85	87	89	91
30	69	71	73	75	77	79	81	83	85	87	89	91	93

[a] Based on formulations presented in reference 20.

(T_m) in °C is equal to twice the number of A or T bases plus four times the number of G or C bases. Annealing temperature is typically taken to be 5°C below T_m. This provides a reasonable first estimate as GC-rich regions are more stable than AT-rich regions, and estimated annealing temperatures calculated this way are presented in *Table 2*. However, the duplex stability depends more on sequence than on GC content. For example, the duplex TT/AA is more stable than GA/TC (16). Other factors include the size of the expected PCR product (17), helix structure (18), and buffer composition. Ultimately, these should be used only as guidelines and the annealing temperature optimized experimentally for each particular set of primers. Several computer programs for IBM PC and Apple Macintosh are now available commercially to aid in the design of PCR primers (12,17,19). While computer-aided design of PCR primers is a useful tool, experimental validation of oligonucleotides and analysis of the PCR product is essential to verify the utility of the primers.

Some genomic regions are sufficiently well conserved that the sequence is essentially the same for a broad spectrum of organisms. When these conserved regions flank a more variable region, and provided the intervening, variable region is not too large, they are useful for the design of 'universal' primers for inter- and intraspecific phylogenetic screening. A number of pairs of universal primers are described in *Table 3*.

Table 3. Universal primers

Primer sequence[a]	Length[b]	Taxa[c]	Reference
Portion of mtDNA cytochrome *b*			
Primer 1: CCATCCAACATCTCAGCATGAAA			
Primer 2: CCCCTCAGAATGATATTTGTCCTCA	308 bp	M,B,A,R,F	64
Portion of mtDNA 12S rRNA			
Primer 1: CAAACTGGGATTAGATACCCCACTAT			
Primer 2: AGGGTGACGGGCGGTGTGT	387 bp	M,B,A,R,F,I	64
mtDNA control region			
Primer 1: TTCCCCGGTCTTGTAAACC			
Primer 2: ATTTTCAGTGTCTTGCTTT	1 kb	M,F	65
Portion of mtDNA 16S rRNA			
Primer 1: GTGCAAAGGTAGCATAATCA			
Primer 2: TGTCCTGATCCAACATCGAG	366 bp	M,B,F,I,G	[d]
Portion of mtDNA CO3			
Primer 1: AGTAGACCACAGCCCATGAC			
Primer 2: ATCTACGAAGTGCCAGTATCA	700 bp	M,B,F,I,G	[d]
D2 expansion segment from rDNA 28S			
Primer 1: TAAGGGAAAGTTGAAAAGAA			
Primer 2: GTTAGACTCCTTGGTCCGTG	700 bp	M,B,F,I,G	
rDNA, ITS-1 region			
Primer 1: ACACCGCCCGTCGCTACT			
Primer 2: GCGTTCGAAGTGTCGATG	1.1 kb	M,A,I	66
Chloroplas *rbcL* gene			
Primer 1: ATGTCACCACAAACAGA			
Primer 2: TAGTAAAAGATTGGGCC	(1.4 kb)	P	67

[a] All primers listed 5′–3′; Primer 1 is the sense primer and Primer 2 the antisense one.
[b] Approximate length in human sequence (except entry in parenthesies).
[c] M, mammals; B, birds; R, reptiles; F, fish; I, insects; A, amphibians; G, gastropods; P, various plants.
[d] Pennington and Hoelzel, unpublished. The 16S portion begins at human L2585; CO3 begins at human L9236.

6. Reaction conditions

The standard PCR buffer contains 1.5 mM $MgCl_2$, 10 mM Tris–HCl pH 8.4, 50 mM KCl, and 100 μg/ml gelatin. Magnesium concentration affects the reaction such that too little reduces yield and too much increases non-specific amplification. 1.5 mM $MgCl_2$ is often optimal, but varying this concentration between 0.5 and 10 mM can increase the efficiency of the reaction dramatically in some cases. For example, Saiki (14) used two overlapping sets of primers that each amplified 1.8 kb from the human β-globin gene. One set amplified best at 1 mM $MgCl_2$, while the other amplified best at 6 mM $MgCl_2$ (and not at all at 1 mM $MgCl_2$). Deoxynucleotide triphosphates (dATP, dCTP, dGTP, and dTTP) quantitatively bind Mg^{2+}, so any adjustment of the dNTP concentration may require a compensatory adjustment of $MgCl_2$.

dNTPs are usually included at a concentration of 50–200 μM for each nucleotide. Higher concentrations may promote mis-incorporations by the polymerase (8). A concentration of 50 μM is sufficient (enough for the synthesis of approximately 7 μg of DNA), though it is more common to use 200 μM. Neutralized dNTP solutions can be purchased from a number of suppliers (e.g., Pharmacia, United States Biochemicals). Alternatively they can be purchased as lyophilized powders, which are less expensive, but the aqueous solutions must be neutralized with NaOH to pH 7.0–7.5.

KCl is included to facilitate annealing, and it can increase the synthesis rate of *Taq* DNA polymerase by 50–60% (with an optimal effect at 50 mM; reference 20). However, Innis *et al.* (8) have found that KCl above 50 mM is inhibitory to *Taq* DNA polymerase activity. Gelatin is included to help stabilize the enzyme, though the reaction usually works well without it. Bovine serum albumin (BSA) is sometimes used in place of gelatin.

Both template DNA and *Taq* concentration can affect the efficiency of the amplification. About 100 pg of plasmid DNA or 10–100 ng of genomic DNA is usually appropriate; however, this will often depend on the template and quality of the DNA preparation. Too much DNA (or substances co-purifying

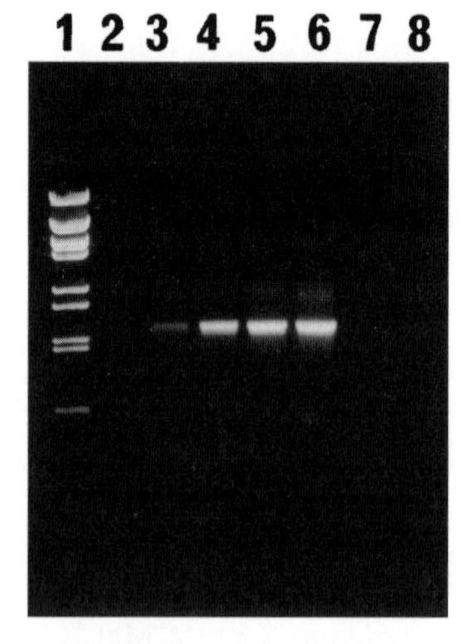

Figure 2. PCR amplification of *Mirounga angustirostris* mitochondrial DNA control region from whole-cell DNA preparations (*Protocol 2*) using the primers described in *Table 3* and target DNA in various concentrations. DNA concentration was determined by visual comparison with known weight standards (see Chapter 9) or by dilution. For all trials DNA was added in 1 μl TE, and one-fifth of the PCR reaction was run on the 0.8% agarose gel. Reaction times were 2 min at 50°C, 2 min at 70°C, and 45 sec at 94°C for 25 cycles. Lane 1, lambda/*Bst*EII size marker; lane 2, negative control; lane 3, amplification from 200 pg whole-cell DNA; lane 4, from 1 ng; lane 5, from 8 ng; lane 6, from 125 ng; lane 7, from 0.5 μg; lane 8, from 1.0 μg. A variety of DNA sources, templates, and extraction techniques show a similar pattern (unpublished observation).

with the DNA) can be inhibitory (*Figure 2*), while it is possible to amplify from as little as one copy of template DNA given the right conditions and a sufficient number of cycles (21). The *Taq* concentration usually varies from 0.5 to 2 units per 50 μl reaction. Too little will limit the amount of product, while too much can produce unwanted non-specific products (14).

To give an example, a reaction might be made up as follows.

- 5 μl 10 × PCR buffer: 15 mM $MgCl_2$, 100 mM Tris–HCl pH 8.4, 500 mM KCl, 1 mg/ml gelatin
- 5 μl 2 mM dNTPs: 2 mM each of dATP, dTTP, dGTP, and dCTP
- 1 μl of each primer (0.25 μM; *c.* 0.5 μg)
- 38 μl H_2O
- 1 unit of *Taq* DNA polymerase: polymerase can be added in 1 × PCR buffer after the initial denaturation step to prolong its activity (*Taq* DNA polymerase has a half-life of 40 min at 95°C), but this is often unnecessary (it will lose about 6% of its activity during an initial denaturation step of 95°C for 5 min, but the remaining activity should be sufficient for the remaining 25–35 cycles). Further, other enzymes with longer half-lives can be used (see *Table 1*).
- 1 μl template DNA (about 10–100 ng genomic DNA added last before the initial denaturation step)

This is overlaid with 50 μl light mineral or paraffin oil (unless a heated lid is available on the PCR machine, in which case no oil is needed). When many samples are being run together using the same primers, it is useful to combine all components with the exception of the DNA in a stock solution, using the proportions indicated above. Then aliquot to separate tubes, and add the DNA to each tube. The reaction mixture is then cycled as described in Section 6.3.

6.1 Additives that improve yield and specificity

Various additives are sometimes included to enhance PCR performance. Non-ionic detergents such as Triton X-100, Tween-20, or Nonidet P-40 (NP-40) at concentrations of 0.1–1% (v/v) will sometimes increase yield, though non-specific amplification is sometimes also increased. They help stabilize the enzyme and, if there is any SDS contamination (left over from DNA extraction), non-ionic detergents will neutralize its inhibitory effect. Just 0.01% SDS contamination will reduce *Taq* DNA polymerase activity to 10%, but the inclusion of 0.5% (v/v) Tween-20 and 0.5% (v/v) NP-40 reverses this effect (20). Non-ionic detergents may also suppress the formation of secondary structures (22).

Specificity of PCR priming can sometimes be increased by the inclusion of dimethylsulfoxide (DMSO) at 2–10% (v/v) (23), formamide at 1–5% (v/v) (24), or tetramethylammonium chloride (TMAC) at 15–100 mM (25). Non-specific priming is especially problematic in GC-rich regions. DMSO is thought to work by reducing secondary structure in the template, and was found to be essential for the amplification of a 75% GC-rich region of the retinoblastoma gene (26). However, 10% (v/v) DMSO has been found to reduce *Taq* DNA polymerase activity by 50%, and can thereby reduce yield (20). The same study showed that up to 10% (v/v) formamide did not affect the activity of *Taq* DNA polymerase. For amplifications of two regions from the human D2 receptor gene, 1.25% (v/v) formamide worked as well as 2.5% and 5%, and no amplification was seen at 10% formamide (24). TMAC was shown to eliminate non-specific priming in amplifications from tumour necrosis factor β and interleukin-1α cDNAs (27). It has previously been used to reduce potential DNA–RNA mismatch (28) and to improve the stringency of hybridization reactions (29).

The use of 7-deaza-2′-deoxyguanosine (dC^7GTP) in place of dGTP can sometimes facilitate the reaction when stable secondary structures in the template DNA interfere with amplification (30). The base analogue dC^7GTP destabilizes secondary structure. The reaction conditions are the same (see Section 6) except that dGTP is replaced in a ratio of 3:1 dC^7GTP:dGTP.

6.2 Contamination

PCR is an impressively sensitive procedure, capable of amplifying from minute concentrations of DNA. This can work to your disadvantage when solutions or equipment become contaminated with non-target DNA. There are a number of common-sense practices that can help to minimize this problem:

- use disposable gloves
- autoclave buffer solution
- add template DNA last to help avoid contaminating stock solutions
- use positive displacement pipettes or use filtered pipette tips and clean pipettes frequently with 0.25 M HCl (or autoclave), as the barrel of the pipette can become contaminated with aerosols of DNA solutions

As an added precaution in some laboratories, PCRs are set up in an enclosed bench-top Perspex (Plexiglass) container which contains a small short-wavelength UV lamp. The lamp is left on at all times when the container is not in use (to degrade contaminating DNA), and a flap or arm holes provides access for pipetting solutions.

There are also several post-hoc methods for eliminating contaminating DNA including DNase treatment (31) and gamma irradiation (32). The simplest and most effective method in our hands is to irradiate the PCR reaction with UV light after all reaction components have been added, with the exception of the template DNA and *Taq* DNA polymerase (33). The sample is irradiated with 254 or 300 nm UV light for 5–20 min. This is especially effective if the contaminating DNA is genomic, as UV light most effectively degrades large fragments. A 5 min exposure in a standard microwave oven at maximum setting will often work for contamination with small fragments.

6.3 Temperature cycle

The PCR reaction typically requires cycling between three temperatures: the first to denature the template strands, the second to anneal the primer to the template, and the third for extension at a temperature optimal for the DNA polymerase. This cycle is usually repeated 25–40 times. After 20–25 cycles the amplification will start to reach a plateau. This attenuation of the initial exponential rate of amplification is usually concomitant with the accumulation of 0.3–1 pM of product (34). It is important to keep the number of cycles down to the minimum required, as non-specific amplification will sometimes increase after the plateau has been reached. The number of cycles required depends in part on the quantity of template. More than 40 cycles are usually only necessary when there are fewer than 1000 initial template molecules (34).

Insufficient denaturation is a common problem leading to the failure of a PCR reaction. It is usually sufficient to heat to 94°C for 30–45 sec; however, some templates may require up to 98°C, especially if the sequence is GC-rich. On the other hand, *Taq* DNA polymerase loses activity at high temperature (with a half-life of about 40 min at 95°C and about 5 min at 98°C), so denaturation times and temperature should be kept to a minimum. It is important to note that some PCR machines control the temperature cycle using thermocouples on the block, and there may be a delay between the times at which the block and the sample tube reach the target temperature. This will require some compensation when inputting the durations for the cycle profile (see Section 2.2).

The temperature of the annealing step will depend to some extent on the base composition and length of the primers (see Section 5). Unless the primers are very short or AT-rich, temperatures in the range 50–60°C are usually best. Temperatures that are too low can permit non-specific priming. A good first choice is 5°C below T_m, if known; otherwise try 55°C. When designing a large number of primers, optimizing annealing temperatures can be time-consuming and may warrant the use of specially designed thermocyclers (e.g. Robocycler (Stratagene)) which produce a temperature gradient across a block dedicated to annealing (samples are robotically moved between fixed temperature blocks). Annealing may require only a few seconds, but this will depend on a variety of factors such as interference from secondary structure and primer concentration. A time within the range 30 sec to 2 min is usually appropriate.

At the optimal temperature for *Taq* DNA polymerase (70–80°C depending on the preparation and the template) extension time depends on the length and concentration of the target sequence. Estimations of the rate of DNA polymerase synthesis vary from 35 to 150 nucleotides/sec, depending on the enzyme (*Table 1*), buffer composition, and the nature of the template (8). As a rule of thumb, 1 min per kb is probably ample. However, products of up to 2 kb seldom require more than 1 min, while products over 5 kb may require considerably longer extension times. *Taq* DNA polymerase is active at temperatures outside the optima, though at a reduced level (half to one-third the optimum at 55 °C); therefore for some short templates the extension step can be omitted.

Prior to cycling, a preliminary denaturing period of 2–5 min at 95 °C is often included when amplifying from genomic DNA to increase the proportion of initial single-stranded template. After cycling, an extended extension period for 5–10 min is often included to ensure that all annealed template is fully polymerized.

A typical example of a PCR profile follows:

(a) 5 min at 95°C;
(b) a repeated cycle (25 times) consisting of: 55°C for 1 min, 72°C for 1 min, and 94 °C for 30 sec;
(c) 55°C for 1 min;
(d) 72°C for 10 min;
(e) refrigerate.

In some instances it may be appropriate to change the temperature of the annealing step over the course of the experiment. For example, if homology with the template is poor, non-specific amplification can sometimes be reduced by using a 'step-up' procedure whereby the annealing temperature (and therefore the stringency) is increased after a few initial cycles. Alternatively, when the homology with the template is good, but there is a persistent secondary amplification product, this can sometimes be diminished using a 'step-down' procedure where the stringency of the reaction is gradually reduced every few cycles, starting with an annealing temperature that is about 5–10°C above the T_m and going down to about 5 °C below the T_m. The idea is to enrich for the target sequence during the early stringent cycles.

Another procedure that can reduce non-specific amplification and primer-dimer formation, especially for amplifications from low concentrations of the template (35), is the 'hot start' method (36). This entails preventing the reaction from commencing until the mixture has reached an elevated temperature. One way to do this is physically to separate portions of the reaction mixture with wax (e.g. AmpliWax, Perkin–Elmer), which only melts and allows the reaction components to mix when the temperature gets beyond 70–80°C. Another way is to use a modified *Taq* DNA polymerase (*Taq* Gold, Perkin–Elmer) which needs to be activated with an initial heat shock.

6.4 Long PCR

Amplification using *Taq* DNA polymerase is typically limited to templates of less than 3–4 kb. However, longer amplifications would be extremely useful in population studies, for example by facilitating the analysis of long introns or by permitting the amplification of long segments of the mitochondrial genome, which could then be analysed for restriction fragment length polymorphisms (RFLPs) in agarose gels stained with ethidium bromide (e.g. see ref. 37). One early method for extending the length of the amplified product used the phage T4 gene protein (Gp32; at a concentration of 1 nmol/μl) to prevent non-specific primer annealing during the preparation of the reaction (38). This permitted the amplification of fragments of up to 6.5 kb using an extension time of 5 min. More recently methods have been developed that facilitate much longer amplifications (up to 35 kb from bacteriophage lambda templates; ref. 39), including the amplification of the entire vertebrate mitochondrial genome (about 17 kb; refs 37 and 40). These methods are based on the inclusion of a proof-reading DNA polymerase (such as *Pwo* or *Pfu*, see *Table 1*), in addition to *Taq* DNA polymerase. Used on its own the 3′–5′ exonuclease activity of enzymes like *Pwo* could excessively degrade the primers, but at a low concentration with *Taq* as the dominant DNA polymerase, the proof-reading feature serves to repair errors that could otherwise terminate the reaction. 'Long PCR' methods also depend to varying extents on changes in the reaction mixture. These include the addition of co-solvents to facilitate the separation of DNA strands at lower temperatures, and buffering agents to maintain higher pH levels (41). Extension times are 10–15 min for the longer fragments, and primers are chosen with strong homology and annealing temperatures in the 62–68°C range. Very high quality template DNA seems to be important especially for the longest amplifications. To amplify the entire human mitochondrial genome, Cheng *et al.* (40) used 25 mM Tricine (pH 8.7), 80 mM potassium acetate, 10% (v/v) glycerol, 2% (v/v) DMSO, 1.0 mM magnesium acetate, 75 μg/ml non-acetylated BSA, 0.2 mM of each dNTP, 1.0 U *Tth* DNA polymerase, 0.04 U Vent DNA polymerase, and 0.2 μM of each primer. Commercial kits are also available (e.g. Expand, Boehringer Mannheim).

7. Cloning double-stranded PCR product

PCR product is notorious for being difficult to clone. There are two primary problems related to the activity of *Taq* DNA polymerase. First, *Taq* DNA polymerase tends to add an extra base to the 3′ end of the fragment, and this base is almost exclusively an adenosine (42). Second, there is some indication that the *Taq* DNA polymerase may remain bound to the DNA and thereby inhibit endonuclease activity. The first problem interferes with blunt cloning, while the second interferes with cutting restriction sites designed into the PCR primers, which would allow cloning into compatible sites in a vector molecule. In fact, there are other problems related to the use of restriction sites in the primers. Some enzymes have a low cutting efficiency if the site is close to the end of the molecule (43). Since primer homology to the template should be nearly perfect for at least the first 10–15 bp from the 3′ end, a primer that includes a restriction site may need to be 30–35 bp long.

One way to overcome the problem of an added adenosine at the 3′ end is to use one of the thermostable DNA polymerases that produces primarily blunt ends (see *Table 1*). If *Taq* DNA polymerase is used, another possibility is to fill in with Klenow and, if the 5′ end of the primer has been left with a hydroxyl group, kinase to add a phosphate. This can then be cloned into a phosphatased blunt-ended vector (Chapter 4). *Protocol 3* describes this method. Alternatively, the vector can be modified to include a single 3′-overhang T-residue. This can be accomplished either by incubating a vector that has been cleaved with a blunt cutting enzyme such as *Sma*I or *Eco*RV for 2 h at 70°C with *Taq* DNA polymerase (1 unit/μg plasmid/20 μl volume), standard PCR buffer (see Section 6), and 2 mM dTTP (44), or by the method described in *Protocol 4* (after ref. 45). If background is high using the former method, phosphatase the vector and, if necessary, kinase the insert (see *Protocol 3*).

Crowe and co-workers (46) found that treating PCR product with proteinase K increased restriction digestion, and therefore cloning efficiency (when restriction sites have been built into the primer sequence), by several orders of magnitude (*Protocol 5*).

Protocol 3. End-fill with Klenow

Reagents

- 10 × Klenow buffer: 0.5 M Tris–HCl pH 7.2, 0.1 M $MgSO_4$, 1 mM dithiothreitol (DTT); 500 μg/ml BSA (pentax fraction V); store at –20°C
- 1% low melting point (LMP) agarose gel
- Ethidium bromide
- TAE or TBE buffer (*Appendix 4*)
- 2 mM dNTP and 10 mM dATP
- 100 mM DTT
- T4 DNA polynucleotide kinase
- TE (*Protocol 2*)
- DNA polymerase (Klenow fragment)

Method

1. Separate PCR product in 1% LMP agarose gel in TAE or TBE. Include ethidium bromide in the gel at a concentration of 5 ng/ml. Cut out the band to be cloned in as small a volume as possible. Dilute in 2–3 vol. of TE and incubate at 65°C for 15 min with occasional mixing.
2. Add 15 μl of DNA in LMP agarose to 2 μl of 10 × Klenow buffer (commercial medium-salt restriction enzyme buffers will work as well), 2 μl of 2 mM dNTP (see Section 6), and 1 unit of DNA polymerase (Klenow fragment).
3. Incubate at room temperature for 2 h.
4. If the PCR primers are phosphorylated, then no further preparation is necessary. If the primers have a free 5′ hydroxyl group, they must be

phosphorylated. Add 1 μl of 10 mM ATP, 1 μl of 100 mM DDT, and 1 unit of T4 DNA polynucleotide kinase. Incubate at 37°C for 1 h and then inactivate the enzyme by heating at 65°C for 15 min.

5. Ligate and clone as described in Chapter 4. It is not necessary to purify the DNA from the LMP agarose or reaction components.

Protocol 4. Making a T-vector

Reagents

- Vector DNA
- Blunt-cutting restriction enzyme represented in the polylinker (e.g. *Eco*RV or *Sma*I)
- Phenol:chloroform:isoamyl alcohol (25:24:1 by vol.)
- Chloroform:isoamyl alcohol (24:1 v/v)
- 7.5 M ammonium acetate
- Absolute ethanol
- TE (*Protocol 2*)
- Terminal transferase
- 5 × terminal transferase buffer: 125 mM Tris–HCl pH 6.6, 1 M potassium cacodylate, 1.25 mg/ml BSA, 7.5 mM $CoCl_2$; store at –20°C
- Dideoxythymidine triphosphate (ddTTP)

Method

1. Digest 5 μg of a suitable vector with a blunt-cutting restriction enzyme represented in the polylinker (e.g. *Eco*RV or *Sma*I). Purify by phenol–chloroform extraction (as in steps **6** and **7** of *Protocol 2*) and precipitate with 0.5 vol. of 7.5 M ammonium acetate and 2.5 vol. of 100% ethanol.
2. Take DNA up in TE (*Protocol 2*). Mix DNA with 8 μl terminal transferase buffer, 100 units of terminal transferase, and ddTTP to a final concentration of 10 μM.
3. Incubate at 37°C for 1 h.
4. Purify DNA by phenol–chloroform extraction and ethanol precipitation (see step **1**).
5. Ligate to PCR product using an equimolar ratio of insert to vector (see Chapter 4).

Protocol 5. Proteinase K digestion of PCR product

Reagents

- Phenol:chloroform:isoamyl alcohol (25:24:1 by vol.)
- Chloroform:isoamyl alcohol (24:1 v/v)
- 7.5 M ammonium acetate
- Absolute ethanol
- 10 × buffer: 50 mM EDTA, 100 mM Tris–HCl pH 8.0, 5% (w/v) SDS
- 1 mg/ml proteinase K in H_2O; store at –20°C
- Incubators at 37°C and 68°C

Method

1. Purify PCR product by phenol–chloroform extraction and ethanol precipitation (see *Protocol 4*, step **1**).

Protocol 5. *Continued*

2. Mix 0.1 vol. of 10 × buffer, 0.05 vol. of proteinase K solution, and purified DNA.
3. Incubate at 37 °C for 30 min.
4. Incubate at 68°C for 10 min.
5. Purify by phenol–chloroform extraction and ethanol precipitation as in *Protocol 4*, step **1**. The DNA is now ready for restriction digestion and cloning (Chapter 4).

8. DNA sequencing PCR product

DNA sequencing provides the greatest possible resolution of genetic variation and, due to advances in the technique and methods for automation (see Chapter 10), is now a relatively affordable and quick method of analysis. This section will focus on manual DNA sequencing using the dideoxy termination method, incorporating ^{35}S-labelled nucleotides (47). The major factors determining the success of the procedure are the amount and quality of the DNA template. The ideal template is uncontaminated single-stranded DNA, such as that obtainable from M13 phage. Double-stranded sequencing from plasmid preparations, using alkali to denature the strands, often works well, but can be limited by the nature of the template. However, all these methods of generating template require cloning, and often sub-cloning, of the DNA to be sequenced. Two methods for the generation of single-stranded sequencing template by PCR, and one method for sequencing double-stranded PCR product are described in the following sections. The sequencing reactions described are based on those published for use with T7 DNA polymerase (7). A modified T7 DNA polymerase (Sequenase, USB) is generally used which polymerizes at an increased rate (roughly threefold) following the chemical inactivation of the 3′–5′ exonuclease activity of native T7 DNA polymerase.

Protocol 6. Pouring and running a sequencing gel

Caution: acrylamide solutions are highly toxic. Use two pairs of gloves when handling solutions and prepare solutions in a fume hood.

Equipment and reagents

- Acrylamide–urea solution:[a] 23.8% (w/v) acrylamide, 12.5% (w/v) methylene bis-acrylamide, 8.3 M urea (99.9% pure)
- 10 × TBE–urea solution:[a] 890 mM Tris–borate, 890 mM boric acid, 20 mM EDTA pH 8.0, 8.3 M (50%) urea
- Urea solution:[a] 8.3 M urea (99.9% pure)
- 10% (w/v) ammonium persulfate (AMPS), made up freshly
- *N,N,N′,N′*-tetramethylethylenediamine (TEMED)
- 10% acetic acid, 10% methanol

- Glass plates
- Spacers (usually 0.4 mm thick)
- Sharks-tooth comb
- Repel-Silane (Pharmacia)
- 100 ml syringe
- 3MM filter paper
- Vacuum gel dryer
- Plastic wrap

Method

1. Thoroughly clean glass plates, and siliconize the top plate only with Repel-Silane (do this under a fume hood applying small amounts using clean paper towels and a circular motion; leave to dry for 5–10 min; see Chapter 10). Put plates together with spacers (usually 0.4 mm thick) and seal edges. Sequencing kits are available that facilitate this step by providing clamps to seal the sides and a base for sealing the bottom of the gel (e.g. from Bio-Rad). If this is not available (well worth the investment), seal the edges of the plates with plastic tape.
2. Mix stock solutions according to the required percentage of monomer in the gel (6% is standard for sequencing gels). The following table gives the volume of components in ml for a total volume of 100 ml.

	Volume (ml)			
	4%	**6%**	**8%**	**10%**
Acrylamide–urea	16	24	32	40
Urea	74	66	58	50
10 × TBE–urea	10	10	10	10

3. Add 600 μl AMPS and 60 μl TEMED for each 100 ml of acrylamide mixture; swirl, and quickly pour into gel apparatus using a 100 ml syringe pushed up against the space between the plates. Tilt the plates slightly and keep the flow constant to facilitate pouring. If bubbles form they can often be dislodged by tapping on the plates.
4. Insert a sharks-tooth comb, pushing the flat side into the gel to the appropriate depth, clamp plates with bulldog clips, and allow 1–2 h to polymerize.
5. Remove the comb, pre-heat plates to 50°C prior to loading, and maintain at 50–55°C during the run. Run under 1 × TBE buffer.
6. Insert the sharks-tooth comb teeth down so that the teeth just penetrate the gel. Use a pipette to flush urea out of the wells prior to loading samples.
7. At the end of the run, dismantle the plates (carefully lifting the siliconized plate up from the gel), and soak the gel in 10% acetic acid, 10% methanol for 20 min (to remove urea, which blocks signal). The solution can be floated on top of the gel, to avoid the risk of the gel separating from the plate during total immersion.
8. Allow the gel to air-dry for 10–15 min and then cut a piece of 3MM filter paper to the size of the gel and smooth it over the gel's damp

Protocol 6. *Continued*

surface, Lift the filter paper and gel and lay it out on a flat surface. Cover the gel with plastic wrap.

9. Dry the gel at 80°C in a vacuum gel dryer for 45 min to 1 h. Expose against X-ray film after removing the plastic wrap[b].

[a] These solutions are available commercially from National Diagnostics.
[b] After running sequencing reactions (see protocols 9–11).

8.1 Sample preparation

Prior to sequencing, the PCR product should be purified from nucleotides and primers in the PCR reaction mixture. There are a number of ways to accomplish this. *Protocol 7* describes a method for the gel isolation of fragments (after ref. 48). This is probably the most efficient method, but also the most time-consuming. DNA separated in agarose gels can also be extracted by freezing the fragment and spinning in a microcentrifuge (49) or using commercial kits (e.g. Geneclean (Bio101)). *Protocol 10* in Chapter 5 describes a method for selectively removing small fragments and nucleotides on a spun column (use Sephadex G-200), which is faster but not as clean (columns are also available commercially). *Protocol 1* in Chapter 7 describes a method employing dialysis tubing. A simple precipitation method is described here in *Protocol 8*, which is probably the fastest method, but the purification is incomplete (though often sufficient) and the yield is lower than for the other two methods.

Protocol 7. DNA extraction from LMP agarose

Reagents

- LMP agarose gel
- TBE buffer (see *Appendix 4*)
- Dilution buffer: 50 mM Tris–HCl pH 8.0, 10 mM NaCl, 10 mM EDTA
- Chloroform
- Equilibrated phenol (see *Protocol 2*) pre-warmed to 37°C
- 3 M sodium acetate
- Absolute ethanol (–20°C)
- 65°C water bath

Method

1. Separate PCR sample on LMP agarose gel (use TBE buffer; see *Appendix 4*) and cut out band of interest in as small a slice as possible.
2. Melt get slice at 65°C in a microcentrifuge tube and dilute with 4 vol. of dilution buffer to 1 vol. of gel solution. Incubate at 65°C for a further 15 min with occasional mixing.
3. Add an equal volume of equilibrated phenol (see *Protocol 2*) pre-warmed to 37°C and shake moderately for 2 min.
4. Spin in microcentrifuge at 13 000 × *g* for 3 min and carefully remove the upper phase, saving it.

5. Repeat steps **3** and **4** once with room temperature phenol, and twice with room temperature chloroform.
6. Precipitate with 0.1 vol. 3 M sodium acetate and 2 vol. 100% ethanol at –20°C for 2 h. Spin down, wash, and take up as in steps **11** and **12** of *Protocol 2*.

Protocol 8. Isopropanol precipitation

Equipment and reagents

- 7.5 ammonium acetate
- Isopropanol
- 70% ethanol
- TE (*Protocol 2*)

Method

1. If an oil overlay was used, pipette PCR reaction from beneath oil overlay by following the interface down with the pipette tip. If there is any carry-over of oil, chloroform-extract once.
2. Add 0.5 vol. of 7.5 M ammonium acetate and 1 vol. isopropanol and invert several times.
3. Incubate at room temperature for 5 min.
4. Spin in a microcentrifuge at 13000 × *g* for 10 min, wash with 70% ethanol, dry on bench or under vacuum, and take up in TE.

8.2 Asymmetric PCR

By this method single-stranded DNA is generated during the PCR reaction. Primers are included in a ratio of 50:1 to 100:1, one primer at the usual concentration and one diluted. The best ratio generally needs to be determined by trial and error. Other components are the same as for the standard reaction (see Section 6). For the first 15–25 cycles most of the product will be double-stranded; then, as the limiting primer is depleted, single-stranded DNA from the strand primed by the abundant oligonucleotide will be generated. *Figure 3* shows the approximate relationship between an increasing number of cycles and the generation of double-stranded versus single-stranded DNA (after ref. 50).

The mobility of the single-stranded DNA in agarose will depend on sequence composition; however, it usually migrates close to the double-stranded product (which will stain more intensely) and is often the more slowly migrating band. This is due in part to the tendency of single-stranded DNA to form hairpin loops, which retards its mobility in a gel matrix. The single-stranded band can then be gel-isolated (as described in *Protocol 7*) and sequenced using the limiting PCR primer or an internal primer (an internal

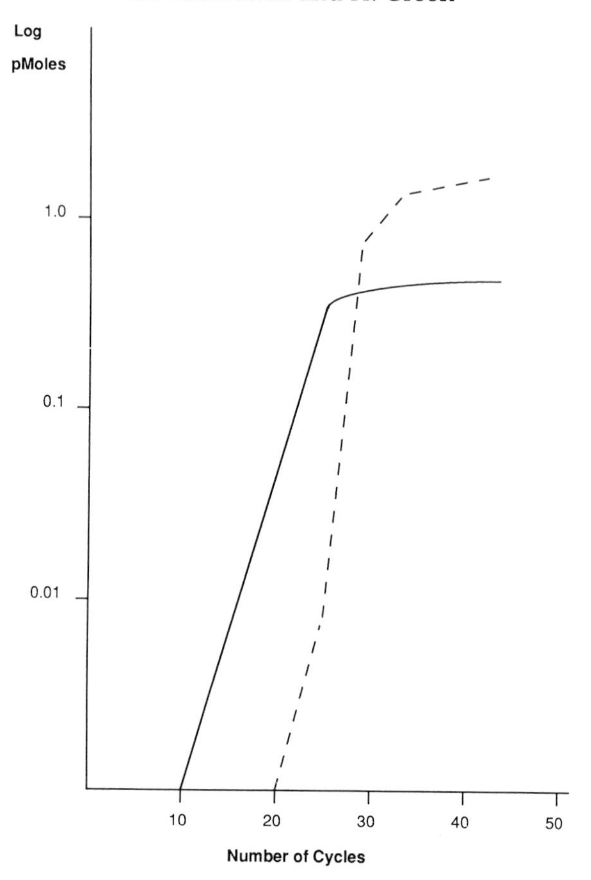

Figure 3. Accumulation of double-stranded (solid line) and single-stranded (broken line) DNA with increasing numbers of cycles (after ref. 50). This is based on the asymmetric amplification of a 242 bp fragment from the HLA-DQ locus using a 100:1 primer ratio, and cycling 43 times (50).

primer often gives better results). Alternatively, the PCR product can be purified as in *Protocol 8* or by spun column (see Chapter 5, *Protocol 10*). The double-stranded DNA, which is retained by these methods, should not interfere with sequencing the single-stranded product.

It will often be necessary to adjust the stringency of the asymmetric PCR reaction by adjusting the annealing temperature and the extension time. An increase in non-specific amplification is a common problem. Specificity and yield can sometimes be improved by first amplifying and gel-isolating the double-stranded PCR product using standard conditions, and then performing an asymmetric PCR from a 500-fold dilution of the purified product. However, a number of researchers have found that in some cases it can be very difficult to produce single-stranded product by asymmetric PCR that can be readily sequenced. Therefore the success of this procedure is to some extent related to the nature of the template.

Protocol 9. Sequencing asymmetric PCR product[a]

Equipment and reagents

- 8 × enzyme buffer: 200 mM Tris–HCl pH 7.5, 80 mM $MgCl_2$, 250 mM NaCl
- Labelling mixture: 6 μM dTTP, 6 μM dGTP, 6 μM dCTP
- Termination mixtures: make up four mixtures (A–D), each including 100 μM dATP, 100 μM dTTP, 100 μM dGTP, 100 μM dCTP, 50 mM NaCl. In addition, Mix A includes 10 μM ddATP, Mix T includes 10 μM ddTTP, Mix G includes 10 μM ddGTP, and Mix C includes 10 μM ddCTP.[b]
- Stop solution: 95% formamide, 15 mM EDTA, 0.05% Bromophenol Blue; 0.05% Xylene Cyanol FF
- 100 mM DTT
- [^{35}S]dATP (12.5 mCi/ml; NEN)
- T7 DNA polymerase
- TE
- Sequencing primer
- 65°C water bath
- 37°C heating block

Method

1. Amplify DNA in reaction conditions described in Section 6 with a 50- to 100-fold dilution of one primer, and cycle 40 times. Separate amplification product from primers and nucleotides by one of the methods described at the beginning of this section.
2. Take DNA to be sequenced up in 7 μl H_2O. All of the DNA amplified in one reaction will usually be required.
3. Prepare labelling solution. For each reaction, mix 1 μl 100 mM DTT, 0.5 μl labelling mixture, 2.0 μl H_2O, and 0.5 μl [^{35}S]dATP (6.25 μCi). Dilute T7 DNA polymerase 1:8 in TE (dilute just before use).
4. Prepare microcentrifuge tubes (or microtitre plate) for termination reactions. For each sample prepare four tubes (wells), one for each termination mix (A, T, G, and C). Label tubes or plate and pipette 2 μl termination mix into each tube (well).
5. Add 2 μl enzyme buffer and 1 μl sequencing primer (at a concentration of 20 pmol, or about 120 ng) to DNA and incubate at 65°C for 5 min in a water bath.
6. Remove tube(s) from water bath and add 4 μl labelling solution and 2 μl diluted DNA polymerase to each reaction and incubate at room temperature for 2–5 min.
7. Pre-heat termination mixes on 37°C heating block for 1 min. Add 4 μl from reaction solution to each of four termination mixtures, mix well with pipette, and incubate at 37°C for 2–5 min.
8. Stop reactions at room temperature by adding 4 μl stop solution to each. These can now be run on a sequencing gel (*Protocol 6*) or frozen and stored at –20°C for several weeks.

Protocol 9. *Continued*

9. Boil samples for 2 min and put on ice for 30 sec immediately prior to loading on sequencing gel.

[a]These reaction mixtures are for use with T7 DNA polymerase (after ref. 7). For methods for sequencing with *Taq* DNA polymerase see ref. 8.
[b]This ratio of dNTP:ddNTP (10:1) is appropriate for sequencing up to 400 bases; to read longer sequences increase the ratio to 30:1 (see ref. 7).

8.3 Biotinylated primers and magnetic beads

The use of streptavidin-coated magnetic beads (available from Dynal), in conjunction with biotinylated PCR primers can provide an uncontaminated single-stranded DNA template for sequencing. This solid-phase method of template purification and sequencing depends on the strong interaction between biotin and streptavidin, allowing the separation of a single strand of DNA (*Figure 4*). A PCR reaction is performed using oligonucleotides specific for the DNA of interest, one of which has been previously biotinylated at the 5′ end. Once cycling is finished, an aliquot of streptavidin-coated magnetic beads is added to the reaction tube, mixed, and incubated.

The biotinylated DNA, both PCR product and unincorporated oligonucleotides, will bind strongly to the streptavidin-coated beads. A small magnet is applied to the outside of the tube, pulling the beads to the wall of the tube, and the supernatant and mineral oil are aspirated. The beads are then resuspended in alkali to denature the double-stranded product. The magnet is then re-applied, to pull down what is now single-stranded DNA, allowing the

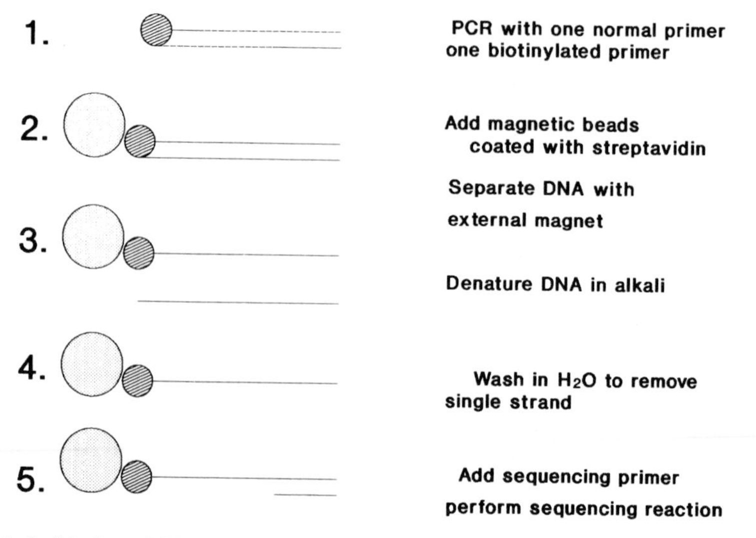

Figure 4. Solid-phase PCR sequencing.

complementary strand to be aspirated. Three washes are performed, each time holding the DNA at the side of the tube with the magnet. The single-stranded template is now ready to be sequenced using a slight modification of the standard dideoxy method. This solid-phase method of template preparation takes approximately 15 min per sample tube from the end of the PCR to the beginning of the sequencing reaction.

As originally described, the template DNA from the PCR was extracted, and the single-stranded DNA was separated and sequenced using γ-^{32}P-end-labelled primer in a dideoxy termination reaction (51). Schofield *et al.* (52) modified this procedure to allow direct separation of the single strand, and sequencing by incorporating an unlabelled primer and ^{35}S-labelled nucleotides. This method also allows direct sequencing of multiple PCR products from agarose gels (53) and clear reading of GC compressions, both of which have been problematic with other methods of template preparation.

Protocol 10. Separation and sequencing of single-stranded DNA on magnetic beads

Equipment and reagents

- Streptavidin-coated magnetic beads (10^8 beads/ml; Dynabeads M-280 (Dynal))
- 0.15 M NaOH made fresh
- Isopropanol
- TE
- Strong magnet
- Sequencing reagents as in *Protocol 9*

Method

1. Amplify by PCR in the reaction conditions described in Section 6, but include one primer that has been labelled with biotin at the 5′ ends, at a 1 μM final concentration. A typical reaction volume is 20–30 μl. If necessary, overlay with an equal volume of light mineral oil.
2. Add 30 μl of resuspended magnetic beads to the PCR and incubate at room temperature for 15 min. It is sometimes necessary to combine two identical PCRs depending on the yield of the PCR and the efficiency with which the primer was biotinylated. The magnetic beads can be introduced under the oil overlay. If you do not wish to perform the optional step **3**, proceed to step **4**.
3. (Optional) Separate the reaction from the oil by pipetting from below the oil (follow the interface down with the tip of the pipette). Precipitate with isopropanol as described in *Protocol 8*, take pellet up in 20 μl TE, and add 30 μl magnetic beads. This removes most of the biotinylated primer (which competes with the biotinylated PCR product) and can substantially improve the quality of the resulting sequence.
4. Pull the beads to the side of the tube using a strong magnet and

Protocol 10. *Continued*

slowly remove the supernatant. Be careful to avoid removing or disturbing any of the beads. A glass pipette drawn out to a very fine tip over a flame is useful for this purpose.

5. Add 20 μl of 0.15 M NaOH and mix well with the pipette. Incubate at room temperature for 5 min.
6. Pull the beads to the side of the tube, carefully remove supernatant, and wash once with 0.15 M NaOH.
7. Wash twice with 40 μl doubly distilled sterile H_2O, ensuring thorough mixing.
8. Resuspend the beads in 7 μl doubly distilled sterile H_2O. At this point the samples can be stored almost indefinitely at −20°C.
9. Proceed with sequencing by adding 2 μl buffer and 1 μl primer (20 pmol, as in *Protocol 9*) and incubating for 2–5 min at 65°C in a lead pot or block immersed in a water bath. It is often best to use a sequencing primer that is internal to the original PCR primer.
10. Remove samples still in lead pot or block and leave on bench for 20–30 min to cool gradually (allow the temperature to reach 35°C).
11. Sequence as described in *Protocol 9*.
12. Prior to loading the samples on a sequencing gel, resuspend the beads by gentle flicking, boil for 2–3 min, put on ice for 30 sec, pull beads to the side of the tube with a strong magnet, and load the supernatant on to the gel.

8.4 Sequencing double-stranded PCR product

Sequencing from double-stranded DNA is usually restricted to 200–300 bp in each direction. Therefore, if PCR products are to be screened for variation by direct sequencing, the region to be amplified is often no longer than about 500 bp (though using internal primers to amplify from 1–2 kb PCR product sometimes gives better results). However, when small fragments are denatured they re-anneal quickly, which interferes with direct sequencing. Bachman and co-workers (22) and Winship (54) have found that the addition of DMSO or non-ionic detergents and a modification of the concentrations of deoxy and dideoxy nucleotides in the sequencing reagents facilitate the sequencing of very short (less than 250 bp) double-stranded PCR products. The method given in *Protocol 11* is based on the annealing step described by these authors and the sequencing protocol described above. Internal primers gave better results than terminal primers for sequencing a 350 bp region from a 1 kb amplification product.

Protocol 11. Sequencing double-stranded PCR product

Reagents

- Sequencing reagents as in *Protocol 9*

Method

1. Purify PCR product from nucleotides and primers by one of the methods described earlier *(Protocols 7* and *8;* Chapter 5, *Protocol 10;* Chapter 7, *Protocol 1*). Use 0.5 pmol DNA (about 120 ng, typically one-tenth of the PCR reaction) in 6 μl.
2. Add 6 μl DNA to 3 μl sequencing buffer and 1 μl primer (20 pmol as in *Protocol 9*). Boil for 3 min and snap-cool on dry ice for 20 sec to 1 min.
3. Proceed with sequencing as in *Protocol 9.*

9. PCR-based DNA fingerprinting

DNA fingerprinting was originally developed as a means to exploit the high levels of variable number tandem repeat (VNTR) variation in minisatellite loci, using an RFLP technique (see Chapter 9). Since the development of PCR, the process of analysing these hypervariable loci has been further facilitated, and new VNTR markers have been exploited, microsatellites in particular (see Chapter 7). More recently, variation that reflects both point-mutational and VNTR-type changes has been screened using PCR methods that amplify numerous loci simultaneously. These are the subject of the brief review in this section. Their strength is that they offer a fast and simple method for revealing high levels of variation. Their weakness is in the interpretation of results. For all of these methods, statistical interpretation is limited by an inability to resolve separate loci, and the fact that different types of evolutionary change (e.g. point-mutational versus VNTR) evolving at very different rates are represented, and indistinguishable based on the observed pattern of bands (see Appendix 3).

Among the PCR-based DNA fingerprinting methods, probably the most extensively used are RAPDs (Section 9.1), and AFLP (Section 9.2); these are reviewed in more detail below. Other methods include DNA amplification fingerprinting (DAF; ref. 55) and arbitrarily primed PCR (AP-PCR; ref. 56). Three of these methods, DAF, RAPD, and AP-PCR, are all variations on the same theme. In each case, a single primer is used to amplify genomic DNA at random. The main distinction between DAF and RAPD is that DAF uses shorter primers (5–10 bp versus 10–12 bp), and thereby amplifies a larger number of fragments. AP-PCR is based on PCR amplification with just one larger (15–25 bp) primer (at 10 μM concentration).

AP-PCR also differs in its temperature cycling profile. An initial two cycles are run at very low stringency (e.g. 5 min at 40°C, 5 min at 72°C, 1 min at 94°C), allowing numerous non-specific amplification products to be generated, followed by 10–15 high stringency cycles (e.g. 1 min at 60°C, 2 min at 72°C, 1 min at 94 °C; these conditions are suitable for the Kpn-R primer described in ref. 56: CCAAGTCGACATGGCACRTGTATACATAYAGTAAC). The Mg^{2+} concentration is kept high (4 mM) to enhance the stability of the primer–template interactions (56). The result is a complex of 10–20 bands that are generated when, under low stringency, the primer by chance anneals on either strand over an amplifiable range. The main drawback with DAF, AP-PCR, and RAPD (see below) is that the resulting pattern of bands is very sensitive to variations in reaction conditions, DNA quality, and the PCR temperature profile.

Burt *et al.* (57) describe a method that combines random amplification with DNA sequencing, avoiding the difficulties of interpretation with RAPD and related techniques, but requiring considerably more time and effort. They refer to the procedure as 'sequencing with arbitrary primer pairs' (SWAPP). They use two different random 10-mers in the amplification, and choose resulting bands that are not amplified by either primer on its own to gel-isolate and sequence.

9.1 Random amplified polymorphic DNA (RAPD) PCR

This is by far the most commonly used variant of single-primer, random-amplification protocols. Examples applied to minke whale (after ref. 58) and elephant seal DNA are shown in *Figure 5*. RAPD PCR employs one 10–12 bp oligonucleotide (though longer primers are sometimes used; see below) to amplify random segments of DNA. A 10-mer will anneal to random sequence approximately every million base pairs. RAPD works because a small number of fragments (usually 5–10) will be amplified when the oligonucleotide anneals on each strand over a length range that can be readily amplified by PCR (usually less than 3–4 kb). When the intervening DNA varies in tandem repeat number, the amplified fragment will be of variable length in the population. If sequence variation at the priming site affects the annealing, then a fragment may not amplify in some individuals. Under carefully controlled reaction conditions, these two factors are likely to explain most of the observed variation. However, the generation of reproducible and comparable banding patterns is very dependent on template quality (59) and reaction conditions (60).

Bielawski *et al.* (60) found that variation in template concentration and in annealing, extension, and denaturing time could all affect the number and pattern of bands amplified. Micheli *et al.* (59) found that template quality is also important, and recommend using only DNA that has been spooled. Bielawski *et al.* (60) describe the following optimized protocol (applied to

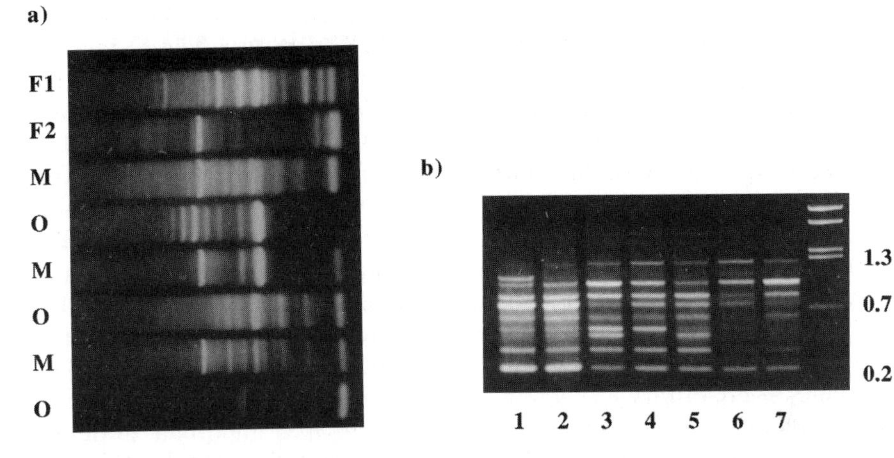

Figure 5. RAPD-PCR amplifications using 1.5 units *Taq* DNA polymerase, 150 ng (a) or 50 ng (b) template DNA, 200 μM dNTPs, 1.5 mM $MgCl_2$, 10 mM Tris–HCl, 50 mM KCl, and 0.4 μM primer DNA. (a) Northern elephant seal DNA amplified from two prospective fathers (F1 and F2) and three mother–offspring pairs (M, O) using the primer 5′-GAGGGTGGCGGTTCT. The primer was based on the core sequence of the M13 repeat region (62). Cycling conditions were: 42°C for 2 min, 70°C for 3 min, and 94°C for 40 sec for 40 cycles. (b) Seven minke whale DNA samples amplified using the primer 5′-AGGTCACTGA. The cycling conditions were: 36°C for 1.5 min, 70°C for 2 min, and 94°C for 45 sec for 40 cycles. Fragment sizes are indicated in kilobases.

various vertebrate species). Using a random decamer they denature at 94°C for 30 seconds, anneal at 36°C for 30 sec, and extend at 72°C for 2 min. They omit the usual initial denaturation step of 5–10 min and repeat the cycle 45 times. The reaction conditions were: 10 mM Tris–HCl pH 8.3, 50 mM KCl, 2.0 mM $MgCl_2$, 0.1% gelatin, 0.2 mM each dNTP, 0.5 μM of one decamer primer, and 1 U of *Taq* DNA polymerase in a 25 μl reaction volume. Variations in DNA concentration affect the number and position of bands resolved, therefore a standardized concentration across all samples is recommended. The number and resolution of bands can sometimes be improved by including 1 nM/μl Gp4 (ref. 38, and see Section 6.4).

Another factor affecting interpretation is the fact that PCR will not necessarily amplify both alleles evenly in RAPD amplifications. This 'dominance effect' (61) can make certain types of analysis very difficult, especially paternity testing and kinship assessment. An example is shown in *Figure 5a*. A paternity test for the northern elephant seal comparing two prospective fathers (F1 and F2) with three mother–offspring pairs is shown. Tests using minisatellite markers established F1 as the most likely father, but as seen in the figure, not all paternal or maternal bands are represented in the banding profile of the offspring. This example also illustrates how RAPD patterns can be generated using longer primers based on hypervariable repetitive loci. In this case, a 15-mer based on the core of the M13 repeat (62) was used.

9.2 Amplified fragment length polymorphisms (AFLPs)

AFLP is based on the selective amplification of sets of restriction fragments from genomic DNA (63). DNA is cut with restriction enzymes and double-stranded adaptors are then ligated to the fragments. PCR primers designed to match the sequence of the adaptors are used to amplify the fragments. The number of fragments amplified is intentionally limited by including nucleotides at the 3′ end of the primer that extend into the unique sequence of the DNA fragment (by at least 3 bp on each primer for the complex genomes of most plants and animals), so only a subset of the fragments will be amplified (those matching the designed 3′ primer extension at the 5′ end of the amplified sequence). The number of amplified fragments can be further limited by using two restriction enzymes, one that cuts frequently and one that cuts rarely (63). To avoid 'double banding' artefacts, it is important to design primers such that the 5′ residue is a guanine. Trials varying DNA concentration over a 1000-fold dilution range suggested no effect on the AFLP pattern except at very low concentrations (due to the consequent faintness of some bands).

For amplifications from higher plants and animals, Vos *et al.* (63) found that a two-step amplification strategy reduced background. This involved an initial low resolution amplification using primers with only one base pair extending into the unique sequence, followed by re-amplification using primers with three base pairs extending into the unique sequence. Further discussion and detailed protocols can be found in ref. 63.

10. Conclusion

PCR, when conducted with proper controls against contamination, offers a uniquely efficient means for the generation of bulk DNA from a specific locus. This DNA can then be analysed for variation in a variety of ways. New methods for the amplification of very long DNA segments (Section 6.4) will facilitate the analysis by RFLP (see Chapter 5) of regions such as nuclear gene introns and whole vertebrate mitochondrial genomes. Methods using radiolabelled probes (Chapter 5) or, for mitochondrial DNA, isolated mitochondrial DNA separate from nuclear DNA (Chapter 3), can be considerably more time-consuming. PCR-amplified DNA will be present in sufficient quantity to analyse using ethidium bromide staining on agarose gels, or for smaller fragments, silver staining on polyacrylamide gels. However, for PCR products over about 10 kb in length, the template DNA must be very pure, and for some sequences it may be difficult to amplify the target sequence alone without non-specific or other background bands.

A variety of non-specific amplification methods (such as RAPD and AFLP) offer a fast and easy means for the amplification of hypervariable banding patterns from small samples of DNA (Section 9). The advantage of these methods is that a lot of information is gained quickly; the disadvantages have to do with reproducibility and interpretation and are discussed in more detail in Section 9.

While there are pros and cons for each method of analysis, DNA sequencing PCR-amplified DNA (Section 8) remains an excellent choice for the routine screening of variation in natural population for two main reasons. First, the data for analysis will be from a defined locus and complete information on variation will be available, including, for example, such factors as the relative frequency of synonymous and non-synonymous substitutions. Variation due to insertions, deletions, and variation in tandem repeat number can be assessed directly by this method. Each of these can be important factors in the interpretation of the meaning of the results, since the expected rate of change over evolutionary time can vary greatly for these different mechanisms of change. Second, recent advances in methodology (especially the automated methods discussed in Chapter 10) have made this procedure faster and much more cost-effective.

References

1. Mullis, K. B. and Faloona, F. (1987). *Methods in Enzymology*, **155**, 335.
2. Huang, S., Hu, Y., Wu, C., and Holcenberg, J. (1990). *Nucleic Acids Res.* **18**, 1922.
3. Leung, D. W., Chen, E., and Goeddel, D. V. (1989). *Technique* **1**, 11.
4. Williams, J. G., Kubelik, A. R., Livak, K. J., Rafalski, J. A., and Tingey, S. V. (1990). *Nucleic Acids Res.* **18**, 6531.
5. Hoelzel, A. R. (1990). *Trends Genet.* **6**, 237.
6. Yap, E. P. H. and McGee, O. J. (1991). *Nucleic Acids Res.* **19**, 1713.
7. Tabor, S. and Richardson, C. C. (1987). *Proc. Natl Acad. Sci. USA* **84**, 4767.
8. Innis, M. A., Myambo, K. B., Gelfand, D. H., and Brow, M. A. (1988). *Proc. Natl Acad. Sci. USA* **85**, 9436.
9. Lawyer, F. C., Stoffel, S., Saiki, Myambo, K. B., Drummond, R., and Gelfand, D. H. (1989). *J. Biol. Chem.* **264**, 6427.
10. Higuchi, R. (1989). In *PCR technology* (ed. H. A. Erlich), p. 31. Stockton Press, New York, NY.
11. Rappolee, D. A., Wang, A., Mark, D., and Werb, Z. (1988). *Science* **241**, 708.
12. Rychlik, W. and Rhoads, R. E. (1989). *Nucleic Acids Res.* **17**, 8543.
13. Williams, J. F. (1989). *BioTechniques* **7**, 762.
14. Saiki, R. K. (1989). In *PCR technology* (ed. H. A. Erlich), p. 7. Stockton Press, New York, NY.
15. Suggs, S. V., Wallace, R. B., Hirose, T., Kawashima, E. H., and Itakura, K. (1981). *Proc. Natl Acad. Sci. USA* **78**, 6113.
16. Breslauer, K. J., Frank, R., Blocker, H., and Markey, L. A. (1986). *Proc. Natl Acad. Sci., USA* **83**, 3746.
17. Lowe, T. L., Sharefkin, J., Yang, S. Q., and Dieffenbach, C W. (1990). *Nucleic Acids Res.* **18**, 1757.
18. Rychlik, W., Spencer, W. J., and Rhoads, R. E. (1989). *Nucleic Acids Res.* **18**, 6409.
19. Bridges, C. G. (1990). *Comput. Applic. BiosSci.* **6**, 124.

20. Gelfand, D. H. (1989). In *PCR technology* (ed. H. A. Erlich), p. 17. Stockton Press, New York, NY.
21. Jeffreys, A. J., Neumann, R., and Wilson, V. (1990). *Cell* **60**, 473.
22. Bachmann, B., Wolfgang, L., and Hunsmann, G. (1990). *Nucleic Acids Res.* **18**, 1309.
23. Smith, K. T., Long, C. M., Bowman, B., and Manos, M. M. (1990). *Amplifications* **5**, 16.
24. Sarkar, G., Kapelner, S., and Sommer, S. S. (1990). *Nucleic Acids Res.* **18**, 7465.
25. Chevet, E., Lemaitre, G., and Katinka, D. (1995). *Nucleic Acids Res.* **22**, 3343–3345.
26. Brookstein, R., Lai, C.-C., To, H., and Lee, W.-H. (1990). *Nucleic Acids Res.* **18**, 1666.
27. Hung, T., Mak, K., and Fong, K. (1990). *Nucleic Acids Res.* **18**, 4953.
28. Wood, W. J., Gitschier, J., Lasky, L. A., and Lawn, R. M. (1985). *Proc. Natl Acad. Sci. USA* **82**, 1585.
29. Jacobs, K. A., Rundersdorf, R., Neill, S. D., Dougherty, J. P., Brown, E. L., and Fritsch, E. F. (1988). *Nucleic Acids Res.* **16**, 4637.
30. McConlogue, L., Brow, M. A., and Innis, M. A. (1988). *Nucleic Acids Res.* **16**, 3360.
31. Furrer, B., Candrian, U., Wieland, P., and Luthy, J. (1990). *Nature* **346**, 324.
32. Deragon, J.-M., Sinnett, D., Mitchell, G., Potier, M., and Labuda, D. (1990). *Nucleic Acids Res.* **18**, 6149.
33. Sarkar, G. and Sommer, S. S. (1990). *Nature* **343**, 27.
34. Innis, M. A. and Gelfand, D. H. (1990). In *PCR protocols* (ed. M. A. Innis, D. H. Gefland, J. J. Sninsky, and T. J. White), p. 3. Academic Press, San Diego, CA.
35. Chou, Q., Russell, M., Birch, D. E., Raymond, J., and Bloch, W. (1992). *Nucleic Acids Res.* **20**, 1717.
36. D'Aquila, R. T., Bechtel, L. J., Videler, J. A., Eron, J .J., Gorczyca, P., and Kaplan, J. C. (1991). *Nucleic Acids Res.* **19**, 3749.
37. Nelson, W. S., Prodohl, P. A., and Avise, J. C. (1996). *Mol. Ecol.* **5**, 807.
38. Schwarz, K., Hansen-Hagge, T., and Bartram, C. (1990). *Nucleic Acids Res.* **18**, 1079.
39. Barnes, W. M. (1994). *Proc. Natl Acad. Sci. USA* **91**, 2216.
40. Cheng, S., Higuchi, R., and Stoneking, M. (1994). *Nature Genet.* **7**, 350.
41. Cheng, S., Fockler, C., Barnes, W. M., and Higuchi, R. (1994). *Proc. Natl Acad. Sci. USA* **91**, 5695.
42. Clark, J. M. (1988). *Nucleic Acids Res.* **16**, 9677.
43. Jung, V., Pestka, S. B., and Pestka, S. (1990). *Nucleic Acids Res.* **18**, 6156.
44. Marchuk, D., Drumm, M., Saulino, A., and Collins, F. S. (1991). *Nucleic Acids Res.* **19**, 1154.
45. Holton, T. A. and Graham, M. W. (1991). *Nucleic Acids Res.* **19**, 1156.
46. Crowe, J. S., Cooper, H. J., Smith, M. A., Sims, M. J., Parker, D., and Gewert, D. (1991). *Nucleic Acids Res.* **19**, 184.
47. Sanger, F., Nicklen, S., and Coulson, A. R. (1977). *Proc. Natl Acad. Sci. USA* **74**, 5463.
48. Perbal, B. (1988). *A practical guide to molecular cloning.* John Wiley, New York, NY.
49. Tautz, D. and Renz, M. (1983). *Anal. Biochem.* **132**, 14.

50. Gyllensten, U. (1989). *PCR technology* (ed. H. A. Erlich), p. 45. Stockton Press, New York, NY.
51. Hultman, T., Stahl, S., Hornes, E., and Uhlen, M. (1989). *Nucleic Acids Res.* **17**, 4937.
52. Schofield, J. P., Vaudin, M., Kettle, S., and Jones, D. S. C. (1989). *Nucleic Acids Res.* **17**, 9498.
53. Green, A., Roopra, A., and Vaudin, M. (1990). *Nucleic Acids Res.* **18**, 6163.
54. Winship, P. R. (1989). *Nucleic Acids Res.* **17**, 1266.
55. Caetano-Anolles, G., Brassam, B. J., and Gresshoff, P. M. (1991). *BioTechnology* **9**, 553.
56. Welsh, J. and McClelland, M. (1990). *Nucleic Acids Res.* **18**, 7213.
57. Burt, A., Carter, D. A., White, T. J., and Taylor, J. W. (1994). *Mol. Ecol.* **3**, 523
58. Hoelzel, A. R. (1994). *Rep. Int. Whaling Comm.* **44**, 487.
59. Micheli, M. R., Bova, R., Pascale, E., and D'Ambrosio, E. (1994). *Nucleic Acids Res.* **22**, 1921.
60. Bielawski, J. P., Noack, K., and Pumo, D. E. (1995). *BioTechniques* **18**, 856.
61. Postlethwait, J. H. *et al.* (1994) *Science* **264**, 699.
62. Vassart, G, Georges, M., Monsieur, R., Brocas, H., Lequarre, A. S., and Christophe, D. (1987) *Science* **235**, 683.
63. Vos, P., Hogers, R., Bleeker, M., Reijans, M., van de Lee, T., Hornes, M., Frijters, A., Pot, J., Peleman, J., Kuiper, M., and Zabeau, M. (1995) *Nucleic Acids Res.* **23**, 4407.
64. Kocher, T. D., Thomas, W. K., Meyer, A., Edwards, S. V., Paabo, S., Villablanca, F. X., and Wilson, A. C. (1989). *Proc. Natl Acad. Sci. USA* **86**, 6196.
65. Hoelzel, A. R., Hancock, J. M., and Dover, G. A. (1991). *Mol. Biol. Evol.* **8**, 475.
66. Palumbi, S. R. (1996). In *Molecular systematics*, 2nd edn (ed. D. M. Hillis, C. Moritz, and B. K. Mable), p. 205. Sinauer, Sunderland, MA.
67. Olmstead, R. G., Michaels, H. J., Scott, K. M., and Palmer, J. D. (1992). *Ann. Missouri Bot. Garden* **79**, 249.

7

Microsatellites

CHRISTIAN SCHLÖTTERER

1. Introduction

Microsatellites are highly polymorphic DNA markers with discrete loci and co-dominant alleles. They have a wide range of applications which previously were difficult to address experimentally. Microsatellites are almost uniformly distributed over the entire genome, which is what makes them so useful for genome mapping projects (1,2). Their high variability has made them a marker of choice in behavioural ecology, as they allow the determination of paternity and kinship (3,4). Finally, they are becoming increasingly important in studying the structure of natural populations. This chapter provides information on how to get started with microsatellite cloning and analysis, and provides some background information on the interpretation and analysis of the data.

1.1 Microsatellites—a definition

Microsatellites are short tandemly repeated sequence motifs consisting of repeat units of 1–6 bp in length (5). They are highly abundant in eukaryotic genomes, but also occur in prokaryotes at lower frequencies. They seldom include more than about 70 repeat units and are interspersed throughout the genome. For example, in mammals it has been estimated that the most common motif (GT/AC) occurs on average every 30 kb. This implies that most microsatellites are embedded in single copy DNA, thus facilitating the unambiguous scoring of alleles (see below).

1.2 Microsatellite polymorphisms

Microsatellite mutations are changes in repeat numbers caused by an intramolecular mutation mechanism called DNA slippage. The most common mutations are changes of a single repeat unit, which allows microsatellite mutations to be interpreted as a very good approximation of a stepwise mutation process (6); a large body of theoretical treatment of this mutation process is available (7–11). The observed mutation rates range from 10^{-3} to 10^{-6} (5,12). The allelic state of microsatellites is scored by polymerase chain reaction (PCR) analysis. Primers flanking the repeat are used for PCR amplification (13,14). Sizing of the PCR products on high resolution gels allows the

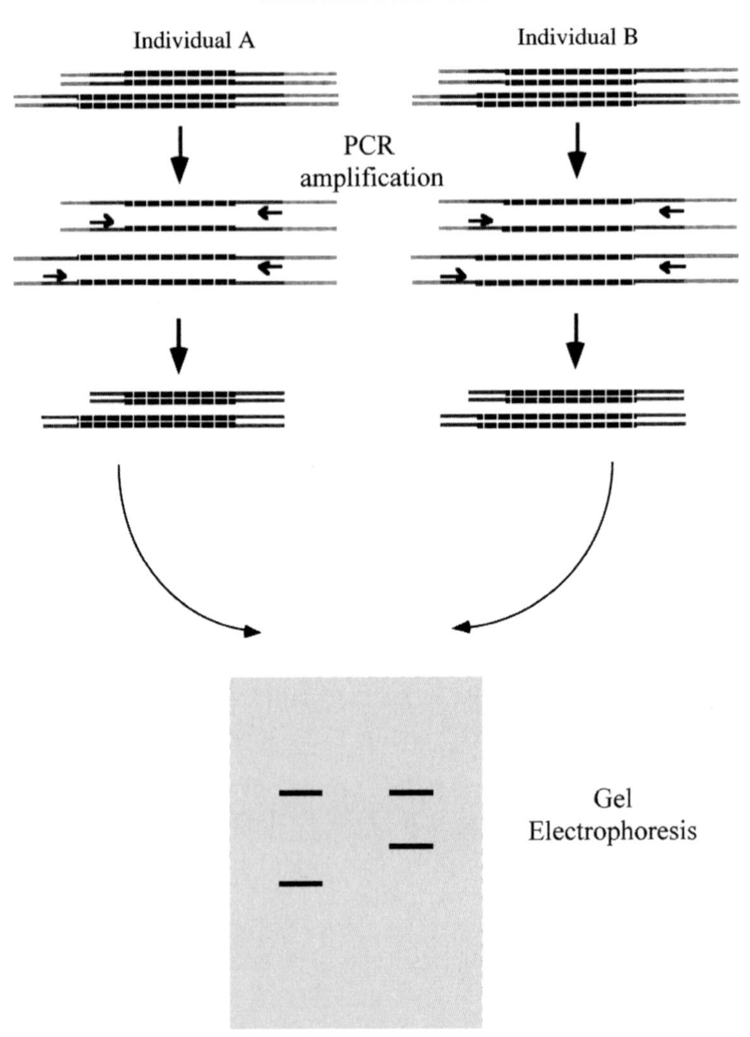

Figure 1. PCR amplification of a microsatellite locus. PCR primers are designed to flank the microsatellite repeat. As individuals A and B differ in microsatellite repeat numbers for one allele, the PCR yields two different products. These PCR products can be separated by gel electrophoresis.

determination of the number of repeats and any variation in repeat number between the different alleles (*Figure 1*).

2. Isolation of microsatellite loci

Before a specific microsatellite locus can be amplified, one needs sequence information for the flanking DNA to allow the design of specific primers. For

Table 1. Comparison of two alternative microsatellite isolation protocols

Protocol	Advantages	Disadvantages	When to use
Standard procedure (*Protocol 1*; ref. 14)	easy and fast	low efficiency for trinucleotide and tetranucleotide repeats	for a few (<30) dinucleotide microsatellites
Enrichment (*Protocol 2*; ref. 15)	more efficient	laborious; sometimes leads to over-representation of a small number of clones	for many microsatellite loci; trinucleotide and tetranucleotide microsatellie loci; when limited amount of DNA is avaliable for preparing library

some model organisms, an increasing number of microsatellites and their flanking sequences have been entered into computer databases (such as EMBL and GenBank; see Appendix 5). Furthermore, there have been several attempts to initiate specific microsatellite databases where primer information for a wide range of species can be found (see below and Appendix 4). To date many organisms lack sufficient numbers of entries in the databases. Therefore, the isolation of microsatellite loci and their flanking regions still remains the first step in their analysis. Two different protocols are given for the isolation of microsatellites from genomic DNA (see *Table 1*). *Protocol 1* provides an outline strategy for establishing a partial genomic library sufficient for the isolation of microsatellite loci. It is recommended for isolation of a limited number (<30) of dinucleotide loci. Given that all molecular methods are established and are working well, this protocol may provide microsatellite loci within 14 days. The details of cloning and library screening protocols are given in Chapter 4.

Protocol 1. Standard protocol for isolation of microsatellites (13)

Equipment and reagents

- Restriction enzymes (*Alu*I, *Hae*III, and *Rsa*I) and appropriate buffers
- 1% agarose gel
- 1 × TAE buffer
- Ethidium bromide
- DNA as size standard
- UV transilluminator
- Dialysis tubing (Sigma) boiled twice in 1 × TE
- Pre-cut and dephosphorylated sequencing vector (Chapter 4)
- 3 M sodium acetate
- 100% ethanol
- DNA sequencing equipment and reagents (see Chapters 6 and 10)
- LB medium and LB agar (see Appendix 4)
- Competent cells (Chapter 4)
- Equipment and reagents for colony/plaque lifting (Chapter 4)
- Microsatellite probes[a]
- X-ray film (e.g. Fuji RX)
- Nylon membranes (e.g. Hybond N+, Amersham)
- T4 DNA ligase

Protocol 1. *Continued*

- Hybridization buffer (Church buffer): 0.5 M sodium phosphate pH 7.0, 7% sodium dodecyl sulfate (SDS)
- Wash solution: 0.1% SDS + SSC, with SSC concentration depending on GC content of probe[a]
- 5 × ligase buffer: 250 mM Tris–HCl pH 7.6, 50 mM $MgCl_2$, 5 mM rATP, 5 mM DTT, 25% (w/v) polyethylene glycol 8000
- Purified genomic DNA

A. *Generation of a partial library*

1. Digest 10 μg of genomic DNA with *Alu*I, *Hae*III, and *Rsa*I.
2. Separate on a 1% agarose gel (at 3 V/cm) in 1 × TAE buffer alongside size standards and stain with ethidium bromide.
3. Recover the 300–500 bp fraction from the gel by electro-elution:
   (a) Excise the 300–500 bp fraction from the gel (avoid lengthy exposure of samples to UV radiation).
   (b) Transfer excised agarose into dialysis tubing (Sigma) that has been boiled twice in 1 × TE.
   (c) Add some 1 × TAE and remove air bubbles carefully before sealing both ends of the tube (use clips or make a knot).
   (d) Electrophorese in a gel chamber until DNA has left the gel (check in UV light).
   (e) Reverse polarity for 15 sec.
   (f) Remove buffer from tubing, taking care that no agarose is transferred.
   (g) Precipitate with 0.1 vol 3 M sodium acetate and 2 vol 100% ethanol (see Chapter 6, Protocol 2).
4. Clone into sequencing vector (use molar excess of insert; see Chapter 4).
5. Transform competent cells with ligation mix (Chapter 4).
6. Grow transformed cells on LB plates (overnight).[b]
7. Colony/plaque lift (Chapter 4).
8. Denature DNA bound to nylon membrane by NaOH treatment (Chapter 4).
9. Neutralize nylon membrane (Chapter 4).
10. UV cross-link DNA to filter (Chapter 4).

B. *Screening for clones carrying a microsatellite*

1. Label 20–50 ng of microsatellite polymer[a] by random priming (see Chapter 5; random primers may be left out). Take care that the labelled nucleotide is contained in the polymer.
2. Pre-hybridize filters at 65°C for 30 min in Church buffer.
3. Boil labelled polymers for 5 min and add directly to pre-hybridized filters.

4. Hybridize overnight at 65°C in Church buffer.
5. Wash filters at 65°C in 0.1% (w/v) SDS + SSC, using an SSC concentration appropriate for the GC content of the probe.[c]

[a]For generation of polymers see *Protocol 2.* Dinucleotide polymers are commercially available (e.g., from Pharmacia)
[b]Screening of dinucleotides is greatly facilitated if bacteria/plaques are well separated, as time-consuming re-screening can be avoided.
[c]Appropriate SSC concentrations are as follows: 0% GC, 5 × SSC; 33% GC, 2 × SSC; 50% GC, 1 × SSC; 66% GC, 0.1 × SSC; 100% GC, 0.1 × SSC + 25% formamide (v/v).

As explained above, screening a partial library works well for isolating small numbers of dinucleotides. However, to isolate many dinucleotide loci or trinucleotide/tetranucleotide repeats, which are less abundant, it is advisable to construct a library that is enriched for the respective microsatellite motif (*Protocol 2*). The essential difference from *Protocol 1* is that only DNA fragments that have been pre-selected for the presence of a microsatellite motif are cloned into the sequencing vector. Hence, fewer clones need to be screened to detect sufficient numbers of microsatellite loci. *Table 1* compares *Protocols 1* and *2* to show the relative merits of each approach.

Protocol 2. Construction of a partial library enriched for microsatellites (14)

Equipment and reagents

- SauL oligonucleotide A: 5′-GCGGTACC-CGGGAAGCTTGG-3′
- SauL oligonucleotide B: 5′-GATCCCAAGC-TTCCCGGGTACCGC-3′
- 10 mM rATP
- 10× Phosphorylation buffer (760 mM Tris–HCl, pH 7.6, 100 mM $MgCl_2$, 50 mM DTT)
- Polynucleotide kinase (New England Biolabs)
- 0.5 M EDTA pH 8.0
- 3 M sodium acetate
- Absolute and 80% ethanol
- Linkers
- T4 DNA ligase (Gibco BRL)
- Equipment and reagents for gel electrophoresis (*Protocol 1*)
- PCR machine
- PCR buffer (10×): 166 mM $(NH_4)_2SO_4$, 677 mM Tris–HCl pH 8.8, 0.1% (v/v) Tween 20, 15 mM $MgCl_2$
- Small nylon filters[a] (e.g. cut from Hybond-N, Amersham)
- Hybridization buffer (Church buffer): see *Protocol 1*
- Wash solution (see *Protocol 1*)
- 0.25 M KOH
- 50 mM KOH, 0.01% (w/v) SDS
- 1 M Tris–HCl pH 4.8
- 50 mM Tris–HCl pH 7.5
- 0.25 M HCl
- 60°C incubator

Method

A. *Linker preparation*

Two single-stranded oligonucleotides are hybridized to each other to produce a double-stranded linker that can be ligated to double-stranded DNA fragments. Note that one primer has to be phosphorylated to allow ligation of the linker and the DNA fragment.

Protocol 2. *Continued*

i. Phosphorylation of SauL B

1. Mix the following:
   - 5 μg SauL B
   - 2 μl 10 × phosphorylation buffer
   - *x* μl double-distilled H_2O (to make a total volume of 20 μl)
   - 5 μl 10 mM rATP
   - 5 U polynucleotide kinase
2. Incubate at 37 °C for 30 min.
3. Add 1 μl of 0.5 M EDTA pH 8.0.
4. Precipitate with 0.1 vol. 3 M sodium acetate and 4 vol. of 100% ethanol and resuspend in 5 μl of water.

ii. Annealing

1. Incubate 5 μg of SauL A and 5 μg of SauL B (phosphorylated) in 0.1 M NaCl at 60 °C for 1 h.
2. Proceed directly to ligation (*Protocol 2C,* part *ii*).

B. *Making target filters*

i. Preparation of microsatellite probes

1. Using complementary primers (17–20 bp) consisting entirely of the repeat motif, the primers can be extended in a PCR-like reaction containing:
   - 1 μM of each primer
   - 0.2 mM of each nucleotide
   - PCR buffer
   - 0.05 U *Taq* DNA polymerase per μl of reaction volume

   Perform 35 PCR cycles consisting of 1 min at 39 °C , 1 min at 72 °C, and 1 min at 94 °C.
2. Long targets are probably important in both the kinetics and selectivity of hybridization selection, so synthetic arrays greater than 200 bp may be gel-purified before use.[b]
3. After ethanol precipitation, re-dissolve at a high concentration (about 0.2 mg/ml).

ii. Spotting microsatellite probes on to a nylon membrane[a]

1. Before proceeding to make the target filter, remove about 20% of the mixture (at least 50 ng) for use as a hybridization probe mix.
2. To the remaining target DNA mix, add 0.15 vol. of 1 M KOH; leave at room temperature for 5 min, then add 0.25 vol. of 1 M Tris–HCl pH 4.8. Keep on ice.

3. Spot the denatured target DNA, 2 μl at a time, on to a small piece (about 2–3 mm on a side) of nylon hybridization filter (e.g. Hybond-N, Amersham). Use both sides of the filter alternately, and allow to dry between each application. If using more than one target mix to select from the same PCR library, or to identify a blank filter, two filters distinguished by shape can be used in the same hybridization.
4. When all the DNA has been applied, allow to dry, and fix the DNA to the filter using UV irradiation to both sides.

C. *Enrichment procedure*

i. Genomic DNA digest

1. Digest 10 μg genomic DNA (if possible, isolated from a single individual) with *Mbo*I.
2. Separate DNA fragments on a 1% agarose gel (at 3 V/cm).
3. Electro-elute a 300–700 bp fraction (see *Protocol 1*).
4. Ethanol precipitate by adding 0.1 vol. 3 M sodium acetate and 3 vol. 100% ethanol).

ii. Re-amplification

1. Ligate 5 μg of linkers using about 500 ng DNA fragments and T4 DNA ligase (overnight at 15°C) (see Chapter 4).
2. Size-select the 300–700 bp fraction away from linker dimers by gel electrophoresis on a 1.2% agarose gel.
3. Gel purify and ethanol precipitate the DNA (see above).
4. Re-amplify the library in 3 parallel 50 ml reactions (cycling conditions: 25–30 cycles of 95°C for 1 min, 67°C for 1 min, 70°C for 2 min) using primer SauL A to make about 1 μg in total.
5. Ethanol precipitate by adding 0.1 vol. 3 M sodium acetate and 3 vol. ethanol and re-dissolve in 5 μl of water.

iii. Hybridization selection

1. Pre-hybridize the target filter(s) (see *Protocol 2B* above) for 1 h in a 1.5 ml microcentrifuge tube with 1 ml of Church buffer at 65°C. Include a blank filter (no target DNA) in the same tube, as a control. Remove Church buffer, and replace with 100 μl fresh buffer.
2. To about 1 μg PCR library (see *ii.* above) in a volume of approximately 5 μl, add 1 μl 0.25 M KOH. After 5 min at room temperature, place on ice and add 5 μl of 1 M Tris–HCl pH 4.8, followed by 0.5 μl 0.25 M HCl. Add to the solution with the filters, mix, and cover with paraffin oil. Incubate overnight at 65°C.
3. Remove the paraffin oil and hybridization solution, and wash the filters for approximately 30 min, four times in the microcentrifuge tube with

Protocol 2. *Continued*

1 ml wash of solution,[c] followed by four washes of 25 ml in a universal container.

4. After washing, blot off excess moisture from each filter. Place each filter, including the blank, in a separate microcentrifuge tube. To each, add 50 μl of 50 mM KOH, 0.01% SDS. Leave at room temperature for 5 min, and remove the 50 μl to a fresh tube.
5. Add 50 μl of 50 mM Tris–HCl pH 7.5 with 0.01% SDS to each tube containing a filter. Leave at room temperature for 5 min, then pool this solution with the corresponding 50 μl from alkaline extraction (step **4**).
6. To the pooled extracts (100 μl), add 10 μl of 10 μM SauL A linker primer (as a harmless carrier), 10 μl of 3 M sodium acetate and mix. Precipitate the recovered DNA with 400 μl ethanol. After centrifugation, wash the pellets (which will be small) with 80% ethanol, dry under vacuum, and re-dissolve in 20 μl water.
7. Re-amplify enriched DNA as described in *Protocol 2C*, part *ii*,[d] digest with *Mbo*I (to remove linkers), size-select on a gel (to remove linker fragments), and clone into a cloning vector (cut with Bam H1). The resulting library can be re-screened using the probe mix and sequencing hybridizing clones (see Chapter 4 for cloning and Chapters 6 and 10 for sequencing).

[a] The nylon filters will be used for hybridization selection in a 100 μl volume, and therefore must be small (2–3 mm on a side); the volume of target DNA used must also be as small as possible. In general, use about 200 ng of each microsatellite probe; these can be combined into target mixtures. For example, 200 ng of each of six tetrameric target sequences can be mixed to give about 1.2 μg DNA, which should be in a volume of 10 μl or less. If necessary, concentrate the target DNA mixture.
[b] Alternatively di- and tri-nucleotide polymers can be generated by incubating the primers with Klenow polymerase. Interestingly, this generates DNA fragments of homogeneous size, eliminating the need of a gel purification step; just purify using Centricon 30 (Amicon) (15).
[c] For washing conditions see *Protocol 1*.
[d] Enrichment efficiency may be tested by amplifying enriched and original DNA, running equimolar DNA amounts next to each other on a gel, blotting the gel, and hybridizing the membrane with the respective microsatellite probe (see *Protocol 2B*, step **1**).

All microsatellite loci isolated by *Protocols 1* or *2* are randomly isolated and nothing is known about their chromosomal location. For most applications in molecular ecology this does not matter, but for some purposes it may be advantageous to analyse microsatellites from a well characterized genomic region. P1 clones or cosmids are among the major tools used in genome projects and provide rapid access to genomic regions of defined origin (17). Microsatellites isolated from single P1 clones, with up to 100 kb of insert DNA, can be used to study linkage disequilibrium, influence of chromosomal position on microsatellite variability, etc. The method is outlined in *Protocol 3* (after ref. 18). The cloning strategy is based on the low vector backbone:insert ratio (19) and the fact that the vector backbone contains no long microsatellite stretch. Therefore, the complete P1 clone can be digested and sub-cloned, which is a much simpler and faster method than other protocols (20). Three restriction enzymes are used separately and the digests are pooled prior to cloning, allowing a more representative cloning of the insert DNA.

Protocol 3. Isolation of microsatellites from P1 clones (18)

Reagents

- P1 DNA
- Restriction enzymes (*Alu*I, *Hae*III, and *Rsa*I)
- Appropriate restiction enzyme buffer(s)
- 3 M sodium acetate
- 80% and 100% ethanol

Method

1. Digest 500 ng P1 DNA for each restriction enzyme (*Alu*I, *Hae*III, and *Rsa*I) separately.
2. Pool the three different digests.
3. Ethanol precipitate (see Protocol 1).
4. Ligate into sequencing vector (see Chapter 4).
5. Follow *Protocol 1B*.

2.1 Choosing the 'right' microsatellites

2.1.1 Repeat type

Dinucleotide repeats are, in general, the most common repeat types. In mammals GT/AC is the most frequent dinucleotide repeat, whereas in plants AA/TT and AT/TA are most common (21,22). The abundance of these dinucleotide repeats makes them an easy target for microsatellite isolation (*Protocol 1*). The major concern about the use of dinucleotide repeats is their high propensity to generate stutter bands (see below) during PCR amplification, sometimes making the assignment of alleles difficult. Trinucleotide and tetranucleotide repeats are not as common, but tend to generate fewer stutter bands and are therefore easier to score. However, the stutter bands may also be very helpful in discriminating non-specific from specific amplification products, as well as in the exact scoring of allele length in neighbouring lanes. Thus, the practical usefulness of a locus tends to depend on the exact pattern of stutter bands that it produces; it is important that the major band can always be clearly scored.

2.1.2 Repeat length

The first systematic studies in humans suggested that longer microsatellites are, on average, more polymorphic than shorter ones (23). However, later studies failed to confirm this trend (24); hence it is not advisable to exclude

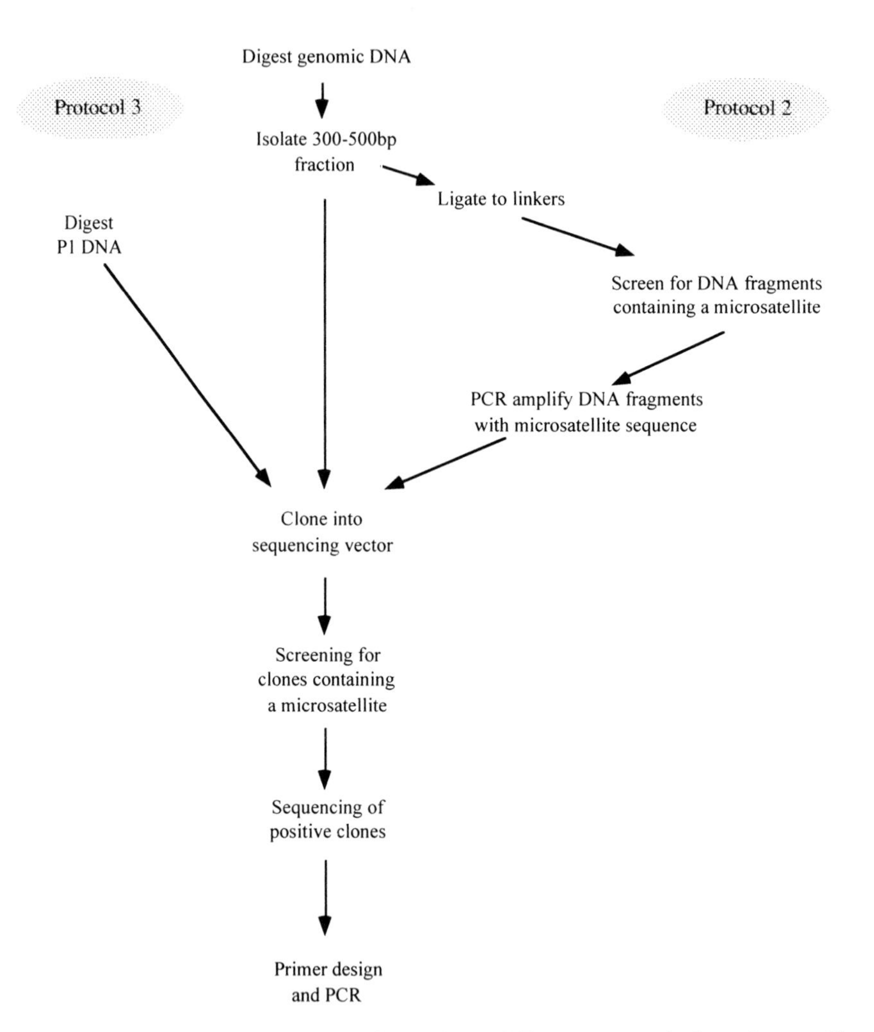

Figure 2. Schematic representation of the three different protocols for microsatellite isolation.

shorter microsatellites from further analysis. In addition, it is currently not possible to rule out the possibility that the average length of microsatellites is species-specific. Interruptions of the microsatellite repeat by mutations in single repeat units may result in lower variability (25–26). Microsatellites consisting of two different repeat types exhibit less length variation, but changes

a)

GTGTGTGTGTGTGTGTGTGTGTGT
CACACACACACACACACACACACA

b)

GTGTGTGTGTGTGGGTGTGTGTGT
CACACACACACACCCACACACACA

c)

GTGTGTGTGTGTGTGAGAGAGAGA
CACACACACACACACTCTCTCTCT

Figure 3. (a) Perfect dinucleotide repeat. (b) Imperfect dinucleotide repeat; the point mutation is indicated by an arrow. (c) Composite microsatellite; the transition from GT/CA to GA/CT repeats is indicated.

in the relative frequencies of each repeat unit may provide some additional insight in population structures due to less homoplasy (28).

3. Microsatellite sizing and detection

Current microsatellite analysis relies on size determination of the entire PCR product consisting of the microsatellite stretch and flanking regions. If information on the sequence of the flanking region is available, the number of repeats may be calculated by subtraction of the flanking nucleotides and dividing the remaining base pairs by the size of the repeat unit. Sizing of PCR products is commonly achieved by polyacrylamide gel electrophoresis.

3.1 What gel to use?

The most common separating technique is gel electrophoresis. In general, denaturing gel systems are preferable to native gels, as heteroduplex molecules generated during the late PCR cycles of heterozygous individuals will result in a third band (sometimes also a fourth). This may cause an incorrect assignment of alleles. Irrespective of the separation technique, in population analyses it is mandatory to size microsatellites alleles with high precision. As a rule of thumb the separation capacity of the gel should be at least half the size of the repeat unit. For example, a dinucleotide repeat should be separated on a gel system allowing size discrimination of single base pairs. Therefore, for most microsatellites, sizing of PCR products on agarose gels is not appropriate as they provide too little resolution. Even the use of special agarose matrices, like Metaphor (FMC), is not advised, as the problem of heteroduplex bands still remains. The most commonly used gel type is a 6% denaturing polyacrylamide gel. Recently, capillary electrophoresis has been established for microsatellite sizing as well (29); a glass capillary filled with a special matrix is used to separate DNA fragments according to their size. Shorter running times, the major advantage of capillary gel electrophoresis, will become fully effective as soon as devices have been constructed which allow parallel use of several capillaries.

3.2 PCR product sizing

The traditional method of sizing microsatellites is to run a sequencing ladder next to the PCR-amplified microsatellites. By comparing this with the sequence reaction, the absolute length of the PCR product can be determined. Alternatively, a commercial sizing ladder (e.g., from Research Genetics) may be used. Note that for the correct calculation of size it is important to consider the priming site of the sequencing primer in the vector. In capillary electrophoresis and some automated sequencers, PCR products are sized using an internal size marker in the same lane (see Chapter 10). This approach automatically corrects for between-lane differences which may occur within a gel. Finally, PCR products may be sized by running a 'cocktail' of known alleles next to the PCR reaction. This approach may be preferable when PCR products are detected by DNA–DNA hybridization (see below).

3.3 How to detect microsatellites after PCR amplification

3.3.1 Radioactive detection

The original and most sensitive approach for microsatellite detection is based on radioactivity. Two different methods for labelling the PCR product are available: incorporation of labelled nucleotides and end-labelling one of the PCR primers (*Protocol 4*). The major difference between them is that both DNA strands are labelled by incorporation of labelled nucleotides during DNA synthesis, while labelling a single PCR primer results in only one marked DNA strand. For many templates the end-labelling of one primer is preferred because differential electrophoretic mobility of the two strands may result in two separate bands on a denaturing gel (22; see Chapter 8).

Protocol 4. Microsatellite detection with radioactively end-labelled PCR primers[a]

Equipment and reagents

- Polynucleotide kinase (New England Biolabs)
- dNTPs (2 mM each)
- Primers 1 and 2 (20 μM each)
- *Taq* DNA polymerase

- 10 × phosphorylation buffer: 700 mM Tris–HCl pH 7.6, 100 mM $MgCl_2$, 50 mM DTT
- [γ-^{32}P]ATP (5000 μCi/mmol) (or [γ-^{33}P]ATP, which provides good exposure times with reduced handling risk)
- 10 × PCR buffer (supplied with DNA polymerase)
- PCR machine
- Loading buffer: 95% formamide, 20 mM EDTA, 0.05% Bromophenol Blue (w/v), 0.05% Xylene Cyanol FF (w/v)
- 6% sequencing matrix: 8 M urea, 19:1 acrylamide:bis-acrylamide
- 37 °C and 95 °C incubators

Method

A. *End-labelling*

1. Combine the following:[b]
   - 0.5 pmol primer
   - 0.5 μl 10 × phosphorylation buffer
   - 2 U (<0.5 μl) polynucleotide kinase
   - 3 μl [γ-^{32}P]ATP or [γ-^{33}P]ATP

   Make up to 5 μl with H_2O.
2. Incubate for >15 min at 37 °C.
3. Incubate for 5 min at 95 °C.
4. Spin down.
5. Add to PCR master mix.

B. *Preparation of master mix*

1. While the labelling reaction (*Protocol 4A*) is proceeding, combine the following (where *n* denotes the number of PCR reactions):
   - $n \times$ 1 μl dNTPs (2 mM each)
   - $n \times$ 0.5 μl primer 1 (20 μM)
   - $n \times$ 0.5 μl primer 2 (20 μM)
   - $n \times$ 1 μl of 10 × PCR buffer (including Mg^{2+})
   - $n \times$ 5.9 μl double-distilled H_2O
   - $n \times$ 0.1 μl *Taq* DNA polymerase (5 U/μl)
2. Mix gently.

C. *PCR*

1. Add 9 μl of master mix (including labelled primer) to 1 μl of DNA (10–100 ng).
2. Perform 25–35 PCR cycles under conditions appropriate for the primers.
3. After PCR add 1 vol. loading buffer.
4. Heat for 2 min at 95 °C and load on to gel.

[a] This protocol is designed for up to ninety 10 μl PCRs.
[b] This will be sufficient for up to 90 PCRs using ^{32}P or 30 reactions using ^{33}P; adjust quantities as necessary for different numbers of reactions (the precise amount of end-labelling mix used is not critical).

3.3.2 Non-radioactive detection

i. Silver staining

Silver staining provides a cheap and sensitive alternative to radioactive detection of nucleic acids. However, sensitivity is too low for reliable detection of a sequencing ladder on a gel. Furthermore, strand-specific staining is not possible, resulting in the problems mentioned above. As silver staining involves several washing steps, the use of a supporting sheet (Gel-Fix polyester sheets; Serva) on which the gel is covalently bound is highly recommended. Alternatively, use binding silane (Pharmacia) to fix the gel to one glass plate.

Protocol 5. Silver staining

Equipment and reagents

- 10% ethanol, 0.5% acetic acid
- 0.1% $AgNO_3$
- 0.75% Na_2CO_3
- 1.5% NaOH, 0.01% $NaBH_4$, 0.15% formaldehyde (this buffer must be made freshly before use)

Method

(During all steps the gel should be submerged while gently rocking.)

1. Incubate for 3 min in 10% ethanol, 0.5% acetic acid.
2. Incubate for 10 min in 0.1% $AgNO_3$.
3. Wash twice in water.
4. Stain for 20 min in 1.5% NaOH, 0.01% $NaBH_4$, 0.15% formaldehyde. Staining time depends on gel thickness and is a trade-off between background and sensitivity.
5. Discard staining solution and add 0.75% Na_2CO_3.
6. Discard 0.75% Na_2CO_3 after 10 min and photograph the gel.

ii. Blotting and hybridization

The detection methods mentioned above detect all PCR products, irrespective of whether they carry a microsatellite motif or not. Thus, non-specific PCR amplifications will result in many bands being visible on the gel. Sometimes it may be difficult to discriminate between non-specific PCR products and microsatellite alleles. By transferring PCR products on to a nylon membrane (*Protocol 6*; see also Chapter 5) and subsequently hybridizing with the respective microsatellite motif, specific detection of microsatellite alleles is possible (30). Furthermore, the blotting procedure permits the use of any of the non-radioactive detection methods that are currently available (e.g. chemoluminiscence, Enhanced Chemiluminscence (ECL), colorimetry; see Chapter 5). The major drawback of the hybridization approach is the higher cost, as membranes and non-radioactive detection kits are still fairly expensive. On the other hand, a potential advantage of the hybridization approach is that it allows separation of several PCRs in a single lane provided they differ in repeat type. PCR products overlapping in their size range can then be detected by subsequently hybridizing the filter with specific probes for the different repeat types.

Protocol 6. Capillary blotting of polyacrylamide gels

Equipment and reagents

- UV transilluminator or UV cross-linker
- Nylon membrane (e.g. Hybond N+, Amersham)
- Whatman 3MM filter paper
- 2 × SSC
- Non-radioactive labelling and detection chemicals (see Chapter 5)

Method

1. Cut membrane to correct size (covering the expected fragment size range).
2. Separate glass plates after electrophoresis.
3. Put dry membrane on top of the gel region containing all alleles.
4. Put four layers of Whatman 3MM paper on top of the membrane.
5. Cover with a glass plate.
6. Put a weight (about 2 kg) on the glass plate.
7. Blot for >1 h (depending on fragment size).
8. Wash in 2 × SSC.
9. UV cross-link (see instructions provided by membrane supplier for protocol).

Several blotting techniques are currently available. The most simple and straightforward approach is capillary blotting after electrophoresis, as described in *Protocol 6*. This method does not require any special equipment; however, the relatively low transfer efficiency reduces sensitivity. Two other techniques, electroblotting (Chapter 5) and direct blotting, require specialized equipment. Direct blotting requires the purchase of a specialized electrophoresis set-up (e.g. GATC 1500 system; GATC). It relies on the transfer of separated DNA fragments on to the nylon membrane during electrophoresis (30). This technique offers the advantage of multiple use of the same gel and excellent resolution of large PCR fragments. However, the correct handling of the direct blotting equipment requires some training.

iii. Fluorescent dyes on automated sequencers

A relatively recent development is the automated detection of microsatellites during gel electrophoresis. To allow detection, the amplified PCR products are labelled with a fluorescent dye (either by incorporation during PCR or by using an end-labelled PCR primer). When activated by laser light, this dye emits a signal which can be detected. By comparing the migration of the PCR product with a length marker, accurate sizing is possible. This frequently used approach is discussed in Chapter 10.

4. Multiplex PCR

High sample throughput with many loci analysed is one key factor for a large population survey. Therefore, there is a high demand for improvements that reduce the amount of pipetting work. This has led to the development of multiplex PCR approaches, which are based on the simultaneous amplification of several microsatellite loci in a single PCR tube. When PCR primers are designed to amplify microsatellite alleles in non-overlapping size ranges, multiplex PCRs can be analysed in single gel lanes. The use of fluorescent dyes on an Applied Biosystems sequencer permits the analysis of overlapping allele sizes if different colours are used (see Chapter 10). For some species (e.g., cattle and horses), multiplex PCRs have been developed which amplify more than five loci simultaneously.

However, the establishment of multiplex conditions can be highly labour-intensive and often requires the re-design of PCR primers. Furthermore, if the overlapping of alleles from different loci needs to be avoided, the allele range of each locus should be well known. Consequently, this effort is often reserved for commercially important species. The decision whether to develop multiplex primer sets is a trade-off between the labour investment to establish the multiplex PCR conditions and the time saved by running fewer gels. For most applications, the development of a highly elaborate multiplex PCR is probably not advisable. Combining two or three loci in a single PCR reaction is a more realistic, feasible alternative.

The most critical step in the establishment of multiplex PCRs is to choose the correct PCR conditions and primer combinations. In addition to varying Mg^{2+} concentration and annealing temperature, a polymerase enhancer (Perfect Match; Stratagene) has been found to be useful to increase specificity in a multiplex PCR. This step can be avoided by amplifying the microsatellite loci separately and subsequently pooling the PCR products. Analysing the pooled microsatellite PCRs on a single gel still provides considerable time savings.

5. Microsatellite statistics

5.1 Theory

Microsatellites provide an excellent marker system for testing paternity and kinship selection. The co-dominance of microsatellites and their high variability allows recognition of the paternal allele by comparing the genotypes of mother and offspring. Based on allele frequencies in the population, paternal exclusion probabilities may then be calculated. Mutation rates ranging from 10^{-3} to 10^{-6} are high enough to generate sufficient polymorphisms to provide high resolution, but low enough to calculate kinship coefficients in social groups.

Microsatellite analysis becomes more difficult as soon as the individuals to be studied are separated for longer time spans, because new mutations will have been acquired. To extract information about population history from observed allele frequencies, it is important to understand the mutation mechanism underlying microsatellite variability.

In vitro and *in vivo* studies indicate that the most common mutation of a microsatellite allele is either the gain or loss of a single repeat unit (16,31). These length alterations are probably generated by DNA slippage, an intramolecular process. It is not possible to make any predictions about whether the next mutation will generate a longer or shorter microsatellite allele, and it is impossible to determine whether two alleles are identical because there is a genuine relationship or whether they are only 'identical by state' (*Figure 4*).

Mutation rates of microsatellites are the result of a balance between the constant generation of new mutations by DNA slippage, and their removal by DNA mismatch repair (32). Higher mutation rates are observed in cells deficient for one mismatch repair system (32), and mismatch repair efficiency is size-dependent, explaining the slightly higher mutation rates of tetranucleotides in humans (despite their lower *in vitro* slippage rates, but see ref. 33). Currently, it is unclear whether the large variances in microsatellite mutation rates, even for microsatellites with the same repeat unit, are a reflection of different efficiencies of the repair mechanism at different chromosomal locations, or just a sampling artefact due to small numbers of observed mutations.

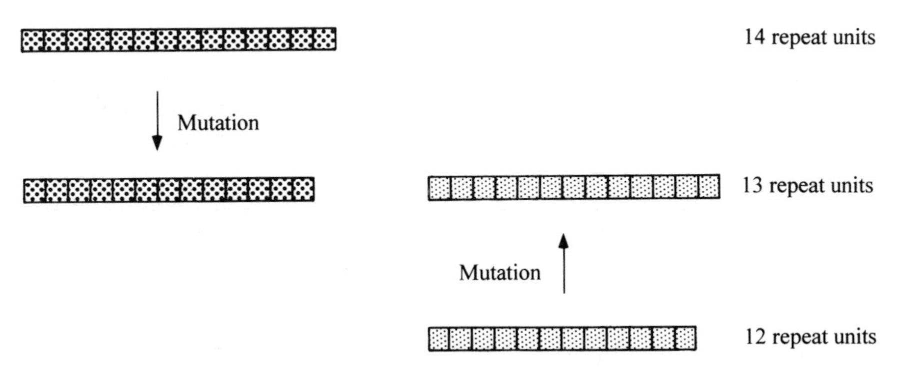

Figure 4. Illustration of 'identity by state'. Two microsatellite alleles with 13 repeat units are shown, derived from different ancestral alleles. Thus, despite their identical lengths, the two 13-repeat alleles are less closely related to each other than, e.g., the 14- and 13-repeat alleles. This phenomenon is called homoplasy.

Interestingly, the microsatellite mutation pattern very closely fits the stepwise mutation model (6) which has been studied intensively by population geneticists in order to explain patterns of allozyme variation. On this basis it is possible to introduce genetic distance, a widely used measurement in population genetics, to microsatellite analysis. These genetic distances are called R_{st}, delta mu, or D_{sw} (see below and Appendix 5) (7,8,10).

5.1.1 Open questions

Even though the stepwise mutation model is probably a very good approximation of the mutation mechanism of microsatellites, there is evidence that some additional mechanisms need to be considered. Interspecies comparisons of microsatellite allele frequencies showed that no large size differences between orthologous microsatellites are observed (34). Therefore, it is highly likely that microsatellites have upper and lower boundaries within which they may vary. If these boundaries are an intrinsic propensity of microsatellites, they need to be considered in the algorithms for calculating genetic distance. As a further concern, it has been put forward that the mutation process of microsatellites is not random, but biased towards longer alleles (35), which would alter interpretation of observed allele frequencies. Finally, some indications have been reported that in addition to the stepwise mutations, a further mutation process may be occurring at lower frequencies, altering microsatellite length by multiple repeat units (36).

5.2 Gel interpretation

The most common picture of a PCR-amplified microsatellite allele is not a single band, but a ladder of bands. Usually, the most intense band is observed at the expected size of the allele. The additional bands, called stutter bands, are usually smaller than the original allele (but see below) and differ in length by multiple repeat units. Most likely, stutter bands are the result of *in vitro* DNA slippage during PCR amplification. The decreasing intensity of stutter bands on the gel is a reflection of the probability distribution of DNA slippage events, with larger size deviations being less likely. Rates of *in vitro* DNA slippage have been shown to decrease with repeat unit length: more stutter bands occur for dinucleotide repeats than for trinucleotide or tetranucleotide repeats (5).

In addition to stutter bands, many microsatellites show an additional band above the expected allele size. This is the result of the terminal transferase activity of *Taq* DNA polymerase, which adds an A to the PCR product. However, this terminal transferase activity is polymerase- and PCR primer-dependent.

The microsatellite banding patterns described above are typical, but secondary products of various types can produce more complex patterns. *Figure 5* shows several microsatellite patterns with interesting deviations from the consensus pattern.

All microsatellite patterns shown in *Figure 5* are highly consistent, allowing a reliable scoring of the alleles if the gels are analysed manually. For reliable automated scoring of all microsatellite loci, it would be highly desirable to have an algorithm to correct for the patterns specific to a particular microsatellite. Currently, the algorithms available only score the most intense band together with the stutter bands.

5.3 Null alleles

As explained above, microsatellite detection is PCR-based. For correct amplification it is mandatory that both PCR primers match the flanking regions of all alleles. If a deletion or point mutation(s) in the primer binding site of a specific allele interferes with priming, that allele will not be amplified. Such alleles are called 'null alleles' (37,38). The unequivocal identification of null alleles in population surveys is problematic. Normally, the presence of null alleles is suspected if a surplus of homoyzgous individuals is observed. However, such deviation from Hardy–Weinberg equilibrium may be for other reasons, such as population sub-division. By studying a large set of microsatellites, population subdivision may be ruled out if heterozygote deficiency is unique to a particular locus.

The best proof of null alleles would be a pedigree analysis; however, for many species no pedigrees are available. In this case, an alternative strategy would be the design of new primers which should not interfere with the DNA polymorphism causing the null allele.

5.4 Specialized software for microsatellite analysis

F_{st} and related measures of population differentiation are based on the assumption of relatively low mutation rates. This assumption clearly does not hold for microsatellites. Therefore, new measurements have recently been introduced that are taking the microsatellite mutation processes into account (see Appendix 5).

MICROSAT is a highly versatile software package especially designed for microsatellite analysis (39). It contains sub-routines for calculation of delta mu (10) and the proportion of shared alleles (34). These are both distance measurements that have proven highly useful for microsatellite analysis. In addition, the software package contains other frequently used distance measurements, such as Nei's distance. A bootstrap resampling routine is also incorporated. The distance output format is compatible with the PHYLIP software package, which can be used for converting distance data into trees. The compiled version of the program can be obtained through the WWW (`http://lotka.stanford.edu/research/distance.html`) for several platforms. The original source code and thus the compiled versions contain a rather simple random number generator, which one may prefer to change for some applications.

Software to calculate the distance measurement $D_{sw,}$ which is appropriate

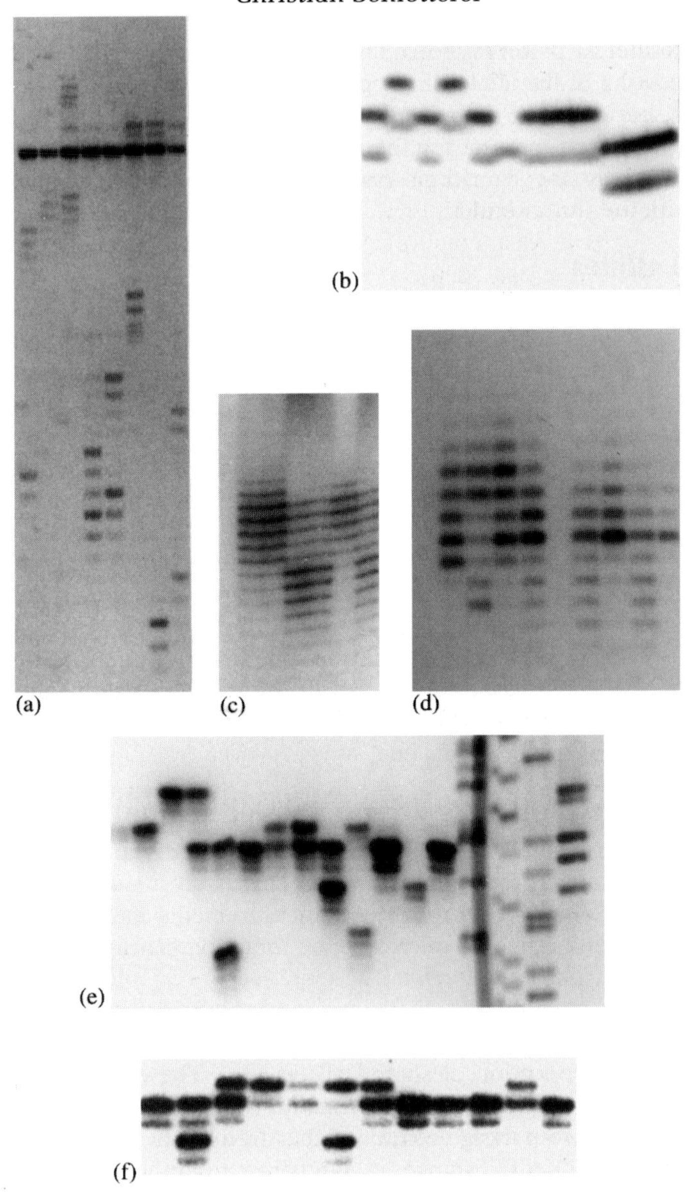

Figure 5. (a) A highly polymorphic dinucleotide microsatellite isolated from pilot whales. In addition to the expected length variation of two base pairs, single base-pair changes are also observed. Note that the most intense band is a PCR artefact. (b) A dinucleotide microsatellite isolated from *Drosophila melanogaster*. This locus shows almost no stutter bands, but has an additional band below the expected allele size. As inbred *Drosophila* lines were typed the second allele was lost by drift during the propagation of the lines. (e) A dinucleotide microsatellite isolated from *D. melanogaster*. This locus shows stutter

for loci with high mutation rates evolving via a stepwise mutation mechanism, is described by M. Shiver (8). The programs are available from the author (e-mail: `mshriver@pgh.auhs.edu`).

A Windows-based program written by L. Excoffier is available from `ftp://anthropologie.unige.ch/pub/comp/win/amova/`. This program provides several useful features for co-variant population data (40). If a matrix of pairwise distances between pairs of microsatellite alleles is provided, the program can be used to calculate R_{st} (7), an F_{st} analogue for microsatellite data.

6. Cross-species amplifications

Despite the relative ease of microsatellite isolation, the efficient characterization of microsatellites is dependent on some experience in molecular biology. Therefore, the finding that polymorphic microsatellites are conserved among closely related species has greatly facilitated the use of microsatellites (26). A table providing primer sequences for some of these loci is given in Appendix 4. Primate research in particular has greatly benefited from the availability of several hundred characterized human microsatellites (41). The range of divergence between two species over which positive amplification of orthologous microsatellites remains possible is highly species-dependent. Some microsatellites have been shown to be conserved over several millions of years, suggesting the possible utility of trial amplifications in even more distantly related species (42–47). *Molecular Ecology* (Blackwell) has a special section devoted to publications describing isolation of new microsatellites and cross-species amplification.

If orthologous primers are used to amplify microsatellites, care should be taken that non-specific amplifications are not incorrectly scored as microsatellite alleles. Some orthologous microsatellites have been shown to differ in the number of flanking nucleotides as well as in the average number of repeat units (26,48). This may cause significant changes in length compared with the allele range in the species for which the microsatellite was identified. Therefore, PCR product length can be a poor indicator of the positive amplification of orthologous microsatellites. The presence of stutter bands is a better indicator, but is not always applicable for trinucleotide and tetranucleotide repeats. Variability of the amplified PCR product makes it highly likely that the amplified product is indeed the orthologous microsatellite. However, the best confirmation is hybridization of the PCR product with the microsatellite motif.

---

bands which are separated by only a single base pair. (d) A dinucleotide microsatellite isolated from pilot whales. This locus exhibits stutter bands both above and below the most intense band (c) A dinucleotide microsatellite isolated from *Latimeria chalumnae*. The very pronounced stutter bands almost disguise the fact that all individuals are heterozygous. (f) The classic example of an easy to read microsatellite (from *Globicephala melas*), with no additional base added.

Researchers not working with primates or species closely related to other model organisms would benefit from accesible information on microsatellite primers isolated from a diversity of species. To accommodate these needs databases for microsatellites are required. So far, several attempts have been made including a newsgroup (`micro-sat@sfu.ca`) and a database (URL: `gopher://nmnhgoph.si.edu:70/11/.lms /.microsat`) and these are worth contacting.

Null alleles are a concern for homologous microsatellite amplification, and it has been suggested that orthologous primers may be even more susceptible. Since only a few studies have explored the origin of null alleles, it is difficult to predict what happens in the case of cross-species amplifications. Assuming that in the most cases only a single point mutation is responsible for the loss of the priming site, null alleles are the consequence of a sequence polymorphism at the primer-binding site. Hence, it is hard to conceive why, *a priori*, the 'source' species should have less polymorphism than a closely related species. Therefore, there is no reason to conclude that orthologous primers should harbour more null alleles. However, if null alleles are caused by multiple mutations, then orthologous primers should be used more carefully.

7. Microsatellites from 'difficult' templates

As a PCR-based technique, microsatellites can be used to analyse samples including only minute DNA quantities or highly degraded DNA. For example, microsatellites have been amplified from skeletal remains (49), museum specimens (50), hair roots (51), and faeces (52–54). Large museum collections are available for numerous species, facilitating the determination of allele distributions going back in time, and this information could be highly relevant to understanding the population dynamics of an endangered species. Collection of DNA from faeces (*Protocol 7*) is becoming widespread due to the fact that faeces is a readily available animal material that can be collected completely non-intrusively (without requiring permission for sampling and transport), and from species that are difficult to trap or encounter in the wild.

Protocol 7. DNA extraction from faeces

Equipment and reagents

- Faeces, collected as fresh as possible in 2 vol. ethanol
- Diatomaceous earth (Sigma Cat. no. D-5384)
- Lysis buffer: 50 mM Tris–HCl pH 7.0, 25 mM EDTA pH 8.0, 1.25% Trition X-100, 5 M guanidine thiocyanate (Sigma)
- Rotation wheel
- Forceps
- DNA binding solution: 1 g size fractionated diatomaceous earth[a] resuspended in 100 ml 50 mM Tris–HCl pH 7.0, 25 mM EDTA pH 8.0, 5 M guanidine thiocyanate

Note: to avoid generation of poisonous HCN gas, dispose of buffers containing guanidine thiocyanate after mixing them with 0.1 vol. 10 M NaOH (check with your safety officer).

Method

1. Briefly dry 50–100 mg faeces in air at RT.
2. Add 1 ml lysis buffer.
3. Vortex until faeces is completely suspended.
4. Incubate overnight at room temperature.
5. Spin for 10 min at 16 000 × *g*.
6. Remove 700 μl supernatant.
7. Add 700 μl of DNA binding solution and vortex briefly.
8. Incubate for 2 h on rotation wheel at room terperature.
9. Sediment diatomaceous earth particles (2 min at 16 000 × *g*).
10. Re-dissolve pellet in 1 ml of 70% ethanol.
11. Sediment diatomaceous earth particles (2 min at 16 000 × *g*).
12. Repeat steps **10** and **11**.
13. Remove ethanol.
14. Dry pellet at 56 °C.
15. Elute DNA with 120 μl of TE by incubating re-dissolved pellet for 10 min at 56 °C.
16. Use 5 μl of supernatant for PCR.

[a] Add 5 g of diatomaceous earth to a 500 ml graduated cylinder filled with double-distilled water. Sediment for 3 h and remove supernatant.

Systematic studies with bonobo (*Pan paniscus*) faeces have shown some potential risks associated with PCR amplification from faeces (54). By re-extracting the same faeces sample and performing multiple PCRs, it has been shown that some heterozygous individuals falsely appear to be homozygous following PCR. Similar observations have been made for hair samples. Furthermore, not all faeces samples were found to yield sufficient DNA for PCRs. Therefore, it is advisable to collect multiple independent samples from each individual.

8. Perspective

Future research on microsatellite evolution will provide a better theoretical framework for the correct interpretation of microsatellite polymorphisms. New screening methods, like DNA chips (55), will be developed for microsatellites that will allow the measurement of repeat size and composition directly, instead of sizing PCR products. Finally, the increasing number of microsatellites isolated will incease the likelihood of finding sequences from a species related to your study species, and facilitate studies in molecular ecology and evolution.

Acknowledgements

I am grateful to J. Armour and D. Dawson for providing enrichment protocols and critically reading the enrichment protocol given in this chapter. Furthermore, I should like to thank W. Miller, D. Tautz, R. Hoelzel, and M. Haberl for comments on the manuscript.

References

1. Dietrich, W.F. *et al.* (1996). *Nature* **380**, 149.
2. Dib, C. *et al.* (1996). *Nature* **380**, 152.
3. Bruford, M.W. and Wayne, R.K. (1993). *Curr. Opin. Genet. Dev.* **3**, 939.
4. Schlötterer, C. and Pemberton, J. (1994). In *Molecular ecology and evolution: approaches and applications* (ed. B. Schierwater, B. Streit, G.P. Wagner, and R. DeSalle), p. 203. Birkhäuser Verlag, Basel.
5. Tautz, D. and Schlötterer, C. (1994). *Curr. Opin. Genet. Dev.* **4**, 832.
6. Ohta, T. and Kimura, M. (1973). *Genet. Res.* **22**, 201.
7. Slatkin, M. (1995). *Genetics* **139**, 457.
8. Shriver, M.D., Jin, L., Boerwinkle, E., Deka, R., Ferrell, R.E., and Chakraborty, R. (1995). *Mol. Biol. Evol.* **12**, 914.
9. Goldstein, D.B., Ruiz Lineares, A., Cavalli-Sforza, L.L., and Feldman, M.W. (1995). *Genetics* **139**, 463.
10. Goldstein, D.B., Ruiz Lineares, A., Cavalli-Sforza, L.L., and Feldman, M.W. (1995). *Proc. Natl Acad. Sci. USA* **92**, 6723.
11. Zhivotovsky, L.A. and Feldman, M.W. (1995). *Proc. Natl Acad. Sci. USA* **92**, 11549.
12. Schug, M.D., Mackay, T.F.C. and Aquadro, C.F. (1997). *Nature Gent.* **15**, 99.
13. Tautz, D. (1989). *Nucleic Acids Res.* **17**, 6463.
14. Rassmann, K., Schlötterer, C., and Tautz, D. (1991). *Electrophoresis* **12**, 113.
15. Armour, J.A.L., Neumann, R., Gobert, S., and Jeffreys, A.J. (1994). *Hum. Mol. Genet.* **3**, 599.
16. Schlötterer, C. and Tautz, D. (1992). *Nucleic Acids Res.* **20**, 211.
17. Hartl, D.L., Nurminsky, D.I., Jones, R.W., and Lozovskaya, E.R. (1994). *Proc. Natl Acad. Sci. USA* **91**, 6824.
18. Schlötterer, C., Vogl, C., and Tautz, D. (1997), *Genetics*, **146**, 309.
19. Sternberg, N. (1990). *Proc. Natl Acad. Sci. USA* **87**, 103.
20. Rowe, P.S.N., Francis, F., and Goulding, J. (1994). *Nucleic Acids Res.* **22**, 5135.
21. Powell, W., Machray, G.C., and Provan, J. (1996). *Trends Plant Sci.* **1**, 215.
22. Langercrantz, U., Ellegren, H., and Andersson, L. (1993). *Nucleic Acids Res.* **21**, 1111.
23. Weber, J.L. (1990). *Genomics* **7**, 524.
24. Beckmann, J.S. and Weber, J.L. (1992). *Genomics* **12**, 627.
25. Goldstein, D.B. and Clark, A.G. (1995). *Nucleic Acids Res.* **23**, 3882.
26. Schlötterer, C., Amos, B., and Tautz, D. (1991). *Nature* **354**, 63.
27. Kunst, C.B. and Warren, S.T. (1994). *Cell* **77**, 853.
28. Estoup, A., Tailliez, C., Cornuet, J.-M., and Solignac, M. (1995). *Mol. Biol. Evol.* **12**, 1074.

29. Butler, J.M., McCord, B.R., Jung, J.M. and Allen, R.O. (1994). *BioTechniques* **17**, 1062.
30. Schlötterer, C. (1993). *Nucleic Acids Res.* **21**, 780.
31. Levinson, G. and Gutman, G.A. (1987). *Mol. Biol. Evol.* **4**, 203.
32. Strand, M., Prolla, T.A., Liskay, R.M., and Petes, T.D. (1993). *Nature* **365**, 274.
33. Chakraborty, R., Kimmel, M., Stivers, D. N., Davison, L. J., and Deka, R. (1997). *Proc. Natl. Acad. Sci. U.S.A.* **94**, 1041.
34. Bowcock, A.M., Ruiz-Lineares, A., Tonfohrde, J., Minch, E., Kidd, J.R., and Cavalli-Sforza, L.L. (1994). *Nature* **368**, 455.
35. Rubinsztein, D.C, Amos, W., Leggo, J., Goodburn, S., Jain, s., Li, S.-H., Margolis, R.L., Ross, C.A., and Ferguson-Smith, M.A. (1995). *Nature Genet.* **10**, 337.
36. Di Rienzo, A., Peterson, A.C., Garza, J.C., Valdes, A.M., Slatkin, M., and Freimer, N.B. (1994). *Proc. Natl Acad. Sci. USA* **91**, 3166.
37. Pemberton, J.M., Slate, J., Bancroft, D.R., and Barrett, J.A. (1995). *Mol. Ecol.* **4**, 249.
38. Callen, D.F. *et al.* (1993). *Am. J. Hum. Genet.* **52**, 922.
39. Minch, E., Ruiz-Linares, A., Goldstein, D., Feldman, M., and Cavalli-Sforza, L.L. (1995).
40. Michalakis, Y. and Excoffier, L. (1996). *Genetics* **142**, 1061.
41. Morin, P.A., Wallis, J., Moore, J.J., and Woodruff, D.S. (1994). *Mol. Ecol.* **3**, 469.
42. Angers, B. and Bernatchez, L. (1996). *Mol. Ecol.* **5**, 317.
43. Coltman, D.W., Bowen, W.D., and Wright, J.M. (1996). *Mol. Ecol.* **5**, 161.
44. FitzSimmons, N.N., Mority, C., and Moore, S.S. (1995). *Mol. Biol. Evol.* **12**, 432.
45. Kijas, J.M.H., Fowler, J.C.S., and Thomas, M.R. (1995). *Genome* **38**, 349.
46. Pépin, L., Amigues, Y., Lépingle, A., Berthier, J.-L., Bensaid, A., and Vaiman, D. (1995). *Heredity* **74**, 53.
47. Primmer, C.R., Møller, A.P., and Ellegren, H. (1996). *Mol. Ecol.* **5**, 365.
48. Garza, J.C., Slatkin, M., and Freimer, N.B. (1995). *Mol. Biol. Evol.* **12**, 594.
49. Hagelberg, E., Gray, I.C., and Jeffreys, A.J. (1992). *Nature* **352**, 427.
50. Ellegren, H. (1991). *Nature* **354**, 113.
51. Morin, P.A., Wallis, J., Moore, J.J., and Woodruff, D.S. (1994). *Science* **265**, 1193.
52. Deuter, R., Pietsch, S., Hertel, S., and Müller, O. (1995). *Nucleic Acids Res.* **23**, 3800.
53. Constable, J.J., Packer, C., Collins, S.A., and Pusey, A.E. (1995). *Nature* **373**, 393.
54. Gerloff, U., Schlötterer, C., Rassmann, K., Rambold, I., Hohmann, g., Fruth, B., and Tautz, D. (1995). *Mol. Ecol.* **4**, 515.
55. Southern, E.M. (1996). *Trends Genet.* **12**, 110.

8

Detection of genetic variation by DNA conformational and denaturing gradient methods

MICHAEL DEAN and BROOK G. MILLIGAN

1. Introduction

Single-stranded conformation polymorphism (SSCP) is one of the most widely used and practical approaches for the detection of mutations in DNA and for the detection and analysis of DNA variation (1). The principal advantage of the method is that it is very rapid to perform and can be carried out using equipment available in most molecular biology laboratories. If properly optimized the method is highly sensitive, and different mutations within a fragment can often be distinguished on the same gel (2). A number of variations of SSCP have been developed that allow analysis of RNA or for the assay to be carried out non-radioactively, as reviewed in ref. 3. SSCP is often combined with heteroduplex analysis to increase further the sensitivity of the method. Despite the denaturation of the sample prior to loading the SSCP gel, there is often the formation of a significant fraction of double-stranded DNA. This is visible at the bottom of the SSCP gel. Since heteroduplexes can occasionally be resolved from homoduplexes, the appearance of heteroduplexes on the SSCP gel can give additional information on the presence of variants.

In SSCP analysis the sample is amplified by the polymerase chain reaction (PCR) in the presence of a radiolabelled nucleotide (for example, ^{32}P or ^{33}P). The product is denatured to generate single-stranded molecules and loaded on to a non-denaturing gel. Single-stranded molecules are resolved on the gel and DNA sequence alterations between the primers appear as fragments of altered mobility due to differential folding of the single strand. The gel composition and running conditions can be varied to alter the temperature or degree of cross-linking, or to include glycerol or sucrose. These variations change the type of conformations seen and increase the sensitivity of detection (4). *Figure 1* displays analysis of a large number of point mutations created in a plasmid vector and analysed by SSCP. Each sample has a pattern distinct from the wild-type, and most mutants are distinct from each other.

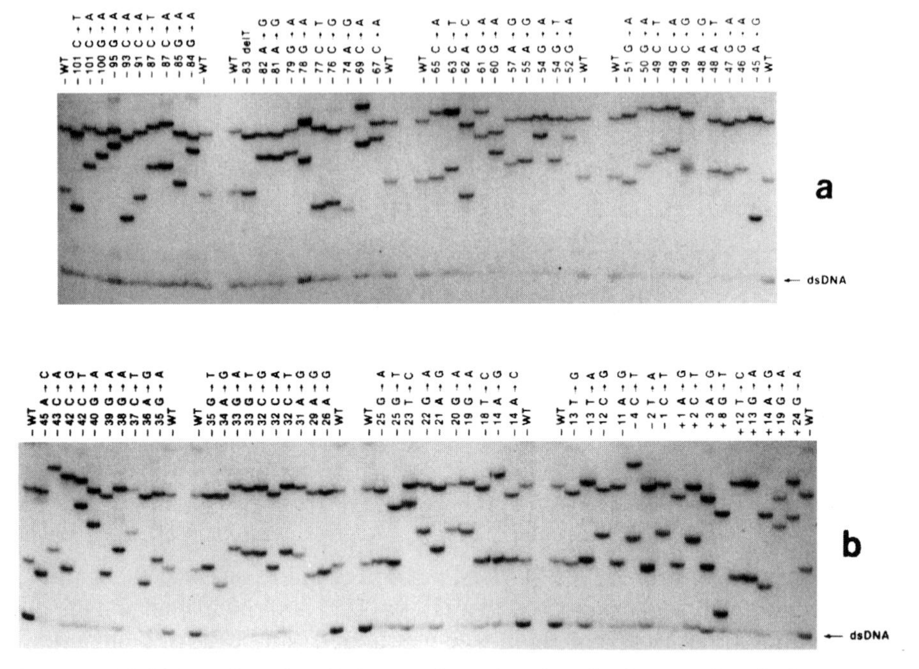

Figure 1. SSCP detection of multiple mutations. A series of point mutations in the mouse β-globin gene promoter (kindly provided by Rick Myers) were cloned into a plasmid. Each plasmid was amplified with flanking primers and assayed on an SSCP gel. The position and nature of each point mutation is indicated above the lane. WT, wild type; dsDNA, double-stranded DNA. From ref. 2, with permission from John Wiley and Sons, Inc.

2. Optimization of PCR

Optimum conditions for the PCR are essential to generating easily interpretable SSCP gels (see also Chapter 6). Important variables to consider are the concentration of Mg^{2+} in the reaction and the parameters of the PCR cycles (temperatures and number of cycles). In addition, several researchers have found that a number of additives such as dimethylsulfoxide (DMSO), spermidine, non-ionic detergents, and tetramethylammonium chloride (TMAC) improve the yield of the PCR product. We have recently found that a step-down PCR program gives clean product from a number of primers using identical Mg^{2+} concentrations (5). In this procedure, the annealing is initially carried out at a high temperature, perhaps above the melting temperature (T_m) of the primers, then in subsequent cycles the annealing temperature is gradually lowered. In this fashion, the reaction should initiate at the temperature that is optimum for the fully base-paired product. In subsequent cycles, at lower temperatures, a sufficient mass of the expected product is generated, which should out-compete any unwanted side reactions. In essence the primers are allowed to find their own optimum temperture to start priming; this avoids the need for extensive optimization of Mg^{2+} concentrations. A version of this procedure is given in *Protocol 1*.

Protocol 1. PCR set-up and optimization

Equipment and reagents

- 10 × PCR buffer (Perkin-Elmer, or see chapter 6)
- dNTPs at 200 μM each
- Primers 1 and 2 (1 OD_{260} unit/ml)
- *Taq* DNA polymerase (Perkin-Elmer)
- Genomic DNA (100 ng/μl)
- Mineral oil[a]
- PCR machine
- Small (10–20 cm) 8% acrylamide or 2% agarose gels for visualizing PCR products
- 100–500 V power supply
- UV light box with camera for photographing gels

Method

1. Make a cocktail of all of the following reagents:
   - 2.5 μl 10 × PCR buffer
   - 2.5 μl 200 μM each dNTP
   - 1.0 μl primer 1
   - 1.0 μl primer 2
   - 16.8 μl water
   - 0.2 μl *Taq* polymerase (5 U/μl)
2. Add 1 μl (100 ng) of genomic DNA to the PCR tube then add 24 μl of the cocktail and one drop of mineral oil.[a]
3. Spin briefly in a microcentrifuge and place in the PCR machine.
4. Perform either a standard PCR program (a) or a step-down program (b):
   (a) Standard PCR:
      (i) 5 min at 95°C;
      (ii) 30 cycles of 0.5 min at 95°C, 1.0 min at 55°C, 2.0 min at 72°C;
      (iii) final extension of 72°C for 10 min.
   (b) The modified step-down program is:
      (i) 3 min at 94°C;
      (ii) two cycles of 35 sec at 94°C (denaturing), 35 sec at 65°C (annealing), 75 sec at 72°C (extension);
      (iii) two cycles of 35 sec at 94°C, 35 sec at 64°C, 75 sec at 72°C;
      (iv) continue reducing the annealing temperature by 1°C at a time, carrying out two cycles at each temperature until you reach 59°C;
      (v) 22 cycles of 35 sec at 94°C, 35 sec at 50°C, 75 sec at 72°C;
      (vi) 5 min at 72°C.

Protocol 1. *Continued*

3. Following the PCR, run 10 μl of the reaction on an 8% acrylamide gel or a 2% agarose gel. Compare the results from the step-down procedure with two or three different Mg^{2+} concentrations with standard cycling conditions. The conditions giving the highest yield and cleanest product are used for SSCP analysis.

[a] If the PCR machine has a heated lid, mineral oil is not necessary.

3. SSCP analysis

3.1 SSCP sample preparation

Once optimum conditions for the PCR have been established, a scaled-down reaction (10 μl) is performed in the presence of a radiolabelled nucleotide, typically ^{32}P. Other labelled isotopes can be used, e.g. ^{33}P or ^{35}S. Alternatively, the primers can be end-labelled using [γ-^{32}P]ATP. (Note that both primers need to be labelled to visualize the two strands. This is because in SSCP it is important to visualize both strands, as conformers can be observed on either strand.). Amplified samples are run on a long, thin acrylamide gel.

Protocol 2. Preparing PCR samples for SSCP

Equipment and reagents

- dNTPs (200 μM each)
- 10 × PCR buffer (Perkin-Elmer)
- Primers 1 and 2 (each at 1 OD_{260} unit/ml)
- [α-^{32}P]dCTP (3000 Ci/mmol)
- *Taq* DNA polymerase (Perkin-Elmer)
- Microtitre plate
- Loading dye (95% formamide, 0.1 % Bromophenol Blue, 0.1% xylene cyanol 10 mM NaOH)

Method

1. The SSCP reaction is identical to the test reaction except that the concentration of dNTPs is reduced by one-third (to 70 μM), 0.1 μl of [α-^{32}P]dCTP is added to the reaction, and the final volume is 10 μl. Mix the following:
   - 1.0 μl genomic DNA (100 ng)
   - 1.0 μl 10 × PCR buffer
   - 0.3 μl 200 μM each dNTP
   - 1.0 μl primer 1
   - 1.0 μl primer 2
   - 0.1 μl [α-^{32}P]dCTP
   - 5.4 μl water
   - 0.2 μl *Taq* DNA polymerase (5 U/μl)

2. Mix 2 μl of the PCR product in a well of a microtitre plate with 8 μl of loading dye. Alternatively add 30 μl of loading dye to the entire sample.
3. Heat the plate to 95°C for 2 min.
4. Load 2–3 μl of the sample on to the gel. It is also useful to load a lane of undenatured DNA, to display the double-stranded DNA fragments.

Protocol 3. Preparing and running an SSCP gel

Equipment and reagents

- 10% ammonium persulfate, freshly made
- *N,N,N',N'*-tetramethylethylenediamine (TEMED)
- 35 cm sequencing gel box with 0.4 mm spacers and shark's-tooth combs
- 2000–3000 V power supply
- Loading dye (*Protocol 2*)
- 40% acrylamide/bis-acrylamide (75:1; 1.3% cross-linking) solution: dissolve 39.5 g acrylamide and 0.53 g bis-acrylamide in a total volume of 100 ml of distilled water
- Whatman 3MM filter paper
- Plastic wrap
- Autoradiography film

Method

1. Clean a set of plates for a 35 cm, 0.4 mm thick gel.
2. Prepare one of the gel formulations listed in *Table 1* and use to prepare the gel. For 75 ml of gel solution add 500 μl of fresh 10% ammonium persulfate and 50 μl of TEMED. Allow the gel to polymerize for 30 min, and pre-run in the gel box in a 4°C cold-room for at least 5 min.
3. Load 2–3 μl of sample (Protocol 2) and allow to run at 4°C until the Bromophenol Blue has reached the bottom.
4. Carefully separate the plates, remove the gel on to a piece of Whatman 3MM filter paper, and cover with plastic film. Dry the gel and expose to X-ray film for 2–24 h.

3.2 Optimization of SSCP detection

The conditions of the SSCP gel can be varied to produce gels in which the samples form alternate conformations, i.e. different secondary structures at the strands will form depending on the nature of the gel. The number of possible conformations is so large that there is no theoretical basis for choosing conditions. Clearly the more conditions, run the greater the sensitivity; however, there are limits to the number of gels that can be run. A single gel can give 60–90% efficiency of detection, and can be used as an initial screening method. Two to four gels can be employed to reach close to 100% detection. The variables that have been employed are as follows:

- temperature: gels can be run at 20–40 W constant power overnight at room temperature, or at 50 W for 3–5 h in a 4°C cold room

- cross-linking: different ratios of acrylamide:bis-acrylamide can be used to give differing degrees of cross-linking. 19:1 acrylamide:bis-acrylamide gives 5% cross-linking, 37.5:1 gives 2.6% cross-linking, and 75:1 gives 1.3% cross-linking; alternatively, 0.5 × MDE (Mutation Detection Enhancer; Avitech Diagnostics) matrices can be used
- additives: 5–10% glycerol or 10% sucrose

Each of these variables can alter the conformation of single-stranded molecules. SSCP was initially run on 5% acrylamide gels with 5% bis-acrylamide cross-linker (5% cross-linking) at either room temperature or 4°C with or without glycerol. More recently researchers have shown that gels with higher percentage of acrylamide and lower cross-linking can detect more mutations (6). Sucrose has also proven to be a useful additive (4). MDE is an alternative matrix that has been employed in place of acrylamide (4). *Table 1* gives gel recipes for several different conditions that produce optimum results (2).

Unfortunately it is not possible to predict how a given change in condition will affect the mobility of a specific fragment. However, it can be observed in *Figure 2* that differences in the percentage of acrylamide and the extent of cross-linking affect the mobility of different mutations. For instance, the –45A→C mutation can be distinguished from the wild type much more readilly on 10% acrylamide (1.3% cross-linking; *Figure 2a*) than for the other conditions. In contrast, the –14A→G and –14A→C variants are more easilly resolved on the 5% acrylamide (2.6% cross-linking) gel (*Figure 2c*).

The size of the PCR product can also affect sensitivity. In general, the smaller the product the higher the sensitivity, with the optimum being 200–300 bp. SSCPs have been observed on products as small as 50 bp and as large as 1 kb. Longer products can be cleaved with restriction enzymes to yield a series of bands that can be assayed simultaneously.

Table 1. Three gel formulations

Final concentration	Stock concentration	Amount per 250 ml
10% acrylamide (1.3% cross-linking)	40%	62.5 ml
1 × TBE	10 ×	25 ml
0.5 × MDE	2 ×	62.5 ml
10% glycerol	100%	25 ml
1 × TBE	10 ×	25 ml
10% acrylamide (1.3% cross-linking)	40%	62.5 ml
10% sucrose		25 g
1 × TBE	10 ×	25 ml

All solutions are to made up to final volume (250 ml) with distilled water.

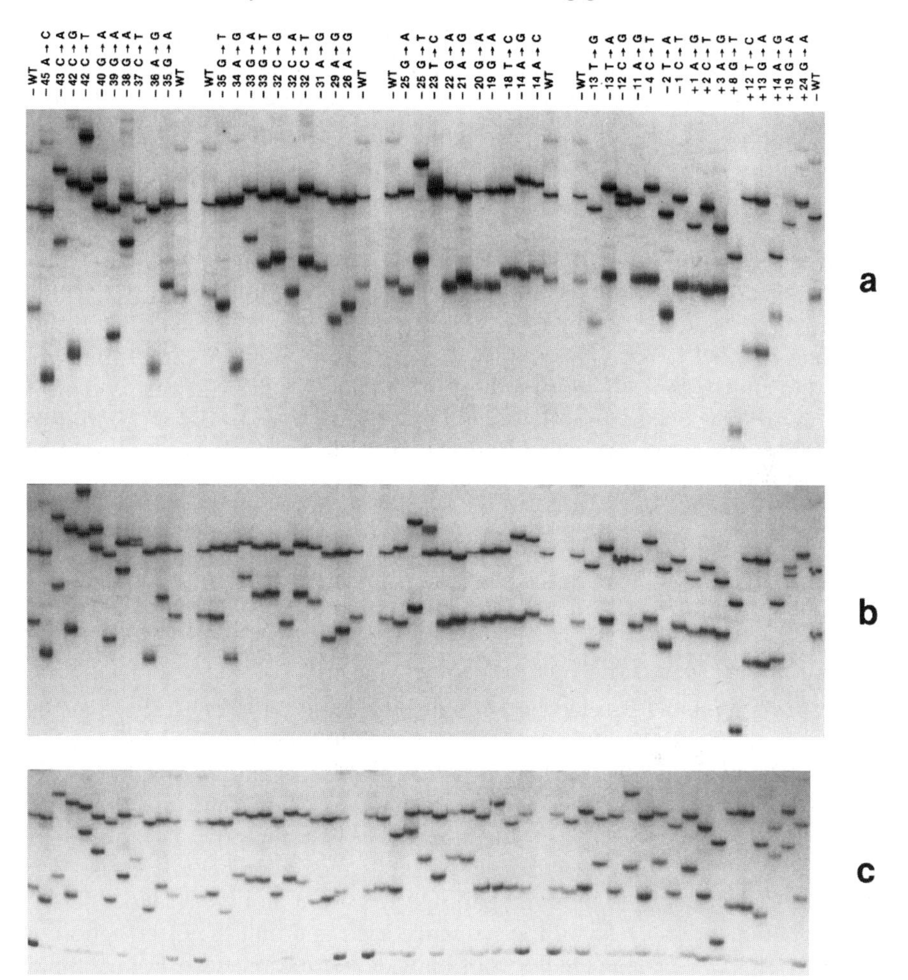

Figure 2. PCR–SSCP analysis of mutations in a 193 bp fragment of the mouse β-globin gene on three different polyacrylamide gel matrices. (a) 10% acrylamide with 1.3 % bis-acrylamide cross-linking. (b) 5% acrylamide with 2.6 % cross-linking. (c) 5% acrylamide with 2.6 % cross-linking. All gels were run in a 4°C cold room at 50 W constant power in 0.8 × TBE buffer. Reprinted from Figure 3 in ref. 3, with permission from Academic Press, Inc.

3.3 Interpretation of SSCP results

The interpretation of the results of an SSCP gel requires some experience. However, a few controls can assist in the analysis. Some double-stranded DNA often re-forms after the denaturation step, producing bands near the bottom of the gel. It is helpful to include a undenatured sample to help in identifying these bands. Double-stranded heteroduplex molecules, formed when a mutant and wild-type strand anneal, can also indicate sequence alterations. These heteroduplexes migrate just above the double-stranded DNA. While heteroduplexes are most prominent in the case of insertions or deletions, they can be seen with some point mutations (7).

Since the exact migration of SSCP conformers can vary from gel to gel, it is useful to include any available control samples at least once on each gel. While it is expected that the two strands will migrate differently, this is not always the case. Similarly, in a heterozygote, not all four strands are always resolved. The important point to look for is that a mutant sample should give a clearly different pattern from a wild-type sample. In addition, a heterozygote should display approximately equal intensity in all four bands.

Occasionally PCR artefact bands occur that can confuse the interpretation. If a band appears only in samples that amplified better than the rest, it is probably an artefact. In samples that amplify well, or are overloaded, alternative conformations can appear. These may be the result of two conformers partially annealing. Diluting the samples will chase the double-stranded DNA into single strands and remove some of these additional bands. Where possible the segregation of altered bands should be demonstrated within a pedigree to confirm it. The final proof of the alteration comes from sequencing.

4. Denaturing gradient gel electrophoresis

Denaturing gradient gel electrophoresis (DGGE) is another effective means of detecting unknown mutations in DNA and is therefore useful in the detection and analysis of DNA variation in natural populations. Unlike SSCP techniques, the method does require specialized equipment; however, it is possible to survey larger DNA fragments using the method and to detect a greater proportion of the variation present. DGGE is based on the twin notions that the physical melting properties of a double-stranded DNA molecule depend in detail on the sequence itself and that the electrophoretic mobility of a DNA molecule depends on its denaturation state. For example, two molecules identical except for a single nucleotide position will have similar but not identical melting properties. As a result, they will be denatured under slightly different conditions, e.g., different concentrations of a chemical denaturant or different temperature. If these molecules are subjected to electrophoresis through a gradient of increasing denaturant concentration, they will become denatured at slightly different positions along their path through the gel. Because the migration rate of these molecules depends on their denaturation state, the two will also exhibit slightly different patterns of migration rate as they begin to be denatured. The net result of this is that they will travel to different positions along the gradient gel after electrophoresis for the same amount of time. Conversely, two fragments completely identical in sequence will be denatured at the same point in their path through the gel, will exhibit the same pattern of migration rate, and will travel to the same position. Thus, a set of fragments, perhaps derived from different individuals within a population, may be readily differentiated into genotypic classes, the members of which are identical in sequence to each other but different from members of other classes.

4.1 DGGE equipment

Because the DGGE technique is unlike most other DNA electrophoresis methods in that it relies on samples experiencing non-uniform denaturing conditions through the gel, it requires some unusual equipment. Specifically, the chemical conditions in the gel must be maintained in such a way as to denature DNA in one portion while not doing so in another. One of the easiest means of accomplishing this is to incorporate a chemical gradient into the gel when casting it and to elevate the gel temperature so that one end of the gradient leads to fragment denaturation while the other does not. Because gel temperature is an important determinant of denaturing conditions it must be accurately maintained throughout electrophoresis. To do this, we immerse the gel in a temperature-controlled aquarium; the main portion of the tank forms the anode, while an isolated buffer chamber in contact with the upper end of the gel forms the cathode. This set-up is illustrated in *Figure 3* and

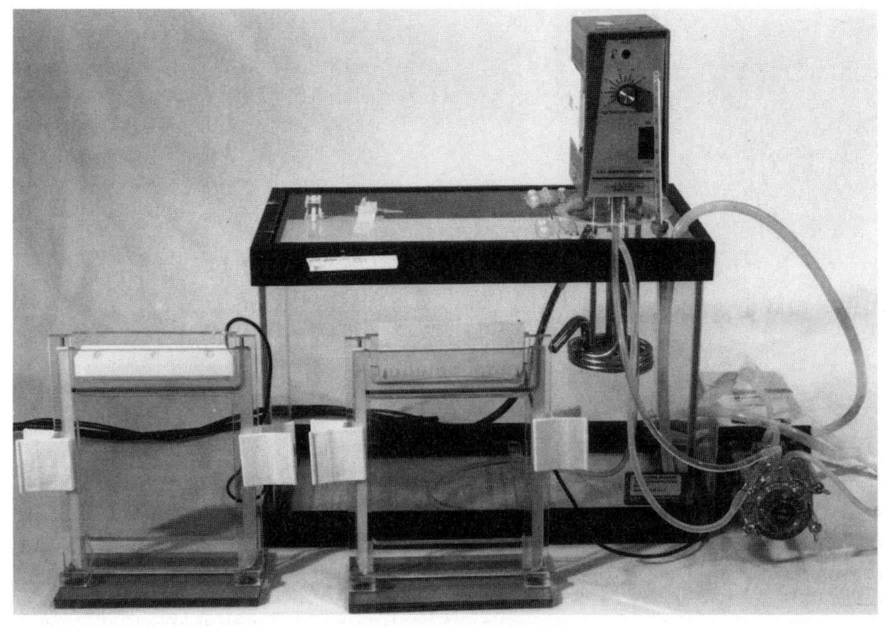

Figure 3. Denaturing gradient gel electrophoresis apparatus. The large aquarium contains 20 litres of buffer which is maintained at a constant temperature by the circulation heater. The two vertical gel rigs are immersed in the tank while electrophoresis is in progress. The one on the left illustrates the use of a single-tooth comb for perpendicular gels, while the one on the right illustrates the use of a shark's-tooth comb for parallel gels. The peristaltic pump on the right circulates buffer into the upper electrode tank of each gel rig.

described more fully in refs 8 and 9. While commercially fabricated equipment is available from at least two sources (CBS Scientific and Bio-Rad), all the components are readily obtainable individually. We routinely use a homemade version with equivalent results.

To establish the chemical gradient in the gel, a gradient maker is required in addition to the electrophoresis equipment. Although gradients of various shapes may be poured using different types of equipment, a linear gradient is most useful if intermediate concentrations must be interpolated based on position within the gradient. As mentioned below, this is important when optimizing the gel conditions for an individual fragment of interest. The linear gradient former we use (*Figure 4*) consists of two chambers of approximately 20 ml linked at the bottom through a valve. One chamber, the mixing chamber, has an outlet port. Equal volumes of two different solutions are placed into each chamber and continually mixed as the solution is allowed to leave the mixing chamber; the result is a linear gradient ranging in composition from that of the solution initially introduced into the mixing chamber to that of the

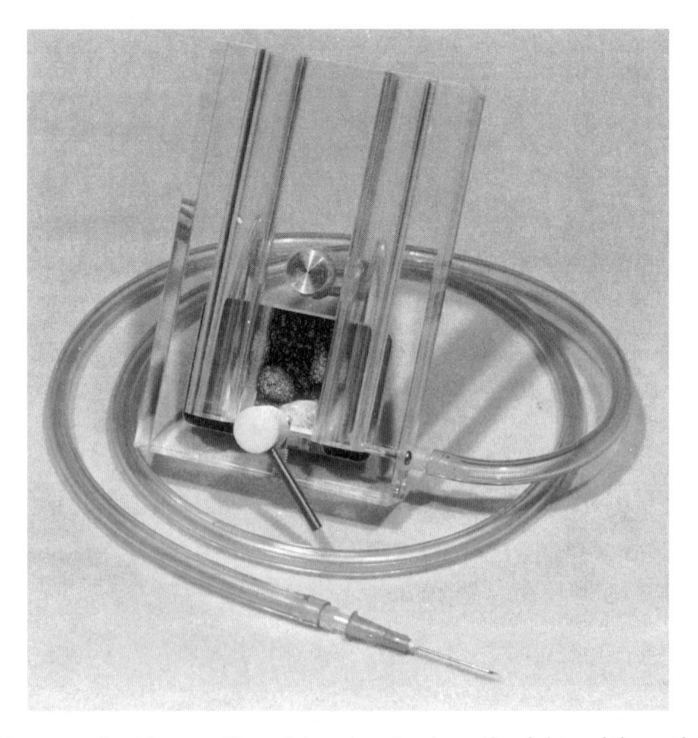

Figure 4. Linear gradient former. The mixing chamber is on the right and the outlet port is connected to a syringe needle with flexible tubing. Flow through this tubing into the gel rig can be controlled by a tubing clamp. A magnetic stirrer bar is placed within the mixing chamber and the gradient former is placed on a magnetic stirrer plate to maintain consistent mixing.

solution introduced into the other chamber. Thorough mixing of the solution within the mixing chamber is, of course, important as is the lack of mixing later, e.g., within the gel casting mould.

4.2 Establishing DGGE conditions

The application of DGGE techniques to the survey of DNA variation among samples is based on electrophoresing the samples in parallel to the denaturant gradient, as described in the example above. This enables one to recognize fragments of distinct DNA sequences from a single gel. Such gels are referred to as parallel DGGE gels because the gradient is parallel to the direction of travel and all samples on the gel are electrophoresed for the same amount of time and experience the same gradient.

The results of running a parallel gel, however, depend on the electrophoretic conditions used. The appropriate conditions vary from DNA fragment to DNA fragment because of the sequence specificity of melting properties. Furthermore, the conditions that are appropriate for resolving differences between certain DNA variants may not resolve other variants from the same genomic fragment. Thus, before undertaking a survey it is necessary to determine the electrophoretic conditions empirically.

Two other types of DGGE gel are useful for efficiently evaluating different electrophoretic conditions. The first is a perpendicular gel (*Figure 5*) in which the gradient is perpendicular to the direction of migration. A single sample is loaded across the entire width of the gel, and hence across the entire gradient. As this sample travels through the gel, different portions of it will encounter different denaturing conditions but these conditions are constant throughout the run. As a result, the portion of sample loaded at one end of the gel will encounter only non-denaturing conditions throughout the run, whereas the portion loaded at the other end will encounter only fully denaturing conditions. Perpendicular gels, therefore, provide an indication of how fragment mobility is influenced along a continuum of denaturing conditions. Typical perpendicular gels yield a sigmoidal banding of DNA with the end travelling further corresponding to the low-denaturant end of the gradient. A rapid transition between the migration rate characteristic of double-stranded DNA and a much lower rate characteristic of denatured DNA occurs at some intermediate concentration of denaturant. This intermediate concentration should be used as the mid-point in parallel gradients for surveying many samples. Linear gradients are most useful for identifying the correspondence between denaturant concentration and fragment mobility, information that is necessary for translating the results of a perpendicular gradient gel into useful parallel gradients.

While perpendicular gradients are very informative about the effect of denaturant concentration on fragment mobility, they provide little indication of how electrophoresis time can influence the discrimination between DNA

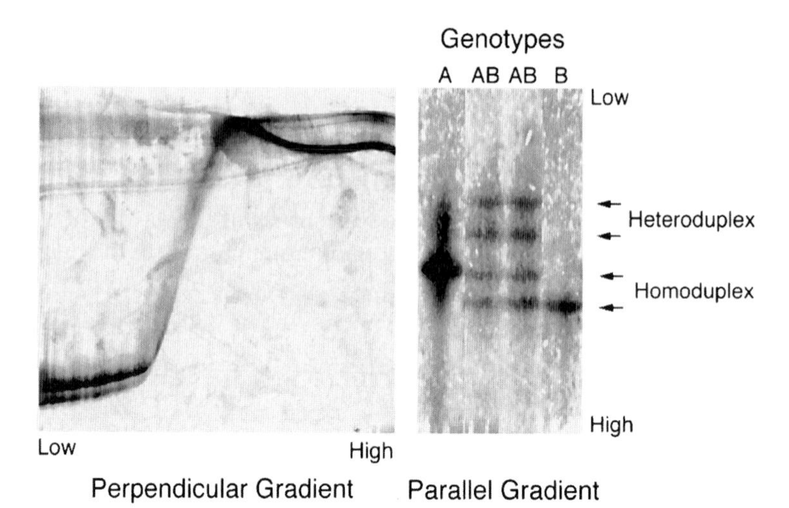

Figure 5. Denaturing gradient gels. For the perpendicular gradient gel (left) a single sample was loaded across the entire width of the gel. For the parallel gradient gel (right) a distinct sample was loaded in each of the four lanes; the two outer samples are homozygotes for different alleles, while the two inner samples are heterozygotes for the same two alleles. In both cases the direction of electrophoresis is from top to bottom.

fragments with distinct nucleotide sequences. In practice, two fragments may be distinguishable if electrophoresed for one period of time, but not if electrophoresed for another.

The final type of DGGE, the travel gel, is useful for determining the electrophoresis time once the gradient has been established. In travel gels the direction of electrophoresis is parallel to the denaturing gradient, and different samples are electrophoresed for different times. Those samples run for only a short period of time only encounter relatively low concentrations of denaturant and, therefore, are never denatured. In contrast, those samples electrophoresed for longer periods of time will eventually encounter denaturing conditions and migrate more slowly. Ideally, samples should be electrophoresed until they become denatured. This seems to provide the greatest resolution between fragments of different sequence. Because analysis of samples using DGGE requires several steps in addition to the survey, an overview of the entire process is given below:

(a) Run a single sample on a 0–80% denaturant gradient oriented perpendicular to the direction of migration.

(b) Identify a suitable gradient, centred on the transition from undenatured to denatured conditions, that spans the full transition.

(c) Run several samples, preferably ones known or suspected to contain distinct DNA sequences, on the new gradient oriented parallel to the direction of migration. Load each set of samples repeatedly to determine how long to run future samples. It is often adequate to load samples every hour with electrophoresis times ranging from 4 h to 10 h.

(d) Identify an electrophoresis time that leads to clear denaturation of the samples. This is readily apparent as a dramatic slowing of the migration rate as electrophoresis time increases. If the samples are known to represent distinct sequences, the efficacy of the gradient for distinguishing them can be immediately determined and the electrophoresis time chosen accordingly. The same is true if previously unknown variation is uncovered at this point.

(e) Run the population samples for the established time through gels containing the new gradient oriented parallel to the direction of migration. It is important to maintain consistency from gel to gel with respect to the gradient, the electric field applied, the electrophoresis time, and gel temperature.

(f) As variants are detected it is important to re-organize them on new gels in such a way that all adjacent samples are of the same genotype. This provides clear verification of the identity of different samples.

(g) For each variant, electrophorese a second independent PCR amplification to verify that the variation is not due to imprecise replication during the amplification process.

(h) A set of standardized markers should be constructed from the samples that exemplify different genotypes. These should be included on every gel to provide constant genotypic references for interpreting the unknown samples. Note that normal DNA size standards are useless for this purpose, because they do not exhibit appropriate denaturing characteristics.

(i) Sequence a subset of each genotypic class to verify the within-class uniformity and the between-class distinctness. Sequences from several individuals from each class from each population should be sufficient.

Although two different types of gels, parallel and perpendicular gradients, are required during the course of DGGE analysis, much of the procedures are the same for both types. Essentially, they differ only in how the gel moulds are constructed and in the number of samples loaded.

Protocol 4. Pouring DGGE gels

Equipment and reagents

- Glass plates (approximately 18 cm × 22 cm), matched pair for front and back
- Ethanol
- Three or four 1.0 mm spacers
- Gel pouring gasket (available from CBS Scientific)
- Shark's-tooth comb
- Bulldog clips
- Gradient maker with stirrer bar (available from CBS Scientific or Bio-Rad)
- Tygon tubing with clamp and 21-gauge syringe needle

Protocol 4. *Continued*

- 20 × DGGE running buffer: 800 mM Tris pH 7.4, 400 mM sodium acetate, 20 mM EDTA pH 7.4
- High-denaturant stock solution corresponding to upper end of gradient (e.g. 80% denaturant stock solution: 6.5% 37.5:1 acrylamide:bis-acrylamide, 32% formamide, 5.6 M urea, 1 × DGGE running buffer);[a] store in amber bottles at 4°C
- Low-denaturant stock solution corresponding to lower end of gradient (e.g. 0% denaturant stock solution: 6.5% 37.5:1 acrylamide:bis-acrylamide, 1 × DGGE running buffer); [a] store in amber bottles at 4°C
- 10% ammonium persulfate (store at 4°C)
- TEMED
- Mylar-based tape

Method

A. *Assembling moulds for perpendicular gels*

1. Clean both glass plates first with H_2O then with ethanol. Be sure plates are thoroughly dry and clean, as dirty plates can ruin the gradient.
2. Form a gel mould sealed with spacers on three sides and with a comb containing a single wide tooth on the fourth (top) side. Leave a small space, just large enough to allow entry of the syringe needle used in pouring the gel, along one side adjacent to the top comb. In pouring the gel, this space should be oriented upwards; i.e., the gel should be rotated 90° relative to the normal orientation so that the top of the gel mould is on the side. It is also useful to place the side spacer adjacent to the opening at a slight angle to allow for easier migration of air bubbles toward the exit opening. Be sure the glass plates are clamped firmly against the spacers and that a good seal is formed. Agarose can be used as a sealant if leakage is a problem.
3. Keep the gel mould upright by resting it on the bulldog clips clamping the side opposite the small side opening, if necessary.

B. *Assembling moulds for parallel gels*

1. Clean both glass plates as above.
2. Form a sealed open-topped mould between the two glass plates. This may be accomplished with three spacers placed along the sides and bottom of the mould, or with two spacers along each side and a special gasket (available from CBS Scientific) designed to fit around three sides of the uneared glass plate. Be sure the glass plates are clamped firmly against the spacers and that a good seal is formed.
3. If required a convenient stand for keeping the gel mould upright can be made by clipping bulldog clips at the very bottom of each side.

C. *Pouring gradient gels*

1. Set the gradient maker, with a tiny stirrer bar in the chamber nearest the outlet, on a cool stirrer plate. Close the valve between the two chambers. Attach tygon tubing terminated with a 21-gauge syringe needle to the outlet. Clamp the tubing closed.

2. Place the assembled gel mould below the gradient maker. For parallel gradient gels, orient the mould with the open side upwards; for perpendicular gels orient the mould with the small opening upwards. Insert the syringe needle[a] between the plates.
3. Each of our gels holds 32.5 ml. Make up two small beakers of equal volume (16.5 ml to give a slight surplus) of high- and low-denaturant stocks, respectively. Add 165 ml of 10% ammonium persulfate to each beaker. Add 10 ml of TEMED to each beaker. Swirl gently. Now polymerization has begun, so do not waste time.
4. Turn on the stirrer plate. Add the contents of the high denaturant beaker to the gradient chamber closest to the outlet. Slightly open the valve between the two chambers to fill the passage with solution. This is important, because an air bubble trapped in the valve will hinder good gradient formation. Close the valve. Add the contents of the low-denaturant beaker to the other chamber.
5. Simultaneously open the valve between the chambers and loosen the clamp on the tygon tubing. Adjust the tubing clamp to control the rate of flow so that there is good mixing in the gradient maker while preventing the fluid from rushing into the gel form.
6. When the gel mould is full, clamp the tubing to prevent further flow.
7. For parallel gels, which have a completely open top on the mould, place the flat side of the comb approximately 8 mm into the gel. This now forms the fourth side of the mould.
8. Clean the gradient maker immediately to prevent polymerization of acrylamide inside the valve.
9. The gel should be usable in 2 h, but may be stored at 4°C for at least two days if moistened with buffer and wrapped in plastic.

[a] Note that stock solutions with other denaturant concentrations may be needed to establish the correct gradient. They may be constructed by adjusting the formamide and urea concentrations appropriately, for example, by mixing appropriate volumes of the 0% and 80% denaturant stocks.

Protocol 5. Electrophoresis of DGGE gels

Equipment and reagents

- DGGE tank full of 1 × DGGE running buffer (see *Protocol 4*) equilibrated to 60°C
- Polymerized DGGE gels (*Protocol 4*)
- Plastic or stainless steel bulldog clips
- Mylar-based tape

Method

1. Prepare the gel for electrophoresis by exposing the top and bottom of the gel. In the case of parallel gels, insert the shark's-tooth comb into

Protocol 5. *Continued*

the top so that the teeth make solid contact with, but do not deeply penetrate, the top of the gel.

2. Tape the sides of the gel with Mylar-based tape. This prevents diffusion of the denaturants during electrophoresis and helps maintain the gradient.
3. Clamp the gel in place and immerse it inside the electrophoresis tank. Be sure to use only plastic or stainless steel clips to prevent corrosion.
4. Fill the upper buffer reservoirs and adjust the buffer flow rate into the reservoirs. The exact rate of flow is not important as long as buffer flow is continuous.
5. Check the electrophoresis set-up by applying the electric field (150 V for our gels, which are approximately 50 cm in field path length) to the gel. The current flowing should be consistent from gel to gel (80 mA for our gels). An unusual amount of current may indicate an improper set-up, such as a lack of isolation between the upper electrode and the rest of the tank or an improperly sealed gel. Be sure to turn off the electric field prior to loading the gel.
6. Load samples on to the gel. For parallel gels, each well formed by the shark's-tooth comb receives a distinct sample. In contrast, for perpendicular gels, a single sample is loaded in a single well across the entire width of the gel.
7. Electrophorese the sample(s). Note that it is important to maintain consistency from gel to gel in both the voltage and the length of time. This is much more important with gradient gels than it is for other types of electrophoresis under constant conditions.
8. Visualize DNA by silver staining (*Protocol 6*).

Although DNA may be detected in DGGE gels based on ethidium bromide staining, silver staining is much more sensitive (*Protocol 6*; see also Chapter 7). Furthermore, because detection is based on the *in situ* precipitation of metallic silver, much as is done with photographic negatives, the developed gel provides a permanent record of the electrophoresis experiment.

Protocol 6. Silver staining of polyacrylamide gels

Reagents

- Wash solution: 10% acetic acid in H_2O
- Impregnating solution: 0.1% (w/v) silver nitrate, 0.15% (v/v) formalin (prepare this solution just prior to staining)
- Developer: 3% (w/v) sodium carbonate, 2 p.p.m. (w/v) sodium thiosulfate,[a] 0.15% (v/v) formalin

Method

1. Soak gel in wash solution for 20 min.
2. Rinse gel in H_2O three times for 1 min each time to remove excess acetic acid.
3. Impregnate the gel with the impregnating solution for 30 min.
4. Rinse the gel in H_2O for 1 min.
5. Develop the gel with developing solution until the desired band intensity is obtained. Note that bands can intensify rapidly so have stop solution (wash solution) ready prior to beginning development. It may also be advisable to remove the developer just before the desired degree of development has occurred, to allow for the delay involved with transfering solutions.[b,c]
6. Stop development by soaking the gel in wash solution.
7. Dry gel against blotter paper. The image is formed by precipation of metallic silver, so the gel can serve as a permanent record.

[a] Sodium thiosulfate is conveniently added as 1 ml per litre of developer of a 0.2% (w/v) solution.
[b] Removal of solutions from the developing tray is easily accomplished with an aspirator.
[c] Cooling the developer slows the reaction and may ease its control.

4.3 Optimization of DGGE detection

Although detection of nucleotide sequence variation among DNA samples is readily accomplished with DGGE, careful attention to electrophoretic conditions is necessary and optimization of those conditions is required. The main variables that influence discrimination of different sequences are:

- the nature of the gradient itself
- the temperature of the gel
- the electric field applied
- the length of time allowed for electrophoresis

All of these variables should be controlled as much as possible to maintain consistency between gels. The effect of variation in either the gradient or the temperature is largely to displace the normal position of denaturation, although it may also influence the resolving power of the gel. The effect of variation in the electric field is largely to alter the time required to reach a position of denaturation. As a result, if necessary the electric field may be changed somewhat (e.g., by at least 10%) as long as the time is compensated in such a way as to maintain a constant field time. Thus, although control of these factors is important, and much more so than with most other electrophoretic techniques, extreme measures beyond normal laboratory care are not required. Because resolution of distinct DNA sequences depends on the proper interaction of four factors, namely time, temperature, electric field, and denaturant gradient, careful optimization is necessary for each DNA fragment. To simplify matters, we have standardized on a constant temperature of 60°C and a constant electric field of 3 V/cm (150 V for a field path length of approximately 50 cm). In our experience increasing the field strength to 9 V/cm (with appropriate changes in electrophoresis time) has little apparent effect on the outcome, although resolution of distinct bands may depend on the field strength (10). One perpendicular gel and one travel gel usually suffice to identify rapidly the necessary interaction between the other two factors, gradient composition and time.

It is, however, important to keep questioning whether each apparently identical pair of samples is in fact identical. This may be accomplished in two ways. First, it is important regularly to run samples more than once in different orders to make sure that each sample is identical in mobility to the ones adjacent to it. Generally, it is easy to recognize a continuous series of adjacent and identical samples; this is much more difficult if identical samples are interspersed with a few samples of distinct sequence. Thus, the most effective means of checking identity is to rearrange the samples so that all those from a putative genotype class are adjacent to each other. Very rapidly this process both reduces the number of samples that must be compared and identifies the genotypic classes.

Second, it is important to be attentive to even slight variation in mobility. This may signal the occurrence of distinct sequences that have been ineffectively resolved. In our experience, slight modifications of the electrophoretic conditions resolve the fragments as distinct, and sequencing verifies their differences.

Proper primer design can also contribute to enhanced detection of sequence variation among DNA samples by overcoming a limitation inherent in the denaturation process. All DNA fragments are composed of one or more domains, regions of adjacent nucleotide pairs with similar denaturation characteristics; one of those domains necessarily is most stable. Differentiating distinct DNA sequences based on denaturation characteristics tends to be difficult if the differences are localized within the most stable domain. To overcome this difficulty and to increase resolution of differences among samples, it is advisable to create an artificial 'most stable domain' that is constant across all samples. This may be accomplished by synthesizing one of the primers used in amplification reactions with a long GC tail at the 5′ end (a 'GC clamp'). The purpose of this segment is simply to provide a common most stable domain that is never denatured under the electrophoretic conditions encountered by the sample. As a result, essentially all differences among samples in denaturation conditions, and hence sequence, are resolvable. As long as the GC clamp accomplishes this purpose and does not interfere with the amplification reaction, any sequence is suitable; for consistency we have used the following which is derived from, but not identical to, primer

3 in Sheffield *et al.* (39): CGCCCGCCGCGCCCCGCGCCCGCCCCGC-CGCCCCCGCCCC. Occassionally, to stay within the constraints of commercial oligonucleotide synthesis services, this GC clamp can be shortened by one to several nucleotides at the 5′ end with no loss of effectiveness.

5. Thermal gradient gel electrophoresis

The denaturing gradient techniques described above are based on exposing DNA fragments to an increasing concentration of a chemical denaturant as they migrate through the gel matrix. The same effect may be obtained by exposing DNA fragments to an increasing concentration of any denaturant. Thus, thermal gradient gel electrophoresis (TGGE) is based on establishing a thermal gradient along the gel, and exposing DNA fragments to increasing temperature as they migrate. While the physical set-up is quite different from that used for chemical gradients, the theory and applications are essentially the same, as is the importance of maintaining consistency of running conditions from gel to gel. In fact, the two methods are virtually identical in terms of their ability to detect DNA variation and in their applicability to population studies.

5.1 TGGE equipment

Thermal gradients are relatively easy to establish across the surface of a gel. Perhaps the simplest means is to fabricate a pair of aluminum blocks that fit against the surface of the glass plates used to contain the gel and that contain a channel across each end through which water may be circulated (11). The desired temperature gradient can be established simply by adjusting the temperature of the circulated water. Although thermal gradients are normally oriented parallel to the direction of electrophoresis, and therefore can be used in the same way as travel gels and parallel chemical gradients discussed above, so long as the thermal plates are designed appropriately they can also be oriented perpendicular to the direction of electrophoresis. Thus, an appropriate set of thermal plates and associated temperature control equipment (e.g., water baths, pumps, etc.) can duplicate the capabilities of the chemical denaturants discussed above. As a result, essentially all of the points raised earlier concerning the overall process involved with denaturing gradients, the utility of gradients in different orientations, the optimization of electrophoresis conditions, and the interpretation of results apply equally to thermal gradients as to chemical gradients. However, both the composition of the gels and the electrophoresis differ slightly.

5.2 Interpretation of parallel gradient gels

Interpretation of parallel gradient gels is straightforward but requires attention to the fact that the PCR amplification reaction can potentially (though not necessarily) generate a diversity of products, even neglecting any spurious products derived from extraneous regions of the genome. Because of the continual denaturation and re-annealing that occurs during amplification, all possible homoduplexes and heteroduplexes are commonly generated. In the case of fragments amplified from a single locus of a diploid genome, two homoduplexes are formed, one from each of the two homologous alleles, and two heterduplexes are formed, each combining one strand from each homologous allele. If the individual is heterozygous at this locus, the two homoduplexes will be composed of exactly complementary strands whereas the two heteroduplexes will be composed of mismatched strands. As a result of the mismatch, the heteroduplexes will be denatured under less stringent conditions than the homoduplexes; as a result of the differences between alleles, the two homoduplexes (and the two heteroduplexes typically) will be denatured under different conditions. In contrast, if the individual is homozygous at this locus, all four amplification products will be identical and composed of exactly complementary strands; hence, all four will be denatured under identical conditions. After electrophoresis, heterozygous samples typically exhibit a four-banded pattern, while homozygous samples exhibit a one-banded pattern (*Figure 5*).

Protocol 7. Pouring TGGE gels

Equipment and reagents

- Equipment for constructing gel mould (*Protocol 4*)
- 10 × TGGE running buffer: 900 mM Tris pH 8.1, 900 mM sodium borate, 20 mM EDTA pH 8.1
- 1 × TGGE gel solution: 6.5% 37.5:1 acrylamide:bis-acrylamide, 4 M urea, 23% formamide, 0.5 × TGGE running buffer
- 10% ammonium persulfate (store at 4°C)
- TEMED

Method

1. Clean both glass plates first with H_2O then with ethanol. Be sure plates are thoroughly dry and clean; dirty plates can ruin the gel.
2. Form a sealed open-topped mould between the two glass plates. This may be accomplished with three spacers placed along the sides and bottom of the mould, or with two spacers along each side and a special gasket (available from CBS Scientific) designed to fit around three sides of the uneared glass plate. Be sure the glass plates are clamped firmly against the spacers and that a good seal is formed.
3. If required, make a convenient stand for keeping the gel mould upright by clipping bulldog clips at the very bottom of each side.
4. Each of our gels holds 32.5 ml. To 33 ml of 1 × TGGE gel solution, add 330 μl of 10% ammonium persulfate and 20 μl of TEMED. Swirl gently. Now polymerization has begun, so do not waste time.

5. Pour the gel solution into the gel mould quickly but smoothly to prevent bubbles.
6. When the gel mould is full, place the flat side of the comb approximately 8 mm into the gel. This now forms the fourth side of the mould.
7. The gel should be usable in 2 h, but may be stored at 4°C for at least two days if moistened with buffer and wrapped in plastic.

Protocol 8. Electrophoresis of TGGE gels

Equipment and reagents

- TGGE gel rig (*Protocol 7*)
- TGGE thermal plate and associated temperature control equipment
- Polymerized TGGE gel (*Protocol 6*)
- 0.5 × TGGE running buffer (*Protocol 6*)

Method

1. Prepare the gel for electrophoresis by exposing the top and bottom of the gel and inserting the shark's-tooth comb into the top so that the teeth make solid contact with, but do not deeply penetrate, the top of the gel.
2. Clamp the gel in place with the thermal plate firmly mounted against the glass. Fill the electrode tanks with 0.5 × TGGE running buffer.
3. Apply the electric field to be used for running the gel and adjust the thermal gradient to the desired temperature range (typically 18–44°C or 28–32°C). Allow the gel to equilibrate for 45–60 min during this pre-electrophoresis period. Be sure to turn off the electric field prior to loading the gel.
4. Load samples on to the gel; each well formed by the shark's-tooth comb receives a distinct sample.
5. Subject the sample(s) to electrophoresis.[a]
6. Visualize DNA by silver staining (*Protocol 6*).

[a] It is important to maintain consistency from gel to gel in both the voltage and the length of time. This is much more important for gradient gels than it is for other types of electrophoresis under constant conditions.

The situation is greatly simplified for haploid genomes, e.g., those in the organelles, that are homoplasmic. In such cases samples only exhibit single bands because the amplication results in a uniform population of identical and fully complementary strands (31).

Clearly the situation is greatly complicated for either polyploid genomes or for amplifications that target multiple loci. The principles for interpretation remain the same; the difference is simply that many more combinations of fully and partially complementary strands may be formed. Certainly in complex situations, and even in simple diploid cases, homozygote samples greatly ease the problems of interpretation by providing markers for the homoduplexes within heterozygote samples.

In some cases multiple loci can be evaluated simultaneously without confusion if the alleles derived from each locus differ sufficiently in denaturing conditions from alleles at other loci. Fragments will then migrate to different regions of the gel depending on the locus from which they are resolved, and within each region patterns characteristic of homozygotes and heterozygotes will be resolved. There is, however, the possibility of observing bands involving heteroduplexes composed of strands from different loci, so this strategy probably cannot be extended to more than a few loci. A more practical approach of obtaining data on multiple loci is to combine the products of several separate amplifications known to be denatured under suitably different conditions. In this way, complex heteroduplexes are avoided, yet multiple loci may be scored simultaneously.

6. Applications and discussion

Since the invention of the PCR–SSCP and DGGE techniques, many applications of the method have been demonstrated. SSCP has been used to identify alterations in tumour DNA samples, mutations in human disease genes, as well as in animal models (1,12–15). SSCP can be applied to RNA, or to reverse-transcribed RNA (cDNA) either to increase sensitivity or to assay multiple exons simultaneously (16). Multiplex assays, in which several exons are amplified at once, have also been demonstrated (17).

SSCP has also been applied to complex loci with many alleles, such as the HLA locus (18). Since different alleles usually give distinct patterns, five to ten alleles can usually be resolved on the same gel. SSCP can also be readily applied to the mapping of genes in interspecies backcrosses, where differences between the parental species can be easily detected in a 200–500 bp PCR product. SSCP alterations are often found in the 3′ untranslated regions and introns of genes (19,20). The 3′ untranslated regions are rarely disrupted by introns, and such polymorphic sequence-tagged sites can allow a gene to be mapped genetically (in families), as well as physically (on yeast artificial chromosome clones, and radiation hybrids (21).

Since SSCP can be used to scan large numbers of samples rapidly, the method lends itself to the analysis of both nuclear and mitochondrial DNA from population samples. SSCP has been applied to the analysis of an insecticide resistance gene in the beetle *Hympothenumus hampei* (22), variation in a human fungal pathogen (*Coccidiodes inunitis*, 23), analysis of the major histocompatibility complex (MHC) locus in whales and moose (24,25), and variations in the hepatitis C virus (26,27). The DGGE method has also been applied in population biology reserach (28–31). These methods can be used to differentiate individuals and sub-species, develop genetic markers for family and evolutionary studies, and identify alterations in candidate genes for diseases and phenotypic traits.

7. Other methods

Although SSCP, TGGE, and DGGE are excellent methods for many applications, other methods may be warranted for some applications (32). Other established methods to consider are chemical cleavage and RNase protection (reviewed in ref. 33), which both rely on cleavage of heteroduplex products at the site of the mismatch. These methods can be applied to larger DNA fragments (up to 1–2 kb), and give information on the position of the mutation, and the nature of the alteration. A chemical cleavage protocol employing fluorescently labelled primers (fluorescence-assisted mismatch analysis, FAMA) and resolution on an automated sequencer allowed for high efficiency scanning of the human serpin C1 inhibitor gene (34). Because of the high resolution of the sizing of the products, and information on the nature of the mutation from the pattern of chemical cleavage, the many alterations can be characterized without the need for sequencing.

Recently methods for identifying unknown mutations relying on enzymatic detection (enzyme mismatch cleavage (EMC), mismatch repair enzyme cleavage; 35,36) or reversed phase liquid chromatography (37) have been developed that may prove useful in population-based studies. Bacteriophage resolvases cleave branched DNA structures, recognize mismatched bases in double-stranded DNA, and cleave the DNA at the mismatch. The EMC method takes advantage of this characteristic to detect individuals who are heterozygous at a given site. Radiolabelled DNA is cleaved by the enzyme at the site of mismatch in heteroduplex DNA and digestion is monitored on a gel. Like other cleavage methods, the presence and the position of an alteration is revealed. Improvements of the method allow up to 98% of alterations to be detected (38).

References

1. Orita, M., Suzuki, Y., Sekiya, T., and Hayashi, K. (1989). *Genomics*, **5**, 874.
2. Glavac, D. and Dean, M. (1993). *Hum. Mutat.*, **2**, 404.
3. Glavac, D. and Dean, M. (1995). *Methods Neurosci.*, **26**, 194.
4. Ravnik-Glavac, M., Glavac, D., and Dean, M. (1994). *Hum. Mol. Genet.*, **3**, 801.
5. Hecker, K. H. and Roux, K. H. (1996). *BioTechniques*, **20**, 478.
6. Ravnik-Glavac, M., Glavac, D., Chernick, M., di Sant'Agnese, P., and Dean, M. (1994). *Hum. Mutat.*, **3**, 231.
7. White, M. B., Carvalho, M., Derse, D., O'Brien, S. J., and Dean, M. (1992). *Genomics*, **12**, 301.
8. Myers, R. M., Maniatis, T., and Lerman, L. S. (1987). In *Methods in enzymology*, Vol. 155, p. 501. Academic Press.

9. Myers, R. M., Sheffield, V. C., and Cox, D. R. (1988). In *Genome analysis: a practical approach* (ed. K. Davies), p. 95. IRL Press, Oxford.
10. Abrams, E. S., Murdaugh, S. E., and Lerman, L. S. (1995). *Nucleic Acids Res.*, **23**, 2775.
11. Wartell, R. M., Hosseini, S. H., and Moran, C. P. (1990). *Nucleic Acids Res.*, **18**, 2699.
12. Claustres, M., Laussel, M., Desgeorges, M., Giansily, M., Culard, J.-F., Razakatsara, G., and Demaille, J. (1993). *Hum. Mol. Genet.*, **2**, 1209.
13. Soto, D. and Sukumar, S. (1992). *PCR Methods Applic.*, **2**, 96.
14. Dean, M., White, M. B., Amos, J., Gerrard, B., Stewart, C., Khaw, K.-T., and Leppert, M. (1990). *Cell*, **61**, 863.
15. Claustres, M., Gerrard, B., Kjellberg, P., Desgeorges, J., and Dean, M. (1992). *Hum. Mutat.*, **1**, 310.
16. Danenberg, P. V., Horikoshi, T. M., Volkenandt, M., Danenberg, K., Lenz, H. J., Shea, C. C. L., Dicker, A. P., Simoneau, A., Jones, P. A., and Bertino, J. R. (1992). *Nucleic Acids Res.*, **20**, 573.
17. Kozlowski, P., Sobczak, K., Napierala, M., Wozniak, M., Czarny, J., and Krzyzosiak, W. J. (1996). *Nucleic Acids Res.*, **24**,1177.
18. Carrington, M., Miller, T., White, M., Gerrard, B., Stewart, C., Dean, M., and Mann, D. (1992). *Hum. Immunol.*, **33**, 208.
19. Nielsen, D. A., Dean, M., and Goldman, D. (1992). *Am. J. Hum. Genet.*, **51**, 1366.
20. Glavac, D., Ravnik-Glavac, M., O'Brien, S. J., and Dean, M. (1994). *Hum. Genet.*, **93**, 694.
21. Poduslo, S., Dean, M., Kolch, U., and O'Brien, S. (1991). *Am. J. Hum. Genet.*, **49**, 106.
22. Borsa, P. and Coustau, C. (1996) *Heredity*, **76**, 124.
23. Burt, A., Carter, D. A., Koenig, G. L., White, T. J., and Taylor, J. W. (1996). *Proc. Natl Acad. Sci USA*, **93**, 770.
24. Murray, B. W., Malik, S., and White, B. N. (1995). *Mol. Biol. Evol.*, **12**, 582.
25. Ellegren, H., Mikko, S., Wallin, K., and Andersson, L. (1996). *Mol. Ecol.*, **5**, 3.
26. Kurosaki, M., Enomoto, N., Marumo, F., and Sato, C. (1995). *Arch. Virol.*, **140**, 1087.
27. Enomoto, N., Kurosaki, M., Koizumi, K., Asahina, Y., Sakuma, I., Murakami, T., Yamamoto, C., Marumo, F., and Sato, C. (1994). *Nippon Rinsho*, **52**, 1707.
28. Lessa, E. P. (1992). *Mol. Biol. Evol.*, **9**, 323.
29. Norman, J. A., Moritz, C., and Limpus, C. J. (1994). *Mol. Ecol.*, **3**, 363.
30. Taylor, M. F. J., Shen, Y., and Kreitman, M. E. (1995). *Science*, **270**, 1497.
31. Strand, A. E., Milligan, B. G., and Pruitt, C. M. (1996). *Evolution*, **50**, 1822.
32. Cotton, R. G. H. (1996). *Am. J. Hum. Genet.*, **59**, 289.
33. Cotton, R. G. H. (1993). *Mutat. Res.*, **258**, 125.
34. Verpy, E., Biasotto, M., Brai, Misiano, G., Meo, T., and Tosi, M. (1996). *Am. J. Hum. Genet.*, **59**, 308.
35. Mashal. R. D., Koontz, J., and Sklar, J. (1995). *Nature Genet.*, **9**, 177.
36. Youil, R., Kemper, B. W., and Cotton, R. G. H. (1995) *Proc. Natl Acad. Sci. USA*, **92**, 87.
37. Underhill, P. A., Jin, L., Zemans, R., Oefner, P. J., and Cavalli-Sforza, L. L. (1996). *Proc. Natl Acad. Sci. USA*, **93**, 196.
38. Youil, R., Kemper, B. W., and Cotton, R. G. H. (1996) *Genomics*, **32**, 431.
39. Sheffield, V. C., Cox, D. R., Lerman, L. S., and Myers, R. M. (1989). *Proc. Natl Acad. Sci. USA*, **86**, 232.

9

Multilocus and single-locus DNA fingerprinting

MICHAEL W. BRUFORD, OLIVIER HANOTTE, JOHN F. Y. BROOKFIELD, and TERRY BURKE

1. Introduction

Over the last ten years DNA fingerprinting has been and continues to be applied in many diverse areas of biological science, and has now become a standard technique for rapidly screening genetic variation in animals, plants, and other eukaryotes. This chapter outlines the techniques of both multilocus and single-locus DNA fingerprinting using minisatellite sequences and outlines the various uses of these techniques in population genetics, evolutionary studies, and behavioural ecology.

1.1 What is DNA fingerprinting?

Five years ago the technique of DNA fingerprinting (1,2) was relatively easy to define, since it almost exclusively involved the use of the 'minisatellite' family of sequences. Since that time the term 'DNA fingerprinting' has been used to describe several other multilocus techniques, including polymerase chain reaction (PCR)-based methods such as randomly amplified polymorphic DNA (RAPD) analysis (3) and amplified fragment length polymorphism (AFLP) DNA analysis (ref. 4; see also Chapter 6). Of the 1500+ publications appearing between 1992 and 1996 having DNA fingerprinting in the title or keywords, only about 30% utilize 'classical' DNA fingerprint loci, i.e. minisatellites and other variable number of tandem repeat (VNTR) loci. This is perhaps a regrettable development for several reasons. Firstly, there is some merit in using a generic term such as DNA fingerprinting to describe a cohesive set of techniques. Secondly, many of the non-VNTR techniques do not have the prerequisite property of a DNA fingerprinting technique, i.e. the ability to reveal individual-specific patterns in most cases. Finally, the PCR-based methods described above yield patterns that differ fundamentally in their genetic properties from classical DNA fingerprints. Features such as the almost complete dominance associated with the amplification of RAPD and

AFLP fragments (see Chapter 6) make their range of applications rather different from that of DNA fingerprints, and in some ways the only common feature among the techniques is the generation of a multi-fragment, bar-code-like pattern. We would therefore suggest that the term 'DNA fingerprinting' be restricted to individual-specific DNA patterns, and that 'DNA profiling' be used in other cases.

Classical DNA fingerprinting, both multilocus and single-locus, continues to be applied in a large number of studies across a wide range of disciplines, including behavioural ecology (predominantly, but not exclusively, in birds), population genetics, animal breeding, and conservation biology. In this chapter, we will deal exclusively with the 'minisatellite' and other VNTR sequences, since other techniques are covered elsewhere in the book. The 'minisatellite' family of sequences consist of arrays of up to several hundred 15–60 bp units often widely scattered throughout the chromosomes of many (if not all) eukaryotes. These tandem repeat units often contain a common 'core' sequence. A number of different core sequences have been identified and different polycore sequence constructs are available to detect large numbers of loci in many different species. There has been much recent debate regarding which VNTR-based approach is the most appropriate (multilocus, single-locus minisatellites, or microsatellite loci). Several factors have to be taken into consideration such as cost, availability of probes, time, application, etc., and there are no strict rules (see further discussion in Sections 2.1 and 3.1 and Chapter 7). It is clear, however, that minisatellite loci are often the best choice for pedigree and parentage analysis, where precise estimates of relatedness are not required.

The final product of the multilocus DNA fingerprinting procedure is a pattern of bands, resembling a bar-code. In contrast to some of the more recently developed PCR-based multilocus systems, such as RAPD, this pattern is often effectively specific to the individual (typically 10–25% of the bands are shared between any two individuals by chance), except in extreme cases of inbreeding, or in monozygotic twins. The bands comprising the fingerprint pattern are inherited in a Mendelian manner (on average half of the bands are derived from each parent), and usually show high somatic and germ-line stability. The practical procedures involved in multilocus DNA fingerprinting are discussed in Section 2.

1.2 DNA fingerprinting applied to population biology

DNA fingerprinting has rarely been effectively applied to analysing differences between populations. This is primarily due to the fact that the minisatellite loci that are detected exhibit levels of variation too great to yield meaningful results when comparing most wild populations. The characteristics of these sequences have instead lent themselves to intra-population studies, where there is a need to identify individuals or clones (2,5) or close relatives (6), and in studies of reproductive behaviour and mating success (for examples see refs 7–19). It is in this context that multilocus DNA fingerprinting still represents the most cost-effective and sensitive methodology available.

Some studies (20–25) have shown that population level comparisons can be carried out, especially where effective population sizes are sufficiently small to result in the reduction of individual variability, and in these cases population-specific patterns may occur. These situations are unusual in nature and such analyses are mainly confined to 'bottlenecked' populations and those which are naturally very small. However, with the recent emergence of the application of molecular techniques in conservation biology, such studies are increasing rapidly (26,27).

One of the major problems of applying multilocus DNA fingerprinting to population analyses is that it is often impossible to identify which fingerprint bands derive from the same locus; for example, bands that appear to be shared by individuals (i.e. bands that co-migrate) are not always identical alleles of the same locus. This problem can be overcome, and population level analysis tackled more judiciously, by using single-locus minisatellite probes or microsatellite-based typing systems (16,18,25,28–33; Chapter 7). These probe or primer sequences can be isolated from the species of interest, and applied to individuals from different populations in the knowledge that the same loci are being examined. Cloning, isolating, and characterizing single-locus minisatellites has proven to be problematic and microsatellites are now the preferred single-locus marker in the great majority of laboratories. However, single-locus minisatellites have a specific set of properties, not least their extremely high levels of heterozygosity and allelic diversity (often in excess of 0.95 and 25 per locus, respectively), which make them especially valuable in certain studies. The techniques involved in cloning and isolating these sequences will be described in Section 3.

2. Multilocus DNA fingerprinting

2.1 Applications

Multilocus DNA fingerprinting was first described by Jeffreys *et al.* (1,2) and has since transformed many studies in population biological research (for reviews see refs 34 and 35). Applications have mainly centred around the analysis of reproductive success in wild populations of vertebrates (examples are found in refs 5–19); however, population/social group comparisons have been carried out in some species (20–22,26,27,36–38). DNA fingerprinting has also been used to distinguish between commercial crop cultivars (5,39) and between sexually reproducing and selfing berry species (40) and to establish paternity in apples (14). While confirming parentage in studies of reproductive behaviour involves testing hypotheses of close relatedness (10), behavioural ecologists would often like to identify more distant relatives, and this is where multilocus DNA fingerprinting becomes less useful. For example, as described by Lynch (41) and Brookfield (42), even second-degree relatives cannot always be distinguished statistically from unrelated individuals in cases where band-sharing coefficients for non-relatives are much above zero. This problem can be overcome by using a set of single-locus markers, where the allele frequencies in the base population can be measured and relatedness can be directly quantified (43).

The following section describes some of the procedures used to produce a 'legible' fingerprint. The techniques described often represent specific modifications of standard molecular methods, and where these are important they will be highlighted explicitly. However, many of the routine techniques are described in much more detail in general laboratory manuals (44,45).

2.2 Extracting DNA from blood and tissue

Clean, high molecular weight DNA is a prerequisite for DNA fingerprinting as degraded DNA does not sustain the integrity of restriction fragments required to produce clear band patterns. Consequently, appropriate methods of sample storage in the field are of extreme importance, as storage can greatly affect the quality of DNA extracted. Blood is a convenient and popular source of DNA but as mammalian red cells are not nucleated, the method of storage and extraction differs between mammals (46) and other vertebrates. For example, avian blood can be collected and stored in three volumes of sterile isotonic buffer such as 1 × SSC (0.15 M NaCl, 15 mM trisodium citrate, pH 7), 10 mM EDTA pH 7.4. This blood/buffer solution can be stored at 4°C for up to three months, but should be frozen as soon as possible (tubes should be shaken to resuspend the cells immediately before freezing). Alternatively blood is stored at room temperature in a lysis buffer such as 0.1 M Tris–HCl pH 8.0, 0.1 M EDTA, 10 mM NaCl, 0.5% sodium dodecyl sulfate (SDS) (36) or 8 M urea, 0.4 M NaCl, 0.2 M Tris–HCl pH 8.0, 20 mM EDTA, 0.5% SDS (47) or in >10 volumes 100% ethanol (avoid polystyrene tubes). Avian and reptilian blood can also be preserved for fingerprinting by drying smears on microscope slides. Other tissues should be frozen if possible, but can be stored at ambient temperatures in several buffers, such as 20% dimethylsulfoxide (DMSO) in saturated NaCl (48) or 4 M urea, 0.2 M NaCl, 0.1 M Tris–HCl pH 8.0, 10 mM EDTA, 0.5% SDS (47). Plant DNA appears to be stable *in situ* until the tissue itself begins to degrade, and can be extracted from dried material (see Chapter 2 for plant DNA extraction techniques).

The extraction of high molecular weight, double-stranded DNA is a standard technique in all molecular biology laboratories, and the methods usually employed are perfectly appropriate for DNA fingerprinting (see Chapter 2). Most laboratories still extract DNA using phenol–chloroform treatment (see ref. 44 for details); however, a safer and quicker alternative involves precipitating the denatured protein using high salt concentrations (49). The DNA resulting from this method is relatively protein-free (though it may appear a little discoloured) and restriction digests work equally well as with phenol-extracted DNA. *Protocol 1* is specifically for mammalian and other vertebrate blood, though homogenized animal tissue such as muscle or liver can also be used (see Chapter 2, Section 3.8).

Protocol 1. DNA extraction from mammalian, avian, reptilian, and amphibian blood and from animal tissue

Equipment and reagents

- EDTA- or sodium heparin-coated tube
- EL buffer: 0.155 M NH_4Cl, 10 mM $KHCO_3$, 1 mM EDTA, pH 7.4
- KL buffer: 10 mM Tris, 2 mM EDTA, 0.4 M NaCl, pH 8.2
- TE buffer: 10 mM Tris, 0.1 mM EDTA, pH 7.4
- 1 × TNE buffer: 50 mM Tris–HCl, 100 mM NaCl, 5 mM EDTA, pH 7.5
- Proteinase K (20 mg/ml)
- 20% or 25% (w/v) SDS
- 6 M NaCl
- 100% and 70% ethanol
- Sterile scalpel
- Mortar and pestle
- Liquid nitrogen

A. *DNA extraction from mammalian blood*

1. Collect 1–10 ml blood and place in an EDTA- or sodium heparin-coated tube.[a]
2. Add 3 vol. cold EL buffer.
3. Mix by inversion and place tube on ice for 15 min, mixing occasionally.
4. Centrifuge for 10 min at 8000 × *g*.
5. Remove supernatant carefully; retain pellet.
6. Wash again with EL buffer and repeat suspension and centrifugation once or twice until there is no sign of haemoglobin in the pellet.
7. Resuspend pellet in 3 ml KL buffer.
8. Add 100 μl of proteinase K (20 mg/ml) and 150 μl of 20% SDS. Mix carefully. The solution will be viscous. Incubate overnight at 37°C or at room temperature for 3 days. More proteinase may be added, if necessary, to break up the pellet.
9. Add one-third to half a volume of 6 M NaCl to the solution.
10. Shake very hard for approximately 15 sec.
11. Centrifuge for 15 min at 1700 × *g* at room temperature.
12. Remove the supernatant; if there is a 'foamy' layer on the top of the supernatant be careful not to remove any of the foam (otherwise take as little of the foam as possible and repeat steps **10–12**).

Protocol 1. *Continued*

13. The DNA solution may not necessarily be viscous but should be clear. Add 2–2.5 volumes of 100% ethanol.
14. Mix by inversion until DNA precipitate appears.
15. Spool DNA and rinse in 70% ethanol a few times, then transfer into 0.5–1 ml TE buffer (DNA dissolves better in buffer than in unbuffered water). Store the DNA at 4°C and allow to dissolve overnight.

B. *DNA extraction from avian/reptilian/amphibian/fish blood*

1. Place 50–100 μl blood[b] into a 15 ml tube.
2. Add 3 ml of 1 × TNE, 300 μl of 1 M Tris–HCl pH 8.0, 5 units of proteinase K, and 80 μl of 25% SDS.
3. Incubate overnight at 37°C or for a minimum of 1.5 h at 50°C. Proceed as in *Protocol 1A*, steps **9–15**.

C. *DNA extraction from animal tissue (muscle, liver, skin, etc.)*

1. Defrost approximately 50 mg of tissue and cut into small pieces using a sterile scalpel blade in a Petri dish (on ice/dry ice).
2. (Optional) Place tissue pieces into a mortar, cover with liquid nitrogen, and gently grind with a pestle. Repeat until the tissue has been powdered.
3. Proceed as in *Protocol 1B*.

[a] Vacutainers are not recommended for taking blood because blood drawn very rapidly may contain a reduced proportion of white cells as these tend to cling to the vein walls.
[b] Blood stored in ethanol should be briefly dabbed dry on tissue.

2.3 DNA fingerprint electrophoresis

2.3.1 Preparation of DNA samples for electrophoresis on agarose gels

Most of the techniques involved at this stage are fairly standard (see Chapter 5 and ref. 44). However, there are some specific steps required for fingerprinting which are designed to maximize the resolution of bands in what can be a very complex pattern.

i. Assessment of condition and concentration of DNA samples

Whole genomic DNA with even a small amount of associated protein does not always dissolve totally or evenly in solution. Most DNA samples prepared for fingerprinting will fall into this category unless extreme care is taken. If the concentration of such samples is measured in the conventional manner (UV spectroscopy or fluorometry) readings often never 'settle down', and repeatability can be poor. The best method of getting a reasonable estimate of DNA concentration and condition at the same time is by electrophoresis.

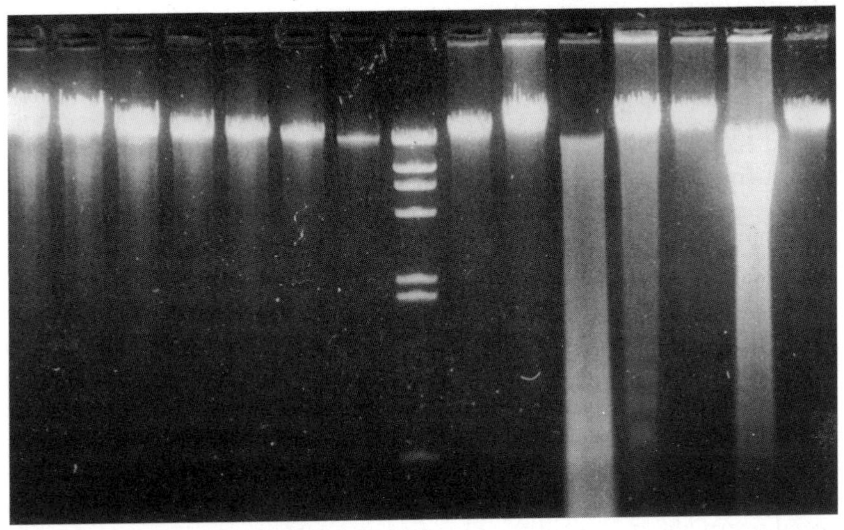

Figure 1. Example of a typical genomic DNA concentration gel. The first seven samples are phage lambda DNA concentration standards (2, 1.5, 1, 0.75, 0.5, 0.25, and 0.125 μg from left to right), followed by 0.5 μg lambda DNA digested with *Hin*dIII. The last seven samples are typical genomic DNA preparations. Note that samples 3 and 6 are degraded. The concentration of intact genomic DNA can be estimated by comparison with the standards.

Figure 1 shows a typical genomic DNA checking gel. Intact, high molecular weight genomic DNA is resolved into one band—any DNA seen below this band has undergone degradation (see sample 3). Very low molecular weight material might be RNA (seen in non-blood tissue preparations), which should be removed by digestion with DNase-free RNase A (most easily by including 1 μg in the restriction digest mix; stock solution is prepared by boiling for 5 min at 1 mg/ml in 200 mM sodium acetate, pH 5.0). Comparison with a molecular weight marker (such as phage lambda DNA digested with *Hin*dIII) is useful to assess the level of degradation. Concentrations can be estimated by eye with serially diluted lambda DNA (e.g. 2 μg–0.125 μg). To overcome the problem of uneven concentration within a DNA sample, it can be heated to 55°C and passed through a 1 ml micropipette tip to homogenize the solution. A small amount (e.g. 2 μl) of the DNA can then be taken (the narrower aperture of the 200 μl pipette tip should be increased by cutting a few millimetres off the end) and loaded on the gel. Any standard Ficoll/glycerol-based loading buffer is adequate (see Chapter 5). The gel in *Figure 1* is 0.7% agarose in 1 × TBE (pH 8.8) running buffer (0.089 M Tris–HCl, 0.089 M boric acid, 2 mM EDTA, pH 8.8), stained with 0.5 μg/ml ethidium bromide. When the DNA solutions have been made homogeneous, this method provides a good, rapid approximation of the concentration, and can deal with a great many (up to 100) samples simultaneously by using up to four rows of wells in a 20 cm × 20 cm gel.

ii. Digestion and preparation of DNA samples for fingerprints

Once the concentrations of genomic DNAs have been assessed, 5 μg of each sample is digested with a restriction endonuclease having a four-base recognition site. Since these enzymes cut frequently, most DNA will be reduced to small fragments, and the large fragments remaining are likely to be repetitive sequences such as minisatellite loci. The enzyme is selected to produce a maximally informative fingerprint pattern; this will depend on the species being studied. For this reason it is often appropriate, when starting work on a species not previously studied, to test a number of enzymes to establish a repeatably scorable fingerprint system, preferably with more than one multilocus probe. The enzymes most commonly used in higher vertebrates are *Hin*fI, *Hae*III, *Mbo*I, and *Alu*I, though methylation-sensitive enzymes (including *Hin*fI) (50) are best avoided, especially if a variety of tissue sources is being used. Enzymes vary greatly in the time for which they are active, and in some cases complete digestion is best achieved by allowing the reaction to proceed overnight at 37°C in the presence of 4 mM spermidine trihydrochloride and the manufacturer's buffer (usually a 10 × concentrate), using at least two units of enzyme per microgram of DNA. However, shorter digestion times can be adequate if samples are agitated occasionally, and more enzyme used. In order to reduce (sample) background in the fingerprint, and salt- or ribonucleotide-related aberrant running effects, it is beneficial to clean and precipitate digested DNA samples by phenol–chloroform extraction, ethanol precipitation, and washing in 80% ethanol (7). Once dried, the samples are dissolved in a standard volume of distilled water, so that re-digestion can be carried out if necessary. Once all samples have been digested to completion, there are a number of methods which can be employed to assess sample concentration accurately enough to permit the even loading of gels.

iii. Fluorometry

DNA fluorometers (e.g. DyNA Quant™ 200) are now being used in an increasing number of laboratories, and are very useful for measuring the concentration of digested DNA (51), although in our experience they are much less useful when looking at whole genomic DNA. As digested fragments are usually completely dissolved, repeatable, stable, and accurate readings can be obtained from a small aliquot of the digest (at a concentration in the range 10–500 ng/ml using 1.0 μg/ml Hoechst 33258 dye in 1 × TNE), and the sample loadings can be adjusted accordingly. Automated fluorometers which can take readings from samples loaded into microtitre plate wells are very time-saving.

iv. Electrophoresis

A whole genomic DNA sample of accurately known concentration (obtained by careful extraction followed by dialysis in 4 mM Tris–HCl, 1 mM EDTA,

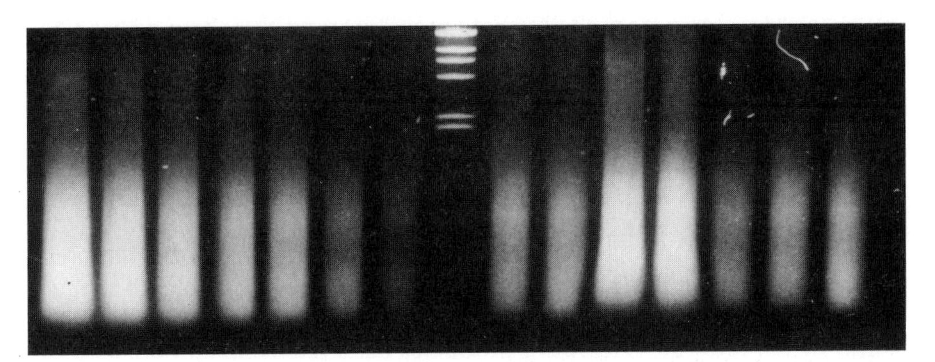

Figure 2. Example of a typical gel for checking restriction digests. The first seven samples are concentration standards of avian genomic DNA digested with *Alu*I (2, 1.5, 1, 0.75, 0.5, 0.25, and 0.125 μg), followed by 0.5 μg lambda/*Hin*dIII DNA. The last seven samples are avian genomic DNA samples digested with *Alu*I. All samples are fully digested, and the concentrations are estimated by comparison with the standards.

pH 7.4 and spectrophotometric measurement (44)) is digested, and serially diluted to make concentration standards. These are electrophoresed alongside the fingerprint samples; concentrations are assessed by eye and adjusted accordingly. A further check can be made by electrophoresing a small (1–2 μl) aliquot (in 10 μl 1 × loading buffer) of the sample to be loaded on the fingerprint gel to ensure that there is an equal concentration in all samples. Electrophoresis is also the only way to identify any DNA samples that have not been fully digested by the restriction enzyme (these are not suitable for fingerprinting). These are readily spotted due to the presence of whole genomic DNA or a distribution of restriction fragments atypical of the species–enzyme combination (see *Figure 2*).

2.3.2 Running the gel

Due to the resolution required for fingerprint analysis, DNA samples often have to be electrophoresed for a considerable number of volt hours (Vh) in order for the DNA fragments to migrate a sufficient distance. To this end, different laboratories have arrived at many different electrophoretic systems, including buffer replacement and/or re-circulation, and a variety of different Tris–acetate and Tris–borate running buffers. A compromise between a faster rate of electrophoresis (usually Tris–acetate) and high buffering capacity (Tris–borate) is 1 × TBE, pH 8.8 (0.089 M Tris, 0.089 M borate, 2 mM EDTA, pH 8.8) (10). For fingerprint gels we most often use 1.0% agarose (Sigma, low EEO). Buffers may also contain ethidium bromide (but see below). It is important to avoid overheating during the run, as this results in uneven running across the gel. We suggest a maximum of 2 V/cm between electrodes; our gel tanks hold 2.5 litres of buffer (for a 20 cm × 20 cm gel) or 4 litres of buffer (for a 20 cm × 30 cm gel). Loading buffer should include the same solution as the running buffer (although it is stored as a 5× or 10× concentrate); most of the standard combinations of Ficoll/glycerol and Bromophenol Blue/xylene cyanol give adequate results.

Molecular weight markers, such as lambda/*Hin*dIII DNA, are usually run in the outside lanes, and running times are calculated with reference to known size marker fragments. As migration is proportional to volt hours, the required end point of a run can be conveniently determined by examining the gel during the run. One option that facilitates later fragment size estimation is to add a small amount (e.g. 5 ng) of internal marker fragments to each sample (see refs. 10 and 131). The membrane is then probed with radiolabelled marker DNA once the fingerprint probe has been removed, and the autoradiographs overlaid to allow the accurate interpolation of molecular weight.

i. Ethidium bromide

There has been much debate about the use (and possible over-use) of ethidium bromide, and its effect on DNA mobility (52). In the presence of ethidium bromide, DNA samples at higher concentrations run faster, and this can seriously undermine band-sharing comparisons if there is a significant discrepancy between amounts of DNA loaded in different lanes. Ethidium bromide is also a powerful mutagen, which poses questions of environmental and user safety, so its use should be minimized. Ethidium bromide is usually used at a concentration of 0.5 μg/ml in gel and running buffer, and under these conditions it is advisable to add to the loading buffer 1 μg for each microgram of digested DNA to avoid 'smiling' bands. However, gels can be run equally well without any ethidium bromide (using a loading buffer dye such as xylene cyanol to give an idea of the migration—it runs equivalent to approximately 4 kb duplex DNA), and can be stained with ethidium bromide (0.5 μg/ml) in a small (re-useable) volume of 1× TBF electrophoresis buffer at the end of the run. The use of standardized running conditions and internal markers eliminates the need for ethidium bromide altogether, and this is now standard practice in many laboratories.

2.3.3 Blotting DNA fingerprint gels

Once the DNA samples in the gel have run the desired distance, they are immobilized on a membrane by a process of capillary or vacuum blotting (see Chapter 5). The protocols used for vacuum blotting tend to follow those recommended by the manufacturers of the membranes and apparatus. With vacuum blotting, the extreme rapidity of the gel treatment and blotting is ideal for DNA fingerprinting, as the probability of lateral diffusion of DNA molecules during the process is greatly reduced. The choice of membrane has also been a matter of debate—nitrocellulose possibly gives marginally better results than other membranes, but can only be re-probed a few times before it disintegrates. Nylon membranes, despite some serious initial problems with inconsistent hybridization, seem now to be the most popular choice, and can be re-probed as many as 10 times. Basic capillary blotting techniques (44,45; see also Chapter 5) are adequate for fingerprinting and are modified in order to decrease the opportunity for lateral diffusion of the DNA. Modifications include rapid soaking times (2 × 15 min for denaturation and neutralization steps) and changing paper towels on the blot every 5 min for the first 15 min, and then every 15 min for the next hour. The blot can then be left overnight with sufficient towels (or the towels can simply be changed every 15 min for another 2 h). An alternative is to use 10 Quickdraw blotting sheets (Sigma) and leave for 3 h.

i. Alkali blotting

A quick and convenient modification of the capillary and vacuum blotting procedure when using nylon membranes involves the use of an alkaline transfer medium. Here, one can blot immediately after depurination, using 0.4 M NaOH as a denaturing transfer buffer. This labour-saving method is routine practice in some laboratories.

ii. Gel drying

Blotting is not the only method by which DNA can be immobilized for hybridization. The technique of gel drying followed by in-gel hybridization has been applied to DNA fingerprinting with considerable success (see refs 53–55 for examples). The method involves vacuum drying the gel on to Whatman 3MM paper, followed by immersion in a denaturing solution and then a neutralizing solution (54). Either non-isotopic or radioactively labelled probes can be used, but it would appear that so far only short oligonucleotide sequences are effective (see Section 2.4), as longer molecules (such as the Jeffreys probes 33.6 and 33.15 (1)) may be too large to circulate through the dried gel matrix.

2.4 Multilocus DNA fingerprinting probes

Several sequences that detect mostly different sets of hypervariable loci have been characterized and are generally available to the scientific community. These fall into two distinct classes: the 'minisatellite' sequences, which directly detect sets of tandem repeats containing particular 10–20 bp GC-rich 'core' regions (e.g. refs 1, 56, 57), and 'simple sequence' oligonucleotide repeats such as $(GATA)_4$ (58), $(CAC)_5$ (59), and $(TG)_n$ (60). Even some randomly generated 14-mer oligonucleotide repeat sequences detect polymorphic sequences in the human genome (61). While many probes are now available, usually one only needs to test a few before finding a system that works in the species of interest. The four most commonly used multilocus probes are 33.6, 33.15 (1,2), M13 hypervariable region (57), and α-globin 3′ hypervariable region (56). These sequences have been demonstrated to detect large sets of independently segregating loci in a number of species (7,8,10,20,46,62–64). DNA probes are usually available from the laboratories of origin; however, M13 is available from most molecular biology suppliers and probes 33.6 and 33.15 are available from Zeneca Cellmark Diagnostics and also in pSPT plasmid derivatives to enable the production of highly sensitive RNA probes (65).

2.4.1 Labelling

Two types of label are used in fingerprinting laboratories; the most common practice is radiolabelling with [α-^{32}P]dCTP, though non-isotopic methods (e.g. refs 66 and 67) are likely to become more popular, especially now that chemiluminescent methods have been developed and are commercially available (Chapter 5).

i. Radiolabelling

The 'Jeffreys probes' 33.6 and 33.15 were originally made available as recombinant M13mp8 and M13mp19 plasmids, respectively. The procedure for the radiolabelling involves primer extension of the single-stranded plasmid followed by gel excision of the radioactive fragment; it is detailed in ref. 68. Random oligonucleotide priming, which involves less exposure to ^{32}P, has, however, been found to be more convenient, and can be applied to any minisatellite probe. This technique routinely yields specific activities of 10^9 d.p.m./μg. A generalized procedure is outlined in *Protocol 2*—this is adapted from the protocols in refs 69 and 70. A more detailed description of a similar oligolabelling procedure is provided in Chapter 5.

Protocol 2. Radiolabelling DNA probes

Equipment and reagents

- Solution O: 1.25 M Tris–HCl, 0.125 M $MgCl_2$, pH 8.0; store at 4°C
- Solution A: 1 ml solution O, 18 μl 14.3 M β-mercaptoethanol, 5 μl each of dATP, dTTP, and dGTP (each at 0.1 M in 3 mM Tris–HCl, 0.2 mM EDTA, pH 7.0); store at −20°C
- Solution B: 2 M Hepes titrated to pH 6.6 with 4 M NaOH; store at 4°C
- Solution C: random hexanucleotides (Boehringer) evenly suspended in 3 mM Tris–HCl, 0.2 mM EDTA pH 7.0; suspend 50 optical density (OD) units in 550 μl; store at −20°C
- Oligolabelling buffer, prepared by mixing solutions A, B, and C in the ratio 2:5:3; store at −20°C (this solution is good for 3 months with repeated freeze–thawing)
- Bovine serum albumin (BSA), enzyme grade (BRL) at 10 mg/ml in ultra-pure water; store at −20°C
- Stop mix: 20 mM NaCl, 20 mM Tris–HCl pH 7.5, 2 mM EDTA, 0.25% (w/v) SDS
- DNA in solution or in low melting point (LMP) agarose (Sea Plaque; FMC Bioproducts), excised from a gel with water added at a ratio (w/v) of 1.5 water:1 gel

Method

1. Use DNA in solution or in LMP agarose. Boil an aliquot of the DNA solution after adding H_2O to a total volume of 10 μl, or all of it in the case of DNA isolated in agarose, for 5 min prior to the labelling reaction in order to denature the DNA (and melt the gel).
2. If the DNA is to be labelled immediately, microcentrifuge it briefly and

transfer the aliquot to be labelled to an ice bath to allow it to cool to 0°C (2 min). Hold it at that temperature (especially in the case of mixes containing gel). For DNA in LMP, store the remainder of the gel at –20°C. Before subsequent reactions, re-boil any previously denatured DNA for 5 min prior to cooling at 0°C and holding it at that temperature.

3. Carry out the labelling reaction at room temperature or 37°C (better for probes in agarose) by adding the reagents in the following order:
   - H_2O to a total volume of 20 μl
   - 3 μl oligolabelling buffer
   - 0.6 μl BSA
   - 10–20 ng DNA
   - 0.6 μl (1 unit/μl) DNA polymerase I (Klenow fragment; Amersham)
   - 1.5 μl (3000 mCi/mmol) [α-^{32}P]dCTP

4. Incubate overnight at room temperature, or for 2 h at 37°C.

5. Stop reaction with 65 μl stop mix.

6. For a crude assessment of the incorporation of ^{32}P (optional), precipitate the DNA[a] and use a ^{32}P monitor to check for a high level of counts in the pellet.[b] Dissolve the pellet in 500 ml water; the solution can be stored for up to 5 days at –20°C. When required, incubate in a heating block at 95°C or boil for 5 min to denature the DNA, and add to the hybridization solution.

[a] A carrier pellet such as 1–2 μg glycogen (Boehringer) or 5 μg tRNA can be used to assist the precipitation of this small amount of DNA by the addition of 50 μl 7.5 M ammonium acetate, pH 4.8 and 2.5 vol. 100% ethanol, mixing at room temperature, and microcentrifuging for 10 min.

[b] When using single-locus probes at high stringency (Section 3.4.2), we do not generally find it necessary to purify the probe or measure the incorporation. The entire denatured reaction mix is simply added to the hybridization solution

ii. Non-isotopic labelling

Several different types of non-isotopic label are now being used routinely in fingerprinting laboratories: biotin (66), alkaline phosphatase (71), digoxigenin (72, Chapter 5), and horseradish peroxidase (chemiluminescence, ref. 73). The labelling protocols are many and varied, all seem to produce acceptable results, and some of the hybridization times are rapid. One problem with the dye-based systems is that the degree of staining cannot easily be controlled, and stripping membranes for re-probing may be problematic. Chemiluminescent labelling methods potentially provide a system largely analogous to ^{32}P-based ones and therefore circumvent these problems. The labelling systems are available as commercial kits and are supplied with detailed instructions.

2.4.2 Hybridization

The following three main types of hybridization apparatus seem to be in common use:

- Perspex (Plexiglass) hybridization chambers immersed in a shaking water bath
- polythene bags floating in a shaking water bath
- tubular bottles attached to a rotisserie in a temperature-controlled oven

All have advantages and disadvantages. Hybridization chambers (our preferred method for multilocus fingerprinting) give good repeatable results. Hybridization bags use very much less solution but uniform hybridization and adequate washing can be difficult to achieve. Rotisserie bottles also use very little solution and the temperature control is very accurate, but membranes must be rolled up to fit into the bottle and, even when nylon mesh spacer sheets are used to separate each membrane, we have found that background hybridization is not easily washed off the parts of the membrane that overlap (this is, however, our preferred method when using single-locus probes at high stringency, for which it works very well). Protocols for hybridizations involving nitrocellulose membranes have mainly used SSC buffer and Denhardt's solution for blocking non-specific hybridization (7,8,44). Higher sensitivities were achieved by using polyethylene glycol 6000 (BDH) in hybridizations (1,7). *Protocol 3*, however, is particularly suitable for nylon membranes. Stringency (i.e. buffer concentration and temperature of hybridization and washing) should ideally be optimized for each probe–species combination, and is usually arrived at by trial and error. Although in early studies hybridizations included denatured salmon sperm DNA as a competitor (1,7), this was later found to bind preferentially to minisatellites (57), dramatically reducing the signal. We have since omitted competitor DNA of any kind from our multilocus experiments (but see Section 3.4 concerning single-locus analysis). *Protocol 3* is adapted from refs 74 and 75 (see also Chapter 5) and the temperatures are merely suggestions.

Protocol 3. Hybridization of DNA fingerprint probes

Equipment and reagents

- 1 M phosphate buffer (per litre): 134 g of Na_2HPO_4 (mol. wt 142); adjust to pH 7.2 with *c.* 7 ml of 88% orthophosphoric acid and add 45 ml water to give 1 mol of phosphate ions (use 167.8 g if using $Na_2HPO_4{\cdot}2H_2O$)
- Wash 1: 0.25 M phosphate buffer; 1% (w/v) SDS
- Washes 2 and 3: 2 × SSC, 0.1% (w/v) SDS
- Pre-hybridization/hybridization solution: 0.25 M phosphate buffer, 1 mM EDTA, 7% (w/v) SDS, 1% BSA (Sigma, type V). This can be stored at –20°C in a partially made form without the BSA; the BSA is most efficiently dissolved simply be floating on the surface. Optionally, de-gas immediately before use.
- Washes 4 and 5: 1 × SSC, 0.1% (w/v) SDS

- 2 × SSC
- 3 × SSC
- 0.4 M NaOH
- Nylon membrane carrying DNA fragments, and blank membrane (*c.* 25 cm^2)
- Temperature controlled incubator

A. *Pre-hybridization*

1. Wet the membrane by floating on a bath of 3 × SSC and immerse once fully soaked. Prepare a blank piece of membrane (*c.* 25 cm^2) the same way.
2. Place in hybridization vessel together with the blank in an appropriate volume of pre-hybridization solution and incubate under agitation at 62°C for 1–3 h.

B. *Hybridization*

1. Boil the probe solution as described in step **6** of *Protocol 2.*
2. To ensure thorough mixing, remove the membranes from the pre-hybridization solution and add the probe solution to it (or use fresh solution, pre-warmed to 62°C). Replace membranes and incubate overnight at 62°C.

C. *Washing*

Note: keep all solutions at 62°C throughout this stage.

1. Remove the membranes from the hybridization solution (which can be stored for re-use), and immerse in pre-warmed wash 1. Incubate under agitation for 15 min.
2. Immerse in wash 2 and incubate for 15 min. Measure radioactive counts on fingerprint membrane and compare with blank membrane.
3. Wash in the other solutions as necessary—stop when the blank gives only a background reading.
4. Briefly immerse membrane in 2 × SSC to remove any SDS. Blot off excess solution and cover membrane in plastic film to retain moisture (probes can be stripped from the membrane more effectively if it stays damp).

D. *Stripping probes off membranes for re-use*

To reprobe nylon filters, old probe must be removed first unless sufficient time has elapsed for the previous probe's activity to decay.

1. Incubate membranes in 0.4 M NaOH for 15 min at 45°C.
2. Incubate membranes in 0.1 × SSC, 0.01% SDS for 45 min at 45°C. Repeat procedure if necessary. Alternatively, boil a solution of 0.1% SDS (w/v) and then immerse the membrane at 80°C and allow to cool

Protocol 3. *Continued*

to room temperature. Nitrocellulose can also be re-probed at least a couple of times, and stripping is best accomplished by immersing the membranes in water that has just been boiled (*c.* 90°C) and leaving to cool down. This procedure may well have to be repeated a couple of times. It is often desirable to check that stripping is complete by exposing the stripped membrane to X-ray film.

In-gel hybridization can be done with either ^{32}P-labelled probes (53,54) or non-isotopically labelled probes (55) and excellent results have been obtained. Gel drying followed by in-gel hybridization to chemiliminescent probes (see Section 2.4.1) might potentially represent the ideal combination of convenience and safety. Most workers including ourselves, however, follow the general approach described in this chapter. This is because the most popular multilocus probe sequences are probably too large to use for in-gel hybridization, though short synthetic oligonucleotide probes and appropriate protocols have been developed, and also because blotting and radiolabelling are currently the most usual procedures in the molecular genetic laboratories where training and advice are sought.

2.4.3 Autoradiography

Exposure times for DNA fingerprints can take from 2 h to 2 weeks (at −80°C) depending on the levels of radioactivity on the membrane, whether none, one, or two intensifying screens (such as manufactured by Cawo) are used in the cassette, and the type of film used. There are no hard and fast rules, and again trial and error is the only way to adjust for these variables. Generally speaking, the best resolution is gained without the use of intensifying screens, but this is often impractical due to the low level of radioactivity on the membranes. A short exposure using two screens, followed by a longer exposure with one or no screen, is probably the best approach.

2.5 Analysis of DNA fingerprints

In all applications of fingerprinting involving species not previously studied, it is important where possible to establish that the bands being detected in the fingerprint occur independently of each other, i.e. that minisatellite alleles are represented by single bands, and that there is no linkage disequilibrium among the loci from which they derive (but see refs 76 and 132). To establish that the detected bands are indeed unlinked, a segregation analysis can be performed on a relatively large pedigree (7,8,10,46,62,64); this analysis is discussed in the next section. In many species, it may be impossible to obtain families to allow segregation analysis (76), and in such cases cases an alternative approach is more appropriate in which the distribution of the band sharing values for known relatives and non-relatives are determined empirically (see refs 11 and 77 and *Figure 5* for examples).

2.5.1 Segregation analysis

Figure 3 and *Table 1* illustrate how the DNA fingerprints should be analysed in a segregation analysis, which should typically include one or more large families of two parents and eight or more offspring (64). Each parental band is scored and given an identity. Bands that are shared (see Section 2.6.2) between the parents and bands that cannot be scored reliably in the offspring because of their proximity to an intense band in the other parent cannot be used. The transmission of each band into the offspring can then be scored as a '1' where present and '0' where absent (see *Figure 3b*). Paternal and maternal bands are analysed separately. The pattern of segregation is then analysed, preferably using a simple computer program (64), and can be interpreted as follows.

(a) Where there is consistent co-segregation (always 11 or 00, 111 or 000, etc.) then the bands are either tightly linked or one locus has been cleaved internally by the restriction enzyme. Using the same probe with a different enzyme might avoid this problem. In such a case, only one of the linked bands would be retained for the analysis of segregation among loci.
(b) Where there is consistent segregation (always 10 or 01) then both alleles of the same locus are being detected, and only one of the alleles is retained for further analysis.
(c) When bands are transmitted to all of the offspring in the pedigree, one can conclude that the parent is probably homozygous at that locus. Such bands are omitted from further analysis.

The precise probabilities of obtaining consistent apparent linkage, allelism, or homozygosity by chance can be easily calculated for any given number of offspring and analysed bands.

After these adjustments to the data set, the segregation pattern of the remaining set of bands can be further analysed to check for Mendelian inheritance and any evidence of incomplete linkage. The frequency with which bands are transmitted should follow a binomial distribution, with a mean of 0.5 (see *Table 1*). We use a computer program (64) to handle the data as a 1010, etc., matrix in order to identify the linked and allelic bands (for removal), to calculate an overall band transmission frequency, and to compare the observed distribution with that expected for different mean levels of linkage. The significance of any deviation between the distributions has to be found by simulation due to the non-independence among the data points.

The number of independently segregating loci, L, detected on the fingerprint is given by the equation $L = n - a - b$, where n is the total number of bands, a is the number of allelic pairs, and b is the total number of bands in tightly linked groups minus the number of tightly linked groups. Occasionally,

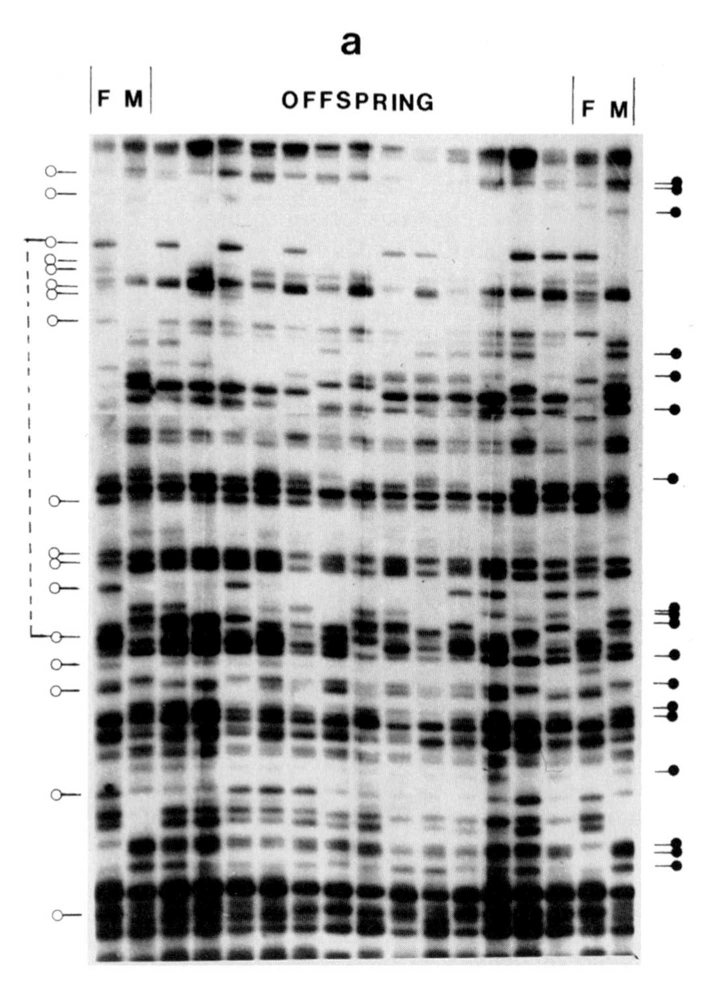

Figure 3. Example of a DNA fingerprint segregation analysis. (a) DNA fingerprint (*Hae*III digest, probe 33.15) of a 12-offspring pedigree in the chicken. Scored maternal bands are indicated by open circles and paternal bands by filled circles. Allelic bands are connected by a dotted line. No bands are linked on this fingerprint. (b) Matrix of segregation of parental bands in the offspring (1 = present; 0 = absent). (c) Segregation data. See T*able 1*.

a band may be seen in the fingerprint which is not attributable to either parent. If this only occurs once or twice in an offspring, then it is likely that one of the alleles has mutated (e.g. refs 8 and 11). Typical mutation rates in fingerprints are of the order of 10^{-3} per fragment per gamete (1,7,8,78).

2.5.2 The band-sharing coefficient and individuality

The main index of similarity used to describe DNA fingerprints from two individuals is the band-sharing coefficient. This gives the probability that a band

b

Female bands	Offspring 1	2	3	4	5	6	7	8	9	10	11	12	13
A	1	1	1	1	1	1	1	1	1	1	1	1	1
B	0	0	1	1	0	0	1	1	0	0	0	0	1
C	1	0	1	0	1	0	0	1	1	0	0	1	1
D	0	1	0	1	1	0	1	0	1	1	0	1	0
E	0	1	0	1	1	1	1	0	1	1	0	0	1
F	0	1	1	0	0	0	0	0	1	1	1	0	1
G	0	0	1	1	0	0	0	1	1	0	1	0	1
H	1	1	1	1	1	1	1	1	1	1	1	1	1
I	0	0	0	0	1	0	1	1	1	1	0	0	0
J	0	1	1	1	1	0	0	0	1	0	0	1	0
K	1	1	1	1	0	0	1	0	1	0	1	1	0
L	1	1	1	1	0	1	0	1	0	1	1	0	0
M	0	0	0	1	0	0	0	0	0	1	1	0	1
N	0	1	0	1	0	1	1	0	0	1	1	0	0
O	1	1	1	1	1	0	0	0	0	0	0	1	0
P	0	1	0	1	0	1	0	1	0	1	1	0	1
Q	0	0	1	1	1	0	0	0	0	0	0	1	0
R	1	1	1	1	1	1	1	1	1	1	1	1	1
S	1	1	1	1	1	1	1	0	1	1	0	1	0
T	0	1	0	0	1	1	1	1	0	1	1	1	1
Male bands													
a	0	0	1	0	0	0	1	1	0	0	0	0	1
b	1	0	0	1	0	1	0	0	0	1	1	0	1
c	0	1	1	1	1	0	0	0	1	0	0	1	1
d	1	1	1	1	1	1	1	1	1	1	1	1	1
e	1	0	0	0	0	1	0	0	1	1	1	1	0
f	1	1	0	0	0	1	0	1	1	1	1	0	1
g	1	1	1	1	1	0	0	1	1	1	1	0	0
h	1	1	1	1	1	1	1	1	1	1	1	1	1
i	1	1	1	1	1	1	1	1	1	1	1	1	1
j	0	0	0	1	1	1	0	0	1	0	0	1	0
k	1	1	1	1	1	1	1	1	1	1	1	1	1
l	1	1	1	1	1	1	1	1	1	1	1	1	1
m	1	0	0	0	1	0	1	1	0	0	1	0	1
n	1	0	1	1	1	0	1	0	0	0	0	1	0
o	1	1	1	0	0	0	1	1	0	1	0	0	1
p	1	1	1	1	0	1	0	0	0	0	1	1	1
q	0	0	1	1	0	1	0	1	0	1	1	1	0
r	1	1	0	0	0	0	1	1	0	1	1	0	1
s	0	1	1	1	1	1	1	0	0	1	1	0	0
t	1	1	1	1	1	1	1	1	1	1	1	1	1
u	1	1	0	0	1	0	1	0	1	0	1	0	1
v	1	0	0	0	0	1	0	0	0	1	1	1	0
w	1	0	1	0	1	1	0	1	1	0	0	1	0

will be shared for a certain molecular weight range within the fingerprint. Shared bands are those that have migrated to the same distance on the gel (within 0.5 mm) and have no more than a two-fold intensity difference, i.e. the difference between a homozygote and heterozygote. There are two ways of calculating band sharing, namely,

$$x = ((N_{ab}/N_a) + (N_{ab}/N_b))/2 \qquad \text{(refs 2 and 7)}$$

and

$$s = 2N_{ab}/(N_a + N_b) \qquad \text{(refs 9 and 79)}$$

where N_{ab} is the number of bands of similar intensity and electrophoretic mobility in individuals a and b, N_a is the total number of bands in individual a which could be scored, if present, in individual b and N_b is the total number of bands in individual b which could be scored, if present, in individual a. In this

Table 1. Segregation analysis of the chicken family in *Figure 3*. Transmission of single fragments and pairwise combinations. All but one of each set of co-segregating fragments or pair of apparently allelic fragments in a parent are excluded so that each scored locus is represented only once[a]

	No. of single fragments or pairwise combinations transmitted to[c]							
	Female offspring				Male offspring			
	Single fragment		Pair (++ or −−)		Single fragment		Pair (++ or −−)	
No of offspring[b]	Obs.	Exp.	Obs.	Exp.	Obs.	Exp.	Obs.	Exp.
0	0	0.0	0	0.0	0	0.0	0	0.0
1	0	0.0	2	0.6	0	0.0	5	0.4
2	2	0.3	4	3.3	0	0.0	3	2.6
3	0	1.1	9	12.3	0	0.2	6	9.6
4	3	2.7	30	30.6	3	0.8	29	24.1
5	3	4.6	65	55.1	3	2.1	37	43.4
6	5	5.8	67	73.5	4	3.8	58	57.8
7	5	5.5	78	73.5	9	5.0	65	57.8
8	7	3.9	50	55.1	3	5.0	39	43.4
9	1	2.1	30	30.6	2	3.8	21	24.1
10	1	0.8	8	12.3	0	2.1	10	9.6
11	0	0.2	5	3.3	0	0.8	2	2.6
12	0	0.0	3	0.6	0	0.2	1	0.4
13	0	0.0	0	0.0	0	0.0	0	0.0
Total	27		351		24		276	

[a]The data are combined for probes 33.6 and 33.15 and were first presented in ref. 49. The mean transmission frequency is 0.493.
[b]Number of offspring to whom single fragment or pairwise combination is transmitted.
[c]Obs. observed; Exp., expected.

way, any weakly hybridized bands that although apparent in individual a (or b) would have been obscured, if present, by a stronger band in individual b (or a) are excluded. In the formulae that follow, x is used as the band-sharing coefficient, i.e. the probability that a band in an individual is in a second, random individual. It can be shown that s is an under-estimate of this quantity, though not to a large degree. Under the assumption that shared bands are always identical alleles from the same locus, and that all bands have equal population frequencies, the band-sharing coefficient for any given band is related to the allele frequency q by the formula $x = 2q - q^2$ (2,4). As band-sharing coefficients are proportions, summary statistics are arrived at by working with angular transformed data. This transformation expresses the proportion p as an angle Θ, where $\Theta = \arcsin\sqrt{p}$ (80). Lynch (79) considers that 'provided that large numbers of loci are examined, it seems reasonable in hypothesis testing to treat S_{xy} as a normally distributed variable' ($S_{xy} = s$).

Having established that bands are independent markers, we can conservatively find the mean probability that all n bands in an individual's fingerprint are present in a second random individual as x^n. This figure is usually extremely small; for example, for an individual with 30 scorable bands in a population in which the mean band sharing coefficient is 0.2, the probability of another individual having the same pattern is $0.2^{30} = 1.1 \times 10^{-21}$.

i. Parentage analysis

The probability, I, of the non-detection of a misassigned pair of parents (as might arise through undetected conspecific nest parasitism, adoption, or misidentification of an offspring sample) is found as $I = (1 - (1 - x)^2)^n$ (assuming that the actual parents are unrelated to the putative parents) (6); the exclusion probability for false parentage is $1 - I$. The mean probability of the non-detection of a misassigned father, assuming the mother is correctly assigned and the actual father is unrelated to the putative father, is given by $I' = x^m$, where m is the number of paternal-specific bands in the offspring; the exclusion probability for false paternity is $1 - I'$. The mean value of m is expected to be $n(1 - x)/(2 - q)$. These formulae use only paternal-specific bands in the determination of paternity. More complex analyses also allow the incorporation of information from the mother (42). A practical example of paternity assignment is seen in *Figure 4*. Here a family of dunnocks was analysed to determine which of two resident males was the true father of each of four offspring (8). Contrary to expectations from behaviour, the subdominant male fathered three out of four chicks (D, E, and F).

An alternative approach to analysing parentage is exemplified in *Figure 5* (from ref. 81). This method can be used to discriminate offspring whose bands do not appear in their behavioural parents due to mutation or extra-pair paternity. Usually, the presence of more than two or three such 'novel' fragments indicates that the offspring is unrelated to one or both parents. A low band-sharing coefficient with either parent can be used as further evidence for extra-pair paternity (81).

ii. Band-sharing coefficient between relatives

Under the assumption that shared bands represent identical alleles of the same locus and that all alleles have the same frequency, q, the expected mean band sharing between full siblings is

$$x = (4 + 5q - 6q^2 + q^3)/4(2 - q) \qquad \text{(ref. 6)},$$

that between parents and offspring is

$$x = (1 + q - q^2)/(2 - q),$$

that between second-order relatives having a direct line of descent (such as aunt–nephew, but not double first-cousins) is

$$x = (1 + 5q - 5q^2 + q^3)/2(2 - q),$$

and that between third-order relatives is

$$x = (1 + 13q - 13q^2 + 3q^3)/4(2 - q).$$

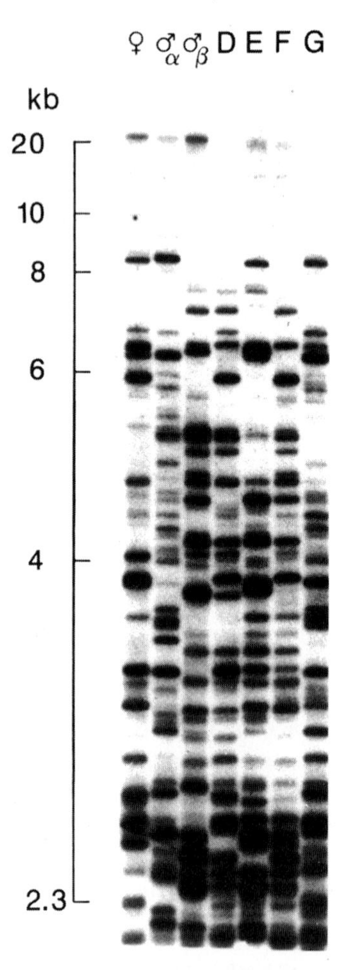

Figure 4. DNA fingerprints of a family of dunnocks (see refs 8 and 82). Analysis of both the presence of unique paternal bands in the offspring and of father–offspring band-sharing coefficients confirmed that the beta male fathered chicks D, E, and F.

Even if it is assumed that bands shared by unrelated individuals are never allelic, and instead represent the chance co-migration of alleles at different loci, the mean probability of band sharing between both kinds of first order relative will still be approximately $(1 + q - q^2)/(2 - q)$.

In order to draw conclusions about relationships from fingerprint data, the deviation of observed band-sharing coefficients from these expected values from various hypothesized levels of relatedness can be examined using goodness-of-fit tests (7,10,80). If non-independence among bands (perhaps detected using more than one probe) has been established through a segregation study, the probability obtained using a statistical analysis that assumes

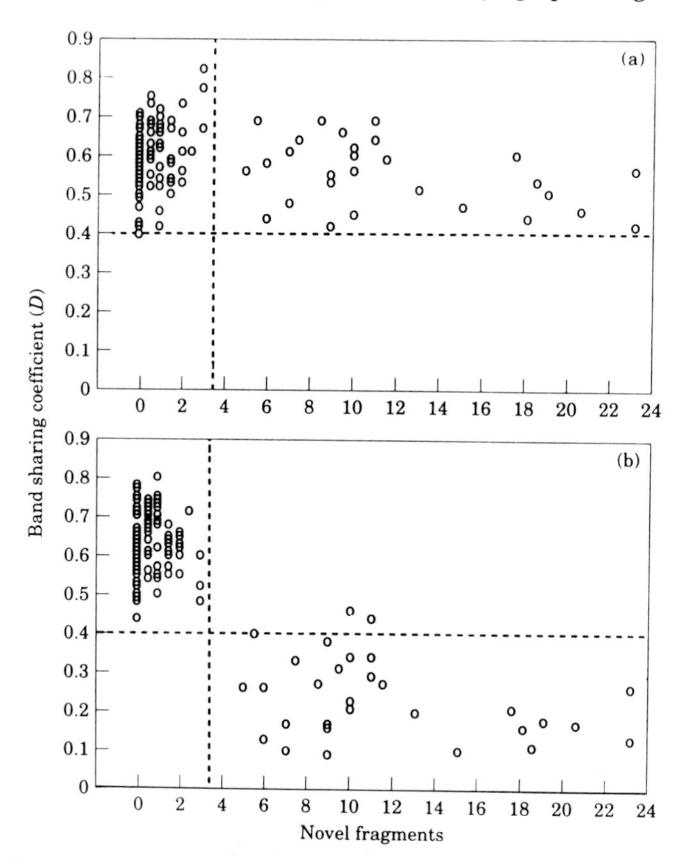

Figure 5. The relationship between band-sharing coefficients with putative parents (a, mother; b, father) and the number of unassigned or 'novel' fragments in the offspring fingerprint. Note that unlike with the putative mothers, most putative fathers who had more than two novel fragments also had a band-sharing coefficient of less than 0.4, indicating that they are not the true father. Reproduced from ref. 81.

independence can be corrected by adjusting the probability as $P' = P^{(H/B)}$, where H is the number of haplotypes represented by B bands (10). The degree of relatedness can rarely be ascertained precisely below the level of first-order relatives (such as father–son or sister–sister). A more accurate estimation would be possible if the allelic contributions in the population were known, unless the band-sharing coefficient in the base population is essentially zero (41, 42). In practice, it is usually found that first-order relatives are more significantly similar than non-relatives, while this is the case for only a proportion of comparisons between second-order relatives.

Multilocus DNA fingerprinting is usually inappropriate for distinguishing between populations. This is because the extreme variation that is usually detected yields too much intra-population variation for any inter-population comparisons to be made, reflecting the fact that the resolved pattern generally contains just one of the alleles at each of a minor proportion of the loci detected by a particular probe (79), and that these loci have very variable and generally high mutation rates (78). Exceptions to this general situation arise where either a small number of founders or severe bottlenecking lead to the population being relatively highly inbred, sometimes producing what is effectively a population-specific pattern. Such situations have been found in several wild populations (21,36–38), but are probably rare. However, where applicable they may reveal intra- and inter-population variation previously undetected using other systems, and Gilbert *et al.* (21) provide an example of how data can be handled. In a theoretical study, Lynch (79) puts the similarity index into the context of standard population genetic parameters, and shows that, while the index does not provide a good estimate of the mean homozygosity of a population, it is expected to be correlated with it. Provided that the fingerprints represent a large number of loci, the comparison of means and sampling variances allows the identification of populations that lack large amounts of genetic variation, such as some endangered populations (21,24,27).

3. Single-locus DNA fingerprinting

3.1 Applications

Multilocus fingerprinting is the method of choice, in most cases, for accurately testing paternity and maternity (e.g. refs 6–8,34,82) It also sometimes enables the determination of relative values of genetic variability within and between populations (21,38). However, there are technical, theoretical, and statistical difficulties that make the use of multilocus DNA fingerprinting difficult or inappropriate in studies of complex mating systems (such as lekking or social species) or in attempts to unravel the genetic structure within and between populations:

(a) Multilocus fingerprinting can be expensive and labour-intensive when a large number of samples is being analysed.
(b) The scoring of a fingerprint gel is time-consuming.
(c) The relative migration of fragments in non-adjacent lanes can be difficult to compare.
(d) Comparisons are impossible between samples on different gels without accurate interpolation using internal size markers.
(e) It is only rarely possible to assign specific restriction fragments to a particular locus and thus identify alleles, determine genotypes, and measure allele frequencies.
(f) As stated in Section 1.2, apparent co-migration of non-homologous fragments and variation in the number of loci that are represented in the scorable region of the fingerprint increase the variance of the estimate of similarity between individuals.

The use of hypervariable single-locus probes overcomes most of these difficulties. Identification of specific loci and alleles is straightforward, and comparison between individuals on the same gel is easier, though several size standards or even internal markers would still be needed for the accurate identification of alleles between gels and sometimes within gels, especially at the more variable loci. Moreover, the presence of less variable loci in species showing highly variable multilocus fingerprints potentially allows the choice of loci showing different levels of heterozygosity to suit the problem being tackled. Probes used singly in succession or together in a cocktail can be as powerful as multilocus analysis using one probe (30,35,83). The high level of specificity of single-locus minisatellite probes and their use at high stringency results in a strong and clean hybridization signal. Furthermore, they are more easily used and the results are obtained more rapidly than with multilocus probes. Importantly, single-locus probes can be extremely sensitive and, under optimal conditions, can detect loci at digest concentrations of less than 100 ng (83). The obvious way to obtain single-locus minisatellite data is to clone them in the species under study (see below). However, note that alternative approaches are possible through using probes cloned from related species, by testing probes cloned in unrelated species (see Section 3.4.4), or by identifying specific loci in multilocus patterns (see 35).

3.2 Cloning minisatellite loci

Until recently, minisatellite loci have proven very difficult to clone, and the effort required seemed justifiable only in humans, genetic model species, and commercially important animals and plants. The difficulties lie in the instability of tandemly repeated sequences in classical cloning systems (83,84). A strategy which overcame many of these problems has been designed and applied in humans (29,85). This strategy is an efficient way of cloning minisatellite sequences in species in which genuine multilocus fingerprints can be obtained. It has now been applied to a considerable number of animal species (e.g., see 16,18,25,30,86–90). Size-selected DNA fragments, rich in minisatellites, are cloned into charomid (85) cosmid vectors chosen to accept inserts within a particular size range (*Figure 6*). The cloning strategy is summarized in *Figure 7* and described below.

3.2.1 Multilocus DNA fingerprinting

Initially, it is necessary to characterize the multilocus DNA fingerprint in the species of interest. Total genomic DNA of several unrelated individuals is digested with the restriction enzyme *Mbo*I (or an isoschizomer such as *Nde*II). This enzyme recognizes a 4 bp sequence (5′-GATC-3′) included within the 6 bp recognition sequence of *Bam*HI (5′-GGATCC-3′) which is present in the polycloning site of the charomid vector (*Figure 7*). A multilocus DNA fingerprint is then obtained following the procedures described in Sections 2.2–2.4.

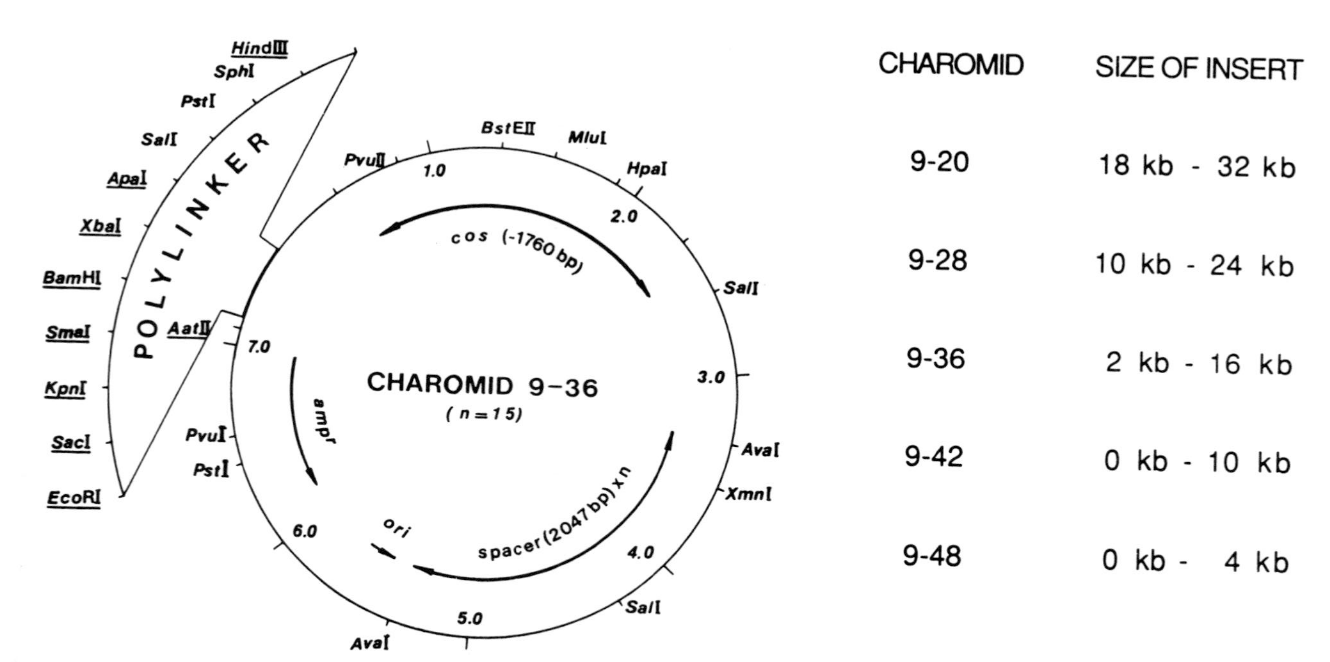

CHAROMID	SIZE OF INSERT
9-20	18 kb - 32 kb
9-28	10 kb - 24 kb
9-36	2 kb - 16 kb
9-42	0 kb - 10 kb
9-48	0 kb - 4 kb

Figure 6. (a) Simplified restriction map of a charomid vector (28,59). Nine restriction sites potentially useful for cloning are present (underlined), but only *Bam*HI contains the 4 bp recognition site of the enzymes used in DNA fingerprinting (*Mbo*I, *Sau*3AI, and *Nde*II). Depending on the number of copies of the 2 kb spacer, the size of the charomid ranges from 5.3 kb (no spacer) to 52 kb (23 copies of the spacer). (b) Different size ranges of genomic DNA can be inserted into the charomid vectors, but the total size of the recombinant molecule must be between 38 kb and 52 kb to be packaged efficiently into bacteriophage lambda particles.

The importance of this excercise is that it provides a means of checking that the enzyme to be used in preparing for the cloning does not produce a profile rich in satellite DNA classes which hybridize to polycore probes, and which might completely obscure any underlying minisatellite pattern (as in the Japanese quail in ref. 7). Such sequences are also likely to yield large numbers of positively hybridizing but unwanted clones in the genomic library. Unfortunately, *Mbo*I is the only four-base recognizing endonuclease compatible with the six-base charomid cloning sites. It is also necessary to choose an appropriate size fraction which contains significant numbers of minisatellite loci having the desired level of variability required for the study envisaged.

3.2.2 Preparation of the genomic DNA

Minisatellite loci characteristically have many alleles of widely different sizes (83,91), so to ensure that many loci are represented in the genomic fraction, DNA from 10–20 individuals should ideally be pooled to form the source of DNA for cloning. Choosing individuals of the heterogametic sex will ensure the equal representation of sex-linked loci.

3.2.3 Extraction and digestion of genomic DNA

Equal amounts of genomic DNA from 10–20 individuals are pooled (ideally 50–100 μg each, extracted as in *Protocol 1*), and re-extracted once with phenol–chloroform, and once with chloroform (45). If DNA is limiting then many more samples can be pooled. Very much less DNA could, in principle, be used, but would be more difficult to work with in subsequent steps. The sample is then dialysed (45) against four changes of 2 litres of TE (10 mM Tris–HCl, 1 mM EDTA, pH 7.5) over 36 h at 4°C. The concentration of a small aliquot of the sample is confirmed by UV absorbance at 260 nm (1 OD_{260} unit ≈ 50 μg/ml). The ratio of OD_{260}:OD_{280} should be about 1.8:1. If not, then re-extract and re-dialyse the DNA. Digest the DNA to completion using *Mbo*I as described in Section 2.3.1, except that only 0.5 units of this enzyme per microgram of DNA is adequate, as it should be a very pure DNA sample. Check a small amount (0.5 μg) of the sample on a gel to ensure it is fully digested. Purify and recover the DNA as described in Section 2.3.1, and suspend the DNA in 1.7 ml of TE, 200 μl 10 × loading buffer (see Section 2.3.2), and 100 μl ethidium bromide (5 μg/ml).

3.2.4 Size selection of DNA fragments

The procedure for recovering size-selected DNA fractions is described in *Protocol 4*, and is relatively cheap and simple. Alternative methods include DNA extraction kits (e.g. Qiagen), electrophoresis on to DEAE–cellulose membranes (for fragments of < 5 kb), electro-elution into dialysis bags (for all size fragments, see Chapter 7), and LMP agarose gels (44).

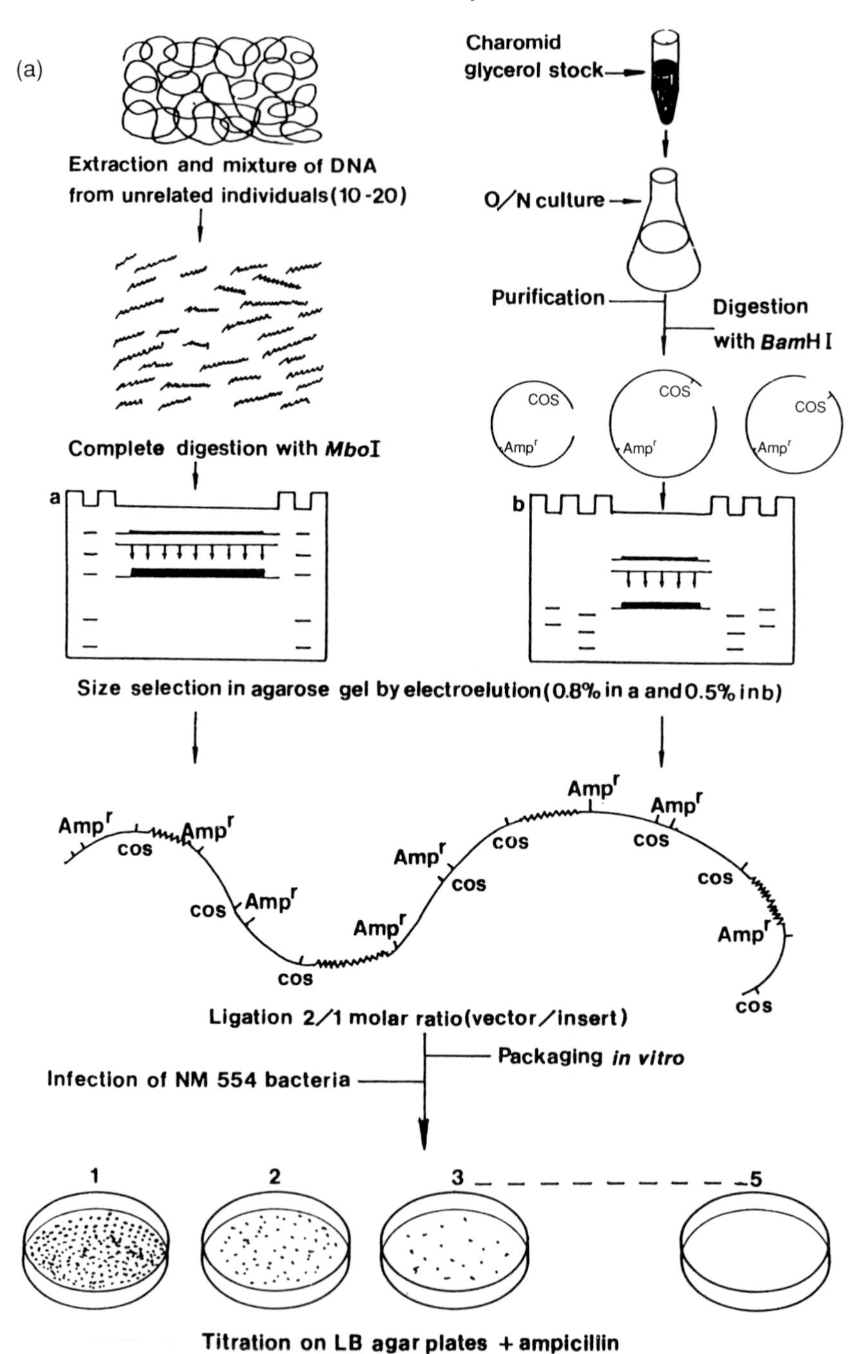

Figure 7. Strategy for cloning minisatelite sequences in charomid vectors. (a) Construction of the library. (b) Establishment of the library on microtitre plates.

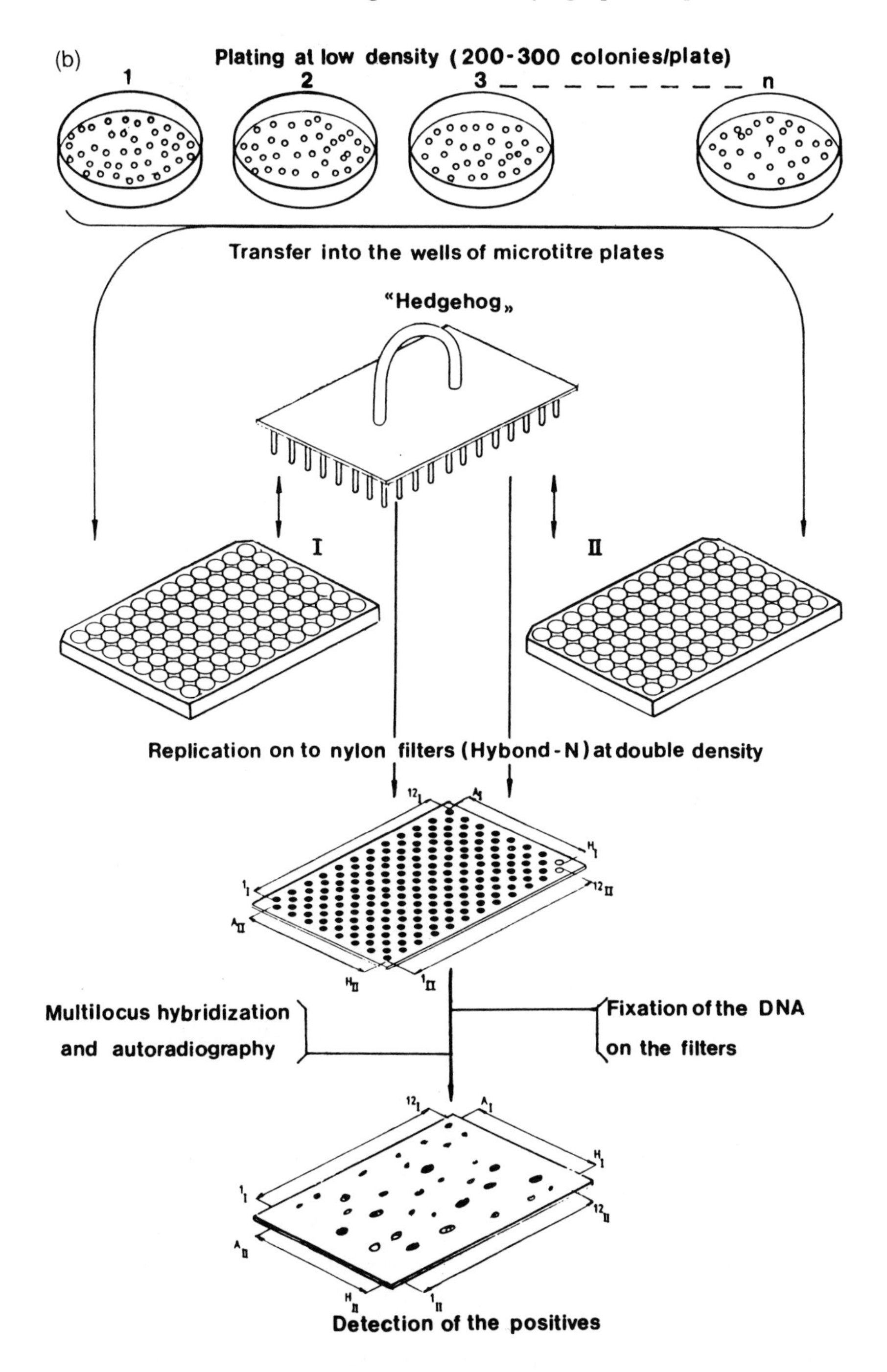
(b)
Plating at low density (200-300 colonies/plate)
1
2
3
n
Transfer into the wells of microtitre plates
«Hedgehog»
I
II
Replication on to nylon filters (Hybond-N) at double density
Multilocus hybridization
and autoradiography
Fixation of the DNA
on the filters
Detection of the positives

Protocol 4. Size fractionation of genomic DNA

Equipment and reagents

- Agarose (Sigma, Type I low EEO)
- 1 × TBE buffer pH 8.8 (see Section 2.3.2)
- Ethidium bromide (see Section 2.3.2)
- Gel mould (20 cm × 20 cm)
- Gel combs (2.5 mm thick) with a long central (14 cm) slot and at least two small (5 mm) slots at each end
- *Mbo*I-digested genomic DNA
- Lambda DNA cut with *Hin*dIII
- 254 nm UV hand lamp
- 1 × and 10 × TBE loading buffer
- Dialysis membrane, prepared by boiling for 5–10 min, storing in 70% ethanol at 4°C, and rinsing well in distilled water before use
- Blunt tweezers (such as manufactured by Millipore)
- 20 ml sterile tube (Sterilin)
- 2 M sodium acetate pH 5.2
- 100% and 80% ethanol
- TE, pH 7.5

Caution!

Several safety precautions should be taken:

- do not remove the membrane from the tank at voltages greater than 100 V
- always wear rubber gloves for insulation (we use two pairs)
- insulate tweezers and scalpel with electrical tape
- do not put hands in the buffer
- do not put tweezers in opposite electrode compartments

Method

1. Prepare agarose gel in the following way (for a 20 cm × 20 cm gel):
   (a) To prevent gel-slot leakage , make a base gel by pouring 75 ml of 2% agarose in 1 × TBE buffer pH 8.8 (see Section 2.3.2), 0.5 μg/ml ethidium bromide (see Section 2.3.2).
   (b) Clamp a comb at each end, 2 mm above the agarose base plate.
   (c) Pour 600 ml of 0.8% agarose in 1 × TBE buffer pH 8.8, 0.5 μg/ml ethidium bromide. When set, transfer the gel into an electrophoresis tank, and fill with TBE containing ethidium bromide until the buffer comes 2 mm above the surface of the gel. It is important later (step **10**) that the tank used will allow a voltage to be applied after removal of the lid.
2. Load 2 μg of lambda DNA cut with *Hin*dIII in 100 μl 1 × TBE loading buffer into the outermost side slots.
3. Load the *Mbo*I-digested DNA into the main slot. Run at 40 V overnight.
4. Cut a sheet of clean dialysis membrane to give two single layers 8 mm longer than the main slot and 8 mm wider than the depth of the gel.
5. Check the migration of the lambda DNA markers using a 254 nm UV hand lamp. If the DNA has migrated far enough, switch off the voltage and mark the limits of the size-selected DNA fraction by using a micropipette to inject the gel with 5 μl of 10 × TBE loading buffer, in line with the edge of the main slot.

6. Using a sharp scalpel, make incisions right dc gel plate. These incisions should be in front (l and at the rear (high molecular weight) of the siz Extend the incisions 1 cm either side of the main not to allow the incisions to extend to the side edges
7. Using two pairs of tweezers (blunt, such as manufact pore), insert one membrane into each incision. The men often curved; insert them such that the convex face of the n faces the loading slot. Care should be taken to avoid placing of the membrane in the lanes containing the marker DNA.
8. Run the DNA on to the dialysis membrane at 200 V.
9. When all of the size-selected fraction is loaded on to the front membrane, reduce the voltage to 100 V and extend the incision to one side of the gel with a scalpel. Gently tease the rear section of the gel backwards 1–2 mm, so that the rear of the membrane is no longer held in place by the gel. Wait a couple of minutes to ensure complete loading.
10. With the current still on, using tweezers at each end of the membrane, pick up the membrane, fold it bilaterally keeping the DNA side upwards, and transfer it into a 20 ml sterile tube as quickly as possible.
11. Pinch a small corner of the membrane by screwing down the lid of the tube, and centrifuge for 2 min at 2500 $\times$ *g*. Wash each side of the membrane with 200 μl of ultra-pure water, centrifuge again at 2500 $\times$ *g* for 2 min. Check with the UV hand lamp that all the DNA is at the bottom of the tube.
12. Remove the dialysis membrane and transfer the solution to one or more 1.5 ml microcentrifuge tubes. Avoid transferring pieces of membrane and agarose. Add 0.1 vol. 2 M sodium acetate (pH 5.2) and 2.5 vol. 100% ethanol. Centrifuge at 12300 $\times$ *g* for 5 min, and aspirate the ethanol solution.
13. Add 200 μl of 80% ethanol, tap the tube, and centrifuge for 5 min. Aspirate the ethanol and air-dry or vacuum-dry the pellet. Resuspend the pellet in a small volume (100–200 μl) of TE pH 7.5.
14. Add 0.1 vol. of 10 $\times$ TBE loading buffer and add 1 μg of ethidium bromide per 5 μl of final volume.
15. Because some low molecular weight fragments aggregate during electrophoresis, it is necessary to repeat the size fractionation. Prepare a second gel but use a comb with a shorter slot (2–4 cm) and repeat steps **1–13**.
16. Resuspend the DNA in 10 μl of TE pH 7.5 or ultra-pure H_2O.
17. Estimate the concentration by running a small aliquot against known amounts of *Mbo*I-digested genomic DNA on a minigel.

3.2 M e

.5 Selection and preparation of the vector

ost cloning vectors can, in theory, be used to clone minisatellite loci. However, bacteriophage lambda and its derivatives are not advisable as the high nstability of tandemly repeated sequences can lead to inviability of the phage during propagation. This protocol uses cosmid vectors known as charomids (59; supplied by the Japanese Cancer Research Resources Bank), which are particularly suitable for the cloning of size-selected DNA ranging from 2 to 45 kb. Members of the charomid family bear different copy numbers (1–23 copies) of a 2 kb spacer fragment linked head-to-tail in tandem arrays. The particular charomid used depends on the genomic size fraction chosen (*Figure 6*). For example, charomid 9-36 is too small (36 kb) to be packaged efficiently without an insert and can be used to clone fragments of 2–16 kb charomids contain a polycloning site with nine different enzyme recognition sequences, including the site recognized by *Bam*HI (*Figure 7*), and carry ampicillin and tetracycline resistance genes.

Because charomids themselves carry tandem repeats, they are unstable and must be propagated in *rec*A$^-$ bacteria (e.g. *Escherichia coli* ED8767 or NM554). Also, during the propagation of the vector, the number of generations of bacterial growth must be kept to a minimum and, after purification of the plasmid and digestion with *Bam*HI, the vector must be size selected. The vector is prepared by the alkaline lysis method (92), the volume of Luria broth used varying with the amount of cosmid desired. The charomid is selected by ampicillin resistance. Once prepared, a small aliquot of the sample is electrophoresed to estimate concentration, and 25 μg or more of the vector is digested with at least 2 units enzyme per microgram of DNA in the presence of the manufacturer's buffer, 4 mM spermidine trihydrochloride, and 0.1 mg/ml RNase A. Incubate at 37°C for 3–5 h. Check the digestion of a small aliquot on a 0.5% gel. Then prepare a 0.5% gel with a comb with a 5 cm slot, and recover the appropriate size fragment (e.g. 36 kb if charomid 9-36 is being used; see *Figure 7*) as in *Protocol 4*, using lambda DNA samples cut separately with *Xho*I and *Hin*dIII as size markers. The vector is then resuspended in 10–20 μl TE, and an aliquot is electrophoresed to estimate concentration as in Section 2.3.1.

3.2.6 Ligation and packaging

Ligation is carried out at high concentrations of DNA to promote the formation of recombinant DNA molecules. A vector:insert molar ratio of 2:1 seems optimal. Before ligation careful estimation of the vector and insert concentrations is important, either by electrophoresis on a 0.5% gel or by fluorimetry (see Section 2.3.1). The principle of *in vitro* packaging has been described elsewhere (44; Chapter 4). Briefly, it is performed in a concentrated lysate of two induced lysogens; one is blocked at the pre-head stage and the other is prevented from forming any head structure. In the mixed lysate, genetic complementation occurs and exogenous DNA is packaged into the phage.

Protocol 5. Single-locus minisatellite ligation and *in vitro* packaging

Reagents

- 10 × ligase mix: 10 μl 1 M Tris–HCl pH 7.5, 2 μl 1 M $MgCl_2$, 2 μl 1 M dithiothreitol (DTT; Sigma) in 0.01 M sodium acetate pH 5.2; 6 μl ultra-pure H_2O
- 10 mM ATP (Sigma)
- T4 ligase (1.5 Weiss unit; Sigma)
- 30 mM spermidine trihydrochloride
- Ultra-pure H_2O
- Packaging mix (e.g. Stratagene Gigapack)

Method

1. Prepare a ligation reaction mix from the following ingredients:
   - vector:insert DNA in a 2:1 molar ratio (e.g. 1.2 μg charomid 9-36: 200 ng 6–16 kb insert)
   - 1.5 μl 10 × ligase mix
   - 1.5 μl 10 mM ATP
   - 1.5 Weiss unit T4 ligase
   - 1.5 μl 30 mM spermidine trihydrochloride
   - ultra-pure water to a volume of 15 μl
2. Mix; incubate the sample and a control sample[a] at 4°C for 2 days, and then add an additional 0.45 μl of 10 mM ATP. Leave for another 1–2 days at 4°C.
3. Assess the success of the ligation by electrophoresing the control and an aliquot of the sample together with undigested control fragments and vector:insert mix. Store at –20°C.
4. For *in vitro* packaging follow the manufacturer's instructions as precisely as possible. Take care not to introduce air bubbles into the packaging mix. Store the packaged ligation at 4°C; do not freeze. Packages can be stored this way for several months.

[a] A control can be carried out, for example, by ligating 1 μg of lambda DNA/*Hin*dIII frgaments.

As previously mentioned, a *rec*A$^-$ bacterial host gives the greatest likelihood of stable propagation of minisatellite sequences. We have used *E. coli* strain NM554 (93) (available from Stratagene), which is also *mcr*A$^-$ and *mcr*B$^-$, two site-specific systems known to digest eukaryotic DNA containing methylated cytosine (93). The protocol below is adapted from standard procedures that can be found in refs 44 or 45 (see also Chapter 4). All solutions should be filtered and sterile.

Protocol 6. Infection and titration

Reagents

- LUB (LB broth): 1% NaCl, 0.5% Difco Yeast Extract, 1% Difco Bacto-tryptone in distilled water
- LUA: LB + 1.5% agar (Difco)
- Phage dilution buffer (per litre): 5.8 g NaCl, 2.0 g $MgCl_2$, 50 ml Tris–HCl pH 7.5, 5 ml 2% gelatin

Method

1. Inoculate 1–2 ml of Luria broth (LUB) containing 0.2% maltose and 10 mM $MgSO_4$ with a single colony of NM554 or another appropriate bacterial host, and place in a shaking incubator overnight at 37 °C together with an uninfected control.
2. Dilute the overnight culture 1:10 in LUB and shake at 37 °C for 3 h.
3. Prepare five 9 cm diameter Petri dishes with LUA medium (agar plates) containing 50 μg/ml ampicillin (stored at 4 °C for 1–2 weeks maximum; do not add to agar above 50^°C).
4. Prepare five tubes by serial dilution as follows.
   (a) Add the following amounts of phage dilution buffer to each tube: tube 1, 80 μl; tube 2, 100 μl; tube 3, 100 μl; tube 4, 99 μl; control tube, 100 μl.
   (b) Add 21 μl packaged ligation to tube 1.
   (c) Perform the serial dilution by adding 1 μl from tube 1 to tube 2. Mix and then add 1 μl from tube 2 to tube 3. Mix and then add 1 μl from tube 3 to tube 4. Mix. The dilution factors for each tube are: tube 1, 5; tube 2, 5×10^2; tube 3, 5×10^4; tube 4, 5×10^6.
5. Add 200 μl of bacterial culture to each tube (from step **2**) which must be in log phase growth.
6. Incubate for 20 min at room temperature without shaking.
7. Add 700 μl of LUB without ampicillin and shake at 37 °C for 1 h.
8. Spread 200 μl of each infection (steps **4–7**) on to a different agar plate. Incubate the five plates at 37 °C to dry, then invert and leave at 37 °C overnight.
9. Choose a plate where the colonies can be easily counted, and determine the approximate titre of the total library suspension: total number of clones × dilution factor × 25. Your library should contain 10^5–10^6 clones. No colonies should grow on the control plate.

3.3 Establishment of library on micotitre plates

3.3.1 Library size

The library size required can be estimated using the following parameters:

(a) the percentage of the genome in the recovered fraction, corrected for efficiency of recovery (this parameter is needed when size selecting for specific sequences and is difficult to estimate, but the percentage of the genome in the size fraction will rarely exceed 1%);
(b) the average size of the selected DNA;
(c) the total genome size of the species of interest.

From these data, it is possible to calculate the number of recombinant colonies that have to be screened in order to find a specific minisatellite with a certain level of probability. This is given by the formula

$$N = \frac{\ln(1 - P)}{\ln(1 - f)}$$

(see ref. 44 and Chapter 7) where P is the desired probability, f is the fractional proportion of the size selected genome in a single recombinant, and N is the number of recombinants. For example, to achieve a 95% probability of representing a specific locus in a library of fragments averaging 12 kb average fragments, with a size fraction of 1% of a total haploid genome equivalent of 10^9 bp, approximately 2500 colonies must be screened (see equation below), or two genome equivalents of this size fraction.

$$N = \frac{\ln(1 - 0.95)}{\ln(1 - (12 \times 10^3/10^7))} = 2495$$

A correction must be made for the proportion of non-recombinants in the library (see *Protocol 6*). A good library has 60–80% recombinants.

Protocol 7. Establishing a microtitre library and replicating on to nylon

Equipment and reagents

- LUA (see *Protocol 6*) plates with 50 μg/ml ampicillin
- Parafilm
- Aluminium foil
- 96-well microtitre plates (e.g. Flow Laboratories)
- Sterile LUB (*Protocol 6*) containing 15% glycerol and 50 μg/ml ampicillin
- Nylon membrane
- 'Hedgehog' microtitre replicator (e.g. Sigma R2508)
- 70% ethanol

Method

A. *Establishing the library*

1. Make 10–20 LUA (see *Protocol 6*) plates with 50 μg/ml ampicillin, and plate out the infected bacteria at a titre determined (from *Protocol 5*) to

Protocol 7. *Continued*

obtain 200–300 colonies per plate. Spread the bacterial suspension evenly over the plate.

2. Grow overnight at 37 °C, until the colonies are 1–2 mm in diameter. Close the plate, sealing with Parafilm and wrap the plates in aluminium foil, storing inverted at 4 °C until required.
3. Using standard 96-well microtitre plates, add 100 μl sterile LUB containing 15% glycerol and 50 μg/ml ampicillin to each well, leaving the bottom right-hand well empty for orientation. This must be done under sterile conditions in a sterile flow cabinet.
4. Transfer the appropriate number of colonies using sterile pipette tips and a 200 μl micropipette set at 20 μl. Mix the bacteria evenly in the medium, number the plates and store at –20 °C.

B. *Replication of the library on to nylon*

1. Cut appropriate sized pieces of membrane. Labelling them and place them on LUA trays containing 50 μg/ml ampicillin.
2. Bathe the prongs of a 'hedgehog' microtitre replicator in 70% ethanol, and then flame off the ethanol to sterilize and leave to cool.
3. Dip the 'hedgehog' replicator in a microtitre plate, agitate gently, and place on the surface of a membrane on the surface of the LUA so that all the prongs are on the membrane. Repeat the process for each plate. If necessary, to save membrane, two plates can be replicated on to one membrane by offsetting the points of contact of the hedgehog prongs with respect to each other (*Figure 6*).
4. Incubate the plates overnight at 37 °C. Once the colonies are clearly visible on the membranes, the DNA can be fixed to the membranes by microwaving (see ref. 94).

3.3.2 Ordering and replicating the library

The transfer of the library into an ordered array brings several advantages. Screening of the library by a single round of hybridization, the easy identification of positives including those that hybridize only weakly, comparison of the sets of positively hybridizing clones detected by different multilocus probes, and the identification of multiply detected clones can all be achieved much more conveniently than on Petri dishes. Once positives have been identified and characterized, the relatively long shelf-life of the plates (at least a year at –20 °C) allows the library to be screened with single-locus probes. This is called 'probe walking' and can generate useful numbers of new probes (30). The major drawback is the labour required for transferring the colonies.

3.3.3 Screening the library

Hybridization is carried out as for multilocus DNA fingerprinting with nylon membranes (*Protocol 3*) with the stringency varying as appropriate for the probe. However, as the charomid contains sequences derived from plasmids pBR322 and pUC18 and a cosmid, pTBE, it is important to ensure that the probe is pure insert sequence. If this precaution is not taken, backgound hybridization can be severe enough to obscure true positives. A competitor DNA consisting of charomid vector DNA and *E. coli* DNA at 1 μg/ml and 2 μg/ml respectively (boiled for 10 min just before use) can be added to the hybridization mix. If this does not reduce the background, then increase the washing stringency. Include a blank piece of membrane in the hybridization (*Protocol 3*) and stop washing once this membrane is at background radiation levels.

After washing, check the filters with a ^{32}P monitor; the counts should be low (usually < 5 c.p.s.), and distributed irregularly on the membranes due to the random distribution of the positives. Expose the membranes overnight initially, and then for 2–8 days (to detect faint positives). The autoradiograph can be oriented using marks from radioactive ink placed on tape next to the membranes. Before re-probing, strip the membranes as in *Protocol 3D* and expose overnight; the autoradiograph should be clear.

3.4 Molecular characterization of positive clones

Inserts purified from positively hybridizing clones do not always detect a recognizable single-locus pattern, and some single-locus patterns do not show useful variation (29,30). For these reasons, each positive clone has to be tested and characterized before application to any study.

3.4.1 Preparation of the insert

A small-scale amplification of a positive clone yields enough insert DNA for the probing experiments required for characterization. The isolated charomid is digested with *Sau*3AI (not *Mbo*I or *Nde*II, which are inhibited by the *dam* methylation of AT dinucleotides which occurs in most strains of *E. coli*, including NM554). The insert is then recovered by electro-elution on to dialysis membrane. Up to 10–20 different sequences can be isolated at one time. For use as a probe, the insert is then labelled by oligolabelling (*Protocol 2*). The isolation protocol is detailed in *Protocol 8*.

Protocol 8. Small-scale preparation of the insert

Materials

- 3 ml LUB (*Protocol 6*) containing 50 μg/ml of ampicillin
- Glycerol, sterilized by autoclaving
- Lysis solution: 25 mM Tris–HCl pH 8.0, 10 mM EDTA pH 8.0, 50 mM glucose, 2 mg/ml lysozyme (added just before use)

Protocol 8. *Continued*

- 0.2 M NaOH, 0.1% (w/v) SDS
- 3 M potassium acetate pH 4.8
- 100% and 80% ethanol
- Propan-2-ol
- 7.5 M ammonium acetate
- Charomid DNA
- *Sau*3AI
- Equipment and reagents for gel preparation and electrophoresis (*Protocol 4*)

A. *Preparation of the insert*

1. Select positively hybridizing clones of different intensities. Thaw the library plates and inoculate 3 ml LUB (*Protocol 6*) containing 50 μg/ml of ampicillin with 5 μl from the required well. Grow culture overnight at 37 °C with strong shaking.
2. Make a permanent stock of recombinant by transferring 200–850 μl of bacterial culture into a sterile tube containing sufficient glycerol (sterilized by autoclaving) to form a 15–20% solution, and store at –70 °C.
3. Fill a 1.5 ml microcentrifuge tube with culture and pellet the bacteria by centrifuging for 30 sec (12 300 × *g*). Drain off the supernatant and refill the microcentrifuge tube with the remainder of the culture, then centrifuge again.
4. Resuspend the pellet in 100 μl lysis solution. Leave at room temperature for 5 min.
5. Add 200 μl 0.2 M NaOH, 1% SDS and mix gently. Incubate on ice for 5 min.
6. Add 150 μl 3 M potassium acetate (pH 4.8), mix gently, incubate on ice for 8 min.
7. Centrifuge at 12 300 × *g* for 10 min. Transfer the supernatant to a new microcentrifuge tube.
8. Add 500 μl propan-2-ol, and mix.
9. Centrifuge for 10 min at 12 300 × *g*, aspirate the supernatant and re-dissolve the pellet in 100 μl of water.
10. Add 50 μl of 7.5 M ammonium acetate, mix gently, and leave at room temperature for 10 min.
11. Centrifuge at 12 300 × *g* for 10 min and transfer the supernatant to a new microcentrifuge tube.
12. Add 300 μl cold 100% ethanol; centrifuge at 12 300 × *g* for 10 min.
13. Pour off the supernatant. Wash the pellet in 100 μl 80% ethanol and centrifuge at 12 300 × *g* for 5 min. Aspirate supernatant and air-dry or vacuum-dry.
14. Dissolve the pellet in 20 μl of ultra-pure water. Store at –20 °C or proceed directly to restriction.

B. *Digestion of charomid and purification of insert*

1. Digest the charomid DNA with 20 units of *Sau*3AI for 3–5 h at 37 °C. As cloned inserts are recovered from gels RNase is not used.
2. Check the restriction by electrophoresing an aliquot (Section 2.2.1), add TBE loading buffer, and load on to a gel suitable for electro-elution (*Protocol 4*) with 5–10 mm slots and without a base gel (see *Protocol 4*). Use an appropriate DNA size marker.
3. Ethanol precipitate the insert and re-dissolve in 20 μl water. Estimate the concentration of an aliquot on a gel, and dilute to a final concentration of 5–10 ng/μl. Oligolabel 10–20 ng of the insert as previously described (*Protocol 2*).

3.4.2 Screening the insert for locus specificity

i. Electrophoresis

In order to test the inserts for variability, they should be hybridized against Southern blots of *Mbo*I-digested genomic DNA from a standard panel of four unrelated individuals, and possibly an individual from a closely related species to test for the presence of possibly homologous sequences. Repeats of this group are loaded together on 20 cm × 20 cm gel with a one lane gap between each panel, and electrophoresed until the 564 bp lambda/*Hin*dIII fragment has migrated to the end of the gel. After DNA transfer, the membranes are cut into strips containing one panel each, and probed.

ii. Hybridization at high stringency

Single-locus probes are hybridized at high stringency to avoid the potential detection of many loci (as in a DNA fingerprint). However, sometimes even this is not enough to prevent the detection of several loci showing close similarity to at least part of the probe sequence. Cross-hybridization might, for example, be due to the presence of an interspersed repeat sequence within the DNA flanking a cloned minisatellite. To avoid this, include whole genomic competitor DNA from the species under examination (preparing by boiling for 20 min to shear and denature it) at a concentration of 5 μg/ml (more may be required for some probes).

The conditions suitable for nylon hybridization are described elsewhere (*Protocol 3*) with the modification of the addition of specific competitor DNA, and, normally, the adjustment of the hybridization temperature to 65°C. Washing is much more stringent than for multilocus fingerprinting. Once hybridization is complete, the membranes are washed at 65°C in 40 mM sodium phosphate pH 7.2, 1% SDS for 10 min, and then in 0.1 × SSC, 0.01% SDS for 10 min. It is important to monitor the radioactivity on the membranes thoroughly, and to ensure that the blank membrane is completely clear.

The results give a characteristic pattern or 'signature' for each probe

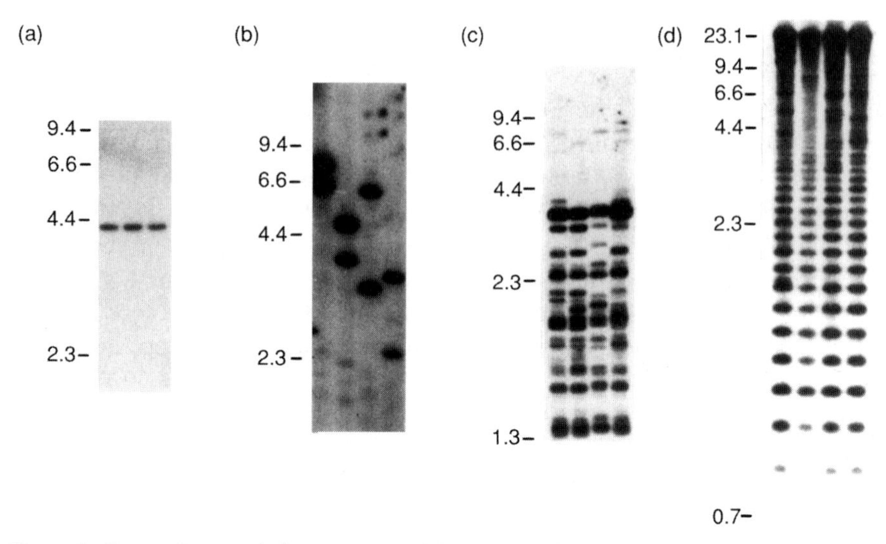

Figure 8. Some characteristic patterns or 'signatures' of probes isolated from positively hybridizing clones detected after screening charomid libraries of different avian species with multilocus probes. House sparrow (*Passer domesticus*) c*Pd*oxx—detection of a satellite sequence. Indian peafowl (*Pavo cristatus*) c*Pcr*30—detection of a monomorphic locus. Chicken (*Gallus gallus*) c*Gga*22—detection of a multi-band pattern corresponding to one or several loci. Ruff (*Philomachus pugnax*) c*Ppu*6—detection of a highly polymorphic minisatellite showing eight alleles in four unrelated individuals.

through which repeated isolation of the same sequence may be detected (29). *Figure 8* shows some patterns detected by inserts of positively hybridizing clones. Once useful variation has been detected with a probe, Mendelian inheritance, heterozygosity, and mutation rate can be ascertained by probing the appropriate blots (see Section 3.5). Therefore, reference membranes of large families and of 20–25 unrelated individuals should be prepared (Section 2.3). Interesting clones should then be amplified on a larger scale.

3.4.3 Probe nomenclature

The charomid system described above has now yielded many single-locus probes in a number of species (e.g., 29,30,87–90), and some of these probes can be used not only in the species of origin, but also in other species (e.g., 18). A standard nomenclature for such sequences including reference to the species of origin has been suggested (30). For example, the prefix c*Pcr*N refers to the *N*th recombinant charomid isolated from *Pavo cristatus*, and c*Gga*N refers to the *N*th isolated from *Gallus gallus*. Plasmid subclones are designated p*Pcr*N, and so on. Where a species' binomial abbreviation is potentially ambiguous, we suggest using additional letters for the new species (so, for example, *Podiceps cristatus* might be abbreviated to c*Pcri*N instead of just c*Pcr*N). 'MS' may be incorporated into the name to indicate that a particular clone is a minisatellite (e.g. c*Gga*MS32).

3.4.4 Minisatellite single-locus probes characterized to date

Many of the minisatellite probes published so far have been cloned at the University of Leicester using the protocol described here. *Table 2* details the probes that have been cloned to date, and that are available for use from the source laboratories.

3.5 Analysis of single-locus DNA fingerprint data

Ideally, single-locus minisatellite data should be analysed using the allele frequencies at each locus. The estimation of these frequencies is possible for minisatellite loci that have a small number of alleles, and here the mathematical tools developed for the study of allozyme polymorphism can be applied (125). Account will need to be taken of the very high and variable rates of mutation to detectably different mobility states, through the loss or gain of minisatellite repeat units (78) that occur at these loci. However, for highly polymorphic loci, accurate determination of the frequency of each allele would require the screening of a very large number of individuals. In practice this may not be feasible. In these cases, mean allele frequency can be determined for each locus, and the data analysed under the same assumptions as in multilocus DNA fingerprinting.

3.5.1 Probabilities of identity, exclusion, and inclusion

The probability of identity is the probability that two individuals in a population will share a genotype by chance. In parentage analysis, the exclusion probability, E, is the probability of detecting an incorrectly assigned parent. The false inclusion probability is the probability of failing to detect an individual that has been incorrectly assigned as a parent ($= 1 - E$). The false exclusion probability (i.e., the probability of excluding a true parent) approximately equals $2m$, where m is the mutation rate per allele.

Assuming that all alleles have the same frequency, q, the probability that two unrelated individuals will share the same genotype at a locus is $q^2(2 - q)$ (83). Similarly, and assuming no mutation, the probability that an offspring's paternal allele is also present in an unrelated individual (the probability of false inclusion) can be estimated as $2q - q^2$ (83).

More precise calculations can be undertaken using the actual allele frequencies. The probability that two unrelated individuals share the same genotype at a given locus is (30)

$$\sum_{i=1}^{n} (q_i^2)^2 + \sum_{\substack{i,j=1 \\ (i>j)}}^{n} (2q_iq_j)^2$$

where q_i is the frequency of the ith of n alleles

Table 2. Available cloned single-locus polymorphic minisatellites (after ref. 35)

Group	Species	No. of probes	Reference
Arachnids	pseudoscorpion (*Cordylochernes scorpiodes*)	2[a]	88
Insect	*Drosophila mauritiana*	1	95
Fish	Atlantic salmon (*Salmo salar*)	6	96,97
		4	98,99
		8[a]	100[c]
	brown trout (*Salmo trutta*)	11	101
	Chinook salmon (*Oncorhynchus tshawytscha*)	1	102
	tilapia (*Oreochromis niloticus*)	2	103
	sea bass (*Dicintrarchus labrax*)	6[a]	unpublished[e]
Amphibian	common toad (*Bufo bufo*)	3[a]	90[c]
Reptile	marine iguana (*Amblyrhynchus cristatus*)	3[a]	unpublished[c]
Bird	chicken (*Gallus gallus*)	60[a]	86[c]
	peafowl (*Pavo cristatus*)	10[a]	30[c]
	ruff (*Philomachus pugnax*)	13[a]	104[c]
	greylag goose (*Anser anser*)	5[a]	unpublished[c]
	Brent goose (*Branta bernicla*)	5[a]	unpublished[c]
	house sparrow (*Passer domesticus*)	16[a]	105,106[c]
	reed bunting (*Emberiza schoeniclus*)	4[a]	18,107[c]
	Eastern bluebird (*Sialia sialis*)	3[a]	unpublished[c]
	pied flycatcher (*Ficedula hypoleuca*)	8[a]	unpublished[c]
	blue tit (*Parus caeruleus*)	9[a]	108
	willow warbler (*Phylloscopus trochilus*)	1	13,109
	starling (*Sturnus vulgaris*)	4[a]	unpublished[c]
	golden parakeet (*Aratiga guarouba*)	2[a]	unpublished[c]
	peregrine (*Falco peregrinus*)	5[a]	110
	merlin (*Falco columbarius*)	3[a]	110
	kestrel (*Falco tinnunculus*)	1[a]	110
	red kite (*Milvus milvus*)	3[a]	111,112
	golden eagle (*Aquila chrysaetos*)	2[a]	110
Mammal	domestic dog (*Canis canis domesticus*)	7[a]	113
	domestic cat (*Felis cattus domesticus*)	2	114
	badger (*Meles meles*)	8[a,b]	unpublished[c]
	grey seal (*Halichoerus grypus*)	2[a]	unpublished[c]
	minke whale (*Balaenoptera acutorostrata*)	6[a]	25,115
	pilot whale (*Globicephala vulgaris*)	1[a]	115[c]
	red squirrel (*Sciurus vulgaris*)	2[f]	unpublished[d]
	mouse (*Mus musculus domesticus*)	2	116,117
	cattle (*Bos taurus*)	36	118
	pig (*Sus scrofa domestica*)	3[a]	119,120
		2	121,122
	human (*Homo sapiens*)	>50	83,123,124
		23[a]	29

[a] Cloned into charomid vector.
[b] All monomorphic, or nearly so, in English badgers.
[c] Available from Terry Burke, Department of Zoology, University of Leicester, Leicester LE1 7RH, UK.
[d] D. T. Parkin, Department of Genetics, Queen's Medical Centre, University of Nottingham, Nottingham NG7 2UH, UK.
[e] P. Benedetti, Department of Biology, Universita Degli Studi di Padova, via Trieste 75, 35121 Padova, Italy.
[f] Unstable loci showing a high germline mutation rate (2.5–3.5%) and somatic instability.

Chakravarti and Li (126, 127) derived formulae to describe the expected mean exclusion probability in cases of misassigned paternity or parentage under the assumption of Hardy–Weinberg equilibrium. In studies in wild bird populations, for example, genetic paternity might be misassigned as a result of a successful extra-pair copulation with the mother, and misassigned parentage might result from conspecific nest parasitism or adoption. They showed that the paternal exclusion probability at a locus with K alleles can be approximated by

$$E(K) = a_1 - 2a_2 + a_3 + 3(a_2a_3 - a_5) - 2(a_2{}^2 - a_4)$$

where $a_1 = \sum_{k=1}^{k} P_i^k$ and P is the frequency of the ith allele.

This can be extended to situations where the mother's genotype is unknown using:

$$PE_i = (1 - a - b)^2$$

where PE_i is the probability of exclusion at the ith locus, and a and b are the frequencies of the alleles found in the offspring.

The combined probability of over all loci (PE_C) can be derived for each method (128),

$$PE_C = 1 - \Pi\,(1 - PE_i).$$

This gives the proportion of randomly selected males which would be excluded based on the population allele frequencies and makes the same assumptions about independence among loci as in the multilocus analysis.

3.5.2 Interpretation

i. Mutation rate

Minisatellite loci can have very high mutation rates. Studies in humans (78) and in the willow warbler (*Phylloscopus trochilus*) (13) have indicated that the mutation rate can be as high as 0.05 per gamete for highly polymorphic minisatellite loci (heterozygosity, 0.97–0.99). Such loci used in parentage studies are subject to a high probability of false exclusion, and results should be confirmed with other single-locus or multilocus data (13,34). Conversely, pedigree analysis using probes isolated in the Indian peafowl (*P. cristatus*) failed to detect allelic mutations for five ministellites showing heterozygosity values ranged between 22 and 78%, suggesting that the mean mutation rate per locus is too low (< 0.004, 95% confidence maximum) (30). In humans it has been demonstrated that mutation rate increases with heterozygosity following the neutral mutation/random drift model, and is practically equal to zero for loci showing heterozygosity values less than 95% (78). The relationship between heterozygosity and mutation rate is dependent on the historical effective population size. The heterozygosity below which mutations are not expected to be observed will therefore vary across species.

ii. Linkage

Although multilocus segregation analyses indicate that in general minisatellite loci sort independently (e.g. refs 3,6,10), clustering of such loci has also been observed (63,129). Linkage between loci should therefore be tested by pedigree analyses where at least one parent is heterozygous at both loci. The most widely used method is the maximum likelihood method or lod score (130):

$$Z(\Theta) = \log_{10}(L(\Theta)/L(0.5))$$

The method consists of estimating the recombination fraction, Θ, and testing whether an observed estimate, $\hat{\Theta}$, is significantly smaller than 50% (null hypothesis of free recombination which corresponds to two unlinked loci) (130). The recombination fraction is estimated as the likelihood of the observations, $L(\Theta)$, which is the probability of the occurrence of the phenotypes of all the individuals. We calculate $Z(\Theta)$ for different values of Θ ($\Theta = 0, 0.001, 0.05, 0.1, 0.2, 0.3, 0.4, 0.5$) and Θ is selected according to which $Z(\Theta)$ is largest. If $\Theta < 0.5$ the lod score will be positive, and a conventional rule is to conclude that loci are linked wherever lod scores are above 3, corresponding to odds of linkage of 1000:1 (124). Conversely, if $\Theta = 0.5$ the lod score is negative, and the convention is to conclude a lack of linkage when $Z(\Theta)$ is less than –2. For values giving a lod score between these limits, no definite conclusions can be drawn, and more data should be collected (125). Two situations can occur: either the linkage phase of the double heterozygote parent is known, i.e. association between the four alleles of the two loci has been determined (through, for example, the analysis of a three-generation pedigree), or the linkage phase is unknown. Where the phase is known, the offspring phenotypes occur independently of each other and the lod score is calculated as (130)

$$Z(\Theta) = \frac{\Theta^r(1-\Theta)^{n-r}}{(0.5)^n}$$

or

$$Z(\Theta) = r\log_{10}(2\Theta) + (n-r)\log_{10}(2(1-\Theta))$$

where r is the number of recombinants and n is the total number both recombinants and non-recombinants in the pedigree. If the phase is unknown, it is generally assumed that each phase is equally likely and the offspring phenotypes can no longer be assumed to be independent (except for $\Theta = 0.5$) and the lod score is calculated as (130)

$$Z(\Theta) = \log_{10}[2^{n-1}(\Theta^r(1-\Theta)^{n-r} + (1-\Theta)^r\Theta^{n-r})]$$

The results from different families can be combined and the overall lod score is the sum of the separate scores. $Z(\Theta)$ may be derived by hand, but it is easier with the aid of a computer program calculating the lod scores for given values of the recombination fraction. Simple examples of calculations of lod scores can be seen in ref. 130.

iii. Sex-linked loci

Minisatellite loci have been shown to be sex-linked in several species (e.g., 29,36,86). Where the locus is linked to the X (or Z) chromosome, all individuals of the heterogametic sex (XY or ZW) will appear as hemizygous (one band only). If the locus is on the Y or W chromosome, the homogametic sex (XX or ZZ) will lack the locus altogether.

iv. 'Null' alleles

The presence of null alleles (in this case alleles actually present but apparently faintly detected or undetected by the probe) has been demonstrated in minisatellite with minisatellite loci isolated in human (29) and in the chicken, *Gallus gallus* (86). These can be detected if during a pedigree analysis an apparently homozygous parent fails to transmit to one or several of the offspring, or if during population censuses some individuals appear to have bands, and others do not.

4. Conclusions

Multilocus and locus-specific DNA fingerprinting can play different but complementary roles in molecular analyses of populations. Analyses of individuality and reproductive success where there are a limited number of progenitors to be screened seem to be tackled most effectively by multilocus DNA fingerprinting. However, where the mating system becomes more complex, and where comparison between populations is required, single-locus probes of the appropriate levels of variation can play an important role. The development of polymorphic single-locus microsatellite markers (see Chapter 7 and Appendix 4) has, however, proved be very important. In many situations microsatellites are possibly more appropriate than minisatellites (particularly where there are small amounts of material available, or with degraded DNA, such that PCR-based methods are necessary). However, when there is good quality DNA available, and a highly polymorphic system is required, minisatellite probes may prove advantageous, especially where microsatellites are not available or are less polymorphic (131).

Acknowledgements

We would like to acknowledge the following people: Iris van Pijlen, Elaine Cairns, Louise Jenkins, Simon Bailey, Eddie Needham, Janine Dann, and Andrew Krupa for advice and assistance in the laboratory; John Armour and Alec Jeffreys for invaluable advice and encouragement; John Brookfield for discussions about statistics; and Mr R. Léonard and Mr T. Dennett for help with the figures.

References

1. Jeffreys, A.J., Wilson, V., and Thein, S.L. (1985). *Nature*, **314**, 67.
2. Jeffreys, A.J., Wilson, V., and Thein, S.L. (1985). *Nature*, **316**, 76.
3. Welsh, J. and McClelland (1990). *Nucleic Acids Res.*, **18**, 7213.
4. Vos, P., Hogers, R., Bleeker, M., Reijans, M., Vandelee, T., Hornes, M., Frijters, A., Pot, J., Peleman, J., Kuiper, M., and Zabeau, M. (1995). *Nucleic Acids Res.*, **23**, 4407.
5. Nybom, H., Rogstad, S.H., and Schaal, B.A. (1990). *Theor. Appl. Genet.*, **79**, 153.
6. Jeffreys, A.J., Brookfield, J.F.Y., and Semeonoff, R. (1985). *Nature*, **317**, 818.
7. Burke, T. and Bruford, M.W. (1987). *Nature*, **327**, 149.
8. Burke, T., Davies, N.B., Bruford, M.W., and Hatchwell, B.J. (1989). *Nature*, **338**, 249.
9. Wetton, J.H., Carter, R.E., Parkin, D.T., and Walters, D. (1987). *Nature*, **327**, 147.
10. Birkhead, T.R., Burke, T., Zann, R., Hunter, F.M., and Krupa, A.P. (1990). *Behav. Ecol. Sociobiol.*, **27**, 315.
11. Westneat, D.F. (1990). *Behav. Ecol. Sociobiol.*, **27**, 67.
12. Rabenold, P.P., Rabenold, K.N., Piper, W.H., Haydock, J., and Zack, S.W. (1990). *Nature*, **348**, 538.
13. Gyllensten, U.B., Jakobsson, S., and Temrin, H. (1990). *Nature*, **343**, 168.
14. Nybom, H. and Schaal, B.A. (1990). *Theor. Appl. Genet.*, **79**, 763.
15. Packer, C., Gilbert, D.A., Pusey, A.E., and O'Brien, S.J. (1991). *Nature*, **351**, 562.
16. Kempenaers, B., Verheyen, G.R., Van den Broeck, M., Burke, T., Van Broeckhoven, C., and Dhondt, A.A. (1992). *Nature*, **351**, 494.
17. Pemberton, J.M., Albon, S.D., Guinness, F.E., Clutton-Brock, T.H., and Dover, G.A. (1992). *Behav. Ecol.*, **3**, 66.
18. Dixon, A., Ross, D., O'Malley, S.L.C., and Burke, T. (1994). *Nature*, **371**, 698.
19. Krokene, C., Anthonisen, K., Lifjeld, J.T., and Amundsen, T. (1996). *Anim. Behav.*, **52**, 405.
20. Jeffreys, A.J., Wilson, V., Kelly, R., Taylor, B.A., and Bullfield, G. (1987). *Nucleic Acids Res.*, **15**, 2823.
21. Gilbert, D.A., Lehman, N., O'Brien, S.J., and Wayne, R.K. (1990). *Nature*, **344**, 764.
22. Kuhnlein, U., Zadworny, D., Dawe, Y., Fairfull, R.W., and Gavora, J.S. (1989). *Theor. Appl. Genet.*, **77**, 669.
23. Heath, D. (1995). *Mol. Ecol.*, **4**, 389.
24. Robinson, N.A., Murray, N.D., and Sherwin, W.B. (1993). *Mol. Ecol.*, **2**, 195.
25. Van Pijlen, I.A., Amos, B., and Burke, T. (1995). *Mol. Biol. Evol.*, **12**, 459.
26. Brock, M.K. and White, B.N. (1992). *Proc. Natl Acad. Sci. USA*, **89**, 11121.
27. Fleischer, R.C., Tarr, C.L., and Pratt, T.K. (1994). *Mol. Ecol.*, **3**, 383.
28. Wong, Z., Wilson, V., Patel, I., Povey, S., and Jeffreys, A.J. (1987). *Ann. Hum. Genet.*, **51**, 269.
29. Armour, J.A.L., Povey, S., Jeremiah, S., and Jeffreys, A.J. (1990). *Genomics*, **8**, 501.
30. Hanotte, O., Burke, T., Armour, J.A.L., and Jeffreys, A.J. (1991). *Genomics*, **9**, 587.
31. Amos, B., Schlotterer, C., and Tautz, D. (1993). *Science*, **260**, 670.
32. Roy, M.S., Geffen, E., Smith, D., Ostrander, E., and Wayne, R.K. (1994). *Mol. Biol. Evol.*, **11**, 553.
33. Paetkau, D. and Strobeck, C. (1994). *Mol. Ecol.*, **3**, 489.

34. Burke, T., Hanotte, O., Bruford, M.W., and Cairns, E. (1991). In *DNA fingerprinting: approaches and applications* (ed. G. Dolf, T. Burke, A. Jeffreys, and R. Wolff), p. 154. Birkhäuser Verlag AG, Basel.
35. Burke, T., Hanotte, O., and Van Pijlen, I.A. (1996). In *Molecular genetic approaches in conservation* (ed. T.B. Smith and R.K Wayne), p. 251. Oxford University Press, New York, NY.
36. Longmire, J.L. *et al.* (1988). *Genomics*, **2**, 14.
37. Faulkes, C.G. and Abbott, D.H. (1990). *J. Zool. Lond.*, **221**, 87.
38. Reeve, H.K., Westneat, D.F., Noon, W.A., Sherman, P.W., and Aquadro, C.F. (1990). *Proc. Natl Acad. Sci. USA*, **87**, 2496.
39. Dallas, J.F. (1988). *Proc. Natl Acad. Sci. USA*, **85**, 6831.
40. Nybom, H. and Schaal, B.A. (1990). *Amer. J. Bot.*, **77**, 883.
41. Lynch, M. (1988). *Mol. Biol. Evol.*, **5**, 584.
42. Brookfield, J.F.Y. (1989). *J. Maths Appl. Med. Biol.*, **6**, 111.
43. Queller, D.C. and Goodnight, K.F. (1989). *Evolution*, **43**, 258.
44. Sambrook, J., Fritsch, E.F., and Maniatis, T. (1990). *Molecular cloning: a laboratory manual*, 2nd edn. Cold Spring Harbor Laboratory Press, Cold Spring Harbor, NY.
45. Berger, S.L. and Kimmel, A.R. (eds) (1990). *Methods in enzymology*, Vol 152. Academic Press, New York, NY.
46. Jeffreys, A.J. and Morton, D.B. (1987). *Anim. Genet.*, **18**, 1.
47. Galbraith, D.A. (1989). *Fingerprint News*, **1**(4), 6.
48. Amos, W. and Hoelzel, A.R. (1991). *Rep. Int. Whaling Comm.*, Special Issue **13**, 99.
49. Miller, S.A., Dykes, D., and Polesky, H.F. (1988). *Nucleic Acids Res.*, **16**, 1215.
50. Kessler, C. and Holtke, H.J. (1986). *Gene*, **47**, 1.
51. Labarca, C. and Paiger, K. (1980). *Anal. Biochem*, **102**, 344.
52. Waye, J.S. and Fourney, R.M. (1990). *Appl. Theor. Electroph.*, **1**, 193.
53. Gontijo, N.F. and Pena, S.G. (1989). *DNA Fingerprint News*, **1**(3), 6.
54. Schaefer, R., Zischler, H., Birsner, U., Becker, A., and Epplen, J.T. (1988). *Electrophoresis* **9**, 369.
55. Nanda, I., Feichtinger, W., Schmid, M., Schroeder, J.H., Zischler, H., and Epplen, J.T. (1990). *J. Mol. Evol.*, **30**, 456.
56. Jarman, A.P., Nicholls, R.D., Weatherall, D.J., Clegg, J.B., and Higgs, D.R. (1986). *EMBO J.*, **5**, 1857.
57. Vassart, G., Georges, M., Monsieur, R., Brocas, H., Lequarré, A.S., and Christophe, D. (1987). *Science*, **235**, 683.
58. Ali, S., Müller, C.R., and Epplen, J.T. (1986). *Hum. Genet.*, **74**, 239.
59. Schäfer, R., Zischler, H., and Epplen, J.T. (1988). *Nucleic Acids Res.*, **16**, 5196.
60. Kashi, Y., Tikochinsky, Y., Genislav, E., Iraqi, F., Beckmann, J.S., Gruenbaum, Y., and Soller, M. (1990). *Nucleic Acids Res.*, **18**, 1129.
61. Vergnaud, G. (1989). *Nucleic Acids Res.*, **17**, 7623.
62. Jeffreys, A.J., Wilson, V., Thein, S.L., Weatherall, D.J., and Ponder, B.A.J. (1986). *Am. J. Hum. Genet.*, **39**, 11.
63. Georges, M., Lathrop, M., Hilbert, P., Marcotte, A., Schwers, A., Swillens, S., Vassart, G., and Hanset, R. (1990). *Genomics*, **6**, 461.
64. Bruford, M.W. and Burke, T. (1994). *Anim. Genet.*, **25**, 382.
65. Carter, R.E., Wetton, J.H., and Parkin, D.T. (1989). *Nucleic Acids Res.*, **17**, 5687.

66. Madeiros, A.C., Macedo, A.M., and Pena, S.D.J. (1988). *Nucleic Acids Res.*, **16**, 10394.
67. Macedo, A.M., Madeiros, A.C., and Pena, S.D.J. (1989). *Nucleic Acids Res.*, **17**, 4414.
68. Wells, R.A. (1988). In *Genome analysis: a practical approach* (ed. K.E. Davies), p. 153. IRL Press, Oxford.
69. Feinberg, A.P. and Vogelstein, B. (1983). *Anal. Biochem.*, **132**, 6.
70. Feinberg, A.P. and Vogelstein, B. (1984). *Anal. Biochem.*, **137**, 266.
71. Edman, J.C., Evans-Holm, M.E., Marich, J.E., and Ruth, J.L. (1988). *Nucleic Acids Res.*, **16**, 6235.
72. Schäfer, R., Zischler, H., and Epplen, J. (1988). *Nucleic Acids Res.*, **16**, 9344.
73. Pollard-Knight, D., Read, C.A., Downes, M.J., Howard, L.A., Leadbetter, M.R., Pheby, S.A., McNaughton, E., Syms, A., and Brady, M.A.W. (1990). *Anal. Biochem.*, **185**, 84.
74. Church, G.M. and Gilbert, W. (1984). *Proc. Natl Acad. Sci. USA*, **81**, 1991.
75. Westneat, D.F., Noon, W.A., Reeve, H.K., and Aquadro, C.F. (1988). *Nucleic Acids Res.*, **16**, 4161.
76. Amos, W., Barrett, J.A., and Pemberton, J.M. (1992). *Proc. R. Soc. Lond. B*, **249**, 157.
77. Hunter, F.M., Burke, T., and Watts, S.E. (1992). *Anim. Behav.*, **44**, 149.
78. Jeffreys, A.J., Royle, N.J., Wilson, V., and Wong, Z. (1988). *Nature*, **332**, 278.
79. Lynch, M. (1990). *Mol. Biol. Evol.*, **7**, 478.
80. Sokal, R.R. and Rohlf, F.J. (1981). *Biometry*, 2nd edn. W.H. Freeman, San Fransisco, CA.
81. Krokene, C., Anthonisen, K., Lifjeld, J.T., and Amundsen, T. (1996). *Anim. Behav.*, **52**, 405.
82. Burke, T. (1989). *Trends Ecol. Evol.*, **4**, 139.
83. Wong, Z., Wilson, V., Jeffreys, A.J., and Thein, S.L. (1986). *Nucleic Acids. Res.*, **14**, 4605.
84. Kelly, R., Bulfield, G., Collick, A., Gibbs, M., and Jeffreys A.J. (1989). *Genomics*, **5**, 844.
85. Saito, I. and Stark, G.R. (1986). *Proc. Natl Acad. Sci. USA*, **83**, 8664.
86. Bruford, M.W., Burke, T., and Hanotte, O. (1994). *Anim. Genet.*, **25**, 391.
87. Wetton, J.H., Burke, T., Parkin, D.T., and Cairns, E. (1995). *Proc. Roy. Soc. Lond. B*, **260**, 91.
88. Zeh, D.W., Zeh, J.A., and May, C.A. (1994). *Mol. Ecol.*, **5**, 517.
89. van Pijlen, I.A., Amos, B., and Burke, T. (1995). *Mol. Biol. Evol.*, **12**, 459.
90. Scribner, K.T., Arntzen, J.W., and Burke, T. (1994). *Mol. Biol. Evol.*, **11**, 737.
91. Nakamura, Y., Leppert, M., O'Connell, P., Wolff, R., Holm, T., Culver, M., Martin, C., Fujimoto, E., Hoff, M., Kumlin, E., and White, R. (1987). *Science*, **235**, 1616.
92. Birnboim, H.C. and Doly, J. (1979). *Nucleic Acids Res.*, **7**, 1513.
93. Raleigh, E.A., Murray, N.E., Revel, H., Blumenthal, R.M., Westanay, D., Reith, A.D., Rigby, P.W.J., Elhai, J., and Hanahan, D. (1988). *Nucleic Acids Res.*, **16**, 1563.
94. Buluwela, L., Forster, A., Boehm, T., and Rabbitts, T.H. (1989). *Nucleic Acids Res.*, **17**, 452.
95. Jacobsen, J.W., Guo, W., and Hughes, C.R. (1992). *Insect Biochem. Mol. Biol.*, **22**,

785–792.
96. Taggart, J.B. and Ferguson A. (1990). *J. Fish Biol.*, **37**, 991–993.
97. Taggart, J.B., Prodohl, P.A., and Ferguson, A. (1995). *Anim. Genet.,* **26**, 13–20.
98. Bentzen, P, Harris, A.S., and Wright, J.M. (1991). In *DNA fingerprinting: approaches and applications* (ed. T. Burke, G. Dolf, A.J. Jeffreys, and R. Wolff) p. 243. Birkhauser Verlag, Basel.
99. Bentzen, P. and Wright J.M. (1993). *Genome*, **36**, 271.
100. Thomaz, D.M.P.F. (1995). Alternative life-history strategies of the atlantic salmon (*Salmo salar* L). Unpublished PhD thesis, University of Leicester.
101. Prodöhl, P.A., Taggart, J.B., and Ferguson, A. (1994). In *Genetics and evolution of aquatic organisms* (ed. AR Beaumont), pp. 263–270. Chapman and Hall, London.
102. Heath, D.D., Iwama, G.K., and Devlin, R.H. (1993). *Nucleic Acids Res.*, **21**, 5782.
103. Harris, A.S. and Wright, J.M. (1995). *Genome* **38**, 177.
104. Lank, D.B., Smith, C.M., Hanotte, O., Burke, T., and Cooke, F. (1995). *Nature* **378**, 59.
105. Hanotte, O., Cairns, E., Robson, T., Double, M.C., and Burke, T. (1992). *Mol. Ecol.*, **1**, 127.
106. Wetton, J.H., Burke, T., Parkin, D.T., and Cairns, E. *Proc. R. Soc. Lond. B*, **260**, 91.
107. Dixon, A., Ross, D., O'Malley, S.L.C., and Burke, T. (1994). *Nature* **371**, 698.
108. Verheyen, G.R., Kempenaers, B., Burke, T., Van den Broeck, M., Van den Broeckhoven, C., and Dhondt, A.A. (1994). *Mol. Ecol.*, **3**, 137.
109. Gyllensten, U.B., Jakobsen, S., Temrin, H., and Wilson, A.C. (1989). *Nucleic Acids Res.*, **17**, 2203.
110. Wetton, J.H. and Parkin, D.T. (1997). *Mol. Ecol.*, **6**, 119.
111. May, C.A., Wetton, J.H., Davis, P.E., Brookfield, J.F.Y., and Parkin, D.T. (1993). *Proc. R. Soc. Lond. B*, **251**, 165.
112. May, C.A., Wetton, J.H., and Parkin, D.T. (1993). *Proc. R. Soc. Lond. B*, **253**, 271.
113. Joseph, S.S. and Sampson, J. (1994). *Anim. Genet.*, **25**, 307.
114. Gilbert, D.A., Packer, C., Pusey, A.E., Stephens, J.C., and O'Brien, S.J. (1991). *J. Hered.*, **82**, 378.
115. Van Pijlen, I.A. (1994). Hypervariable genetic markers and popualtion genetic differentiation in the minke whale *Balaenoptera acutorostrata.* Unpublished PhD thesis, University of Leicester.
116. Kelly, R.G., Gibbs, M., Collick, A., and Jeffreys, A.J. (1991). *Proc. R. Soc. Lond. B*, **245**, 235.
117. Gibbs, M., Collick, A., Kelly, R.G., and Jeffreys, A.J. (1993). *Genomics*, **17**, 121.
118. Georges, M. *et al.* (1991). *Genomics*, **11**, 24.
119. Signer, E.N., Schmidt, C.R., and Jeffreys, A.J. (1994). *Mamm. Genome*, **5**, 48.
120. Signer, E.N., Gu, F., and Jeffreys, A.J. (1996). *Mamm. Genome*, **7**, 433.
121. Coppieters, W., Ven de Weghe, A., Depicker, A., Bouquet, Y., and Van Zeveren, A. (1990). *Anim. Genet.*, **21**, 29.
122. Coppieters, W., Zijlstra, C., Van de Weghe, A., Bosma, A.A., Peelman, L., Depicker, A., Van Zeveren A., and Bouquet, Y. (1994). *Mamm. Genome*, **5**, 591.
123. Nakamura, Y. *et al.* (1988). *Genomics*, **2**, 302.
124. White, R. (1989). *Proceedings of the International Symposium on Human Identification*, pp 1–4. Promega Inc. Madison, WI.
125. Weir, B.S. (1996). *Genetic data analysis II.* Sinauer, Sunderland, MA.
126. Chakravarti, A. and Li, C.C. (1983). In *Inclusion probabilities in parentage testing*

(ed. R.H. Walker, R.J. Duquesnoy *et al.*). American Association of Blood Banks, Arlington, VA.

127. Morin, P.A., Wallis, J., Moore, J.J., and Woodruff, D.S. (1994). *Mol. Ecol.*, **3**, 469.
128. Chakraborty, R., Meagher, T.R., and Smouse, P.E. (1988). *Genetics*, **118**, 527.
129. Royle, N.J., Clarkson, R.E., Wong, Z., and Jeffreys, A.J. (1988). *Genomics*, **3**, 352.
130. Ott, J. (1986). In *Human diseases: a practical approach* (ed. K.E. Davies) p. 19. IRL Press, Oxford.
131. Taggart, J.B. and Ferguson, A. (1994). *Mol. Ecol.* **3**, 271.
132. Hanotte, O., Bruford, M.W., and Burke, T. (1992). *Heredity* **68**, 481.

10

Automated DNA detection with fluorescence-based technologies

VICTOR A. DAVID and MARILYN MENOTTI-RAYMOND

1. Introduction

The use of fluorescence-based technologies in DNA detection systems has truly come of age. Automated DNA sequencing instruments, which were once relegated to core sequencing facilities at a few large institutions, are now found at thousands of laboratories throughout the world. The capabilities of these instruments have been expanded to encompass new technical developments such as microsatellites (1,2), random amplified polymorphic DNA (RAPD) markers (3,4), variable number tandem repeats (VNTRs; 5) and single-strand conformational polymorphism (SSCP; 6). The amenability of this technology to automation has made it especially useful in large-scale sequencing and genotyping projects; however, high sample throughput can be accomplished without expensive robotic systems. DNA storage, polymerase chain reaction (PCR) amplification, DNA sequence reactions, and sample loading can all be accomplished in a format compatible with a standard 96-well plate. This allows for the use of multi-channel pipettes to facilitate sample processing and multi-channel syringes for sample loading. This chapter will assess current applications of fluorescence-based DNA detection systems with emphasis on the two most widely used applications, DNA sequence analysis and microsatellite genotyping.

1.1 Application of automated fluorescence-based technology to analysis of populations

Other chapters document the use of DNA sequence analysis (Chapter 6), RAPD markers (Chapter 6), microsatellite detection (Chapter 7), SSCP (Chapter 8), and VNTRs (Chapter 9) for molecular genetic analysis of populations using radio-isotopic or silver-staining detection methods. For all of these applications DNA can be detected in a fluorescence-based system following labelling with a fluorescent dye (fluorophore). This can be accomplished in many ways. If the DNA is amplified by PCR (7) prior to detection, the PCR products are usually labelled either by utilizing a fluorescently labelled primer or by incorporating fluorescently labelled dUTP in the PCR. For example, in DNA sequence analysis, fluorophores can be incorporated into the sequencing primers or fluorescently labelled dideoxynucleotides can be included in the sequencing reaction. DNA can also be directly labelled by incubation with a fluorescent intercalating dye such as TOTO-1 (8). The fluorophore-labelled DNA is then detected in a variety of ways depending on the type of assay performed and the type of instrument used for the detection. For the applications emphasized above, the DNA is electrophoresed through a matrix and as the DNA passes through a window at a fixed migration distance, the fluorophore is excited by a laser and the fluorescence emission is measured.

1.2 Assessment of automated fluorescence-based detection systems

1.2.1 Advantages of automated fluorescence-based detection systems

Switching to a fluorescence-based detection system eliminates the problems of dealing with a radioactivity-based system such as:

- exposure of investigators to radioactivity
- high cost of radioactive waste disposal
- time-consuming paper work involved in radioisotope use
- short half-lives of radioisotopes, necessitating frequent labelling reactions (fluorescently labelled primers are stable for years if properly stored)

A second advantage is the automation of data analysis. There are no X-ray films to develop and the user is supplied directly with computer output, obviating the need for lengthy manual data analysis and data entry. These technologies are especially powerful when combined with automation of sample preparation and further downstream data processing. They become an integral part of a highly automated system. The sequencing of the entire genome of *Haemophilus influenzae* by automated DNA sequence analysis attests to the power of the technology (9). Similarly, the utility of fluorescence-based microsatellite detection in a large-scale project was demonstrated in the genotyping of 96 affected sibling pairs with 290 marker loci in a search for human type 1 diabetes susceptibility genes (10).

There are many robotic workstations that can automate sample set-up such as the Bio-mek (Beckman Instruments), Microlab (Hamilton Company), and Catalyst (Applied Biosystems Division, Perkin–Elmer). In addition, the Bio-mek and Catalyst workstations can perform the PCR and sequencing reactions. While an expensive workstation may be a requirement for a large core facility, it is not necessary for individual projects. The use of multi-channel pipettes for sample set-up and multi-channel syringes for sample loading, greatly reduces the number of pipetting steps, decreasing the workload and lowering the likelihood of operator error (11).

When using conventional techniques that utilize radioactive isotopes and autoradiography, it is sometimes necessary to perform multiple gel exposures to obtain an interpretable autoradiograph. Fluorescence signals are linear over a large range, thus signals that greatly vary can be scored on the same gel (12). For example, sequencing reactions prepared by multiple investigators using different methods could be electrophoresed on the same run. Likewise, multiple microsatellite loci with greatly variable signal strengths could be run together. Since the investigator does not have to worry about radioactive decay, samples can be lyophilized and stored at –20°C for months prior to analysis. The remainder of samples not used for analysis, can be saved in case the analysis needs to be repeated.

1.2.2 Limitations

A major limitation of fluorescence-based detection systems is the expense. The initial instrument cost is prohibitive for many laboratories. However, as these instruments find their way into more institutions, it is becoming easier for investigators in population genetics to gain access to them. Additionally, due to high instrument throughput, it may only be necessary to use an instrument for a short time to complete a project.

Reagent costs for sequencing kits and custom dye-labelled oligonucleotides are also high. However, many oligonucleotide suppliers offer discounts on commonly used dye labelled sequencing and PCR primers.

Finally, this is a computer-intensive technology that uses sophisticated software which takes a while to learn and may entail some down-time for training. However, for those starting work on a new system or working on a system that will require large sample throughput, a fluorescence-based system offers definite advantages.

1.2.3 Alternatives to automated DNA detection

One alternative to switching to a fluorescence-based system is to upgrade an existing radioactivity-based system. Improvements in a radioactivity-based system can be made by:

- automating sample preparation with a robotic workstation
- reducing detection time by using a beta-imager (13)
- use of image analysis system to semi-automate gel scoring (such as BioMax, Eastman Kodak Company; Molecular Imager System, Bio-Rad Laboratories)

Improvements have been made in chemiluminescent detection systems, which also eliminate the use of radioisotopes (see Chapters 5 and 9). These are offered by many manufacturers. They are compatible with image analysis systems and transfer of DNA to a membrane can be automated by use of a direct blotting electrophoresis system (14). Chemiluminescent methods, however, involve post-electrophoresis development steps not required in fluorescence-based systems. The disadvantage of beta-imagers and image analysis systems are that they are expensive and, unlike an automated fluorescence-based system, the data are not collected as the samples run.

Fluorescence detection systems are available which are not a part of an automated system (Molecular Dynamics, Hitachi). They can be used to scan gels after electrophoresis has been completed on a conventional inexpensive electrophoresis system. They offer the advantages of fluorescence technology at a reduced cost.

1.3 Instruments for automated fluorescence-based detection of DNA

Since the first report of fluorescence-based DNA sequencing in 1986 (15) many companies have developed DNA sequencing instruments. Besides instruments, the companies market software for data analysis and kits for performing reactions to be run on their system. The choice of which instrument to use will probably be influenced by which instrument one has access to at their institution. For those purchasing new instruments, important factors to weigh will be sample throughput, cost, precision, reliability, and availability of technical service. Below, we have noted some of the instruments that are most widely used. For the specific applications discussed later, we have used an Applied Biosystems Model 373 Sequencer, because that is the instrument available in our laboratory. We have tried to include protocols with information helpful to users of other machines.

1.3.1 Applied Biosystems Division, Perkin–Elmer Corporation instruments

Applied Biosystems was the first manufacturer to develop a commercial instrument for automated DNA sequencing. The unique characteristic of the Applied Biosystems technology is that their instruments can distinguish between multiple fluorophores electrophoresed in the same lane on a gel. This allows high sample throughput. For sequencing applications, each sample is loaded in a single lane on a gel rather than four lanes. In other applications, this allows the user to combine samples. By labelling PCR primers for different microsatellite loci with different fluorophores, one can electrophorese multiple microsatellites in a single lane. By further taking advantage of the defined size ranges for the different loci, one can also electrophorese amplified loci of different sizes labelled with the same dye in the same lane. The pooling of multiple loci together has been referred to as multiplex analysis. There have been reports of electrophoresis of up to 24 human microsatellite loci 'multiplexed' in a single lane (11). For microsatellite applications, one dye is used to label a set of molecular weight markers that serves as an internal lane standard in each lane, therefore that dye colour cannot be used to label microsatellite primers. This gives precise band sizing between lanes.

Applied Biosystems has marketed a number of different instrument models for automated fluorescence-based DNA detection. Older models that are still widely in use include the Model 373 and, for longer sequence reads, the Model 373 stretch. Newer models include a second-generation gel-based sequencer (Model 377), with temperature regulation capabilities and improved optics, and Model 310, which replaces gel electrophoresis with capillary electrophoresis. The capillary electrophoresis system has reduced throughput; however, the instrument is considerably less expensive.

1.3.2 Pharmacia Biotech ALF system

Pharmacia Biotech offers an automated sequencer marketed as the ALF system. Software is also available for applications such as microsatellite, RAPD, and SSCP detection. Like the Applied Biosystems 377 sequencer, the gel can be cooled by attachment to an external water bath for SSCP applications. The ALF system utilizes a single dye technology, but accomplishes high sample throughput by minimizing run times. Microsatellite multiplexing in a single lane can still be accomplished by running loci of different sizes together. Alleles are sized based on external lane standards. Pharmacia has also developed an optional solid-phase sequencing system that can be used in sequencing PCR products. One of the primers used in the PCR is labelled with biotin. Following PCR amplification, the PCR product can be bound to a streptavidin-coated comb. The comb containing multiple samples can be sequentially transferred to a tray filled with a solution to remove the non-biotinylated DNA strand, then to a second tray to anneal the sequencing primer and to a final tray for the primer extension reaction for DNA sequencing. The combs can then be directly loaded on to the gel. The advantages of this system are:

- it eliminates the need to purify PCR products
- a large number of samples can be prepared in parallel
- pipetting steps during the DNA sequencing reaction are eliminated
- there is no need to precipitate samples prior to gel loading (16)

1.3.3 LI-COR automated DNA sequencer

LI-COR has developed a single-dye, automated DNA sequencer that uses an infra-red laser to reduce background fluorescence (17). Relative to ABI and Pharmacia systems less expensive glass plates are used and the sensitivity of detection is improved. The system utilizes sophisticated deconvolution software to enhance peak resolution, enabling sequence reads over 1 kb.

2. DNA sequence analysis

2.1 DNA sequencing chemistry

The type of instrument used will determine how the sequencing reactions are performed. Each company sells kits or reagents specifically optimized for use on their instrument. Other independent companies have developed their own sequencing kits intended to be used on one of the available instruments.

All of the sequencing chemistries used in these systems are based on the use of dideoxynucleotides to terminate sequencing reactions (18). There are two approaches to fluorescently labelling the products of the sequencing reactions (*Figure 1*). The first approach is to label the primer used in the sequencing reaction (15). This has been called the dye primer method and can be used on any of the instruments mentioned above. Four separate reactions are set up for the DNA template to be sequenced. Each tube contains a fluorescently labelled sequencing primer and one of the four different dideoxynucleotides to terminate the reaction. Following primer extension, the samples can be electrophoresed in four separate lanes on a sequencing gel in a manner analogous to that of radioactivity-based sequencing. For the Applied Biosystems systems, four sequencing primers each labelled with a different dye are used in four separate reactions. Following primer extension, the samples can be pooled and electrophoresed in the same lane.

The second method used in fluorescence-based DNA sequencing uses fluorescently labelled dideoxynucleotides to terminate the extension of the sequencing primer (19). This has been called the dye terminator method. On the Applied Biosystems system (only), each different dideoxynucleotide is labelled with a different fluorophore, so the sequencing reaction can be performed in a single tube containing an unlabelled primer and all of the four different dideoxynuclotides. In general, DNA fragments of the same size generated in the sequencing reaction are labelled at the 3′ end with the same dye.

Both the dye primer method and the dye terminator method are very robust and work for a variety of DNA templates. There are many factors to consider in deciding which method to use (20). The dye primer method generally gives longer sequence reads and more even signal peak strength, which is especially important in heterozygote determination. The dye terminator method has the advantage that the sequencing reaction can be performed in a single tube, so sample processing is simplified. There is no need for sample pooling prior to electrophoresis and the number of wells or tubes used is reduced by a factor of four. It also is convenient because it does not require the use of a fluorescently labelled primer for the sequencing reaction. When attempting to sequence large DNA fragments using the dye primer method, the sequences from the two strands may not overlap. Gaps between the two sequences can be filled in by designing a primer from the 3′ end of the known sequence, then extending the primer into the gap using the dye terminator method. On the Applied Biosystems systems, to use the dye primer method, you must purchase or synthesize the sequencing primer labelled with four different fluorophores. For template that has been cloned into a vector, synthesis of primers is not necessary, because pre-labelled oligonucleotides are commercially available for the commonly used sequencing primers.

The use of cycle sequencing protocols has dramatically simplified and

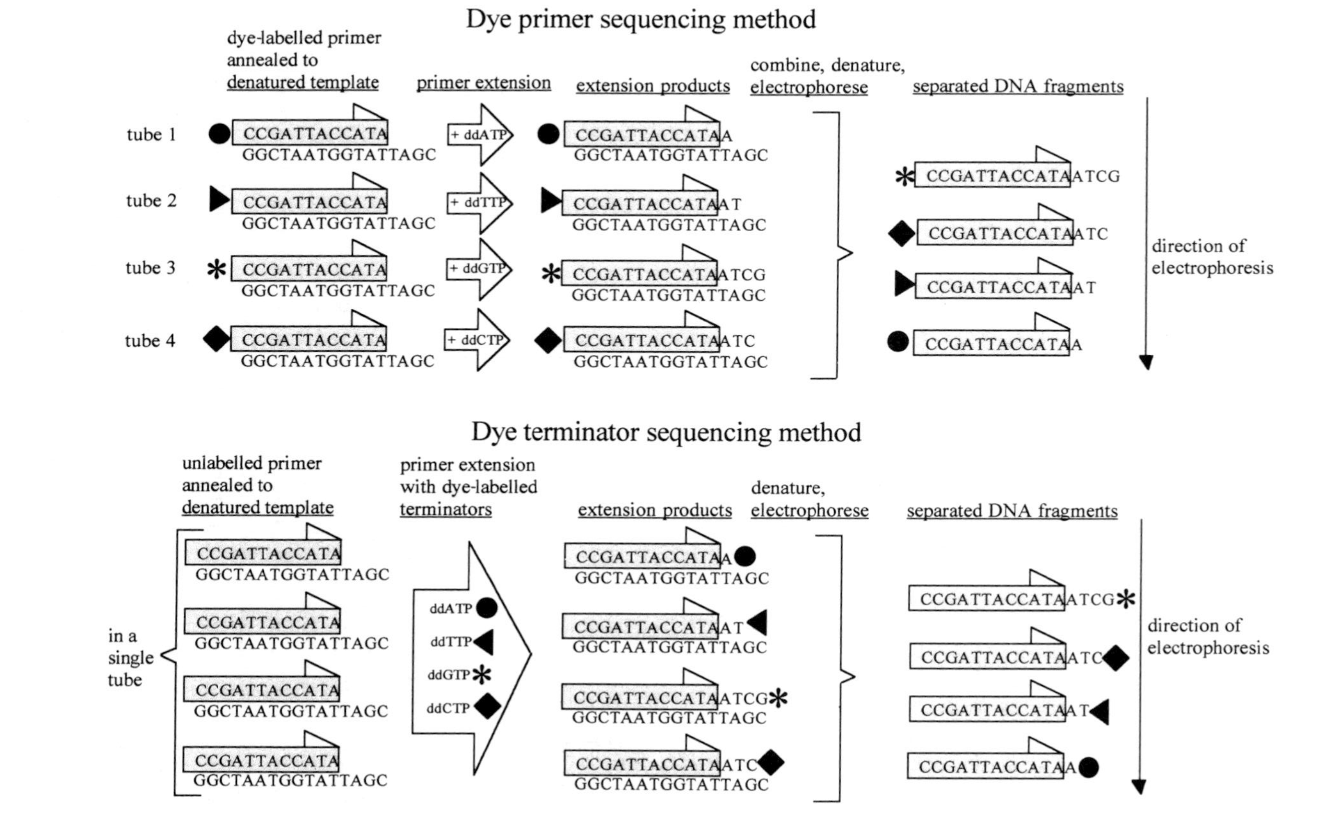

Figure 1. Comparison of the dye primer and dye terminator sequencing methods. The symbols attached to the primer or the dideoxynucleotide represent different dye labels which can be used on the Applied Biosystems systems. For Pharmacia and LI-COR instruments a single dye is used and products generated by the dye primer method are electrophoresed in four separate lanes.

improved DNA sequence analysis (21). This method involves subjecting the sequencing reaction to a thermocycle routine analogous to a PCR. Since only a single primer is present in the sequencing reaction, there is a linear rather than a logarithmic expansion of the product. Cycle sequencing can be used with both dye primer or dye terminator chemistries. It requires less DNA, reduces artefacts due to secondary structure, and simplifies sequencing of double-stranded DNA.

This chapter will emphasize sequencing PCR products, because that is the method most widely used in molecular genetic analysis of populations. Sequencing methods used to depend on the source of the DNA. With tremendous improvements in fluorescence-based DNA sequencing methods, the main difference between sequencing PCR products, single-stranded DNA and plasmids is now the amount of DNA added as a template.

2.2 Isolation of DNA

Fluorescence-based DNA sequence analysis can be performed using DNA from many sources including plasmids, cosmids, M13 templates, and PCR products. The quality of the DNA sequence is dependent upon the amount and purity of the DNA used in the sequencing reaction. For example, residual organics carried over from DNA extraction can cause problems. DNA concentration can be estimated for samples with small amounts of DNA by comparison with a known standard on an agarose gel or by using a DNA DipStick (Invitrogen). Previous chapters have outlined the purification of genomic (Chapter 2) and mitochondrial (Chapter 3) DNA. It has been our experience that commercially available kits from Qiagen work well for preparing plasmid and M13 DNA for fluorescence-based DNA sequencing (but see Chapter 4 for other methods).

2.3 Preparation of PCR products for DNA sequence analysis

2.3.1 Optimization of PCR amplification

PCR products can be directly sequenced or cloned into a suitable vector as described in Chapter 6. It will generally be necessary to clone the PCR product into a vector if you wish to determine DNA sequence differences between different alleles. Procedures for optimization of PCR product formation have been discussed previously in Chapter 6. It is important that a PCR to be used for sequencing generate only a single product, as the addition of multiple DNA templates into a sequencing reaction will result in unintelligible data. Alternatively, the product intended to be sequenced can be separated from the other amplified DNA fragments.

The presence of primer dimers can interfere with sequencing, especially if one of the primers used in the PCR is to be used as the sequencing primer. The use of a 'hot start' method can reduce the formation of primer dimers and generally reduce spurious product amplification (22,23). Variations on the hot start method such as using wax overlays (24), anti-*Taq* polymerase antibodies (TaqStart antibody, Clontech) and modified *Taq* DNA polymerase (*Taq* Gold, Perkin–Elmer) can also be used to reduce primer dimer formation and reduce spurious products. *Taq* Gold is a thermostable DNA polymerase that has no enzymatic activity until activated by an initial heat shock. It is also recommended for use in multiplex PCRs, where multiple PCR primer sets are combined for amplification in one tube. To avoid problems caused by primer dimers, where practical, use a primer in the sequencing reaction that is internal to the primer used in the PCR.

2.3.2 PCR amplification of mitochondrial DNA for DNA sequence analysis

Chapter 3 has already outlined the value of mitochondrial DNA in the molecular genetic analysis of populations. Fluorescence-based DNA sequence analysis offers a rapid method for screening various mitochondrial regions in a large number of individuals. A method that produces PCR products suitable for direct DNA sequencing by the dye primer method is presented in *Protocol 1*. This method is not limited to mitochondrial DNA, but can be used as a general approach to sequencing PCR products. It offers the advantages of the dye primer method (even signal strength and long DNA sequence reads), but avoids the costly procedure of synthesizing custom-labelled sequencing primers (25). The dye terminator method is an alternative for sequencing short PCR products; however, it is only available for the Applied Biosystems instruments. The drawback of direct sequencing of PCR products is that when sequencing genomic regions where an individual is heterozygous for alleles with different DNA sequences, a mixed template is sequenced. By cloning PCR products prior to sequencing, separate alleles may be individually sequenced.

The DNA amplified by the method described in *Protocol 1* is intended to be used in a sequencing reaction using dye primer chemistry. Since it is desirable to use commercially available fluorescently labelled sequencing primers in the sequencing reaction, the amplified PCR product must contain the DNA sequence of a universal sequencing primer. This is accomplished by synthesizing oligonucleotides which at their 5′ ends have the DNA sequence of different universal sequencing primers followed by the sequence for the forward or reverse PCR primer for the target region to be amplified. The PCR product from such an amplification contains the sequence of the target DNA flanked on both ends by DNA sequences of a forward and reverse sequencing primer. A single PCR product can thus be sequenced in both directions.

2.3.3 Purification of PCR products for use in DNA sequence reactions

In most all cases, it will be necessary to remove the dNTPs and the PCR primers from the PCR prior to performing the sequencing reaction. This can be accomplished using an ultrafiltration device such as a Centricon 100

(Amicon; note that as a modification to their protocol, a single wash with 2 ml of water is sufficient) or by various other methods (see Chapter 6). When performing dye primer sequencing, it is sometimes possible to dilute the PCR product tenfold and use it directly in the sequencing reaction without removal of dNTPs and the PCR primers.

Protocol 1. Use of universal-tailed primers to amplify PCR products for DNA sequencing by the dye primer method

Equipment and reagents

- Custom-synthesized oligonucleotides
- DNA thermocycler
- *Taq* DNA polymerase
- DNA for analysis
- PCR reagents (ingredients will vary depending on specific application, see Chapter 6). Use the following 10 × PCR buffer: 100 mM Tris–HCl pH 8.3, 500 mM KCl.

Method

1. Select PCR primer sequences flanking the DNA region to be amplified. Refer to Chapter 6, *Table 3* and Chapter 3, *Table 2* for primer sequences that amplify specific mitochondrial regions.
2. Synthesize universal-tailed primers by including the sequences of the –21 M13 and RP1 M13 primers in the forward and reverse primer sequences respectively from step **1**.
   - –21 M13 primer sequence: 5′-TGTAAAACGACGGCCAGT (append to the 5′ end of the forward primer sequence)
   - RP1 M13 primer sequence: 5′-CAGGAAACAGCTATGACC (append to the 5′ end of the reverse primer sequence)
3. Optimize the PCR as described in Chapter 6. *Table 2* in Chapter 6 shows a simplified method for calculating the primer annealing temperature. For most mitochondrial sequences less than 1 kb long, the following protocol works well in the Perkin–Elmer Model 480 thermocycler under standard PCR conditions and 1.5 mM $MgCl_2$: 30 cycles of
   - 94°C for 30 sec
   - annealing temperature for 90 sec
   - 72°C for 1 min

   If the amplifications do not initially work, try lowering the annealing temperature by 3°C.
4. Perform a 25 μl PCR and remove the PCR primers and dNTPs as described in Section 2.3.3.

Sometimes, even after attempts at optimization of the PCR there are still multiple PCR products present. In such cases, the PCR fragment of interest must be separated from the other DNA fragments prior to sequencing. There are numerous ways to do this. The PCR products are first separated based on size, generally by agarose gel electrophoresis. The band of interest can be excised from the gel and the DNA can be removed by methods such as electrophoresis into a dialysis bag (see Chapter 7) on to DEAE–cellulose paper (26) or using the 'freeze–squeeze' method (27). These methods can also be used instead of expensive ultrafiltration devices for removing dNTPs and PCR primers from the PCR. While these methods are inexpensive to perform, they are time-consuming and sometimes the product recovery is low. A method that we have found to work in fluorescence-based DNA sequencing applications involves the use of a three-stage ultrafiltration device available from Amicon (28). The gel slice is placed in the first stage of the device (Gel Nebulizer), which is a funnel with a small hole in it. When centrifuged, the gel passes through the hole and is homogenized. The agarose is caught in the second stage (Micropure 0.22 μm separator) and the DNA is allowed to pass through to the third stage which is a conventional ultrafiltration membrane that retains the DNA (Microcon 30). The advantage of this method is that it is fast and gives high recovery. The disadvantage is that these devices are expensive.

2.4 DNA sequencing reaction

Conditions will vary greatly for performing the DNA sequencing reactions depending on whether dye primer or dye terminator chemistry is used and on the instrument manufacturer. Sequencing kits have been developed and optimized by each instrument manufacturer and they have detailed protocols for performing the sequencing reactions that will not be repeated here. Improvements to sequencing kits are constantly being made and it is advisable to check with the instrument manufacturer for the latest updates. For example, a genetically engineered *Taq* DNA polymerase (*Taq* FS, Perkin–Elmer) has been developed (29) that has a higher affinity for dideoxynucleotides than *Taq* DNA polymerase. This results in dramatic improvements in both dye primer and dye terminator sequencing (30,31).

Fluorescence-based DNA sequencing reactions typically utilize a cycle sequencing protocol, so access to a thermocycler will be required. Use of an instrument with a heated lid is advantageous, because it eliminates the need for an oil overlay. Removing small volumes of sequencing reactions from under oil is tedious and time-consuming and, if not properly accomplished, can affect the quality of the data.

After performing the sequencing reactions, additional sample processing is generally required before loading the samples on the gel. With dye primer chemistry, the samples are ethanol-precipitated to remove the dye-labelled primers. With the dye terminator chemistry, the dye-labelled dideoxynucleotides must be removed. This can be accomplished by ethanol precipitation or, as we favour, by using a spin column (Centri-Sep; Princeton Separations).

Spin columns are available in a 96-well format (Advanced Genetic Technologies) for larger sequencing projects (32). See Chapter 6 for other methods.

Desiccated sequencing reactions prepared by the dye primer method can be stored at –20°C indefinitely and dried reactions prepared by dye terminator chemistry can be stored for up to a month. The samples are resuspended in a loading buffer and denatured just prior to sample loading.

2.5 Gel preparation

Preparation of the sequencing gel is a critical factor in DNA sequence quality. To reduce gel to gel variation, it is necessary to follow a detailed protocol faithfully. *Protocol 2* details the procedure for pouring a 6% denaturing polyacrylamide DNA sequencing gel used on the Applied Biosystems model 373 DNA sequencer. Depending on the instrument and the application, the type of gel and the plate size will vary. To increase the speed at which the gels can be run, Hydrolink gel matrix (AT Biochem) can be used.

Protocol 2. Preparation of a 6% polyacrylamide sequencing gel

Equipment and reagents

- 24 cm well-to-read glass plates, (i.e. distance from sample well to lasar window is 24 cm) spacers and combs.
- 40% acrylamide:bis-acrylamide solution (19:1; Bio-Rad cat. no. 161-0144) (**warning**: acrylamide is a suspected neurotoxin. Caution should be taken when handling solutions containing acrylamide to prevent personal exposure, such as wearing double gloves and preparing gels in a fume hood.)
- Ammonium persulfate (**note**: ammonium persulfate is a powerful oxidizer and should not be discarded down the sink)
- Urea
- Distilled de-ionized water
- *N,N,N',N'*-tetramethylethylenediamine (TEMED) (**caution**: TEMED is toxic and should be handled with care)
- 10 × TBE: 890 mM Tris–borate, 890 mM boric acid, 20 mM EDTA, pH 8.3 (prepared monthly; also available commercially)
- Alconox detergent (Alconox Inc.)
- Amberlite MB-1A (Sigma)
- 0.2 μm filtration unit (Nalge, cat. no. 121-0020)
- Vacuum source
- Sequencing tape
- Metal clamps

Method

1. It is critical that the glass plates are clean. Any smudges or specks in the region scanned by the laser will be detected as peaks and may obscure data collection for one or more lanes. Dirty plates increase the likelihood of introducing bubbles when pouring the gels. Always handle the plates wearing gloves or only holding the plates on the edges. Clean the plates after each sequence run or before the run if the plates are not clean upon inspection (see step **11**). Scrub the plates with a soft sponge and a small amount of Alconox detergent using hot water, and rinse with de-ionized water. Allow to air-dry in a vertical position. Alternatively, plates can be wiped dry with paper tissues or rinsed with 95% ethanol. Do not wipe dry plates excessively with a tissue, or lint will be left behind. Also note that some brands of

ethanol contain traces of fluorescent compounds that will result in a high background level of fluorescence.

2. Assemble plates and spacers and secure the sides and bottom with sequencing tape. Place two equally spaced clamps on the long sides of the glass plates.
3. In an Erlenmeyer flask combine:
   - 30 ml water
   - 40 g urea
   - 1 g Amberlite
   - 12 ml 40% acrylamide:bis-acrylamide

   Place the flask in a beaker of warm water ($< 50\,^\circ C$) on a stir plate to help the urea dissolve. (Extended heating on a hot plate can cause the urea to break down.) As soon as the urea dissolves (5 min), filter the solution through a 0.2 μm filter unit. Keep the sample under vacuum for 2 min to remove oxygen, which inhibits acrylamide polymerization. Extended degassing may cause the gel to polymerize too rapidly.
4. Transfer the solution to a new Erlenmeyer flask containing 8 ml of 10 × TBE, taking care not to introduce bubbles. Swirl gently, add 400 μl of freshly prepared 10% ammonium persulfate and 45 μl TEMED, and swirl again. Pour the gel immediately (step **5**).
5. Hold the plates at 45° and pour the acrylamide solution between the plates. Bubbles can be eliminated by tapping on the glass plates. If this does not work, insert a clean spacer to remove any bubbles.
6. Insert the comb and secure with three clamps across the top of the plates. For sequencing applications a comb with a single large well is used so that a shark's-tooth comb can be inserted following polymerization. For microsatellite applications a conventional comb with 24, 32, or 36 wells is used.
7. Place the gel in a level horizontal position to polymerize. The gel can be rested on a piece of foam support with the comb end protruding over the edge.
8. Allow the gel to polymerize a minimum of 2 h before using it. Gels can be poured up to 24 h in advance; however, it is preferable to use the gel on the day it is poured. If gels are to be left overnight, cover the comb area with damp paper towels and plastic wrap.
9. Remove the tape and rinse away any acrylamide on the outside of the plates with hot water and a gloved hand. Rinse the plates with deionized water and dry as in step **1** above.
10. Scan the plates prior to electrophoresis to make sure the scan area is clean. If peaks appear when the plates are checked, wipe the scan

Protocol 2. *Continued*

region on both sides of the plates with a damp paper tissue and blow the scan region dry with compressed air.

11. At the end of the run, remove the spacers and manually pry apart the plates. Do not use a metal spatula as this could chip the plates. Discard the gel and rinse with hot water. Rub the plate with a gloved hand, especially around the edges to remove bits of dried acrylamide. To prevent chipping the plates, rest the plates on a foam support (such as comes in commercial shipments of liquid chemicals) in the sink when washing the plates.

2.6 Electrophoresis conditions

Sample loading and electrophoresis conditions will vary depending on the instrument used, the gel matrix, the plate size, and the type of samples sequenced. Careful monitoring of conditions during a sequence run can yield valuable information. The dye primer and dye terminator peaks should appear at the same time for each successive run. Failure to do so can be indicative of gel problems.

2.7 Post-electrophoresis analysis

A major advantage of fluorescence-based sequencing is the automated base-calling feature. At the end of the sequence run, computer files are created for the individual lane scans and a DNA sequence file is generated. An electropherogram for a sample sequenced on an Applied Biosystems Model 373 sequencer using dye primer chemistry is shown in *Figure 2*. The user can see both the peaks at each position as well as the nucleotide assigned by the instrument. The DNA sequence can be edited to remove regions where the peaks are not strong or sharp enough to be reliably called. DNA sequence files can be exported without the need for any manual data entry. At this point the DNA sequence files can be manipulated just as sequence files would be handled had they been generated by conventional DNA sequence analysis. For example, if the sequence files were a part of a sequencing project, they could be reverse-complemented, overlapped, and aligned in order to construct a contiguous consensus DNA sequence.

Both instrument manufacturers and independent vendors (SeqEd, Applied Biosystems; Sequencher, Gene Codes Corp.) offer separate software programs that allow the user to view multiple electropherograms simultaneously. These programs are specifically designed to manage sequencing projects generated on fluorescence-based DNA sequencers. They can perform alignments and generate consensus sequences as described above; however, they have the added advantage that they display the electropherograms as well as

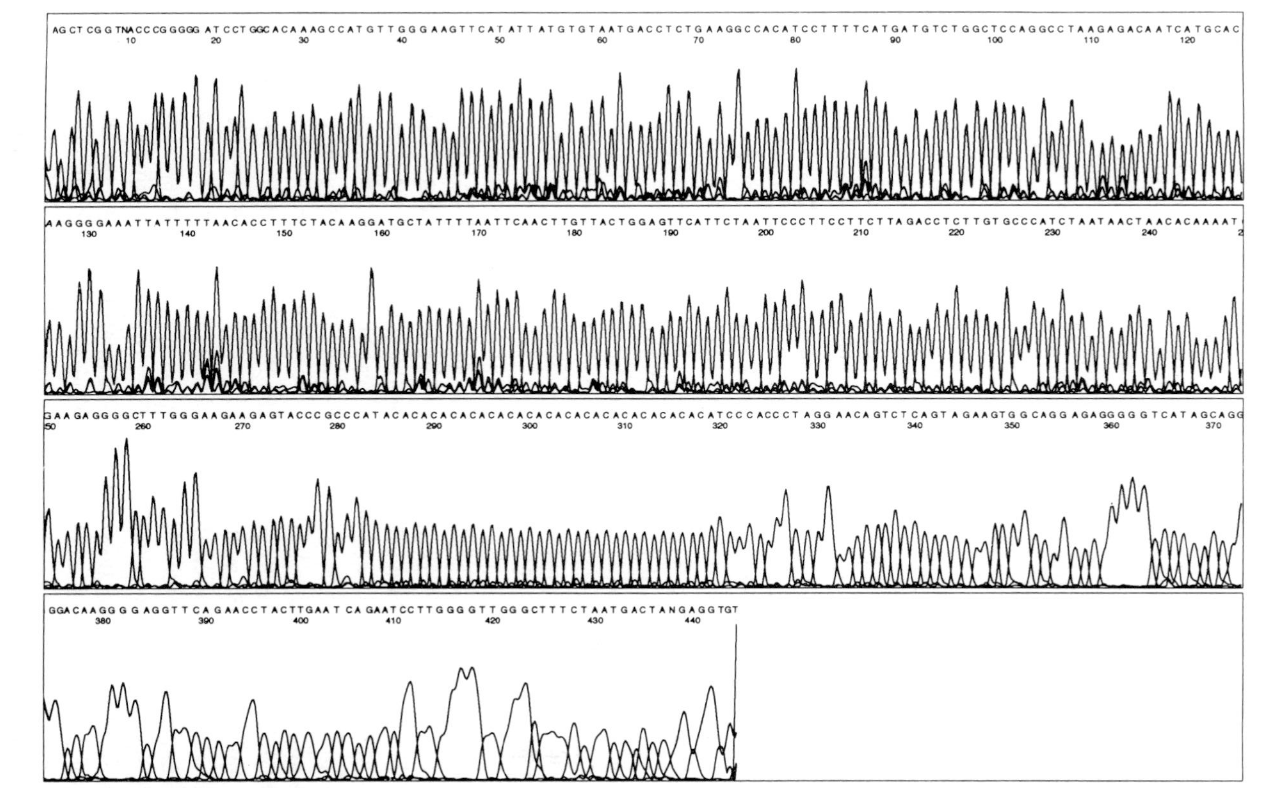

Figure 2. An electropherogram of an M13 single-stranded DNA template sequenced on an Applied Biosystems Model 373 sequencer using dye primer chemistry and a *Taq* FS sequencing kit. Four separate sequencing reactions were performed; the samples were pooled and electrophoresed in a single lane on a gel. Similar plots are generated by instruments marketed by LI-COR and Pharmacia. These instruments require four lanes on a gel for each DNA template sequenced; however, the four lanes are overlaid to generate a single plot like that shown above.

the DNA sequences. The user can view regions of overlap and look at the actual peaks to aid in resolving sequence ambiguities.

2.8 Heterozygote detection

The low background and sharp peak heights obtained on automated fluorescence-based DNA sequencers make it possible to determine if, at any given base position, the template sequenced contained more than one nucleotide. Rather than seeing one normal-sized peak, two smaller peaks would be present at a heterozygous position. This simplifies heterozygote analysis (when the sequence of all genotypes is known), because in most cases it obviates the need to clone separate PCR products in order to sequence the separate alleles. This technique has been used for genotyping individuals at known heterozygous loci (33). Newly developed heterozygote detection algorithms simplify heterozygote detection with less than perfect sequence data (34). The use of a commercial mixture of a mutant *Taq* DNA polymerase and a thermostable pyrophosphatase (*Taq* FS, Applied Biosystems) reduces base-line noise levels and evens out peak heights, further improving the resolution of heterozygote detection (35). During the DNA synthesis one of the products is pyrophosphate, which can cause phosphorolysis or the reversal of DNA synthesis. If this occurs, the terminal fluorescently labelled dideoxynucleotide monophosphate is removed and that termination product is not detected. Phosphorolysis is not random, but occurs preferentially at particular sites, so there may be some regions where the peaks are very low. Addition of a thermostable pyrophosphatase to the cycle sequencing reactions prevents pyrophosphate accumulation.

Alternative fluorescence-based methods that automate detection of mutations, include fluorescence-assisted mismatch analysis (FAMA) (36) and use of a *Taq* Man PCR detection system (Applied Biosystems) (37). These methods are less useful in finding the location and identity of heterozygous positions, but are more applicable to large-scale genotyping of individuals at known mutant sites. FAMA is a method where DNA from a homozygous wild-type individual and a putative mutant are separately amplified with PCR primers that are labelled with strand-specific fluorophores. The two products are then mixed, denatured, and re-annealed. Any mismatches present in the heteroduplex formed between the wild-type and mutant alleles can be cleaved using the chemical mismatch cleavage method (38). The sample can then be electrophoresed on a denaturing gel and the products detected and sized on a fluorescence-based detection system. If mutant alleles are present, peaks will appear as cleavage products shorter than the full-length fragment. The position of the base mutation can also be estimated.

The *Taq* Man PCR detection system can be used to genotype individuals at known polymorphic positions as well as to quantify PCR yield at each successive PCR cycle. An oligonucleotide is designed which is internal to the PCR fragment to be amplified and does not overlap with the PCR primers. A fluorescent reporter dye such as 6-Fam is attached at the 5′ end of the oligonucleotide and a quencher dye such as Tamra is attached distally. The intact oligonucleotide does not fluoresce when both the reporter and quencher dyes are attached; however, if the reporter dye is cleaved from the oligonucleotide, it is no longer under the influence of the quencher and will fluoresce if excited. This double dye-labelled oligonucleotide is included in the PCR. It can bind to the denatured DNA just as the PCR primers bind. It will not fluoresce initially; however, during the PCR, as the DNA synthesis proceeds through the region where the oligonucleotide is annealed, the 5′–3′ exonuclease activity of *Taq* DNA polymerase will degrade the oligonucleotide and separate the reporter dye from the quencher dye. The progress of the PCR can be followed by monitoring the rise in reporter fluorescence, and thereby offers an advance in monitoring quantitative PCR. The drawbacks of the system are that an expensive custom-synthesized double dye-labelled oligonucleotide is required as well as a system for measuring fluorescence. The fluorescence can be monitored by placing the amplified samples in a luminescence spectrophotometer such as the Applied Biosystems *Taq* Man LS-50B or by simultaneous PCR amplification and detection in an Applied Biosystems Prism 7700 system. The system can be used to genotype individuals in regions where a mutation is known (39). One allele-specific oligonucleotide is designed for the wild-type allele and a second allele-specific oligonucleotide is designed for the mutant allele. Both oligonucleotides are labelled with the same quencher, but with different reporters such as 6-Fam and Tet. Under optimized conditions the labelled oligonucleotides will only anneal to the specific allele that they perfectly match. Following PCR amplification of the target region, the genotype of an individual can be determined by which fluorophores were detected.

3. Microsatellite genotyping using fluorescence-based systems

3.1 Selection of microsatellites

As more microsatellites appear in the literature (see Chapter 7 and Appendix 4), investigators may have a choice of which microsatellites to use in their study. To take full advantage of an automated system one will want to electrophorese multiple samples in a single lane. For multiplex analysis, microsatellites should be selected which give strong clean signals with a minimum of stray bands. This could be determined by use of a fluorescently labelled PCR primer and detection of amplified product on a fluorescence-based detection system, using unlabelled PCR primers and detection of amplified products on an agarose or acrylamide gel or by incorporating fluorescently labelled dUTP in a PCR with unlabelled primers followed by detection with a fluorescence-

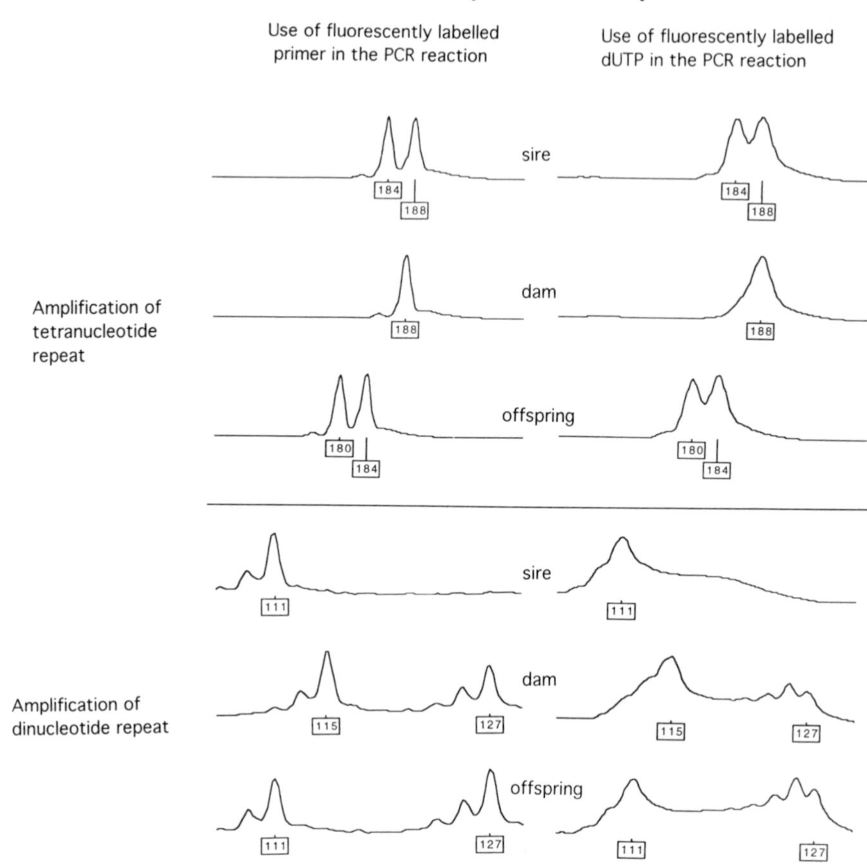

Figure 3. Amplification of dinucleotide and tetranucleotide repeats with either a fluorescently labelled primer or fluorescently labelled dUTP. DNA was amplified in a family of cats using feline dinucleotide or tetranucleotide microsatellite primers. Note that the tetranucleotide repeat peaks have greatly reduced stutter bands compared with the peaks generated by amplification of dinucleotide repeats. The peaks generated by the method using a dye-labelled primer give sharp peaks because only the single dye-labelled strand is detected. The method using fluorescently labelled dUTP gives more complicated patterns because both strands are labelled. The shoulder to the right of the 111 bp peak labelled by the fluorescently labelled dUTP method is fluorescently labelled dUTP. Fluorescently labelled dUTP is acceptable for use with tetranucleotide repeats due to their reduced stutter, but not generally acceptable for accurate scoring of dinucleotide repeats.

based system. The latter two strategies reduce the number of labelled PCR primers that must be synthesized.

There is a choice of which type of microsatellite to use. Besides dinucleotide repeats, trinucleotide and tetranucleotide repeat microsatellites have been determined to be useful genetic markers (40). The only disadvantage of the larger repeat microsatellites is that they have an increased size range, reducing the number of loci that can be multiplexed together. As shown in *Figure 3*, they have reduced 'stutter' or 'shadow' bands (41), which results in easier scoring.

3.2 PCR product labelling strategy

To detect microsatellites in a fluorescence-based system, a fluorophore must be attached to the PCR product. The best ways to accomplish this are by either attaching a fluorophore to one of the PCR primers or by using fluorescently labelled dUTP in the PCR. Both labelling strategies can accommodate multiple fluorophores. On an Applied Biosystems system multiple microsatellite loci can be labelled with different fluorophores and run in the same lane.

3.2.1 Synthesis of fluorescently labelled primers

Fluorescently labelled oligonucleotides can be custom-ordered from many commercial vendors. Pre-labelled PCR primers are available from Applied Biosystems to amplify human and bovine microsatellites. Fluorescently labelled PCR primers can also be synthesized by standard cyanoethyl phosphoramidite chemistry. Applied Biosystems sells 6-Fam, Hex, and Tet amidites and Pharmacia sells carboxyfluorescein amidite which can be added during normal synthesis to the 5′ end of the oligonucleotide (*Protocol 3*). Following purification, the oligonucleotides are ready for use in PCRs. Forty nanomoles of fluorescently labelled oligonucleotide is enough to perform 10 000 PCRs under our standard conditions in *Protocol 4*.

Protocol 3. Preparation of a fluorescently labelled PCR primer

Equipment and reagents

- DNA synthesizer
- Fluorescent dye phosporamidites (6-Fam, Hex, Tet; Applied Biosystems)
- 2.0 M and 0.1 M triethylamine acetate (TEAA) (Applied Biosystems)
- Acetonitrile
- 8% acetonitrile in 0.1 M TEAA
- 20% acetonitrile in water
- Oligonucleotide Purification Cartridge (OPC) (Applied Biosystems)
- Sterile distilled de-ionized water
- 5 ml syringe

Method

1. Modify cycle procedures on DNA synthesizer to accommodate fluorescent dye addition. This may vary depending on the instrument you use; check with the manufacturer for specific recommendations. The following method is essentially as described in Applied Biosystems' *User Bulletin* no. 78.
2. Perform 'Trityl on' 40 nmol synthesis of 5′ dye-labelled oligonucleotide. A single bottle of dye-labelled phosphoramidite is sufficient for 12 couplings. Dye-labelled phosphoramidite should be used within

Protocol 3. *Continued*

4 days of dilution. Extended times on the DNA synthesizer or removal from the instrument for storage result in decreased coupling efficiency.

3. Prepare second oligonucleotide of PCR primer pair without fluorescent label by your normal protocol.
4. Deprotect fluorescently labelled oligonucleotide for 4 h at 55°C. Longer deprotection times may result in breakdown of the dye.
5. Dry the fluorescently labelled oligonucleotide, then resuspend it in 1.0 ml of 0.1 mM TEAA.
6. Load 5 ml of dry acetonitrile into a 5 ml syringe. Attach the syringe to an OPC column and force the liquid through. Repeat with 5 ml of 2 M TEAA.
7. Load the fluorescently labelled oligonucleotide in a 5 ml syringe and slowly (one drop per second) pass it through the OPC column. Collect the eluate in a clean tube and pass it thorough the column a second time.
8. Wash the column first with 5 ml of 8% acetonitrile in 0.1 M TEAA then with 5 ml of water.
9. Slowly elute the fluorescently labelled oligonucleotide from the OPC column with 1 ml of 20% acetonitrile in water. Dry the oligonucleotide and store at –20°C in the dark. Further purification by HPLC is possible but generally not required. The labelling of the oligonucleotide with a fluorescent dye can be confirmed by taking the ratio of the optical density of the sample at 260 nm to the optical density of the sample at the absorbance maximum for the particular dye label. In our experience, as long as a coloured product was eluted from the OPC column, you can assume the fluorescent labelling of the oligonucleotide was accomplished.
10. To quantify the yield, check the optical density at 260 nm. Resuspend the fluorescently labelled oligonucleotide in 200 μl water. The 6-Fam-labelled oligonucleotide should be yellow, the Hex-labelled oligonucleotide should be pink, and the Tet-labelled oligonucleotide should be orange. Add 20 μl of fluorescently labelled oligonucleotide to 980 μl water (1 in 50 dilution). The optical density at 260 nm of the diluted sample should be approximately 0.1–0.2. Calculate the oligonucleotide concentration. One method (26) for calculating oligonucleotide concentration based on the nucleotide composition of the oligonucleotide is:

$$\text{molar concentration} = \frac{(\text{OD of solution at 260 nm}) \times \text{dilution factor}}{15\,000A + 12\,000G + 7300C + 8800T}$$

The calculation is based on the use of a 1 cm path-length cuvette and *A, G, C,* and *T* refer to the total numbers of each individual base in the oligonucleotide.

11. The dye-labelled oligonucleotides (especially Hex-labelled oligonucleotides) are sensitive to light and to freeze–thaw. To avoid frequent freeze–thaw cycles of the concentrated stock, make multiple working stocks. Combine 400 pmol of both PCR primers for a specific microsatellite locus, in a total volume of 40 μl in a 1.5 ml screw-capped tube. This makes a 10 μM PCR primer mix sufficient for 100 PCR amplifications (10 μl reactions; 4 pmol each PCR primer).
12. Store the working stocks and the concentrated stocks at –20°C. For long-term storage (months) it is advisable to dry down the primers and resuspend them as needed.

3.2.2 Use of fluorescently labelled dUTP

Use of fluorescently labelled dUTP in PCR amplification of microsatellite loci obviates the need for the synthesis of a fluorescently labelled PCR primer, as it is simply added to the PCR (*Protocol 5*). Subjecting fluorescently labelled dUTP to PCR amplification conditions results in fluorescent breakdown products, which may obscure products of interest. In such cases, the breakdown products can be removed prior to electrophoresis by ultrafiltration of samples with a Microcon 30 (Amicon). Fluorescently labelled dUTP is available from a number of vendors and is even available labelled with multiple fluorophores (Applied Biosystems) to allow for multiplex analysis of samples labelled with different fluorophores in the same lane. One drawback of labelling with fluorescent dUTP is that both strands of the amplified product are fluorescently labelled. When labelling with a dye-labelled PCR primer, only one of the two amplified strands is labelled. When the PCR product is denatured and electrophoresed on a denaturing acrylamide gel, two dUTP-labelled strands may migrate at different rates through the gel, resulting in multiple peaks being detected. This is further complicated by the presence of stutter bands. Generally, fluorescently labelled dUTP does not work very well with dinucleotide repeats, but it does work with larger repeats, which typically have reduced stutter bands (*Figure 3*).

3.3 PCR amplification of microsatellites

Multiplex analysis can be carried out by individually amplifying multiple microsatellite loci and pooling them following PCR amplification. It can also be accomplished by working out conditions where multiple microsatellite loci can be PCR amplified together in the same tube (for details on optimizing multiplex PCR, see refs 42 and 43). This can be a time-consuming process; however, it is worthwhile in circumstances where a very large number of samples are to be run or, for example, in applications where there may be a limited amount of DNA available (44). Batched analysis of genotypes has been proposed as a powerful tool for population genetics (45). Multiple DNA samples are amplified together in a single PCR; then frequencies for the different alleles in the population are determined based on their relative peak intensities.

It is optimal to maintain samples in a 96-well format as much as possible to streamline sample processing and avoid mistakes. PCR primers are designed so amplified microsatellites range from 100 to 300 bp. With systems that allow multiple fluorophore detection, six microsatellites can be easily be run in the same lane. It is wise to allow a buffer zone of 50 bp between microsatellite loci labelled with the same dye that are electrophoresed in the same lane. For the Applied Biosystems Model 373 system, we amplify 32 different DNA samples individually with PCR primers from six different microsatellites (two each of 6-Fam, Hex, and Tet). Each PCR contains only a single PCR primer pair. As described above, while it is possible to use multiple PCR primers during the PCR, some optimization is required and not all primers work well together. For routine applications, we amplify one microsatellite locus per PCR and pool multiple different microsatellite loci together following PCR amplification for 'post-PCR multiplexing'. The PCRs are performed in two 96-well plates; this represents 192 genotypes per gel (*Figure 4*). Following PCR amplification the six different microsatellite amplifications for a given DNA sample

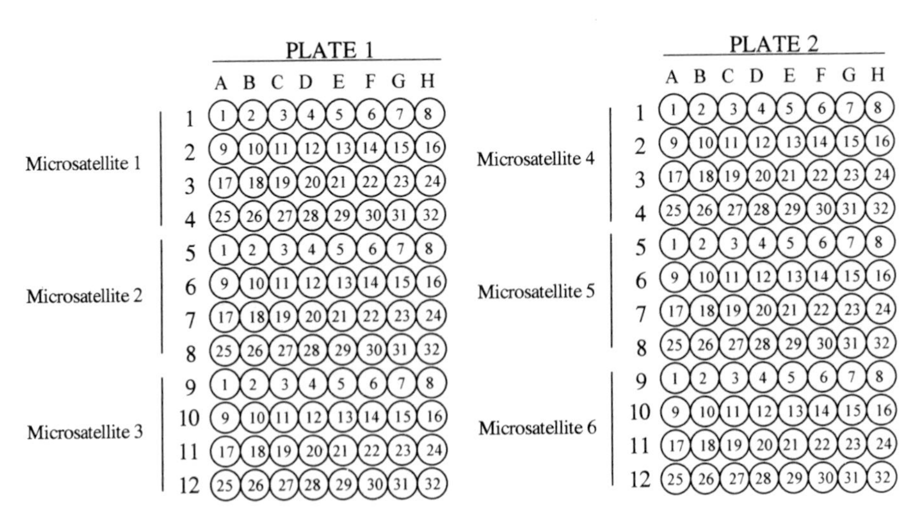

Figure 4. Microsatellite amplification of 32 different individuals with six different microsatellites. The number inside the circle is the number of the DNA sample placed in that position on the 96-well plate. Note that there are six replicates of the each of the 32 samples. The DNA can be added to the wells with a multi-channel pipette. A master mix is prepared containing all the remaining PCR reagents for each of the six different microsatellites to be amplified. The master mix is then added to the rows as indicated.

are pooled, diluted, and electrophoresed in a single lane. It is necessary to determine the optimal dilution empirically for each microsatellite locus.

Protocol 4 outlines a standard protocol for PCR amplification and gel loading used on an Applied Biosystems 373 system that maintains samples in a 96-well format. There are numerous thermocyclers that use a 96-well format, any of which can be used. The protocol presented has been optimized for a Perkin–Elmer Model 9600 thermocycler. For large genotyping projects it is advisable to have PCRs that all work at the same temperature. We design PCR primers using the PRIMER design program (Whitehead/MIT Center for Genome Research, available from `http://www-genome.wi.mit.edu/ftp/pub/software/primer.0.5/`) to have a melting temperature of 55°C. For PCRs that do not work under standard conditions at 2.0 mM $MgCl_2$, we try 1.0, 1.5, 2.5, and 3.0 mM $MgCl_2$. There is a 30 min extension at 72°C at the end of the PCR amplification. *Taq* DNA polymerase can add a non-templated base on to the end of the PCR product (46,47). In cases where some products have the base addition and others do not, it may be difficult to interpret the results, particularly if the data are scored in an automated system. Extended incubations at 72°C (up to 90 min) consistently force the PCR products towards the form with the non-templated base addition (48, and see further discussion in Chapter 6).

Protocol 4. Microsatellite PCR amplification and gel loading (Applied Biosystems 373 System)

Equipment and reagents

- 100 mM dATP
- 100 mM dTTP
- 100 mM dGTP
- 100 mM dCTP
- *Taq* DNA polymerase
- 10 × PCR buffer: 100 mM Tris–HCl pH 8.3, 500 mM KCl
- 25 mM $MgCl_2$
- Centrifuge with adaptors to hold 96-well plates
- DNA thermocycler
- DNA stocks at 12.5 ng/μl
- PCR primer mix containing forward and reverse primers each at 10 μM (one dye-labelled, one unlabelled)
- GS 350 Tamra dye-labelled standard and loading dye (Applied Biosystems no. 401736)
- De-ionized formamide
- PCR tubes (Fisher, Perkin–Elmer, or Robbins) or 96-well 'Cycleplate' PCR plates (Robbins Scientific no. 1038-00-0)[a]
- 6% acrylamide sequencing gel (*Protocol 2*)
- 1 × TBE buffer

Method

A. *Preparation of PCR master mix*

1. Preparation of a PCR master mix will reduce the time it takes to set up PCRs and eliminate an important possible variable in PCR amplification. For example, to make a master mix containing 2.0 mM $MgCl_2$, combine:
   - 30 μl 100 mM dCTP

Protocol 4. *Continued*

- 30 μl 100 mM dGTP
- 30 μl 100 mM dATP
- 30 μl 100 mM dTTP
- 960 μl 25 mM $MgCl_2$
- 1.2 ml 10 × PCR buffer
- 4344 μl water

2. Aliquot 198.8 μl of PCR master mix into each of 30 microcentrifuge tubes; store at –20°C until needed. Each tube has enough PCR master mix for 36 reactions.
3. Dilute stocks of DNA samples to be amplified to 12.5 ng/μl in water; store at 4°C. A panel of 32 different DNA samples (the number of wells across our standard gel) can be stored in rows 1–4 in a 96-well (8 × 12 wells) plate (Beckman, no. 267001). Ideally, the DNA concentrations should be checked on a fluorometer to ensure they are all at the same concentration. For samples where DNA is limited, the amount of DNA can be reduced. The lower limit for DNA will vary depending on the microsatellite. Generally 5 ng of DNA in a 10 μl PCR is the lower limit under our conditions; however, some microsatellites amplify well with as little as 1 ng of DNA.
4. When ready to perform PCRs, thaw one tube of PCR master mix for each microsatellite to be amplified. Add 14.4 μl of 10 μM PCR primer mix and 2.9 μl of *Taq* DNA polymerase (5 units/μl). Mix gently and pulse-spin in a microcentrifuge.

B. *PCR amplification strategy for multiplex analysis of 32 different individuals with six microsatellites*

Note: Follow well-established protocols for preventing PCR contamination (see Chapter 6).

1. Prepare PCR master mix with *Taq* DNA polymerase and PCR primers for the six microsatellite loci to be run on the same gel. The PCR will be performed in a total of 10 μl. Each PCR tube will receive 6 μl of a PCR master mix with added *Taq* DNA polymerase and PCR primers and 4 μl of a DNA sample.
2. To a 96-well PCR plate, add 6 μl of PCR master mix with *Taq* DNA polymerase and PCR primers added. Use of a programmable electronic digital pipette will speed this operation. For example 100 μl can be taken up and added 6 μl at a time to each well. The strategy for amplifying 32 DNA samples with six microsatellites in two 96-well PCR plates is shown in *Figure 4*. The master mix containing the first microsatellite to be amplified is placed in rows 1, 2, 3, and 4 on a 96-well plate. The second microsatellite master mix is placed in rows 5, 6, 7, and 8. Continuing in this manner two 96-well PCR plates are used.

3. Add 4 μl of DNA at 12.5 ng/μl to each well. Use of a programmable eight-channel electronic digital pipette will simplify this operation. If the DNAs were stored in rows 1–4 in a 96-well plate, bring up 24 μl of DNA from row 1 and transfer 4 μl to rows 1, 5, and 9 in each of the two 96-well plates. Transfer from row 2 of the DNA plate to rows 2, 6, and 10 on the PCR plates. Similarly transfer the remaining two rows of DNA samples to the two PCR plates.
4. Cap the tubes or use a rubber plate cover. Pulse-spin the plates to pellet the liquid to the bottom of the tube.
5. Amplification conditions (for a Perkin–Elmer Model 9600):
   - 3 min at 93°C (one cycle)
   - 15 sec at 94°C, 15 sec at 55°C, 30 sec at 72°C (10 cycles)
   - 15 sec at 89°C, 15 sec at 55°C, 30 sec at 72°C (20 cycles)
   - 30 min at 72°C (one cycle)
   - hold at 4°C (can be left overnight)
6. Samples may be stored at –20°C. Samples covered with rubber plate covers may evaporate, but can be resuspended in water.

C. *Sample pooling and loading (six microsatellites per lane)*

1. Prepare dye loading mix by combining 20 μl GS 350 Tamra-labelled standard, 20 μl dye loading buffer, and 120 μl deionized formamide. Larger amounts of the dye loading mix can be prepared and stored for up to 2 weeks at 4°C.
2. Due to the sensitivity of the fluorescence-based detection systems, it is generally necessary to dilute the PCR products prior to loading them on the gel or the signals will be too strong. The optimal dilution factor must be experimentally determined for each microsatellite locus amplified. A starting point is a 10-fold dilution for 6-Fam- and Tet-labelled products and 5-fold for Hex-labelled products. On the Applied Biosystems systems, PCR products should fall between 100 and 3000 fluorescence units. Caution should be used in scoring peaks over 4000 fluorescence units. If peaks are greatly overloaded (equivalent to 8000 fluorescence units), the software may not display the centre of the peaks but only spikes before and after the actual peak. In such cases, a homozygote may be incorrectly scored as a heterozygote.
3. For each of the 32 DNA samples, pool aliquots of the six different microsatellite loci amplified. A programmable eight-channel electronic digital pipette can simplify the operation by consecutively taking up samples from six different rows, then dispensing them to a new 96-well plate. If the optimal dilution factor is 5, 10, or 25, remove 10 μl, 5 μl, or 2 μl respectively of the amplified PCR product. Bring the final volume of the six pooled microsatellites to 50 μl with water.

4. Combine 2 μl of pooled diluted microsatellites with 4 μl of dye loading mix (step **1**) in a 96-well PCR plate. Pulse-spin the plate in a centrifuge to get all the liquid to the bottom of the plate. This should be done within a few hours of when the samples are to be electrophoresed.
5. Mix the samples by manually flicking the bottom of each well. Pulse-centrifuge to get liquid to the bottom of the tube. **Note**: it is absolutely imperative that the samples be mixed well, due to the disparity in density between the pooled microsatellites and the dye master mix containing formamide.
6. When the gel is ready to load, denature the samples for 3 min at 93°C and transfer the samples to an ice bucket. Immediately proceed to load the gel.
7. Load 2 μl of denatured sample per well. If a 32-well comb was used in the acrylamide gel, the samples can be loaded with an eight-channel Hamilton syringe (no. 84503). The syringe can take samples directly from a 96-well plate and will load them into every other well of a 32-well gel. Water is taken up in the syringe and discarded between sample loadings to avoid cross-contamination.
8. If the PCR amplifications are performed in a 96-well format in an instrument that requires an oil overlay for the PCRs, the Hamilton syringe can be used to pool samples. The needles are much better at removing samples from under an oil overlay than a pipette tip. The Hamilton syringe can also be used in step **3** to pool amplified microsatellites, eliminating the cost associated with the use of pipette tips.

[a]We have used PCR tubes from Fisher, Perkin–Elmer, and Robbins and all appear to work equally well. For applications such as this where an entire 96-well plate is required, we use the Cycleplate from Robbins which has all 96 tubes fused together. This obviates the need for costly PCR trays and the manual loading of individual PCR tubes or strips of tubes into trays is no longer required.

Protocol 5. Incorporation of fluorescent dUTP into a PCR product

Equipment and reagents

- 400 μM Tamra-labelled dUTP (Applied Biosystems)
- *Taq* DNA polymerase
- PCR primer mix containing forward and reverse primers each at 10 μM (both unlabelled)
- PCR master mix for 36 samples (*Protocol 4A*, step **1**)
- DNA stocks at 12.5 ng/μl
- Centrifuge with adaptors to hold 96-well plates
- 96-well PCR plates
- DNA thermocycler
- De-ionized formamide
- GS 350 Rox-labelled standard and loading dye (Applied Biosystems no. 401735)
- Microcon 30 (Amicon)

Method

1. Thaw the 36-reaction PCR master mix and add:
   - 14.4 μl of 10 μM PCR primer mix
   - 2.9 μl *Taq* DNA polymerase (5 units/μl)
   - 3.6 μl 400 μM Tamra-labelled dUTP[a]

   Mix gently and pulse-spin in microcentrifuge.
2. Add 6 μl of mix prepared in step **1** to a PCR tube.
3. Add 4 μl of DNA at 12.5 ng/μl.
4. Perform PCR amplification (steps **3–5** in *Protocol 4B*).
5. Tamra dUTP will migrate in the gel at approximately 90–120 bp depending on gel conditions. Fluorescent breakdown products will appear at about 220 bp. If the microsatellites to be run do not have products in these size ranges, they can be run directly on the gel without further purification. Otherwise, the fluorescent dUTP and breakdown products can be removed by ultrafiltration on a Microcon 30 (Amicon).
6. Dilute PCR product 1 in 10 in water. (This is a suggested dilution to start at. The optimal dilution can be determined experimentally.)
7. Prepare dye loading master mix by combining 20 μl GS 350 Rox-labelled standard, 20 μl dye loading buffer, and 120 μl deionized formamide.
8. Combine and mix together 2 μl diluted PCR product with 4 μl of dye loading master mix.
9. When the gel is ready to load, denature the samples for 3 min at 93°C and transfer the samples to an ice bucket. Immediately proceed to load the gel.

[a] The amount of dye-labelled dUTP used will vary depending on the fluorescent label and may need to be experimentally determined. This protocol outlines a protocol which works for Tamra-labelled dUTP.

3.4 Gel electrophoresis conditions

Gel matrix and electrophoresis conditions will vary depending on the instrument used for analysis. On the Applied Biosystems systems, the same gel as described in *Protocol 4* can be used. To reduce electrophoresis time, shorter 12 cm well-to-read plates can be substituted for the 24 cm well-to-read plates used in DNA sequencing. Recommended electrophoresis settings are 2000 V, 40 mA, and 25 W. Under these conditions, the wattage is limiting and it takes about 3 h for the largest fragment in the standard (350 bp) to pass through the laser window. After a single gel run has been completed, a new set of samples can be loaded on the same gel and a second run performed. If amplifications from the first run are especially dirty, a second run on the same gel should not be attempted. Samples on the second run might be inter-mixed with products from the first run and difficult to interpret.

3.5 Post-electrophoresis analysis

Following electrophoresis, the computer generates a gel image showing the bands that were detected. A print-out analogous to an autoradiogram can be obtained. DNA fragment size (in base pairs) can be precisely estimated (11,12,49) for all the peaks in a lane either by the use of internal or external lane standards. The Applied Biosystems system uses an internal lane standard in each lane, which is labelled with a different dye from that used to label the microsatellites. The Pharmacia and LI-COR systems use a single dye so the size standard can either be loaded alone in a lane or electrophoresed in the same lane as the sample, providing there is no overlap between the size standards and the microsatellite amplification products. The software programs recognize the standard peaks and construct a standard curve. The sizes of the product peaks are then estimated based on their migration relative to the known standard. All peaks are labelled, including microsatellite alleles, stutter bands, and stray bands arising from non-specific PCR amplification.

At this point, the investigator can look at the data and manually score the alleles by ignoring stutter bands and stray amplification products. Each allele can be assigned an estimated size based on its migration relative to the standards. This manual operation can be automatically performed by exporting the data to a second computer program such as Genotyper (Applied Biosystems) or Automated Linkage Preprocessor (ALP) (50). ALP was developed for use on Pharmacia systems and is available from A.F. Brown via the internet at `http://www.hgu.mrc.ac.uk/`. The user first defines microsatellite allele ranges and dye label information. The programs then use a filtering algorithm designed to ignore stutter bands and stray amplifications. The results can be viewed as an electropherogram (*Figure 5*) with labelled peaks for manual editing and the final numerical data can be placed into a table. The data can be exported in various formats so that they can be directly imported into other programs such as spreadsheet applications like Microsoft Excel or linkage analysis programs. Genotyper and ALP have the ability to check the data for Mendelian inheritance within pedigrees.

We typically export data to Microsoft Excel where we have a program that converts the allele designations from a fractional value such as 112.46 bp to whole integer or 'bin' designation such as 112 bp. This is not simply a rounding, but rather a grouping of alleles that are likely to contain the same sized microsatellite repeat. For example, fragments sized as 112.46 bp and 112.55 bp in two different lanes on the same gel should both be in the same bin, but simple rounding would put them in bins of 112 and 113 bp respectively. It has been our experience that identical alleles generally migrate within 0.5 bp of

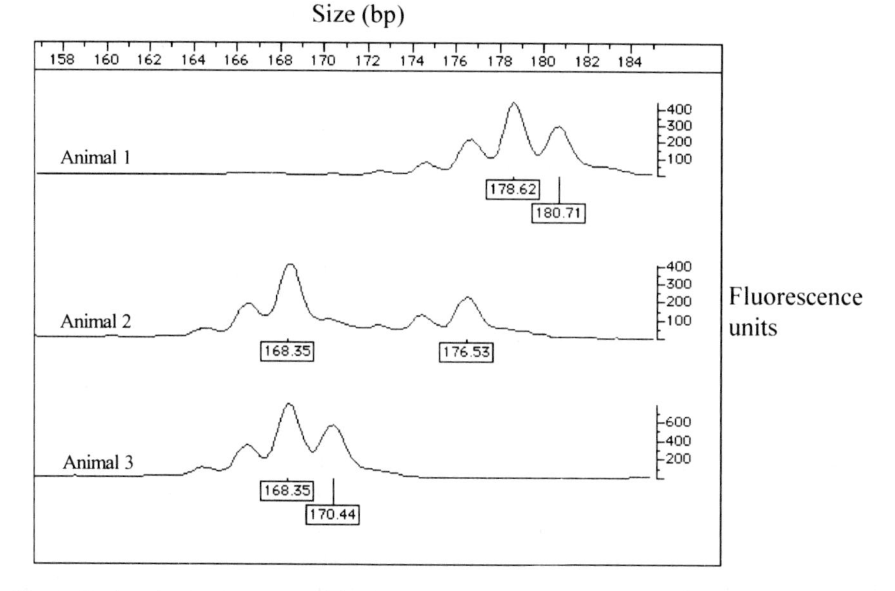

Figure 5. An electropherogram of microsatellite data analysed on an Applied Biosystems 373 system using Genescan and Genotyper software (Applied Biosystems). DNA from three domestic cats was PCR-amplified with primers flanking a feline dinucleotide microsatellite. The peak sizes were determined relative to an internal lane standard using Genescan software. The data were then imported into Genotyper where the peaks were filtered to remove stutter bands. Only the peaks corresponding to alleles remain labelled in the figure. Note that alleles are preceded by smaller stutter peaks. When a peak is preceded by a larger peak, as in animals 1 and 3, both peaks are scored as alleles.

each other on a gel. Larger variations are seen when comparing data from different gels. *Table 1* shows an example of data generated by amplification of a domestic cat microsatellite (51) amplified in a 32-member cat pedigree, which was exported from Genotyper into Microsoft Excel where bin designations were assigned. The first allele of the first sample in *Table 1* (177.73) was used to determine how to round the alleles for binning and to determine a correction factor to apply to all other alleles. The nearest integer value to 177.73 is 178 so this allele was put in bin 178. All other allele values in the table were adjusted by adding a correction value of 0.27 bp to them (178 – 177.73 = 0.27) and then the alleles were rounded to the nearest integer value or bin. A program to do this automatically can be set up in any spreadsheet application; then, as new data are generated they can be pasted into this template for binning. The data in *Table 1* were sorted by allele size and plotted in a histogram (*Figure 6*) to demonstrate the variation observed between alleles in the same bin. It is wise to prepare a histogram for all data generated to ensure that alleles are properly binned.

When it is necessary to compare data across multiple gels, include the same control sample on each gel in order to confirm that bin designations are

Table 1. Results of analysis of 32 different cats with feline microsatellite FCA 347 on a fluorescence-based detectioin system[a b]

Lane[c]	Sample number[d]	Allele 1[e]	Allele 2[f]	Bin 1[g]	Bin 2[h]
2	1	177.73	179.64	178	180
4	2	167.94	171.80	168	172
6	3	171.98	177.74	172	178
8	4	171.76	175.61	172	176
10	5	177.67	179.58	178	180
12	6	172.10	175.90	172	176
14	7	177.73	179.63	178	180
16	8	175.76	179.53	176	180
3	9	167.99	179.62	168	180
5	10	166.24	175.68	166	176
7	11	166.22	179.66	166	180
9	12	166.00	168.14	166	168
11	13	165.97	179.59	166	180
13	14	179.63	0	180	180
15	15	179.62	0	180	180
17	16	179.53	0	180	180
18	17	171.99	179.53	172	180
20	18	171.94	179.66	172	180
22	19	168.12	179.36	168	180
24	20	170.23	172.10	170	172
26	21	171.89	179.57	172	180
28	22	166.25	173.92	166	174
30	23	166.24	172.05	166	172
32	24	171.98	179.62	172	180
19	25	179.75	0	180	180
21	26	160.42	162.10	160	162
23	27	160.42	179.67	160	180
25	28	171.89	179.58	172	180
27	29	160.42	172.05	160	172
29	30	160.63	172.26	160	172
31	31	168.07	171.99	168	172
33	32	160.42	177.65	160	178

[a] Fluorescence-based detection on an ABI Model 373 Sequencer and data analysed using Genescan and Genotyper software packages. The data were exported to Microsoft Excel for binning.
[b] On a typical gel we run a panel of six different microsatellites so six tables such as this one would be generated.
[c] Lane in which the sample was run on the gel. The first lane was not loaded so the first sample was loaded in lane 2. An eight-channel syringe was used to load the gel, so samples in the same row on the PCR plate were loaded in alternate wells on the gel.
[d] Sample number of the cat DNA electrophoresed in this lane. DNA from 32 different cats was electrophoresed on this gel. The table was sorted by cat number.
[e] Size estimated in base pairs for the smaller allele. The size is estimated by comparison with a set of standards electrophoresed in the same lane.
[f] Size estimate of the larger allele amplified. The presence of a zero indicates the absence of a second allele. The absence of the second allele is generally indicative of a homozygote; however, it can also indicate the presence of a 'null' allele which failed to amplify.
[g] The bin designation for the smaller allele. The bins were determined as described in the text.
[h] The bin designation given to the larger allele. All cats where allele 2 was zero were scored as homozygotes.

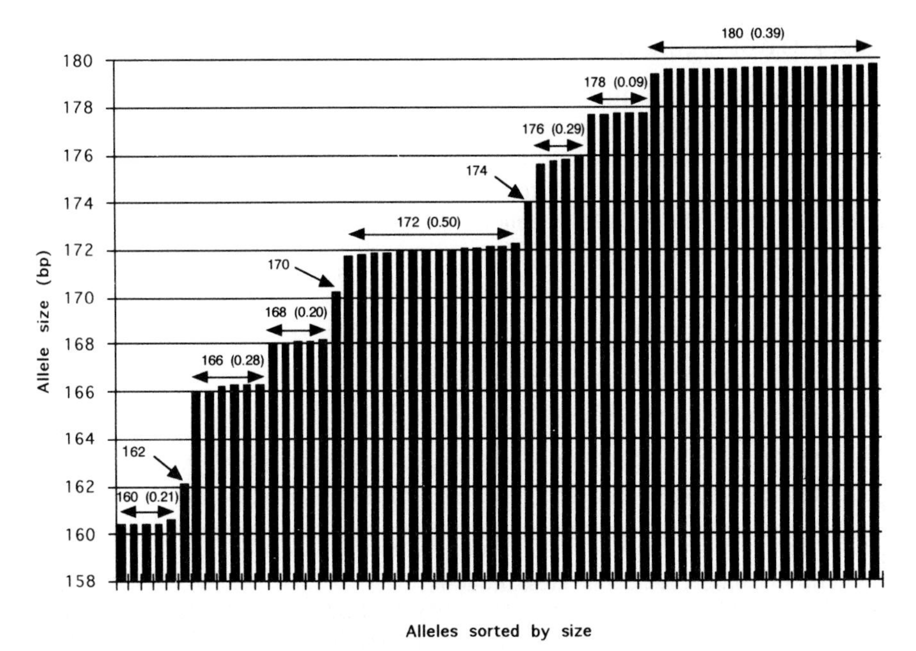

Figure 6. Histogram of allele sizes from *Table 1*. Alleles in *Table 1* were sorted from smallest to largest allele. The range of alleles placed together in the same bin is indicated by the arrowed line. Above the arrowed line is the bin designation and in parenthesis is the range between the largest and smallest allele in that bin.

consistent. The size estimate for a given allele is generally within one microsatellite repeat length of the true size (52). When checking Mendelian inheritance, it is important to be aware that certain 'null' alleles may sometimes fail to be amplified (53). Once recognized, these alleles may in certain cases be tracked through the pedigree.

4. SSCP, RAPD, and VNTR analysis

SSCP, RAPD (54), and VNTR analysis (55) have all been analysed using fluorescence-based systems. SSCP analysis can be performed on an instrument such as the Applied Biosystems Model 373, which does not have a cooling system for the plates (56), or, preferably, on systems such as the Pharmacia and Applied Biosystems Model 377 systems, where the plates can be cooled by circulation of water from an external water bath (57,58). Higher SSCP resolution of PCR-amplified products can be obtained by using PCR primers each labelled with a different fluorophore. Both strands of denatured DNA can then be individually visualized and standardized to in-lane markers (59).

Acknowledgements

We thank Melanie Culver and Janet Ziegle for critical evaluation of the manuscript. The content of this publication does not necessarily reflect the views or policies of the Department of Health and Human Services, nor does mention of trade names, commercial products, or organizations imply endorsement by the US Government.

References

1. Weber, J. L. (1990). *Genomics*, **7,** 524.
2. Tautz, D. (1989). *Nucleic Acids Res.*, **17**, 6463.
3. Williams, J. G., Kubelick A. R., Livak, K. J., Rafalski, J. A., and Tingey, S. V. (1990). *Nucleic Acids Res.*, **18**, 6531.
4. Welsh, J. and McClelland, M. (1990). *Nucleic Acids Res.*, **18**, 7213.
5. Jeffreys, A. J., Wilson, V., and Thein, S. L., (1985). *Nature*, **314**, *67.*
6. Orita, M., Suzuki, Y., Sekiya, T., and Hayashi, K. (1989). *Genomics*, **5**, 874.
7. Mullis, K. B. and Faloona, F. A. (1987). In *Methods in enzymology* (ed. R. Wu), Vol. 155, p. 335. Academic Press, London.
8. Mansfield, E. S. and Kronick, M. N. (1993). *BioTechniques*, **15**, 274.
9. Fleischmann, R. D. *et al.* (1995). *Science*, **269**, 496.
10. Davies, J. L. *et al.* (1994). *Nature*, **371**, 13.
11. Schwengel, D. A., Jedlicka, A. E., Nanthakumar, E. J., Weber, J. L., and Levitt, R. C. (1994). *Genomics*, **22**, 46.
12. Ziegle, J. S., Su, Y., Corcoran, K. P., Nie, L., Mayrand, E., Hoff, L. B., McBride, L. J., Kronick, M. N., and Diehl, S. R. (1992). *Genomics*, **14**, 1026.
13. Dausse, E., Quemeneur, E., and Schwartz, K. (1995). *BioTechniques*, **18**, 426.
14. Pohl, T. and Maier, E. (1995). *BioTechniques*, **19**, 482.
15. Smith, L. M., Sanders, J. Z., Kaiser, R. J., Hughes, P., Dodd, C., Connell, C. R., Heiner, C., Kent, S. B., and Hood, L. E. (1986). *Nature*, **321**, 674.
16. Lagerkvist, A., Stewart, J., Lagerström-Fermér, M., and Landegren, U. (1994). *Proc. Natl Acad. Sci. USA*, **91**, 2245.
17. Middendorf, L. R. *et al.* (1992). *Electrophoresis*, **13**, 487.
18. Sanger, F., Nicklen, S., and Coulson, A. R. (1977). *Proc. Natl Acad. Sci. USA*, **74**, 5463.
19. Tracy, T. E. and Mulcahy, L. S., (1991). *BioTechniques*, **11**, 68.
20. Naeve, C. W., Buck, G. A., Niece, R. L., Pon, R. T., Robertson, M., and Smith, A. J. (1995). *BioTechniques*, **19**, 448.
21. Murray, V. (1989). *Nucleic Acids Res.*, **17**, 8889.
22. Erlich, H. A., Gelfand, D., and Sninsky, J. J. (1991). *Science*, **252**, 1643.
23. D'Aquil, R. T., Bechtel, L. J., Videler, J. A., Eron, J. J., Gorczyca, P., and Kaplan, J. C. (1991). *Nucleic Acids Res.*, **19**, 3749.
24. Hebert, B., Bergeron, J., Potworowski, E. F., and Tijssen, P. (1993). *Mol. Cell Probes*, **7**, 249.
25. McBride, L. J., Koepf, S. M., Gibbs, R. A., Nyugen, P., Salser, W., Mayrand, P. E., Hunkapiller, M. W., and Kronick, M. N. (1989). *Clin. Chem.*, **35**, 2196.
26. Sambrook, I., Fritsch, E. F., and Maniatis, T. (1989). *Molecular cloning: a labora-*

tory manual, 2nd edn, p. 6.22. Cold Spring Harbor Laboratory Press, Cold Spring Harbor, NY.
27. Tautz, D. and Renz, M. (1983). *Anal. Biochem.*, **132**, 14.
28. Krowczynska, A. M., Donoghue, K., and Hughes, L. (1995). *BioTechniques*, **18**, 698.
29. Kalman, L. V., Abramson, R. D., and Gelfand, D. H. (1995). *Genome Sci. Technol.*, **1**, 42.
30. Reichert, F. L., Kalman, L. V., Wang, A. M., Gelfand, D. H., and Abramson, R. D. (1995). *Genome Sci. Technol.*, **1**, 46.
31. Spurgeon, S. L., Reichert, F. L., Koepf, S. M., Chen, S. M., Abramson, R. D., Brandis, J., Kalman, L. V., and Wang, A. M. (1995). *Genome Sci. Technol.*, **1**, 48.
32. Krakowski, K., Bunville, J., Seto, J., Baskin, D., and Seto, D. (1995). *Nucleic Acids Res.*, **23**, 4930.
33. Leren, T. P., Rodningen, O. K., Rosby, O., Solberg, K., and Berg, K. (1993). *BioTechniques*, **14**, 618.
34. Phelps, R. S., Chadwick, R. B., Conrad, M. P., Kronick, M. N., and Kamb, A. (1995). *BioTechniques*, **19**, 984.
35. Chadwick, R. B., Conrad, M. P., McGinnis, M. D., Johnston-Dow, L., Spurgeon, S. L., and Kronick, M. N., (1996). *BioTechniques*, **20**, 676.
36. Verpy, E., Biasotto, M., Meo, T., and Tosi, M. (1994). *Proc. Natl Acad. Sci. USA*, **91**, 1873.
37. Livak, K. J., Flood, S. J. A., Marmaro, J., Giusti, W. and Deetz K. (1995). *PCR Methods Applic.*, **4**, 357.
38. Cotton, R. G. H., Rodriguez, N. R., and Campbell, R. D. (1988). *Proc. Natl Acad. Sci. USA*, **85**, 4397.
39. Lee, L. G., Connell, C. R., and Bloch, W. (1993). *Nucleic Acids Res.*, **21**, 3761.
40. Gastier, J. M., Pulido, J. C., Sunden, S., Brody, T., Buetow, K. H., Murray, J. C., Weber, J. L., Hudson, T. J., Sheffield, V. C., and Duyk, G. M. (1995). *Hum. Mol. Genet.*, **4**, 1829.
41. Hauge, X. Y. and Litt, M. (1993). *Hum. Mol. Genet.*, **2**, 411.
42. Chamberlain, J. S. and Chamberlain, J. R. (1994). In *The Polymerase chain reaction* (ed. K. G. Mullis, F. Ferré, and R. A. Gibbs), p. 38. Birkhäuser Press, Boston, MA.
43. Shuber, A. P., Grondin, V. J., and Klinger, K. W. (1995). *Genome Res.*, **5**, 488.
44. Kimpton, C. P., Gill, P, Walton, A., Urquhart, A., Millican, E. S., and Adams, M. (1993). *PCR Methods Applic.*, **3**, 13.
45. LeDuc, C., Miller, P., Lichter, J., and Parry P. (1995). *PCR Methods Applic.*, **4**, 331.
46. Clark, J. M. (1988). *Nucleic Acids Res.*, **16**, 9677.
47. Hu, G. (1993). *DNA Cell Biol.*, **12**, 763.
48. Smith, J. R., Carpten, J. D., Brownstein M. J., Ghosh, S., Magnuson, V. L., Gilbert D. A., Trent, J. M. and Collins, F. S. (1996). *Genome Res.*, **5**, 312.
49. Frégeau, C. J., and Fourney, R. M. (1993). *BioTechniques*, **15**, 100.
50. He, L., Mansfield, D. C., Brown, A. F., Green, D. K., Morris, S. W., St. Clair, D. M., Muir, W. J., Maclean, A., Wright, A. F., and Blackwood, D. H. (1995). *Am. J. Med. Genet.*, **60**, 192.
51. Menotti-Raymond M. and O'Brien S. J. (1995). *J. Hered.*, **86**, 319.
52. Pritchard, L. E., Kawaguchi, Y., Reed, P. W., Copeman, J. B., Davies, J. L., Barnett, A. H., Bain, S. C., and Todd, J. A. (1995). *Hum. Mol. Genet.*, **4**, 197.

53. Callen, D. F., Thompsom, A. D., Shen, Y., Phillips, H. A., Richards, R. I., Mulley, J. C., and Sutherland, G. R. (1993). *Am. J. Hum. Genet.*, **52**, 922.
54. Grundmann, H., Schneider, C., Tichy, H. V., Simon, R., Klare, I., Hartung, D., and Daschner, F. D. (1995). *J. Med. Microbiol.*, **43**, 446.
55. Tully, G., Sullivan, K. M., and Gill, P. (1993). *Hum. Genet.*, **92**, 554.
56. Ellison, J., Dean, M., and Goldman, D. (1993). *BioTechniques*, **15**, 684.
57. Makino, R., Yazyu, H., Kishimoto, Y., Sekiya, T., and Hayashi, K. (1992). *PCR Methods Applic.*, **2**, 10.
58. Iwahana, H., Fujimura, M., Takahashi, Y., Iwabuchi, T., Yoshimoto, K., and Itakura, M. (1996). *BioTechniques*, **21**, 510.
59. Iwahana, H., Yoshimoto, K., Mizusawa, N., Kudo, E., and Itakura, M. (1994). *BioTechniques*, **16**, 296.

Stain recipes for specific enzymes

BERNIE MAY

Sigma Chemical Company numbers are given in parentheses following each compound. The number to the left of each item is for rapid visual measurement (see Chapter 1, Section 3.3(k)).

Acid phosphatase (ACP)
3 100 mg α-naphthyl acid phosphate (N-7000)
3 50 mg Fast Garnet GBC salt (F-0875)
0.02 M sodium acetate (S-8750) buffer pH 5.0

Aconitase (AC)
16 units IDH (I-2002)
10 mg NADP (N-0505)
3 75 mg aconitic acid (A-7251)—adjust pH to 8.0 with NaOH
MTT (M-2128)/PMS (P-9625)

Adenosine deaminase (ADA)
1 100 mg arsenic acid (A-6756)
3 80 mg adenosine (A-9251)
1 unit nucleoside phosphorylase (N-3003)
2 units xanthine oxidase (X-1875, X-4875)
5 ml 0.01 M sodium phospate (S-0751 and S-0876) buffer, pH 7.0
MTT (M-2128)/PMS (P-9625)
Use: as agar overlay

Adenylate kinase (AK)
3 200 mg glucose (G-5000)
2 100 mg ADP (A-6646)
60 units glucose-6-phosphate dehydrogenase (G-7750)
10 mg NADP (N-0505)
MTT (M-2128)/PMS (P-9625)
3.5 100 units hexokinase (H-5000) (added just before staining)
Use: as agar overlay

Alcohol dehydrogenase (ADH)
2 ml ethanol
20 mg NAD (N-7004)
MTT (M-2128)/PMS (P-9625)

Aldolase (ALD)

3 100 mg fructose-1,6-diphosphate (752–1)
2 50 mg arsenic acid (A-6756)
100 units glyceraldehyde-3-phosphate dehydrogenase (G-0763)
20 mg NAD (N-7004)
MTT (M-2128)/PMS (P-9625)

Alkaline phosphatase (AKP)

3 100 mg β-naphthyl acid phosphate (N-7375)
3 50 mg Fast Garnet GBC salt (F-0875)

Aspartate aminotransferase (AAT)

50 ml AAT buffer (for 1 litre: 0.75 g α-ketoglutarate (K-1750), 2.75 g L-aspartate (A-9256), 1.00 g EDTA (ED2SS), 10.00 g polyvinylpyrrolidone (PVP-40), 15 g NaH_2PO_4 (S-0876), 15 g Na_2HPO_4 (S-0751), in H_2O)
4 1 g Fast Garnet GBC salt (F-0875)
3 200 mg Fast Blue BB salt (optional) (F-0250)
Use: pour on, wait 1 min, pour off; cover with plastic wrap; incubate.

Catalase (CAT)

10 ml 0.01 M sodium phospate (S-0751 and S-0876) buffer pH 7.0
7 ml 0.1 M sodium thiosulfate (S-8503)
5 ml 3% H_2O_2
78 ml H_2O
Use: incubate in above solution for 15 min, pour off, flood tray with 0.1 M KI (P-8256)

Creatine kinase (CK)

3 150 mg phosphocreatine (P-6502)
3 100 mg glucose (G-5000)
2 15 mg ADP (A-6646)
60 units glucose-6-phosphate dehydrogenase (G-7750)
3.5 100 units hexokinase (H-5000)
10 mg NADP (N-0505)
MTT (M-2128)/PMS (P-9625)
Use: as agar overlay

Diaphorase (DIA)

2 25 mg NADH (N-8129)
10 mg MTT (M-2128)
0.5 2 mg 2,6 dichlorophenol-indophenol (D-1878)

Esterase (EST)

2 50 mg β-naphthyl acetate (N-6875)
2 50 mg α-naphthyl acetate (N-8505)
2 50 mg Fast Blue BB salt (F-0250)
Dissolve the naphthyl acetates in 1 ml acetone.

Esterase (EST-F) (fluorescent)

1 30 mg 4-methylumbelliferyl-acetate (M-0883) *or* 30 mg 4-methylumbelliferyl-butyrate (M-7759) in 1 ml acetone
Then add 10 ml R gel buffer

Fructose bisphosphatase (FBP)

3 40 mg fructose-1,6-diphosphate (752–1)
100 units glucosephosphate isomerase (P-9010)
60 units glucose-6-phosphate dehydrogenase (G-7750)

2 50 mg arsenic acid (A-6756)
10 mg NADP (N-0505)
MTT (M-2128)/PMS (P-9625)

Fumarase (FUM)

5 400 mg fumaric acid (F-1506)
500 units malate dehydrogenase (410–13)
20 mg NAD (N-7004)
MTT (M-2128)/PMS (P-9625)
Use: as agar overlay

α-Galactosidase (α-GAL) (fluorescent)

1 10 mg 4-methylumbelliferyl-α-galactosidase (M-7633)
0.1 M citrate buffer pH 4.0

β-Galactosidase (β-GAL) (fluorescent)

1 10 mg 4-methylumbelliferyl-β-galactosidase (M-1633)
10 ml 0.1 M citrate (C-0759) buffer, adjusted to pH 4.0 with Na_2HPO_4

Galactosaminidase (GAM) (fluorescent)

10 ml Na_2HPO_4 (S-0751) or NaH_2PO4 (S-0876) buffer

1.5 20 mg 4-methylumbelliferyl-*N*-acetyl-β-D-galactosaminide (M-9129)

Glucokinase (GK)

3 90 mg glucose (G-5000)

2 50 mg ATP (A-5394)
60 units glucose-6-phosphate dehydrogenase (G-7750)
10 mg NADP (N-0505)
MTT (M-2128)/PMS (P-9625)

Glucose-6-phosphate dehydrogenase (G6PDH)

3 100 mg glucose-6-phosphate (G-7250)
10 mg NADP (N-0505)
MTT (M-2128)/PMS (P-9625)

Glucose phosphate isomerase (GPI)

3 50 mg fructose-6-phosphate (F-3627)
60 units glucose-6-phosphate dehydrogenase (G-7750)
10 mg NADP (N-0505)
(2)MTT (M-2128)/(1)PMS (P-9625)
Use: as agar overlay

α-Glucosidase (α-GLU) (fluorescent)

1 10 mg 4-methylumbelliferyl-α-D-glucoside (M-9766)
10 ml 0.1 M citrate (C-0759) buffer, adjusted to pH 4.0 with Na_2HPO_4

β-Glucosidase (β-GLU) (fluorescent)

10 ml R gel buffer

2 75 mg 4-methylumbelliferyl-β-D-glucoside (M-9766)

Glutamic dehydrogenase (GDH)

3.5 200 mg glutamic acid (G-1626)
20 mg NAD (N-7004) or 10 mg NADP (N-0505)
MTT (M-2128)/PMS (P-9625)

Glutamic pyruvic transaminase (GPT) (fluorescent)
20 ml GPT buffer (for 1 litre: 18 g DL-alanine (A-7502), 1.9 g α-keto-glutarate (K-1750; adjust to pH 7.6 with Tris base (T-1503))

2 20 mg NADH (N-8129)
two drops of lactate dehydrogenase (L-2500 or L-2625)

Glutathione reductase (GR)

2 50 mg glutathione (G-4626) (GSSG)

1 5 mg NADPH (N-1630)

25 1 mg 2,6-dichlorophenol-indophenol (D-1878)
10 mg MTT (M-2128)
Use: as agar overlay

Glucoronidase (GUS)

1.5 10 mg naphthol-α-β-glucoronide (N-1875)

2.5 50 mg Fast Garnet GBC salt (F-0875)
100 ml 0.08 sodium acetate (S-8750) buffer, adjusted to pH 5.5 with HCl

Glyceraldehyde-3-phosphate dehydrogenase (GAPDH)

3 55 mg fructose-1,6-diphosphate (752-1)
100 units aldolase (A-1893)

1 150 mg arsenic acid (A-6756)
20 mg NAD (N-7004)
MTT (M-2128)/PMS (P-9625)

Glycerate dehydrogenase (G2DH)

3 100 mg DL-glyceric acid (G-5626)
20 mg NAD (N-7004)
MTT (M-2128)/PMS (P-9625)
Use: as agar overlay

Glycerol-3-phosphate dehydrogenase (G3P)

3 100 mg α-glycerophosphate (G-2138)
20 mg NAD (N-7004)
MTT (M-2128)/PMS (P-9625)

Guanine deaminase (GDA)

2.5 50 mg guanine (G-0506)
2 units xanthine oxidase (X-1875, X-4875)

2 50 mg arsenic acid (A-6756)
5 ml 0.01 M sodium phospate (S-0751 and S-0876) buffer pH 7.0 and R gel buffer, 1:5 MTT (M-2128)/PMS (P-9625)
Use: as agar overlay

Hexoseaminase (HA) (fluorescent)

1 5 mg 4-methylumbelliferyl-acetyl-glucoseaminide (M-2133)
10 ml 0.1 M citrate (C-0759) buffer, adjusted to pH 4.0 with Na_2HPO_4

Hydroxybutyric dehydrogenase (HBDH)

3 50 mg β-hydroxybutyric acid (H-6501)
20 mg NAD (N-7004)
MTT (M-2128)/PMS (P-9625)

Isocitrate dehydrogenase (IDH)

3 75 mg isocitrate (I-1252)
10 mg NADP (N-0505)
MTT (M-2128)/PMS (P-9625)

Lactate dehydrogenase (LDH)

10 ml 0.5 M lactate (L-1250) pH 7.0
20 mg NAD (N-7004)
MTT (M-2128)/PMS (P-9625)

Leucine aminopeptidase (LAP)

2 20 mg L-leucyl-β-naphthylamide (L-0376)
2.5 50 mg Fast Garnet GBC salt (F-0875)
100 ml 0.05M Tris–HCl extraction buffer pH 7.0

Malate dehydrogenase (MDH)

5 ml 0.5 M malate, pH 7.0 (for 500 ml: 33.5 g DL-malic acid (M-0875), 55.0 g $NaHCO_3$ (S-8875))
20 mg NAD (N-7004)
MTT (M-2128)/PMS (P-9625)

Malic enzyme (ME)

10 ml 0.3 M malate (see MDH)
10 mg NADP (N-0505)
MTT (M-2128)/PMS (P-9625)

Mannitol dehydrogenase (MADH)

3 100 mg mannitol (M-4125)
10 mg NADP (N-0505)
MTT (M-2128)/PMS (P-9625)

Mannose phosphate isomerase (MPI)

2.5 35 mg mannose-6-phosphate (M-8754)
60 units glucose-6-phosphate dehydrogenase (G-7750)
100 units glucose phosphate isomerase (P-9010)
10 mg NADP (N-0505)
MTT (M-2128)/PMS (P-9625)
Use: as agar overlay

α-Mannosidase (α-MAN) (fluorescent)

1 5 mg 4-methylumbelliferyl-α-D-mannopyraniside (M-4383)
10 ml 0.1 M citrate (C-0759) buffer, adjusted to pH 4.0 with Na_2HPO_4

Methylumbelliferyl phosphatase (MUP) (fluorescent)

2 50 mg 4-methylumbelliferyl-phosphate (M-8883)
10 ml 0.1 M citrate (C-0759) buffer, adjusted to pH 4.0 with Na_2HPO_4

Nucleoside phosphorylase (NP)

3 200 mg inosine (I-4125)

2 200 mg arsenic acid (A-6756)
2 units xanthine oxidase (X-1875, X-4875)
MTT (M-2128)/PMS (P-9625)
5 ml 0.01 M sodium phospate (S-0751 and S-0876) buffer pH 7.0
Use: as agar overlay

Octanol dehydrogenase (ODH)
1 ml 2-octanol (O-4750)/ethanol
20 mg NAD (N-7004)
MTT (M-2128)/PMS (P-9625)
Use: mix 2-octanol 1:4 with ethanol; place at 37^°C prior to staining.

Peptidase (PEP)
2.5 50 mg glycyl-leucine (G-2002) *or* leucyl-alanine (L-9250) *or* leucyl-leucyl-leucine (L-0879) *or* leucyl-glycyl-glycine (L-9750) *or* phenyl-alanyl-proline (P-6258), etç.
3 20 mg peroxidase (P-8125)
1 15 mg *O*-dianisidine (D-3252)
1 20 mg amino acid oxidase (rattlesnake venom) (V-7000)
Use: as agar overlay; mix just before staining

Peroxidase (PER)
25 ml 0.1M sodium acetate (S-8750) buffer pH 5.0
20 mg 3-amino-9-ethylcarbazole (A-5754) dissolved in 3 ml of *N,N*-dimethylformamide (D-4254)
0.5 ml 3% H_2O_2

Phosphofructokinase (PFK) (fluorescent)
2 10 mg ATP (A-5394)
2 50 mg fructose-6-phosphate (F-3627)
2 50 mg arsenic acid (A-6756)
500 units triosephosphate isomerase (T-2391)
40 units aldolase (A-1893)
40 units GAPDH (G-0763)
20 mg NAD (N-7004)
1 ml $MgCl_2$ (M-9272)

Phosphoglucomutase (PGM)
4 100 mg glucose-1-phosphate (G-1259)
60 units glucose-6-phosphate dehydrogenase (G-7750)
10 mg NADP (N-0505)
MTT (M-2128)/PMS (P-9625)
Use: as agar overlay

Phosphogluconate dehydrogenase (PGD)
2 80 mg 6-phosphogluconate (P-7627)
10 mg NADP (N-0505)
MTT (M-2128)/PMS (P-9625)

Phosphoglycerate kinase (PGK) (fluorescent)
2 20 mg 3-phosphoglycerate (P-8877, P-0769)

2 20 mg NADH (N-8129)
2 20 mg ATP (A-5394)
10 mg $MgCl_2$ (M-9272)
100 units glyceraldehyde-3-phosphate dehydrogenase (G-0763)

General protein (PRO)
0.05% nigrosin (N-4754)
0.05% Buffalo Black (N-9002)
Use: stain for 10–15 min; rinse in H_2O; destain repeatedly in 1:4:5 acetic acid:H_2O:ethanol

Inorganic pyrophosphatase (PYR)
(a) 10 ml pyrophosphate buffer, pH 7.8: 0.05 M Tris (T-1503), 0.01 M pyrophosphate (P-9146))
1 ml $MgCl_2$ (M-9272)
Use: as agar overlay; incubate for 15 min; scrape off agar
(b) 10 ml ammonium molybdate (M-0878) buffer (per litre: 25 g ammonium molybdate, 40 ml H_2SO_4)
2 50 mg ascorbic acid (A-7631)
Use: as agar overlay

Pyruvic kinase (PK) (fluorescent)
10 ml R gel buffer
1.5 50 mg phosphoenolpyruvate (P-7127)
2 30 mg NADH (N-8129)
10 mg $MgCl_2$ (M-9272)
1 10 mg ADP (A-6646)
1 10 mg EDTA (ED2SS)
1 10 mg fructose-1,6-diphosphate (752-1)
1 100 units lactate dehydrogenase (L-2500 or L-2625)

Shikimic kinase (SK)
3 100 mg shikimic acid (S-5375)
10 mg NADP (N-0505)
MTT (M-2128)/PMS (P-9625)

Sorbitol dehydrogenase (SDH)
3 250 mg sorbitol (S-1876)
20 mg NAD (N-7004)
MTT (M-2128)/PMS (P-9625)

Superoxide dismutase (SOD)
20 mg NBT (N-6876)
6 mg PMS (P-9625)
20 mg NAD (N-7004) (optional)
Use: leave in the light ('negative' stain prevents deposition of dye)

Triosephosphate isomerase (TPI)
0.5 5 mg dihydroxyacetone phosphate (D-7137)
100 units GAPDH (G-0763)

2 50 mg arsenic acid (A-6756)
20 mg NAD (N-7004)
MTT (M-2128)/PMS (P-9625)
Use: as agar overlay

Xanthine dehydrogenase (XDH)

2 60 mg hypoxanthine (H-9377)
20 mg NAD (N-7004)
MTT (M-2128)/PMS (P-9625)
Use: heat substrate until it dissolves (do *not* bring to boil), and allow to cool to room temperature before adding other ingredients.

Restriction enzymes

RICHARD J. ROBERTS

Restriction enzymes are endodeoxyribonucleases that recognize short, specific sequences within DNA molecules and then catalyse double-strand cleavage of the DNA. Three distinct classes of restriction enzymes are known:

(a) *Type I enzymes* recognize a specific sequence in DNA and cleave the DNA chain at random locations with respect to that sequence. They have an absolute requirement for the co-factors *S*-adenosylmethionine and ATP, and during cleavage they hydrolyse the ATP. Because of the random nature of the cleavage, the products are a heterogeneous array of DNA fragments.

(b) *Type II enzymes* also recognize a specific nucleotide sequence, but differ from the type I enzymes in cleaving only at a specific location within or close to the recognition sequence, thus generating a unique set of fragments. At the present time, more than 2900 such enzymes have been characterized, among which more than 180 different sequence specificities occur. In cases where enzymes from different sources recognize the same sequence, then the enzymes are called isoschizomers.

(c) *Type III enzymes* have properties intermediate between those of the type I and the type II enzymes. They recognize a specific DNA sequence and cleave a short distance away from it. However, it is usually difficult to obtain complete digestion. They have an absolute requirement for ATP but they do not hydrolyse it, and although their activity is stimulated by *S*-adenosylmethionine it is not an absolute requirement.

In this Appendix, type II restriction enzymes and their neoschizomers are listed alphabetically by prototype. Also included are all commercially available isoschizomers and their sources are indicated, as are the buffer conditions recommended by the manufacturer. It should be noted that for most restriction enzymes, their activity varies little over a wide range of ionic strength and pH, and the values listed in general have not rigorously been shown to be optimal. The information in this list is taken from Roberts, R. J. and Macelis, D. (1996) *Nucleic Acids Research* **24**, 223–235 and includes updates from REBASE, the restriction enzyme database (URL: `http://www.neb.com/rebase`).

Table 1. Type II restriction endonucleases

Name	Isoschizomer(s)	Recognition sequence	NaCl	Tris	pH	Mg	SH	BSA	other	temp. Source[b]
			Digestion conditions[a]							
*Aat*II		GACGT ↓ C	50 Ka	20 A	7.9	10 A	1			ADEFKLMNOPRS
*Acc*I		GT ↓ MKAC	50 Ka	20 A	7.9	10 A	1			ABDEGJKLMNOPQRS
*Ace*III		CAGCTC(7/11)								
*Aci*I		CCGC(–3/–1)	100	50	7.9	10	1			N
*Acl*I		AA ↓ CGTT	66 K	33 A	7.9	10	1 B		25 °C	I
	*Psp*1406I	AA ↓ CGTT	66 Ka	33 A	7.9	10 A		100		DFMN
*Acy*I		GR ↓ CGYC	100	10	8.0	5	1 B		50 °C	JMR
	*Bsa*HI	GR ↓ CGYC	50 Ka	20 A	7.9	10 A	1	100		NS
	*Hin*1I	GR ↓ CGYC	100	10	8.0	5	1 B	100		DEFO
	*Hsp*92I	GR ↓ CGYC	100	10	8.5	10	1			R
	*Msp*17I	GR ↓ CGYC	100	10	8.5	10	1			I
*Afl*II		C ↓ TTAAG	50	10	7.9	10	1	100		ABJKNOS
	*Bfr*I	C ↓ TTAAG	50	10	7.5	10	1 E			MO
	*Bsp*TI	C ↓ TTAAG	100	50	7.5	10	1 E	100		DF
	*Bst*98I	C ↓ TTAAG	150	6	7.9	6	1			R
	*Msp*CI	C ↓ TTAAG	150	10	7.9	10	1			C
	*Vha*464I	C ↓ TTAAG	50	10	7.6	10	1			I
*Afl*III		A ↓ CRYGT	100	50	7.9	10	1	100		ABMN
*Age*I		A ↓ CCGGT		10 B	7.0	10	1		25 °C	JNR
	*Asi*AI	A ↓ CCGGT		10	7.6	10	1 B		25 °C	I
	*Pin*AI	A ↓ CCGGT	50 K	20	7.4	5				BM
*Aha*III		TTT ↓ AAA								
	*Dra*I	TTT ↓ AAA	50 Ka	20 A	7.9	10 A	1			ABDEFGIJKLMNOPQRS
*Alu*I		AG ↓ CT	10 B	7.0	10	1				ABCDEFGHIJKLMNOPQRST
*Alw*NI		CAGNNN ↓ CTG	50 Ka	20 A	7.9	10 A	1			ABNS
*Apa*I		GGGCC ↓ C	50 Ka	20 A	7.9	10 A	1	100		ABEGIJKLMNOPQRS
	*Bsp*120I	G ↓ GGCCC		10	7.5	10		100		DFN
	*Psp*OMI	G ↓ GGCCC	66 Ka	33 A	7.9	10 A	1			I

*Apa*BI		GCANNNNN ↓ TGC									
*Apa*LI		G ↓ TGCAC	50 Ka	20 A	7.9	10 A	1	100			ADEKNS
	*Alw*44I	G ↓ TGCAC	100 K	10	8.5	10		100			FJMOR
	*Vne*I	G ↓ TGCAC	100	50	7.6	10	1				I
*Apo*I		R ↓ AATTY	100	50	7.9	10	1	100		50 °C	N
	*Acs*I	R ↓ AATTY	100	10	8.0	5	1 B	100		50 °C	DIM
*Asc*I		GG ↓ CGCGCC	50 Ka	20 A	7.9	10 A	1				N
*Asu*I		G ↓ GNCC									
	*Asp*S9I	G ↓ GNCC 100	10	8.5	10	1				I	
	*Bsi*ZI	G ↓ GNCC 100	10	7.5	10	1	100		60 °C	T	
	*Cfr*13I	G ↓ GNCC 66 Ka	33 A	7.9	10 A	0.5	100			DFKO	
	*Fmu*I	GGNC ↓ C									
	*Sau*96I	G ↓ GNCC 50 Ka	20 A	7.9	10 A	1				AJLMNORS	
*Asu*II		TT ↓ CGAA									
	*Bpu*14I	TT ↓ CGAA	50	10	7.6	10	1				I
	*Bsi*CI	TT ↓ CGAA		10	7.5	10	1	100		60 °C	T
	*Bsp*119I	TT ↓ CGAA	100	10	8.0	5	1 B	100			DF
	*Bst*BI	TT ↓ CGAA	50 Ka	20 A	7.9	10 A	1			65 °C	NS
	*Cbi*I	TT ↓ CGAA	50	10	7.5	10	1				J
	*Csp*45I	TT ↓ CGAA	60	10	7.5	7					OR
	*Lsp*I	TT ↓ CGAA	50	10	7.5	10	6 B	100	0.1 T	60 °C	L
	*Nsp*V	TT ↓ CGAA		10	7.5	10	1				ABJKO
	*Sfu*I	TT ↓ CGAA	100	50	7.5	10	1 E				M
*Ava*I		C ↓ YCGRG	50 Ka	20 A	7.9	10 A	1				ABEGJKLMNOPQRS
	*Ama*87I	C ↓ YCGRG	100	10	8.5	10	1				I
	*Bco*I	C ↓ YCGRG	100	50	7.5	10	1			65 °C	AT
	*Bso*BI	C ↓ YCGRG	50	10	7.9	10	1				N
	*Eco*88I	C ↓ YCGRG		10	7.5	10	1 E	100		4 °C	DF
	*Nli*387/7I	CYCGR ↓ G									
*Ava*II		G ↓ GWCC	50 Ka	20 A	7.9	10 A	1				ABEGJKMNPQRS
	*Bme*18I	G ↓ GWCC	100	50	7.6	10	1				I
	*Eco*47I	G ↓ GWCC	100 K	10	8.5	10					FO
	*Hgi*EI	G ↓ GWCC	100 K	10	8.5	10	1	100			D

Table 1. Type II restriction endonucleases

Name	Isoschizomer(s)	Recognition sequence	Digestion conditions[a] NaCl	Tris	pH	Mg	SH	BSA	other temp.	Source[b]
	*Sin*I	G ↓ GWCC		10	7.8	10	1	100		LR
	*Vpa*K11AI	↓ GGWCC								
*Ava*III		ATGCAT								
	*Bfr*BI	ATG ↓ CAT								
	*Eco*T22I	ATGCA ↓ T	100	50	7.5	10	1			AKO
	*Mph*1103I	ATGCA ↓ T	100 K	10	8.5	10		100		F
	*Nsi*I	ATGCA ↓ T	100	10	8.4	10	1			BDEHJLMNRS
	*Ppu*10I	A ↓ TGCAT	66 Ka	33 A	7.9	10 A		100		DFN
	*Zsp*2I	ATGCA ↓ T		10	7.6	10	1			I
*Avr*II		C ↓ CTAGG	50	10	7.9	10	1			N
	*Bln*I	C ↓ CTAGG	100 K	20	8.5	10	1			AKM
	*Bsp*A2I	C ↓ CTAGG	66 K	33 A	7.9	10	1		25 °C	I
*Bae*I		ACNNNNGTAYC[c]								
*Bal*I		TGG ↓ CCA		20	8.5	7	7 B	100		AJKR
	*Mlu*NI	TGG ↓ CCA	66 Ka	33 A	7.9	10 A	0.5			M
	*Msc*I	TGG ↓ CCA	50 Ka	20 A	7.9	10 A	1			BDNOS
*Bam*HI		G ↓ GATCC	150	10	7.9	10	1	100		ABCDEFGHIJKLMNOPQRST
	*Bst*I	G ↓ GATCC								
*Bbv*I		GCAGC(8/12)	50	10	7.9	10	1			IN
	*Bst*71I	GCAGC(8/12)	150	6	7.9	6	1		50 °C	R
*Bbv*II		GAAGAC(2/6)								
	*Bbs*I	GAAGAC(2/6)	50	10	7.9	10	1			N
	*Bbv*16II	GAAGAC(2/6)	50	10	7.6	10	1			I
	*Bpi*I	GAAGAC(2/6)	66 Ka	33 A	7.9	10 A	0.5	100		DF
	*Bpu*AI	GAAGAC(2/6)	100	10	8.0	5	1 B			M
*Bcc*I		CCATC								
*Bce*83I		CTTGAG(16/14)								
*Bcef*I		ACGGC(12/13)								

*Bcg*I		GCANNNNNNTCG[d]	100	10	8.4	10	1		20 A		N
*BciV*I		GGATAC									
*Bcl*I		T ↓ GATCA	100	50	7.9	10	1			50 °C	BCDEFGJLMNOPRS
	*Bsi*QI	T ↓ GATCA	50	10	7.5	10	1	100		60 °C	T
	*Fba*I	T ↓ GATCA	100 K	20	8.5	10	1				AK
	*Ksp*22I	T ↓ GATCA	50	10	7.6	10	1				I
*Bet*I		W ↓ CCGGW									
	*Bsa*WI	W ↓ CCGGW	50	10	7.9	10	1	100		60 °C	N
*Bfi*I		ACTGGG									
*Bgl*I		GCCNNNN ↓ NGGC	100	50	7.9	10	1				ABCDEFGHIJKLMNOPQRST
*Bgl*II		A ↓ GATCT	100	50	7.9	10	1				ABCDEFGHIJKLMNOPQRST
*Bin*I		GGATC(4/5)									
	*AcW*I	GGATC(4/5)	66 Ka	33 A	7.9	10 A	1				I
	*Alw*I	GGATC(4/5)	50 Ka	20 A	7.9	10 A	1				NS
*Bmg*I		GKGCCC									
*Bpl*I		GAGNNNNNCTC									
*Bpu*10I		CCTNAGC(−5/−2)									
*Bsa*AI		YAC ↓ GTR	100	50	7.9	10	1				N
	*Bst*BAI	YAC ↓ GTR	100	10	8.5	10	1 B			25 °C	I
*Bsa*BI		GATNN ↓ NNATC	50	10	7.9	10	1			60 °C	N
	*Bse*8I	GATNN ↓ NNATC	50	10	7.6	10	1			60 °C	I
	*Bsh*1365I	GATNN ↓ NNATC		10	7.5	10	1 E	100			DF
	*Bsi*BI	GATNN ↓ NNATC	100	10	7.5	10	1	100		55 °C	T
	*Bsr*BRI	GATNN ↓ NNATC	50	90	7.5	10					R
	*Mam*I	GATNN ↓ NNATC	100	50	7.5	10	1 E				M
*Bsa*XI		ACNNNNNCTCC									
*Bsb*I		CAACAC									
*Bsc*GI		CCCGT									
*Bse*PI		G ↓ CGCGC	50	10	7.6	10	1			50 °C	I
	*Bss*HII	G ↓ CGCGC	100	10	7.9	10	1	100		50 °C	ABDEJKLMNOPQR
	*Pau*I	G ↓ CGCGC	100 K	10	8.5	10				50 °C	F
*Bse*RI		GAGGAG(10/8)		10 B	7.0	10	1				N
*Bsg*I		GTGCAG(16/14)	50 Ka	20 A	7.9	10 A	1		80 A		N

Table 1. *Continued*

Name	Isoschizomer(s)	Recognition sequence	Digestion conditions[a] NaCl	Tris	pH	Mg	SH	BSA	other	temp.	Source[b]
*Bsi*I		CACGAG(–5/–1)									
	*Bss*SI	CACGAG(–5/–1)	100	50	7.9	10	1				N
	*Bst*2BI	CACGAG(–5/–1)	66 Ka	33 A	7.9	10 A	1			60 °C	I
*Bsi*YI		CCNNNNN ↓ NNGG	50	10	7.5	10	1 E			55 °C	MT
	*Bsc*4I	CCNNNNN ↓ NNGG	100	10	8.5	10	1			55 °C	I
	*Bsl*I	CCNNNNN ↓ NNGG	100	50	7.9	10	1			55 °C	N
*Bsm*I		GAATGC(1/–1)	50	10	7.9	10	1			65 °C	ABDEJLMNOS
	*Bsa*MI	GAATGC(1/–1)	150	6	7.9	6	1			65 °C	R
	*Bsc*CI	GAATGC(1/–1)	100	10	7.5	10	1	100		65 °C	T
	*Mva*1269I	GAATGC(1/–1)	100 K	10	8.5	10		100			F
*Bsm*AI		GTCTC(1/5)	100	10	8.4	10	10	100		55 °C	N
	*Alw*26I	GTCTC(1/5)	50	10	7.9	10	1				DFNR
*Bsp*24I		GACNNNNNNTGG[c]									
*Bsp*1407I		T ↓ GTACA	66 Ka	33 A	7.9	10 A	0.5	100			DF
	*Bsr*GI	T ↓ GTACA	50	10	7.9	10	1	100			N
	*Ssp*BI	T ↓ GTACA	100	10	8.0	5	1 B			50 °C	M
*Bsp*GI		CTGGAC									
*Bsp*HI		T ↓ CATGA	50 Ka	20 A	7.9	10 A	1				AN
	*Rca*I	T ↓ CATGA	50 K	20	7.4	5					BM
*Bsp*LU11I		A ↓ CATGT	100	50	7.5	10	1 E			48 °C	M
*Bsp*MI		ACCTGC(4/8)	150	10	7.9	10	1				N
*Bsp*MII		T ↓ CCGGA									
	*Acc*III	T ↓ CCGGA	100	10	8.5	10	1	10		60 °C	EJKQR
	*Bse*AI	T ↓ CCGGA	100	50	7.9	10	1	100	0.02 T	65 °C	CM
	*Bsi*MI	T ↓ CCGGA	100	50	7.5	10	1			55 °C	AT
	*Bsp*13I	T ↓ CCGGA	200 K	10	7.6	10	1	100		50 °C	I
	*Bsp*EI	T ↓ CCGGA	100	50	7.9	10	1				N
	*Kpn*2I	T ↓ CCGGA	50 K	20	7.4	5				55 °C	BDF

	*Mro*I	T ↓ CCGGA	66 Ka	33 A	7.9	10 A	0.5				MO
*Bsr*I		ACTGG(1/–1)	150 K	10	7.8	10		100		65 °C	N
	*Bse*1I	ACTGG(1/–1)	66 Ka	33 A	7.9	10 A	1			65 °C	I
	*Bse*NI	ACTGG(1/–1)	66 Ka	33 A	7.9	10 A	0.5	100		65 °C	DF
	*Bsr*SI	ACTGG(1/–1)	150	6	7.3	6	1			65 °C	R
*Bsr*BI		CCGCTC(–3/–3)	50	10	7.9	10	1				N
	*Acc*BSI	CCGCTC(–3/–3)	66 Ka	33 A	7.9	10 A	1				I
	*Bst*D102I	CCGCTC(–3/–3)	50	10	7.6	10	1 B				P
*Bsr*DI		GCAATG(2/0)	50	10	7.9	10	1	100		60 °C	N
	*Bse*3DI	GCAATG(2/0)	50	10	7.6	10	1 B			25 °C	I
*Bst*EII		G ↓ GTNACC	100	50	7.9	10	1			60 °C	BDGHJLMNOPRS
	*Bst*PI	G ↓ GTNACC									K
	*Eco*91I	G ↓ GTNACC	100	50	7.5	10		100			F
	*Eco*O65I	G ↓ GTNACC	100	50	7.5	10	1	100			AK
	*Psp*EI	G ↓ GTNACC		10	7.6	10	1				I
*Bst*XI		CCANNNNN ↓ NTGG		100	50	7.9	10	1		55 °C	ABDEFGHJKLMNOQRS
*Cac*8I		GCN ↓ NGC	100	50	7.9	10	1				N
*Cau*II		CC ↓ SGG									
	*Bcn*I	CC ↓ SGG 100	50	7.5	10	1					ADFK
	*Eco*HI	↓ CCSGG									
	*Nci*I	CC ↓ SGG 50 Ka	20 A	7.9	10 A	1					BEJLMNORST
*Cfr*I		Y ↓ GGCCR	66 K	33 A	7.9	10 S		100			F
	*Eae*I	Y ↓ GGCCR		10 B	7.0	10	1				ADKLMNS
*Cfr*10I		R ↓ CCGGY	100 K	20	8.5	3 S			0.02 T		ADFKMO
	*Bse*118I	R ↓ CCGGY	100	50	7.6	10	1			65 °C	I
	*Bsr*FI	R ↓ CCGGY	50	10	7.9	10	1	100			N
	*Bss*AI	R ↓ CCGGY	100 K	20	8.5	3		100	0.04 T	65 °C	C
*Cje*I		CCANNNNNNGT[c]									
*Cje*PI		CCANNNNNNNTC[c]									
*Cla*I		AT ↓ CGAT	50 Ka	20 A	7.9	10 A	1	100			ABKMNPRST
	*Ban*III	AT ↓ CGAT	80 K	10	7.5	7	7 B				O
	*Bsa*29I	AT ↓ CGAT	50	10	7.6	10	1				I
	*Bsc*I	AT ↓ CGAT	100	50	7.8	10	1	100			L

Table 1. *Continued*

Name	Isoschizomer(s)	Recognition sequence	Digestion conditions[a] NaCl	Tris	pH	Mg	SH	BSA	other temp.	Source[b]
	*Bse*CI	AT ↓ CGAT	100	50	7.9	10	1		55 °C	C
	*Bsi*XI	AT ↓ CGAT	50	10	7.5	10	1	100	65 °C	T
	*Bsp*106I	AT ↓ CGAT	100 Ka	25 A	7.6	10 A	0.5 B	10		E
	*Bsp*DI	AT ↓ CGAT	50 Ka	20 A	7.9	10 A	1			N
	*Bsp*XI	AT ↓ CGAT	50	10	8.0	10	10 B	100		G
	*Bsu*15I	AT ↓ CGAT	66 Ka	33 A	7.9	10 A	0.5	100		DF
*Cvi*JI		RG ↓ CY	20 G	8.5	10 A	7 B				Q
*Cvi*RI		TG ↓ CA								
*Dde*I		C ↓ TNAG 100	50	7.9	10	1				ABEGLMNOPR
	*Bst*DEI	C ↓ TNAG 50	10	7.6	10	1			60 °C	I
*Dpn*I		GA ↓ TC 50 Ka	20 A	7.9	10 A	1				ABLMNRS
*Dra*II		RG ↓ GNCCY		10	7.5	10	1			AGMS
	*Eco*O109I	RG ↓ GNCCY	50 Ka	20 A	7.9	10 A	1	100		ABDEFJKLN
	*Pss*I	RGGNC ↓ CY								
*Dra*III		CACNNN ↓ GTG	100	50	7.9	10	1	100		AEGMNS
*Drd*I		GACNNNN ↓ NNGTC		50 Ka	20 A	7.9	10 A	1		N
	*Dse*DI	GACNNNN ↓ NNGTC		66 K	33 A	7.9	10	1 B	25 °C	I
*Drd*II		GAACCA								
*Dsa*I		C ↓ CRYGG	100	50	7.5	10	1 E		55 °C	M
	*Bst*DSI	C ↓ CRYGG	66 Ka	33 A	7.9	10 A	1		65 °C	I
*Eam*1105I		GACNNN ↓ NNGTC 100	10	8.0	5	1 B	100		DFK	
	*Ahd*I	GACNNN ↓ NNGTC	50 Ka	20 A	7.9	10 A	1	100		N
	*Asp*EI	GACNNN ↓ NNGTC		10	7.5	10	1 E			M
	*Ecl*HKI	GACNNN ↓ NNGTC	100	6	7.5	6	1			R
	*Nru*GI	GACNNN ↓ NNGTC		10	7.6	10	1 B		25 °C	I
*Eci*I		TCCGCC								
*Eco*31I		GGTCTC(1/5)	66 Ka	33 A	7.9	10 A	0.5	100		DF
	*Bsa*I	GGTCTC(1/5)	50 Ka	20 A	7.9	10 A	1		55 °C	N

*Eco*47III		AGC ↓ GCT	100	50	7.5	10					BDEFLMNOR
	*Afe*I	AGC ↓ GCT	66 Ka	33 A	7.9	10 A	1				I
	*Aor*51HI	AGC ↓ GCT	50	10	7.5	10	1				AK
*Eco*57I		CTGAAG(16/14)		10	7.5	10		100			DFNS
*Eco*NI		CCTNN ↓ NNNAGG	50 Ka	20 A	7.9	10 A	1				ANS
*Eco*RI		G ↓ AATTC	50	100	7.5	10			0.25 T		ABCDEFGHIJKLMNOPQRST
*Eco*RII		↓ CCWGG	50	50	7.4	6	1				BEJMOS
	*Bsi*LI	CC ↓ WGG		10	7.5	10	1	100		60 °C	T
	*Bst*2UI	CC ↓ WGG	50	10	7.6	10	1			60 °C	I
	*Bst*NI	CC ↓ WGG	50	10	7.9	10	1	100		60 °C	CENS
	*Bst*OI	CC ↓ WGG	50	10	7.9	10	1			60 °C	R
	*Mva*I	CC ↓ WGG	100 K	20	8.5	10	1				ADFKMO
*Eco*RV		GAT ↓ ATC	50	10	7.9	10	1	100			ABCDEGHIJKLMNOPQRST
	*Eco*32I	GAT ↓ ATC	100 K	10	8.5	10					F
*Esp*I		GC ↓ TNAGC									
	*Blp*I	GC ↓ TNAGC	50 Ka	20 A	7.9	10 A	1				N
	*Bpu*1102I	GC ↓ TNAGC	50	50	8.0	10					BDEFK
	*Bsp*1720I	GC ↓ TNAGC	50	10	7.6	10	1				I
	*Cel*II	GC ↓ TNAGC	100	50	7.5	10	1				ALM
*Esp*3I		CGTCTC(1/5)	66 Ka	33 A	7.9	10 A	0.5	100			DF
	*Bsm*BI	CGTCTC(1/5)	100	50	7.9	10	1			55 °C	N
*Fau*I		CCCGC(4/6)		10	7.6	10	1			55 °C	I
*Fin*I		GGGAC									
	*Bsm*FI	GGGAC(10/14)	50 Ka	20 A	7.9	10 A	1			65 °C	N
*Fnu*DII		CG ↓ CG									
	*Acc*II	CG ↓ CG	50	10	7.5	10	1				AJKQ
	*Bsh*1236I	CG ↓ CG	100 K	10	8.5	10	1	100			DEF
	*Bst*UI	CG ↓ CG	50	10	7.9	10	1			60 °C	NS
	*Mvn*I	CG ↓ CG	50	10	7.5	10	1 E				M
	*Sel*I	↓ CGCG									
	*Tha*I	CG ↓ CG		50	8.0	10				60 °C	B
*Fnu*4HI		GC ↓ NGC	50 Ka	20 A	7.9	10 A	1				N
	*Bso*FI	GC ↓ NGC	50	10	7.9	10	1			55 °C	N

Table 1. *Continued*

Name	Isoschizomer(s)	Recognition sequence		Digestion conditions[a] NaCl	Tris	pH	Mg	SH	BSA	other temp.	Source[b]
	*Fsp*4HI	GC ↓ NGC 100	50	7.6	10	1					I
	*Ita*I	GC ↓ NGC 100	50	7.5	10	1 E					M
*Fok*I		GGATG(9/13)	50 Ka	20 A	7.9	10 A	1				AGIJKLMNRS
	*Bst*F5I	GGATG(2/0)	66 Ka	33 A	7.9	10 A	1			65 °C	I
	*Sts*I	GGATG(10/14)									
*Fse*I		GGCCGG ↓ CC	50 Ka	20 A	7.9	10 A	1	100			AKN
*Gdi*II		CGGCCR(–5/–1)									
*Gsu*I		CTGGAG(16/14)		10	7.5	10	1 E	100		30 °C	DFS
	*Bpm*I	CTGGAG(16/14)	100	50	7.9	10	1	100			N
*Hae*I		WGG ↓ CCW									
*Hae*II		RGCGC ↓ Y	50 Ka	20 A	7.9	10 A	1	100			ABDEGJKLMNOPRS
	*Acc*B2I	RGCGC ↓ Y	66 Ka	33 A	7.9	10 A	1				I
	*Bsp*143II	RGCGC ↓ Y	66 Ka	33 A	7.9	10 A		100	0.2 T		F
	*Bst*H2I	RGCGC ↓ Y	66 Ka	33 A	7.9	10 A	1			65 °C	I
	*Lpn*I	RGC ↓ GCY									
*Hae*III		GG ↓ CC 50	10	7.9	10	1					ABCDGHIJKLMNOPQRST
	*Bsh*I	GG ↓ CC 50	10	7.5	10	1	100				T
	*Bsu*RI	GG ↓ CC 100 K	10	8.5	10		100				FI
	*Pal*I	GG ↓ CC 50 Ka	12.5 A	7.6	5 A	0.25 B	5				EP
*Hga*I		GACGC(5/10)		10 B	7.0	10	1				AN
*Hgi*AI		GWGCW ↓ C									
	*Alw*21I	GWGCW ↓ C	100 K	10	8.5	10	1	100			DF
	*Asp*HI	GWGCW ↓ C	100	10	8.0	5	1 B				M
	*Bbv*12I	GWGCW ↓ C	100	50	7.6	10	1				I
	*Bsi*HKAI	GWGCW ↓ C	100	50	7.9	10	1	100		65 °C	N
*Hgi*CI		G ↓ GYRCC									
	*Acc*B1I	G ↓ GYRCC	100 K	10	7.5	10	1				I
	*Ban*I	G ↓ GYRCC	50 Ka	20 A	7.9	10 A	1				AEMNOPRS

	*Bsh*NI	G ↓ GYRCC	100	50	7.5	10					F
	*Eco*64I	G ↓ GYRCC	100 K	10	8.5	10	1	100			DF
*Hgi*EII		ACCNNNNNNGGT									
*Hgi*JII		GRGCY ↓ C									
	*Ban*II	GRGCY ↓ C	50 Ka	20 A	7.9	10 A	1				ABKLMNOPQRS
	*Eco*24I	GRGCY ↓ C	66 Ka	33 A	7.9	10 A	0.5	100			DF
	*Eco*T38I	GRGCY ↓ C	50	10	7.5	10	1				J
	*Fri*OI	GRGCY ↓ C	66 Ka	33 A	7.9	10 A	1				I
*Hha*I		GCG ↓ C 50 Ka	20 A	7.9	10 A	1	100				ABGJKNOPRS
	*Asp*LEI	GCG ↓ C 100	50	7.6	10	1					I
	*Cfo*I	GCG ↓ C	50	8.0	10						BLMRS
	*Hin*6I	G ↓ CGC 100	10	8.0	5	1 B	100				DF
	*Hin*P1I	G ↓ CGC 50	10	7.9	10	1					N
	*Hsp*AI	G ↓ CGC 66 Ka	33 A	7.9	10 A	1					I
*Hin*4I		GABNNNNNVTC									
*Hind*II		GTY ↓ RAC	50 Ka	12.5 A	7.6	5 A	0.25 B				EM
	*Hinc*II	GTY ↓ RAC	100	50	7.9	10	1	100			ABCDEFGHJKLNOPQRS
*Hind*III		A ↓ AGCTT	50	10	7.9	10	1				ABCDEFGHIJKLMNOPQRST
*Hinf*I		G ↓ ANTC 50	10	7.9	10	1					ABDEFGHIJKLMNOPQRST
*Hpa*I		GTT ↓ AAC	50 Ka	20 A	7.9	10 A	1				ABDEFGIJKLMNOPQRST
	*Bst*HPI	GTT ↓ AAC	66 K	33 A	7.9	10	1 B			25 °C	I
*Hpa*II		C ↓ CGG	10 B	7.0	10	1					BDEFGILMNOPQRS
	*Bsi*SI	C ↓ CGG 100	50	7.9	10	1	100		55 °C		C
	*Hap*II	C ↓ CGG	10	7.5	10	1					AK
	*Hin*2I	C ↓ CGG 66 Ka	33 A	7.9	10 A		100				F
	*Msp*I	C ↓ CGG 50	10	7.9	10	1					ABDEFGHIJKLMNOPQRS
*Hph*I		GGTGA(8/7)	50 Ka	20 A	7.9	10 A	1				AFNS
	*Asu*HPI	GGTGA(8/7)	100	50	7.6	10	1				I
*Kpn*I		GGTAC ↓ C		10 B	7.0	10	1	100			ABCDEFGHIJKLMNOPQRST
	*Acc*65I	G ↓ GTACC	100	50	7.5	10					DFINR
	*Asp*718I	G ↓ GTACC	100	10	8.0	5	1 B				MP
*Ksp*632I		CTCTTC(1/4)	66 Ka	33 A	7.9	10 A	0.5				M
	*Bsu*6I	CTCTTC(1/4)	66 K	33 A	7.9	10	1 B			25 °C	I

Table 1. *Continued*

Name	Isoschizomer(s)	Recognition sequence			NaCl	Tris	pH	Mg	SH	BSA	other temp.	Source[b]
					Digestion conditions[a]							
	*Eam*1104I	CTCTTC(1/4)		66 Ka	33 A	7.9	10 A	0.5	100			DEF
	*Ear*I	CTCTTC(1/4)			10 B	7.0	10	1				N
*Mae*I		C↓TAG	500	40	8.0	12	14 B	100			45 °C	M
	*Bfa*I	C↓TAG	50 Ka	20 A	7.9	10 A	1					N
*Mae*II		A↓CGT	220	50	8.8	6	7 B				50 °C	MP
	*Tai*I	ACGT↓	100 K	10	8.5	10					65 °C	FN
	*Tsc*I	ACGT↓	50	50	8.3	10	1 B				65 °C	L
*Mae*III		↓GTNAC	275	20	8.2	6	7 B				55 °C	MP
*Mbo*I		↓GATC	100	50	7.9	10	1					ABCEFGKLNQRST
	*Bsc*FI	GATC	100	10	7.5	10	1	100			55 °C	T
	*Bsp*143I	↓GATC	66 Ka	33 A	7.9	10 A	0.5	100				DF
	*Bsp*KT6I	GAT↓C										
	*Cha*I	GATC↓										
	*Dpn*II	↓GATC	100	10 B	6.0	10	1					N
	*Kzo*9I	↓GATC	50	10	7.6	10	1					I
	*Nde*II	↓GATC	150	100	7.6	10						BDGJM
	*Sau*3AI	↓GATC	100	10 B	7.0	10	1	100				ABDEGHJKLMNOPQRS
*Mbo*II		GAAGA(8/7)		50	10	7.9	10	1				ABDFGJKNOQRS
*Mcr*I		CGRY↓CG										
	*Bsa*OI	CGRY↓CG		50	10	7.9	10	1			50 °C	R
	*Bsh*1285I	CGRY↓CG		50	10	7.5	19		100			DF
	*Bsi*EI	CGRY↓CG		50	10	7.9	10	1	100		60 °C	N
	*Bst*MCI	CGRY↓CG			10	7.6	10	1			50 °C	I
*Mfe*I		C↓AATTG		50 Ka	20 A	7.9	10 A	1				N
	*Mun*I	C↓AATTG			50	8.0	10					BDEFKM
*Mlu*I		A↓CGCGT		100	50	7.9	10	1				ABDEFGIJKLMNOPQRS
*Mme*I		TCCRAC(20/18)										
*Mnl*I		CCTC(7/6)	50	10	7.9	10	1	100				ADEFNQ

*Mse*I		T↓TAA	50	10	7.9	10	1	100		BN
	*Tru*1I	T↓TAA	100 K	10	8.5	10		100	65 °C	F
	*Tru*9I	T↓TAA	100	10	8.0	5	1 B	100		DILMRS
*Msl*I		CAYNN↓NNRTG	50	10	7.9	10	1			N
*Mst*I		TGC↓GCA								
	*Acc*16I	TGC↓GCA	100	10	8.0	5	1 B	100		DI
	*Avi*II	TGC↓GCA	100	50	7.5	10	1 E			M
	*Fsp*I	TGC↓GCA	50 Ka	20 A	7.9	10 A	1			ABJKNOS
*Mwo*I		GCNNNNN↓NNGC	150	50	7.9	10	1			N
*Nae*I		GCC↓GGC		10 B	7.0	10	1			ADEKLMNOR
	*Mro*NI	G↓CCGGC		10	7.6	10	1			I
	*Ngo*AIV	G↓CCGGC	50 Ka	20 A	7.9	10 A				B
	*Ngo*MI	G↓CCGGC	50 Ka	20 A	7.9	10 A	1			NR
*Nar*I		GG↓CGCC		10 B	7.0	10	1			BEJMNOPRS
	*Bbe*I	GGCGC↓C		5	7.1	5	2 B	100		AK
	*Ehe*I	GGC↓GCC		20	8.0	10	1			ADFNO
	*Kas*I	G↓GCGCC	50	10	7.9	10	1	100		N
*Nco*I		C↓CATGG	50 Ka	20 A	7.9	10 A	1			ABCDEFGHJKLMNOPQRST
	*Bsp*19I	C↓CATGG	100	10	8.5	10	1			I
*Nde*I		CA↓TATG	50 Ka	20 A	7.9	10 A	1			ABDEFGJKLMNPRS
	*Fau*NDI	CA↓TATG	66 Ka	33 A	7.9	10 A	1			I
*Nhe*I		G↓CTAGC	50	10	7.9	10	1	100		ABDEFGJKLMNOPRS
	*Ace*II	GCTAG↓C								
	*Asu*NHI	G↓CTAGC								I
	*Pst*NHI	G↓CTAGC	66 Ka	33 A	7.9	10 A	1			I
*Nla*III		CATG↓	50 Ka	20 A	7.9	10 A	1	100		ANOS
	*Cvi*AII	C↓ATG								
	*Hsp*92II	CATG↓	150 K	10	7.4	10				R
*Nla*IV		GGN↓NCC	50 Ka	20 A	7.9	10 A	1	100		NS
	*Bsc*BI	GGN↓NCC		10	7.5	10	1	100	55 °C	T
	*Bsp*LI	GGN↓NCC	66 K	33 A	7.9	10 A		100		F
	*Psp*N4I	GGN↓NCC	66 Ka	33 A	7.9	10 A	1			I
*Not*I		GC↓GGCCGC	100	50	7.9	10	1	100		ABCDEFGJKLMNOPQRST

Table 1. *Continued*

Name	Isoschizomer(s)	Recognition sequence	Digestion conditions[a] NaCl	Tris	pH	Mg	SH	BSA	other	temp.	Source[b]
	*Cci*NI	GC ↓ GGCCGC	66 Ka	33 A	7.9	10 A	1				I
*Nru*I		TCG ↓ CGA	100 K	50	7.7	10					ABCDEGIJKLMNOPQRST
	*Bsp*68I	TCG ↓ CGA	100	10	8.0	5	1 B	100			DF
*Nsp*I		RCATG ↓ Y	100 K	20	8.5	10	1	100	0.1 T		ABKM
	*Bst*NSI	RCATG ↓ Y		10	7.6	10	1 B			25 °C	I
*Nsp*BII		CMG ↓ CKG		10	7.5	10	1				A
	*Msp*A1I	CMG ↓ CKG	50 Ka	20 A	7.9	10 A	1	100			NR
*Pac*I		TTAAT ↓ TAA		10 B	7.0	10	1	100			NO
*Pfl*1108I		TCGTAG									
*Pfl*MI		CCANNNN ↓ NTGG	100	50	7.9	10	1	100			ANS
	*Acc*B7I	CCANNNN ↓ NTGG	50	10	7.6	10	1				IR
	*Esp*1396I	CCANNNN ↓ NTGG	50	10	7.5	10		100			F
	*Van*91I	CCANNNN ↓ NTGG	100	10	8.0	5	1 B	100			DFKM
*Ple*I		GAGTC(4/5)	50 Ka	20 A	7.9	10 A	1				N
	*Mly*I	GACTC(5/5)									
*Pma*CI		CAC ↓ GTG		10	7.5	10	1				AK
	*Bbr*PI	CAC ↓ GTG	100	10	8.0	5	1 B				MO
	*Eco*72I	CAC ↓ GTG	66 Ka	33 A	7.9	10 A	0.5	100			DEFR
	*Pml*I	CAC ↓ GTG		10 B	7.0	10	1	100			N
*Pme*I		GTTT ↓ AAAC	50 Ka	20 A	7.9	10 A	1	100			NP
*Ppu*MI		RG ↓ GWCCY	50 Ka	20 A	7.9	10 A	1				ANO
	*Psp*5II	RG ↓ GWCCY	50	50	8.0	10					BDF
	*Psp*PPI	RG ↓ GWCCY	66 K	33 A	7.9	10	1 B			25 °C	I
*Psh*AI		GACNN ↓ NNGTC	100 K	20	8.5	10	10				AKN
*Pst*I		CTGCA ↓ G	100	50	7.9	10	1				ABCDEFGHIJKLMNOPQRST
*Pvu*I		CGAT ↓ CG	100	50	7.9	10	1	100			ABDEFGKLMNOPQRS
	*Bsp*CI	CGAT ↓ CG	100 Ka	25 A	7.6	10 A	0.5 B	10			E
	*Ple*19I	CGAT ↓ CG	66 Ka	33 A	7.9	10 A	1				I

*Pvu*II		CAG ↓ CTG	50	10	7.9	10	1			ABCDEFGHIJKLMNOPQRST
*Rle*AI		CCCACA(12/9)								
*Rsa*I		GT ↓ AC	10 B	7.0	10	1			ABCDEFGIJLMNOPQRST	
	*Afa*I	GT ↓ AC 66 Ka	33 A	7.9	10 A	5	10		K	
	*Csp*6I	G ↓ TAC	10	7.5	10				DFN	
*Rsr*II		CG ↓ GWCCG	50 Ka	20 A	7.9	10 A	1			BMNS
	*Cpo*I	CG ↓ GWCCG	100 K	20	8.5	10	1			ADFK
	*Csp*I	CG ↓ GWCCG	150	10	7.7	10	1	10	30 °C	EOR
*Sac*I		GAGCT ↓ C		10 B	7.0	10	1	100		ACEFGHJKLMNOPQRST
	*Ecl*136II	GAG ↓ CTC	66 Ka	33 A	7.9	10 A		100		DFN
	*Eco*ICRI	GAG ↓ CTC	50	6	7.5	6	1			R
	*Psp*124BI	GAGCT ↓ C	50	10	7.6	10	1			I
	*Sst*I	GAGCT ↓ C	50	50	8.0	10				BCS
*Sac*II		CCGC ↓ GG	50 Ka	20 A	7.9	10 A	1			ACEGHJKLNOPQRST
	*Cfr*42I	CCGC ↓ GG		10	7.5	10	1 E	100		DF
	*Ksp*I	CCGC ↓ GG		10	7.5	10	1 E			M
	*Mlu*113I	CC ↓ GCGG								
	*Sfr*303I	CCGC ↓ GG		10	7.6	10	1			I
	*Sst*II	CCGC ↓ GG	50	50	8.0	10				BS
*Sal*I		G ↓ TCGAC	150	10	7.9	10	1	100		ABCDEFGHIJKLMNOPQRST
*San*DI		GG ↓ GWCCC	200 Ka	50 A	7.6	20 A	1 B	20		E
*Sap*I		GCTCTTC(1/4)	50 Ka	20 A	7.9	10 A	1			N
*Sau*I		CC ↓ TNAGG								
	*Aoc*I	CC ↓ TNAGG		10	7.5	10	1 E			M
	*Axy*I	CC ↓ TNAGG	50	10	7.5	10	1			J
	*Bse*21I	CC ↓ TNAGG		10	7.6	10	1			I
	*Bsu*36I	CC ↓ TNAGG	100	50	7.9	10	1	100		ENRS
	*Cvn*I	CC ↓ TNAGG	50 K	20	7.4	5				B
	*Eco*81I	CC ↓ TNAGG	50	10	7.5	10	1			ADFKO
*Sca*I		AGT ↓ ACT	100	10	7.4	10	1			ABCDEFGJKLMNOPQRS
	*Acc*113I	AGT ↓ ACT	66 Ka	33 A	7.9	10 A	1			I
	*Eco*255I	AGT ↓ ACT		10	7.5	10		100		F
*Scr*FI		CC ↓ NGG 50 Ka	20 A	7.9	10 A	1			DJMNOS	

Table 1. *Continued*

Name	Isoschizomer(s)	Recognition sequence		Digestion conditions[a] NaCl	Tris	pH	Mg	SH	BSA	other temp.	Source[b]
	*Bss*KI	↓CCNGG 100	50	7.9	10	1	100		60 °C	N	
	*Msp*R9I	CC↓NGG 100	50	7.6	10	1				I	
*Sdu*I		GDGCH↓C	150	10	7.5	3		100			F
	*Bmy*I	GDGCH↓C	66 Ka	33 A	7.9	10 A	0.5				M
	*Bsp*1286I	GDGCH↓C		10	7.9	10	1	100			ADJKNR
*Sec*I		C↓CNNGG									
	*Bsa*JI	C↓CNNGG	50	10	7.9	10	1			60 °C	N
	*Bse*DI	C↓CNNGG	66 Ka	33 A	7.9	10 A	0.5	100			DF
*Sex*AI		A↓CCWGGT	100	10	8.0	5	1 B				M
*Sfa*NI		GCATC(5/9)	100	50	7.9	10	1				IN
	*Bsc*AI	GCATC(4/6)									
*Sfe*I		C↓TRYAG									
	*Bfm*I	C↓TRYAG	66 K	33 A	7.9	10		100			F
	*Bst*SFI	C↓TRYAG	100	50	7.6	10	1			60 °C	I
	*Sfc*I	C↓TRYAG	50 Ka	20 A	7.9	10 A	1	100			N
*Sfi*I		GGCCNNNN↓NGGCC		50	10	7.9	10	1	100		50 °C
ABCDEGIJLMNOPQRST											
*Sgf*I		GCGAT↓CGC	50	10	7.9	10	1				OR
*Sgr*AI		CR↓CCGGYG	66 Ka	33 A	7.9	10 A	0.5				M
*Sim*I		GGGTC(–3/0)									
*Sma*I		CCC↓GGG	50 Ka	20 A	7.9	10 A	1				ABCDEFGHIJKLMNOPQRST
	*Cfr*9I	C↓CCGGG	200 N	10	7.5	5	1				FO
	*Psp*AI	C↓CCGGG	100 Ka	25 A	7.6	10 A	0.5 B	10			E
	*Psp*ALI	CCC↓GGG	66 Ka	33 A	7.9	10 A	1			30 °C	I
	*Xma*I	C↓CCGGG		10 B	7.0	10	1				ADINPRS
	*Xma*CI	C↓CCGGG		10	7.5	10	1 E				M
*Sml*I		CTYRAG									
*Sna*I		GTATAC									

	*Bst*1107I	GTA↓TAC	100 K	10	8.5	10					DFKMN
*Sna*BI		TAC↓GTA	50 Ka	20 A	7.9	10 A	1	100			ACEKLMNPRS
	*Bst*SNI	TAC↓GTA		10	7.6	10	1			65 °C	I
	*Eco*105I	TAC↓GTA	50 L	12.5 A	7.5	5 A	0.5 B	100			DFO
*Spe*I		A↓CTAGT	50	10	7.9	10	1	100			ABDEHJKLMNOPQRST
	*Acl*NI	A↓CTAGT		10	7.6	10	1				I
*Sph*I		GCATG↓C	50	10	7.9	10	1				ABCDEGHIJKLMNOPQRST
	*Bbu*I	GCATG↓C	6	6	7.5	6	1				R
	*Pae*I	GCATG↓C		10	7.5	10	1 E	100			DF
*Spl*I		C↓GTACG	100	50	7.5	10	1	100		55 °C	AK
	*Bsi*WI	C↓GTACG	100	50	7.9	10	1			55 °C	MNO
	*Pfl*23II	C↓GTACG	66 Ka	33 A	7.9	10 A	0.5	100			DF
	*Psp*LI	C↓GTACG	66 Ka	33 A	7.9	10 A	1				I
	*Sun*I	C↓GTACG	50 K	20	7.4	5				55 °C	B
*Srf*I		GCCC↓GGGC	100 Ka	25 A	7.6	10 A	0.5 B	10			EO
*Sse*8387I		CCTGCA↓GG 50	10	7.5	10	1	100			AK	
	*Sbf*I	CCTGCA↓GG	66 Ka	33 A	7.9	10 A	1				I
*Sse*8647I		AG↓GWCCT									
*Ssp*I		AAT↓ATT	50	100	7.5	10			0.25 T		ABCDEFIJKLMNOPQR
*Stu*I		AGG↓CCT	50	10	7.9	10	1				ABEJKLMNPQRST
	*Aat*I	AGG↓CCT	60 K	10	7.5	7	7 B				O
	*Eco*147I	AGG↓CCT	66 Ka	33 A	7.9	10 A	0.5	100			DF
	*Pme*55I	AGG↓CCT		10	7.6	10	1				I
	*Sse*BI	AGG↓CCT	100	50	7.9	10	1				C
*Sty*I		C↓CWWGG	100	50	7.9	10	1	100			BCEJMNRS
	*Bss*T1I	C↓CWWGG	100	10	8.5	10	1			60 °C	I
	*Eco*130I	C↓CWWGG	100	50	7.5	10	1 E	100			DF
	*Eco*T14I	C↓CWWGG	100	50	7.5	10	1				AK
	*Erh*I	C↓CWWGG	100	10	8.5	10	1				I
*Swa*I		ATTT↓AAAT	100	50	7.5	10	1 E			25 °C	M
	*Smi*I	ATTT↓AAAT	100	50	7.6	10	1				I
*Taq*I		T↓CGA 100	10	8.4	10		100			ABCDEFGJLMNOPQRST	
	*Tth*HB8I	T↓CGA 100	50	7.5	10	1			65 °C	AK	

Table 1. *Continued*

Name	Isoschizomer(s)	Recognition sequence			Digestion conditions[a] NaCl	Tris	pH	Mg	SH	BSA	other temp.	Source[b]
*Taq*II		GACCGA(11/9)										
		CACCCA(11/9)										
*Tat*I		WGTACW										
*Tau*I		GCSGC										
*Tfi*I		G↓AWTC	100	50	7.9	10	1			65 °C	N	
*Tse*I		G↓CWGC		100	50	7.9	10	1			25 °C	N
*Tsp*45I		↓GTSAC		10 B	7.0	10	1	100		65 °C	N	
*Tsp*4CI		ACN↓GT										
*Tsp*EI		↓AATT	80	10	8.3	0.3			0.1 T	65 °C	LO	
	*Sse*9I	↓AATT	100	50	7.5	10	1 E	100			DI	
	*Tsp*509I	↓AATT		10 B	7.0	10	1			65 °C	N	
*Tsp*RI		CAGTG(2/–7)		50 Ka	20 A	7.9	10 A	1	100		65 °C	N
*Tth*111I		GACN↓NNGTC		50 Ka	20 A	7.9	10 A	1			65 °C	ADKNPQR
	*Asp*I	GACN↓NNGTC		100	10	8.0	5	1 B				MP
	*Ats*I	GACN↓NNGTC		66 Ka	33 A	7.9	10 A	1				I
*Tth*111II		CAARCA(11/9)										
*Uba*DI		GAACNNNNNNTCC										
*Vsp*I		AT↓TAAT		50	50	8.0	10					BDEFIR
	*Ase*I	AT↓TAAT		100	50	7.9	10	1				AJNOP
	*Asn*I	AT↓TAAT		100	10	8.0	5	1 B				MS
*Xba*I		T↓CTAGA		50	10	7.9	10	1	100			ABCDEFGHIJKLMNOPQRST
*Xcm*I		CCANNNNN↓NNNNTGG			50	10	7.9	10	1			AN
*Xho*I		C↓TCGAG		50	10	7.9	10	1	100			ABCDEFGHJKLMNOPQRST
	*Pae*R7I	C↓TCGAG		50 Ka	20 A	7.9	10 A	1				N
	*Sci*I	CTC↓GAG										
	*Sfr*274I	C↓TCGAG			10	7.6	10	1			50 °C	I
*Xho*II		R↓GATCY			25	7.7	10	1	30			EMR
	*Bst*X2I	R↓GATCY		50	10	7.6	10	1			60 °C	I

	*Bst*YI	R↓GATCY		10	7.9	10	1	100	60 °C	BN
	*Mfl*I	R↓GATCY		10	7.5	10	1			AK
*Xma*III		C↓GGCCG		10	8.2	8			25 °C	BE
	*Bst*ZI	C↓GGCCG	150	6	7.9	6	1		50 °C	R
	*Eag*I	C↓GGCCG	100	50	7.9	10	1			NPS
	*Ecl*XI	C↓GGCCG	100	10	8.0	5	1 B			M
	*Eco*52I	C↓GGCCG	100	10	8.9	3		100		ADFKOR
*Xmn*I		GAANN↓NNTTC	50	10	7.9	10	1	100		DENR
	*Asp*700I	GAANN↓NNTTC	100	10	8.0	5	1 B			M
	*Mro*XI	GAANN↓NNTTC								I

[a]The concentrations of reagents are given in mM and are those recommended by New England Biolabs (NEB) or the first provider listed if not available from NEB. All digests are run at 37 °C unless otherwise indicated.

In each column the abbreviations used are as follows. NaCl: letters indicate replacement of NaCl by KCl (K), potassium acetate (Ka), sodium glutamate (N), potassium glycinate (L) or ammonium sulfate (A). Tris: letters indicate replacement of Tris–HCl by Tris–acetate (A) or glycylglycine-KOH (G).Mg: letters indicate replacement of $MgCl_2$ by magnesium acetate (A) or $MgSO_4$ (S). SH: letters indicate replacement of DTT by DTE (E) or β-mercaptoethanol (B).BSA: numbers indicate the amount of bovine serum albumin in μg/ml. Other: T = % Triton-X-100; A = μM AdoMet.

[b]Commercial sources of restriction enzymes and methylases are abbreviated as follows: A, Amersham Life Sciences-USB; B, Life Technologies Inc, Gibco-BRL; C, Minotech Molecular Biology Products; D, Angewandte Gentechnologie Systeme; E, Stratagene; F, Fermentas MBI; G, Appligene Oncor; H, American Allied Biochemical, Inc.; I, SibEnzyme Ltd; J, Nippon Gene Co. Ltd; K, Takara Shuzo Co. Ltd; L, NBL Gene Sciences Ltd; M, Boehringer-Mannheim; N, New England BioLabs; O, Toyobo Biochemicals; P, Pharmacia Biotech Inc.; Q, CHIMERx; R, Promega Corporation; S, Sigma; T, Advanced Biotechnologies Ltd.

[c]*Bae*I is unusual in that it cleaves on both sides of its recognition sequence as shown below:

5′ ↓NNNNNNNNNNACNNNNGTAYCNNNNNNNNNNNN↓ 3′
3′ ↓NNNNNNNNNNNNNNNTGNNNNCATRGNNNNNNN↓ 5′

This is indicated as (10/15)ACNNNNGTAYC(12/7). Other enzymes that cleave in a similar fashion are:
*Bcg*I: (10/12)GCANNNNNNTCG(12/10),
*Bsp*24I: ((8/13)GACNNNNNNTGG(12/7),
*Cje*I: (8/14)CCANNNNNN(15/9),
*Cje*PI: (7/13)CCANNNNNNNTC(14/8).

A3

Statistical analysis of variation

A. R. HOELZEL and D. R. BANCROFT

The neutral theory predicts that molecular genetic change will accumulate gradually at a constant rate, and that most changes are selectively neutral. This would mean that the accumulation of genetic change can be interpreted to estimate time-dependent genetic differentiation within and between populations. While strict neutrality may not be true in all cases, statistical tests assuming neutrality as a null hypothesis underly most analyses. The purpose of this appendix is to provide only the raw material necessary for a preliminary analysis of molecular variation towards the quantification of measures of diversity and genetic distance. For a more detailed treatment and formulations for the derivation of statistical variance, see references 1 and 2. For further discussion and a review of computer programs available for these types of analyses see Appendix 5.

When DNA variation is measured directly, the formulations are based on the following assumptions.

- nucleotides are randomly distributed in the genome;
- variation arises solely by base substitution;
- substitution rates are the same for all nucleotides;
- all relevant bands or fragments can be detected, and similar bands are not scored as identical.

The first three assumptions do not strictly hold, but it has been argued (1) that small deviations will not significantly alter the results. The last assumption is primarily a methodological consideration and is discussed in the relevant chapters.

Allozyme variation is different in that the measured character is phenotypic. It is necessary to establish that these characters follow the Mendelian rules of segregation during sexual recombination (usually by conducting pedigree analyses), and by testing for correspondence to the Hardy–Weinberg rule (which states that gene and genotype frequencies will remain constant between generations in an infinitely large, random mating population).

1. Allozymes

1.1 Reading allozyme data

The visualization and interpretation of allozyme banding patterns are discussed in detail in Chapter 1. The objective is to identify different alleles, and to determine whether an individual is homozygous or heterozygous. Once the genotypes (for example, AA, aa or Aa) have been counted, gene frequencies can be determined as follows:

$$p = (2N_{AA} + N_{Aa})/2N$$
$$q = (2N_{aa} + N_{Aa})/2N$$

where p is the frequency of the A allele, q is the frequency of the a allele, N is the total number of individuals in the sample, and N_{AA}, N_{aa}, and N_{Aa} are the number of individuals with AA, aa and Aa genotypes, respectively. According to the Hardy–Weinberg rule, the proportion of AA individuals should be p^2, the proportion of aa individuals should be q^2, and the proportion of heterozygotes should be $2pq$. Observed genotype frequencies are usually tested for concordance with the Hardy–Weinberg rule by the chi-squared test, although this will only detect relatively large deviations. The relationship between the number of alleles at a locus, the number of genotypes, and the degrees of freedom in a chi-squared test (given that genotypic classes are not pooled) is given in *Table 1*.

1.2 Variation

Variation is usually measured in terms of polymorphism (P) and heterozygosity (H). P is simply the proportion of polymorphic loci. H is the proportion of heterozygous individuals:

$$H = \Sigma(N_{Aa}/N)/n$$

Table 1. Degrees of freedom in a chi-squared test for concordance with the Hardy–Weinberg equilibrium

Number of alleles at the locus	Number of genotypes	Degrees of freedom
2	3	1
3	6	3
4	10	6
5	15	10
6	21	15
7	28	21
8	36	28
9	45	36
10	55	45

where n is the number of loci including monomorphic loci. When the population is in Hardy–Weinberg equilibrium, heterozygosity can be calculated from allele frequencies at a given locus by

$$h = 1 - \Sigma x_i^2$$

where x_i is the frequency of the ith allele. H is then given as the mean of h over all loci.

1.3 Genetic distance

This is a measure of gene diversity between populations expressed as a function of genotype frequency. There are a number of formulations that have been proposed. Those described by Nei (3) will be given here. Given two populations, X and Y, first consider the probability of identity of two randomly chosen genes at a single locus (j_k):

$$j_x = \Sigma x_i^2$$

and

$$j_y = \Sigma y_i^2$$

where x_i and y_i are the frequencies of the ith alleles at a given locus in populations X and Y respectively. For example, if there are two alleles at this locus with frequencies p and q, then

$$j = p^2 + q^2$$

The probability of identity of a gene at the same locus in populations X and Y is

$$j_{xy} = \Sigma x_i y_i$$

The normalized identity between populations X and Y with respect to all loci is

$$I = J_{XY}/(J_X J_Y)^{1/2}$$

where J_{XY}, J_X, and J_Y are the arithmetic means of j_{xy}, j_x, and j_y respectively taken over all loci. The genetic distance between populations X and Y is then defined as

$$D = -\ln(I)$$

In theory, the relationship between D and time (t) is

$$t = 0.5aD$$

where a is the average rate of detectable change per locus per year (Nei (1) gives an estimate of $a = 10^{-7}$). Note that D is not an estimate of sequence difference.

1.4 Inter-population diversity

This is usually measured using the coefficient of gene differentiation (G_{ST}), which is derived from allozyme data by estimating the average similarity within and between populations (1). The probability of gene identity (j_k) is determined for each population as in Section 1.3, and the gene identity within all populations (J_S) is then given as

$$J_S = \Sigma j_k / s$$

where s is the number of populations. The average gene diversity within populations is given by

$$H_S = 1 - J_S$$

The gene identity for the total population is given as

$$J_T = \Sigma x_i^2$$

where

$$x_i = \Sigma_k w_k x_i$$

and w_k is the proportional weight of the kth population. The inter-population gene diversity is given by

$$H_T = 1 - J_T$$

The coefficient of differentiation can then be defined as

$$G_{ST} = (H_T - H_S)/H_T$$

2. Restriction fragment length polymorphisms (RFLPs)

2.1 Reading RFLP data

The visualization and interpretation of restriction fragments are discussed in detail in Chapter 5. An assumption is made that fragments that migrate the same distance on a gel have been produced by the same restriction sites. When the source DNA is circular (such as the vertebrate mitochondrial genome), then the number of fragments will be equal to the number of restriction sites, and the fragments should add up to the total size of the genome. When a probe for nuclear DNA is used, fragments can be visualized that have only partial homology and extend either side of the probe sequence. Restriction patterns can be compared either by the length of the fragments produced, or by comparing actual restriction sites. The latter is more accurate, but requires the added step of mapping the restriction sites (see Chapter 5). Alternatively, restriction sites can often be determined from the fragment data if a complete sequence of the region in question is available.

2.2 Variation

Restriction patterns can be classified into haplotypes (a contiguous piece of DNA with a given set of restriction sites). A measure of diversity akin to heterozygosity can be derived as a function of the frequency of the different haplotypes

$$h = 1 - \Sigma x_i^2$$

where x_i is the frequency of the ith haplotype. This assumes that the population is in Hardy–Weinberg equilibrium. The nucleotide diversity (π) can be estimated by first measuring the similarity between haplotype patterns.

$$S_{ij} = 2n_{ij}/(n_i + n_j)$$

where n_i and n_j are the number of fragments (or restriction sites) in the ith and jth haplotypes, and n_{ij} is the number of shared fragments (or restriction sites).

$$\pi = (-\ln S_{ij})/r$$

where r is the number of base pairs recognized by the restriction enzyme. If enzymes within different values for r are used, they should be computed separately.

2.3 Genetic distance

In this case genetic distance is usually measured as an estimate of the number of base substitutions per nucleotide separating two populations. Upholt (4) derived the following estimate:

$$p = 1 - \{[(S^2 + 8S)^{1/2} - S]/2\}^{1/r}$$

When restriction sites are known, S can be determined as the mean of S_{ij} as given in Section 2.2, and the proportion of nucleotide differences (p) can then be estimated by

$$p = 1 - S^{1/r}$$

as described by Nei (1). The number of nucleotide substitutions per site (d, after ref. 5) can then be estimated from the equation

$$d = -(3/4)\ln[1 - (4/3)p]$$

When restriction sites are not known, but the total length of the molecule under analysis is fixed (such as with mitochondrial or chloroplast DNA), S can be estimated by comparing fragment lengths, and d can be estimated by

$$d = -(2/r)\ln(G)$$

where G is determined by the following iterative formula (1):

$$G = [S(3 - 2G_1)]^{1/4}$$

The iteration is repeated until $G = G_1$. An initial trial value of $G_1 = S^{1/4}$ is recommended. When comparing populations X and Y, d between populations (d_{XY}) can be corrected for variation within populations by

$$d = d_{XY} - 0.5(d_X + d_Y)$$

2.4 Inter-population diversity

This is esentially the same as for the allozyme data, except that gene identities must be estimated from RFLP patterns. The following derivation is for comparing restriction sites (after ref. 6). The identity probability is estimated as

$$I = (1/l)\{[\Sigma C_i(C_i - 1)]/[n(n-1)]\}$$

where l is the total number of cleavage sites among the n samples, and C_i is the number of cleavage sites at position i among the n sequences. The average conditional identity probability is estimated as

$$J = (1/l)[(\Sigma C_{xi}C_{yi})/(n_x n_y)]$$

where x and y represent two different populations. The within- and between-population gene diversities (H_S and H_T respectively) are then given as

$$H_S = 1 - I$$

and

$$H_T = 1 - \{I/L + J[1 - (1/L)]\}$$

where L is the number of populations. G_{ST} is then given as in Section 1.4.

3. DNA sequences

3.1 Reading sequence data

Sequence data are usually compared using computer programs to determine the best alignment. The identity of two sequences is simply the percentage of shared bases. Deletions and insertions are often scored as a single change, regardless of length.

3.2 Variation

Nucleotide diversity can be estimated by

$$\pi = \Sigma p/n_c$$

where p is the proportion of different nucleotides between DNA sequences, and n_c is the total number of comparisons given by

$$n_c = 0.5n(n-1)$$

where n is the number of individuals sequenced (1).

3.3 Genetic distance

The genetic distance between sequences (d) is often given simply as the percentage difference when d is small (referred to as the uncorrected distance or the p-distance), or corrected for multiple substitutions at a given site as proposed by Jukes and Cantor (7):

$$d = (3/4)\ln[1 - (4/3)k]$$

where k is the percentage difference in base composition. The Jukes–Cantor model corrects for the fact that over increasing time some mutations will have occurred at the same site, but it assumes an equal rate of substitution between all pairs of bases. The Kimura two-parameter model (8) considers transition-type (P) and transversion-type (Q) differences separately:

$$d = (1/2)\ln(1/(1 - 2P - Q)) + (1/4)\ln(1/(1 - 2Q))$$

Further discussion of these two models and of a variety of other measures that account for unequal rates of change is given in ref. 2.

3.4 Inter-population diversity

In this case identity between sequences can be measured directly. Therefore, using the formula for d given in Section 3.3, a measure of gene differentiation can be estimated as

$$g = (d_t - d_s)/d_t$$

where d_t is the mean of all pairwise distances within and between populations, and d_s is the mean pairwise distance for comparisons within populations. This measure will have different statistical properties from G_{ST} because distance measures calculated from DNA sequence data will not take values over the whole range from 0 to 1. Two factors affect the lower bound of sequence similarity estimates:

(a) homologous sequences are being compared, and so genetic identities must exceed some arbitrary value;
(b) unequal base usage could increase the lower bound of similarity.

4. Repetitive DNA

The statistical analysis of single and multi-locus DNA fingerprints is described in Chapter 9. This section will concentrate on the analysis of microsatellite data (see also Chapter 7).

4.1 Reading microsatellite data

Microsatellites are short repeats of simple, usually dinucleotide, trinucleotude or tetranucleotide sequence, that vary in repeat number. They are usually visualized on high percentage polyacrylamide gels after amplification by PCR (see Chapters 7 and 10). During the process of amplification, the polymerase often 'slips' over one or up to a few repeats (especially for dinucleotide repeats). Therefore, there can be numerous artefact bands on the gel. The correct band can usually be distinguished as the darkest band, and the pattern of artefact bands is typically locus-specific, so that artefact bands for all alleles and individuals look similar. These regions are highly variable, so many individuals will be heterozygous. A consequence of the greater variation of microsatellite data compared with allozyme data is that each genotypic class will tend to contain fewer observations for any study of finite population size. This higher level of variation may be desirable for paternity and inter-population studies but will lead to reduced statistical power. In this case, likelihood-ratio tests are preferable to the chi-squared test when interpretating such analyses (9).

4.2 Variation and genetic distance

Since microsatellite polymorphisms are on average more variable than allozymes, large changes in individual reproductive success between cohorts or generations can generate deviations in allele and genotype frequencies. Therefore, population data must be examined for such deviations, especially for studies of highly polygamous species, before allele and genotype frequencies can be reliably estimated for the total population.

Analysis of microsatellites is a single-locus technique and so the resulting data can be analysed by a treatment similar to that used for allozyme variation. However, microsatellite polymorphism is generated by length mutations, probably through the mechanism of DNA slippage (10). Therefore the resulting alleles are discrete, typically over a finite size range. The assumption of an essentially infinite pool of potential alleles does not hold and, given the high rate of change, back-mutation is likely. Further, the size of a new mutant is likely to depend on the size of the allele from which it mutated. These factors together with the fact that the rate of mutation is very high, compromise the accuracy of standard measures of genetic distance and F_{ST} when applied to microsatellite data. New measures that are based on a step-wise mutation model have recently been developed for this purpose (11–13).

Shriver *et al.* (12) showed by computer simulation that their measure of genetic distance, D_{SW}, conforms to linearity with time since divergence better than Nei's distance (Section 1.3) for loci with high mutation rates that evolve via a stepwise mutation mechanism, such as microsatellites. It can be defined as:

$$D_{SW} = d_{xyw} - (d_{xw}\, d_{yw})/2$$

where,

$$d_{xw} = \sum_{i \neq j}\sum x_i x_j \delta_{ij},$$

$$d_{yw} = \sum_{i \neq j}\sum y_i y_j \delta_{ij},$$

$$d_{xyw} = \sum_{i \neq j}\sum x_i y_j \delta_{ij},$$

$$\delta_{ij} = |i - j|,$$

and x_i and y_j are the frequencies of the *i*th and *j*th alleles at a locus in populations X and Y, respectively. R_{ST}, a measure of inter-population diversity comparable to F_{ST}, has been proposed by Slatkin (11) based on similar considerations. A computer program that calculates this measure is reviewed in Appendix 5 and equations are given in ref. 11.

References

1. Nei, M. (1987) *Molecular Evolutionary Genetics*. Columbia University Press, New York.
2. Weir, B.S. (1996) *Genetic Data Analysis II*. Sinauer Press, Sunderland, MA.
3. Nei, M. (1972) *Am. Nat.*, **106**, 283.
4. Upholt, W.B. (1977) *Nucleic Acids Res.*, **4**, 1257.
5. Nei, M. and Li, W.-H. (1979) *Proc. Natl Acad. Sci. USA*, **76**, 5269.
6. Takahata, N. and Palumbi, S.R. (1985) *Genetics*, **109**, 441.
7. Jukes, T.H. and Cantor, C.R. (1969) In *Mammalian Protein Metabolism* (ed. H.N. Munro), p. 21. Academic Press, New York.
8. Kimura, M. (1980) *J. Mol. Evol.*, **16**,111.
9. Hernandez, J.L. and Weir, B.S. (1989) *Biometrics*, **45**, 53.
10. Strand, M., Prolla, T.A., Liskay, R.M. and Petes, T.D. (1993) *Nature*, **365**, 274.
11. Slatkin, M. (1995) *Genetics*, **139**, 457.
12. Shriver, M.D., Jin, L., Boerwinkle, E., Deka, R., Ferrell, R.E. and Chakraborty, R. (1995) *Mol. Biol. Evol.*, **12**, 914.
13. Goldstein, D.B., Linares, A.R., Cavalli-Sforza, L.L., and Feldman, M.W. (1995) *Genetics*, **139**, 463.

Useful graphs and tables

A. R. HOELZEL

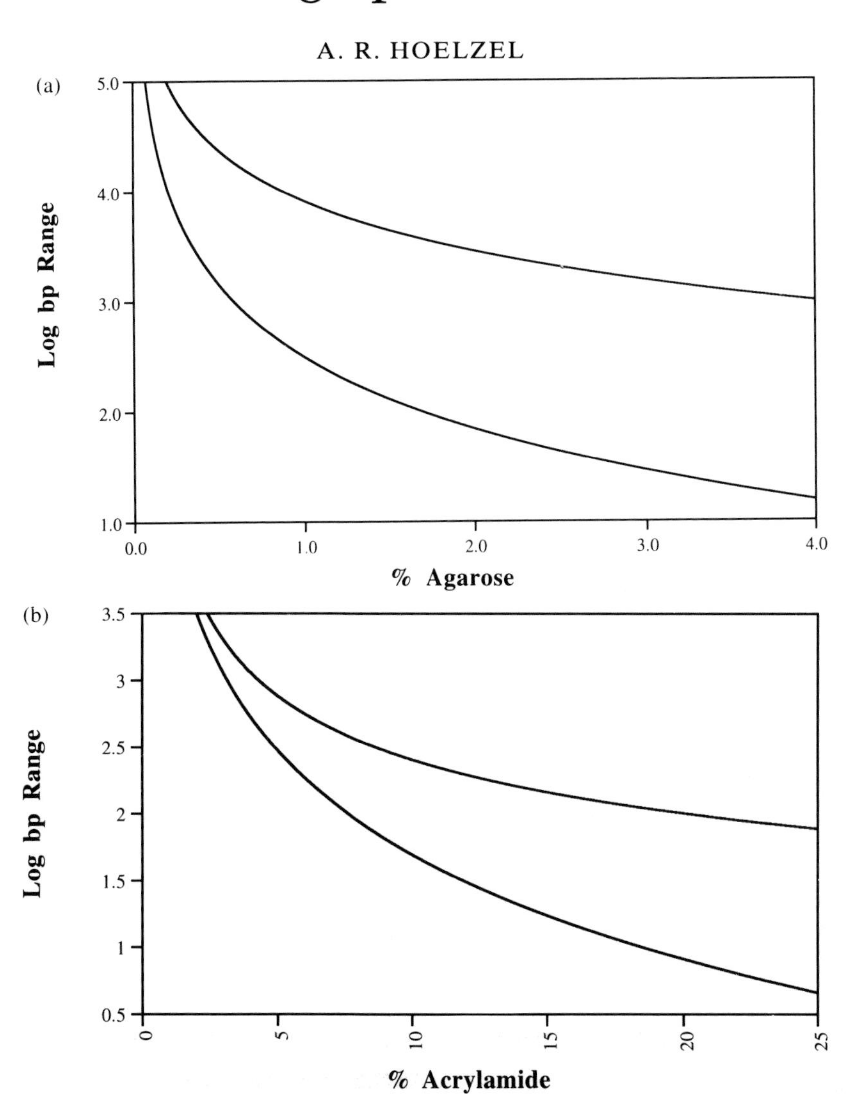

Figure 1. DNA resolution is given as an approximate range indicated between the two graph lines for a given percentage gel, (a) in agarose and (b) in acrylamide.

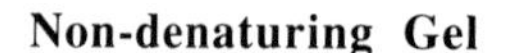

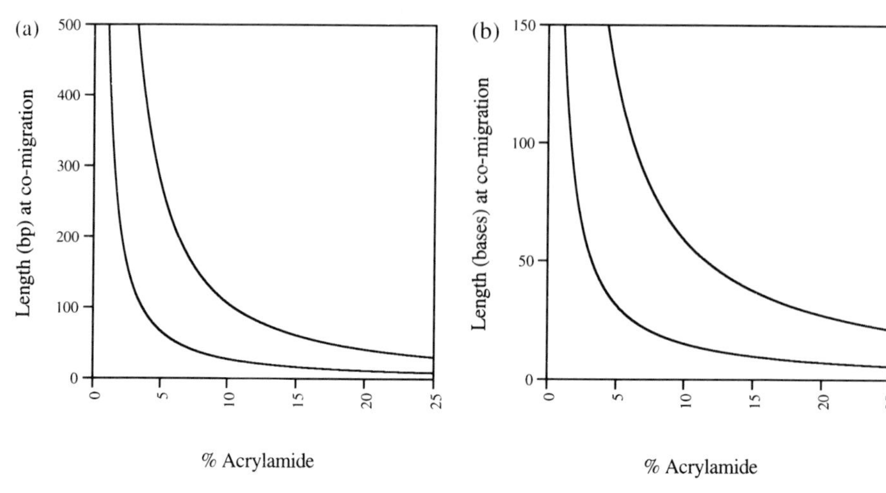

Figure 2. Co-migration of double-stranded (a) and single-stranded (b) DNA with common dye markers in acrylamide gels (Xylene Cyanol FF represented by the upper curve, and Bromophenol Blue, given by the lower curve). Note that over a range of 0.5–1.4% agarose, Bromophenol Blue migrates with 300 bp double-stranded DNA, and Xylene Cyanol FF with 4 kb double-stranded DNA.

Table 1. Size markers for gel electrophoresis (in base pairs)

1 kb ladder[a]	1 kb ladder[b]	λ*Hind*III	λ*Hind*III/*Eco*RI	λ*Bst*EII	pUC18 *Sau*3AI	φX174/ *Hae*III	HMWM[c]
12 216	12 000	23 130	21 226	8454	955	1353	48 502
11 198	10 000	9416	5148	7242	585	1078	38 416
10 180	9000	6557	4973	6369	341	872	33 498
9162	8000	4361	4268	5686	258	603	29 942
8144	7000	2322	3530	4822	141	310	24 776
7126	6000	2027	2027	4324	105	281	22 621
6108	5000	564	1904	3675	78	271	19 399
5090	4000	125	1584	2323	75	234	17 057
4072	3000		1375	1929	46	194	15 004
3054	2000		947	1371	36	118	12 220
2036	1500		831	1264		72	10 086
1636	1000		564	702			8612
1018	750		125	224			8271
517	500			117			
506	250						
396							

[a]Gibco BRL.
[b]Stratagene.
[c]Gibco-BRL High Molecular Weight Marker.

Table 2. Useful buffers and solutions

Solution	Preparation	Storage[a]	Comments
Electrophoresis			
50 × TAE	2 M Tris–acetate, 0.05 M EDTA; for 1 litre combine 242 g Tris base, 57.1 ml glacial acetic acid, 100 ml 0.5 M EDTA pH 8.0, and H_2O to 1 litre	RT	Dilute 50:1 for a 1× running buffer. This buffer offers good resolution in agarose, but needs to be recycled (using a peristaltic pump) during long runs.
10 × TBE	0.9 M Tris–phosphate, 0.02 M EDTA; for 1 litre combine 108 g Tris base, 55.0 g boric acid, 50 ml 0.5 M EDTA pH 8.0, H_2O to 1 litre	RT	Dilute 10:1 for a 1× running buffer for polyacrylamide gels. Use either 1× or 0.5× for agarose gels.
Loading buffer 1	0.25% Bromophenol Blue; 0.25% Xylene Cyanol FF, 15% Ficoll (type 400; Pharmacia) in H_2O	RT	Good general purpose loading buffer for non-denaturing gels.
Loading buffer 2	95% deionized formamide, 20 mM EDTA pH 8.0, 0.05% Bromophenol Blue, 0.05% Xylene Cyanol FF	–20°C	Denaturing gel (e.g. for sequencing) loading buffer. Suitable formamide can be purchased, or it can be deionized by stirring with Dowex XG8 resin.
Ethidium bromide	Make a 10 mg/ml stock solution (e.g. 1 g in 100 ml H_2O), but use a working solution of 50 μg/ml (200:1 dilution) and use one drop (25 μl) for every 50 ml of agarose.	RT	Store in a dark bottle, or wrap bottle with foil. **Caution**: Ethidium bromide is a powerful mutagen—always use gloves when working with solutions, and a mask when weighing the powder.
General purpose			
1 M Tris	For 1 litre dissolve 121.1g Tris base in 750 ml H_2O and add approximately 70 ml concentrated HCl for pH 7.4 or 42 ml for pH 8.0. Allow to cool to room temperature before making final adjustments to pH. Add H_2O to 1 litre.	RT	Sterilize by autoclaving.
0.5 M EDTA pH 8.0	For 1 litre dissolve 186.1 g of disodium ethylenediamine-tetraacetate·$2H_2O$ in 800 ml H_2O. Add approximately 20 g of NaOH pellets to adjust the pH.	RT	Sterilize by autoclaving. Note: EDTA will not dissolve until the pH is close to 8.0.
TE pH 7.4	10 mM Tris–HCl pH 7.4, 1 mM EDTA pH 8.0	RT	DNA storage buffer (store DNA at –20°C). Sterilize by autoclaving.
1 M $MgCl_2$	For 1 litre dissolve 203.3 g of $MgCl_2$·$6H_2O$ in 800 ml H_2O then adjust volume to 1 litre.	RT	$MgCl_2$ is very hygroscopic, so keep bottle well sealed. Sterilize by autoclaving.

Table 2. *Continued*

Solution	Preparation	Storage[a]	Comments
5 M NaCl	For 1 litre dissolve 292.2 g NaCl in 750 ml H_2O, then adjust volume to 1 litre.	RT	Sterilize by autoclaving.
10% SDS	It is sometimes useful to heat the solution to 60°C to facilitate dissolution. Adjust pH to 7.2 with a few drops of HCl.	RT	No need to sterilize. Wear a mask when working with powder.
20 × SSC	For 1 litre dissolve 175.3 g of NaCl and 88.2 g of sodium citrate in 800 ml H_2O. Adjust pH to 7.0 with a few drops of 10 N NaOH. Bring to 1 litre with H_2O.	RT	Sterilize by autoclaving.
Salt–DMSO	Prepare a 20% solution of dimethylsulfoxide (DMSO) in H_2O and saturate with NaCl while mixing with a stirrer bar. Decant from excess salt.	RT	Long-term RT storage solution for tissue samples to preserve high molecular weight DNA. Tissue pieces should be no bigger than 1 cm^2.
3 M Sodium acetate, pH 5.2	For 1 litre dissolve 408.1 g of sodium acetate·$3H_2O$ in 750ml H_2O and adjust the pH to 5.2 with glacial acetic acid. Bring to 1 litre with H_2O.	RT	Sterilize by autoclaving.
1 M DTT	For 10 ml dissolve 1.55 g dithiothreitol (DTT) in 0.01 M sodium acetate (pH 5.2).	–20°C	Sterilize by filtration through a 0.22 μm disposable filter - do not autoclave DTT solutions.
Cloning			
1 M $CaCl_2$	For 500 ml dissolve 135 g $CaCl_2 \cdot 6H_2O$ in 400 ml pure H_2O, then adjust to 500 ml. Store in aliquots.	–20°C	Sterilize by passage through a 0.22 μm filter.
IPTG	For 10 ml dissolve 2 g isopropylthio-β-D-galactoside (IPTG) in 8 ml distilled H_2O, and adjust to 10 ml.	–20°C	Sterilize by passage through a 0.22 μm filter.
X-Gal	Make a 20 mg/ml solution of 5-bromo-4-chloro-3-indolyl-β-D-galactoside (X-Gal) in dimethylformamide. Store in a dark glass bottle.	–20°C	No need to sterilize.
LB	For 1 litre of Luria–Bertani medium (LB) dissolve 10 g Bacto-tryptone, 5 g Bacto Yeast Extract, and 10 g	RT	Sterilize by autoclaving.

	NaCl in 950 ml distilled H_2O. Adjust pH to 7.0 with about 200 μl 5 N NaOH and adjust volume to 1 litre with distilled H_2O.		
LA	To LB, add 15 g/litre Bacto-agar for pouring plates.	RT	Sterilize by autoclaving.
Ampicillin	Make stock solution of 50 mg/ml in H_2O.	–20°C	Use at a working concentration of 20 μg/ml for stringent and 60 μg/ml for relaxed plasmids.
Tetracycline	Make stock solution of 5 mg/ml in ethanol.	–20°C	Use at a working concentration of 10 μg/ml for stringent and 50 μg/ml for relaxed plasmids.
Enzymes			
Pronase	Stock solution: 20 mg/ml in H_2O Reaction buffer: 0.01 M Tris–HCl pH 7.8, 0.01 M EDTA pH 8.0, 0.05% (w/v) SDS	–20°C	Use a working concentration of 1 mg/ml at 37°C. To remove any DNase contamination, self-digest for 1 h at 37°C in 0.01 M Tris pH 7.5, 0.01 M NaCl.
Proteinase K	Stock solution: 20 mg/ml in H_2O Reaction buffer: 0.01 M Tris–HCl pH 7.8, 0. 005 M EDTA pH 8.0, 0.05% (w/v) SDS	–20°C	Use a working concentration of 50 μg/ml at 37–65°C.
10 × CIP buffer	10 mM $ZnCl_2$, 10 mM $MgCl_2$, 100 mM Tris–HCl pH 8.3	–20°C	Use at 10:1 dilution.
10 × Klenow reaction buffer	0.5 M Tris–HCl pH 7.6, 0.1 M $MgCl_2$	–20°C	Use at 10:1 dilution.
10 × T4 DNA ligase buffer	0.5 M Tris–HCl pH 7.6, 0.1 M $MgCl_2$, 100 mM DTT, 500 μg/ml bovine serum albumin (Fraction V)	–20°C	Use at 10:1 dilution. For blunt ligations add hexamine cobalt (III) chloride (HCC) a final concentration of 2 mM[b]

[a]RT, room temperature.
[b]Murray, J.A.H. (1986). *Nucleic Acids Res.*, **24,**10118.

Table 3. Published primers (5′ to 3′) for the PCR amplification of microsatellite loci, with an emphasis on those primers that have been found to amplify from more than one species[a]

Species	P^a	Locus	Primers	Repeat	Ref.
Mammals					
Humans and chimpanzees	2/2	D13S118	CCACAGACATCAGAGTCCTT; GAAATAGTATTTGGACCTGGG	CA	1
	2/2	D13S121	GCTTGAGGTCTCTATGGAAA; TTTCAGAACTCTGTACCAGGA		
	2/2	D13S122	TGGAAACCACCACTCTACTT; TGTGAACCTAGACTGGAATAAA		
	2/2	D13S124	CAAATTCAAATTCTTCCAGC; ACTGTACTCCTGCATGTTAG		
	2/2	D13S193	GCAAGACCCCCATCTCTTAA; CTCACCCCACTCCATGTTC		
	2/2	D13S71	GTATTTTTGGTATGCTTGTGC; CTATTTTGGAATATATGTGCCT		
	2/2	D13S197	TTAATTCCCTGGAGCAGACG; TCAGAGAAGTGGGCATGATG		
Primates: humans, langur, Guinea baboon, orangutan, chimpanzee and seven others where $N < 3$	4/5	D13S159	AGGCTGTGACTTTTAGGCCA; AGCTAGCACCAGGCCACTTT	CA	2
	5/5	D7S503	AAGAATGACTTGGAGTAATG; AGCTAAAAGATAACAGGATA		
	5/5	D1S207	AGCTCAGTCCCAGCACTTCT; AGCTAAGGTCACTGGCAGAG		
	5/5	D17S791	TCCGCCTGAATCTTTCCTGT; GGTTACCTGAGGANCGGATG		
	5/5	D16S420	AGCTTCTATTTCCTGAGGTC; AGCTTGCTTCATGAAATTAT		
			and six more		
Ungulates: addax, bison, cattle, goat, musk ox, sheep, elk, mule deer, caribou, and seven others	12/16	IGFI	GGGTATTGCTAGCCAGCTGGT; CATATTTTTCTGCATAACTTGAACCT	CA	3
	12/16	SRCRSP3	CGGGGATGTGTTCTATGAAC; TGATTAGCTGGCTGAATGTCC		
	12/16	SRCRSP10	ACCAGTTTGAGTATCTTGCTTGGG; AGGAAGTTTATTGGACATGTCTGG		
	12/16	TEXAN2	ACATTGTCATGTGGTTGCTAAC; ACTCTGGGTATGTATATGTGCAAG		
	11/16	SRCRSP2	TGCTGTATCCTGTGTAATATCTT; GCATAAACAGATTATTGTGATGAT		
	11/16	TEXAN4	AGAAGTCCTTGCTTTCACAGAC; TTTTGGTGCTTGATGCTATTAG		
	11/16	TEXAN6	AGGCAGTTACCATGAACCTACC; ATCCTGGTGGGCTACAGTCTAC		
	9/16	SRCRSP9	AGAGGATCTGGAAATGGAATC; GCACTCTTTTCAGCCCTAATG		
	7/16	SRCRSP6	CATAGTTCATTCACAATATGGCA; CATGGAGTCACAAAGAGTTGAA		
	6/16	SRCRSP1	TGCAAGAAGTTTTTCCAGAGC; ACCCTGGTTTCACAAAAGG		
			and 10 more		
Felids: cat, cheetah, puma, lion	4/4	Fca8	ACTGTAAATTTCTGAGCTGGCC; TGACAGACTGTTCTGGGTATGG	CA	4
	4/4	Fca23	CAGTTCCTTTTTCTCAAGATTGC; GCAACTCTTAATCAAGATTCCATT		

	4/4	Fca35	CTTGCCTCTGAAAAATGTAAAATG; AAACGTAGGTGGGGTTAGTGG		
	4/4	Fca45	TGAAGAAAAGAATCAGGCTGTG; GTATGAGCATCTCTGTGTTCGTG		
	4/4	Fca78	TGAACTGAAGTCAGATGCTTAACC; CGGAATCAGCTATTTTTACGG		
	4/4	Fca96	CACGCCAAACTCTATGCTGA; CAATGTGCCGTCCAAGAAC		
	4/4	Fca126	GCCCCTGATACCCTGAATG; CTATCCTTGCTGGCTGAAGG		
	3/4	Fca43	GAGCCACCCTAGCACATATACC; AGACGGGATTGCATGAAAAG		
	3/4	Fca77	GGCACCTATAACTACCAGTGTGA; ATCTCTGGGGAAATAAATTTTGG		
	3/4	Fca90	ATCAAAAGTCTTGAAGAGCATGG; TGTTAGCTCATGTTCATGTGTCC		
Canids: dog, wolf, red wolf,	5/5	CXX172	CCTGTCTCCTGTGGACCAAT; ACATGCAAAAGGACACATTACG	CA	5,6
coyote, golden jackal	5/5	CXX204	CGAGAGCAACATAGGCATGA; CAAAGTGCTGTGGCAGGTC		
	5/5	CXX109	AACTTTAAGCCACACTTCTGCA; ACTTGCCTCTGGCTTTTAAGC		
	5/5	CXX250	TTAGTTAACCCAGCTCCCCCA; TCACCCTGTTAGCTGCTCAA		
	5/5	CXX213	AATATGGGAGAGGAGAAGAGGG; ATGCTTCCTGGTAAGCAATCA		
	5/5	CXX123	AACTGGCCAAACATAAACACG; TTCATTAACCCTTTGCCCTG		
	4/5	CXX225	AGCGACTATTATATGCCAGCG; CTCATTGGTGTAAAGTGGCG		
Pinnipeds: harbour seal, grey	7/7	Pvc19	GGGTGAACAGGATTTATCC; GTGCTAGATAACAATCCTAC	CA	7
seal, hooded seal, harp seal,	7/7	Pvc30	GCATGTGATCTTACAGCAAT; CATGGGTTCTCAATAGAAGA		
Antarctic fur seal, Subantarctic	7/7	Pvc78	GAGTATACCTCCATACTACAC; AGTTGTTCTCCTGACCCAAG		
fur seal, New Zealand fur seal	4/7	Pvc29	GGTTAATTGTGTTGTTTACATCT; AACCAGAAGAATAGAATTAGCAT		
	4/7	Pvc63	CCTGGACTTTGTTTATACCT; GCATGAGTTCATCTAGGGA		
	4/7	Pvc74	CCATCTGTGTCCTCTGATA; CTGATATTCCATGTCTGAGATA		
	2/7	Pvc26	ATTTTCTCCATACCTACATAAT; ATTTGTGATCCCATTTTTGTAA		
Cetaceans: fransiscana, beluga,	26/30	EV37Mn	AGCTTGATTTGGAAGTCATGA; TAGTAGAGCCGTGATAAAGTGC	CA	8
bottlenose dolphin, orca, Baird's	21/30	EV94Mn	ATCGTATTGGTCCTTTTCTGC; AATAGATAGTGATGATGATTCACACC		
beaked whale, fin whale, blue	14/30	EV1Pm	CCCTGCTCCCCATTCTC; ATAAACTCTAATACACTTCCTCCAAC		
whale, and 23 others	14/30	EV14Pm	TAAACATCAAAGCAGACCCC; CCAGAGCCAAGGTCAAGAG		
	14/30	EV104Mn	TGGAGATGACAGGATTTGGG; GGAATTTTTATTGTAATGGGTCC		
	12/30	EV21Pm	CAATAATTGGACAGTGATTTCC; CGCTGAAGGTGTGCCC		
	9/30	EV5Pm	AGCTCCCTTAGACTCAACCTC; TATGGCGAGGGTTCCG	GC	
	8/30	EV96Mn	AAGATGAGTAGATTCACTACACGAGG; CCACTTTTCCTCTCACATAGCC	CA	
	7/30	EV30Pm	GGAATAGAGGTAGGGGTGG; GCTTTTGTTGTGGTCATCC		
			and three more		

Table 3. *Continued*

Species	P^a	Locus	Primers	Repeat	Ref.
Ursids: polar bear, black bear, grizzly bear	3/3	G1A	GACCCTGCATACTCTCCTCTGATG; GCACTGTCCTTGCGTAGAAGTGAC	CA	9,10 11
	3/3	G1D	GATCTGTGGGTTTATAGGTTACA; CTACTCTTCCTACTCTTTAAGAG		
	3/3	G10B	GCCTTTTAATGTTCTGTTGAATTTG; GACAAATCACAGAAACCTCCATCC		
	3/3	G10L	GTACTGATTTAATTCACATTTCCC; GAAGATACAGAAACCTACCCATGC		
	2/3	G10C	AAAGCAGAAGGCCTTGATTTCCTG; GGGGACATAAACACCGAGACAGC		
	2/3	G10M	TTCCCCTCATCGTAGGTTGTA; GATCATGTGTTTCCAAATAAT		
	2/3	G10P	AGGAGGAAGAAAGATGGAAAAC; TCATGTGGGGAAATACTCTGAA		
	2/3	G10X	CCCTGGTAACCACAAATCTCT; TCAGTTATCTGTGAAATCAAAA		
Mustellids: mink, otter, pine marten, stoat, wolverine, badger	2/6	Mvi39	CTCACCAGGCAGGGAGC; AAAACAAGCATTCAGATCACC	CA	12
	1/6	Mvi54	AGAGTCTGTATACCTCCACC; CCCTCCTTGGCTCCGCAC		
	4/6	Mvi57	GAACAGGACCAGCCCTGC; GTTGGAAATGAGGATCTCAC		
	4/6	Mvi87	AACAATAGTAGTGGCAGCAGC; GTCTGTGAAACACTGCAAAGC		
House mouse	1/1	D15Mit16	AGACTCAGAGGGCAAAATAAAGC; TCGGCTTTTGTCTGTCTGTC and 309 others developed for genome mapping	CA	13, 14
Grey red-backed vole	1/1	MSCRB-2	AAGGGTGAGTATGCCAATCA; CCATTAAATGTTCTCAGGGA	CA	15
		MSCRB-3	CATGACCTTCTATTTCTGTCAG; TAAAACCCCAGACAAAGAAGAT and three others		
Birds					
40 passeriformes species from seven families and 19 non-passeriformes species from 14 families	23/59	HrU2	CATCAAGAGAGGGATGGAAAGAGG; GAAAAGATTATTTTTCTTTCTCCC	GT	16, 17
	21/59	HrU5	TCAACAAGTGTCATTAGGTTC; AACTTAGATAAGGAAGGTATAT	GC	
	19/59	HrU6	GCTGTGTCATTTCTACATGAG; ACAGGGCAGTGTTACTCTGC	AAAG	
	19/59	HrU7	GCATTCACAGTGTAGACAATG; GATCACTATGAGTCCCTGGAA	AAAC	
	17/59	HrU3	CACTGGCTCTAGGCTGTCATC; CTGTCCCATGTCAGGCCAGTC	CA	
	12/59	HrU8	CTCCAGATTTGATTGGAATAC; GGCTGGATACTGCCTTCCCAA	TTTG	
	7/59	HrU4	GATCTTGTGAGAGGTTTGAAC; CTTTCTGGAGGCAAACCTTCA	TC	
	5/59	HrU1	TGAACAATGAGAGAACTGATGCAA; CATCACTCAGATGAGAAACTGGAA	GT	
29 passeriformes species from	11/48	FhU2	GTGTTCTTAAAACATGCCTGGAGG; GCACAGGTAAATATTTGCTGGGCC	CA	17

eight families and 19 non-passeriformes species from 14 families	5/48	FhU3	GGATTCCTAGTAATTTAAACTC; ATATTCCCCATAAGATAATGG		
	3/48	FhU4	GGATTCCTAGTAATTTAAACTC; CCTTCCAAACTGAAGAGTAAG		
	2/48	FhU1	TGATCGAAAGACCTGTAAGAT; ATCACGTTAGACCAATACTCTTA		
	2/48	FhU6	ATCTGCTCCTCTGGGCCCTG; GATCCCTGTTCCTGGGTTAC		
Reed bunting and 12 additional species from seven families	8/12	Escu6	CATAGTGATGCCCTGCTAGG; GCAAGTGCTCCTTAATATTTGG	CA	18
	6/12	Escu2	TACAGCAAGGCGGAACTG; TGGGCAAAGATATGGGAAGA	TG	
	6/12	Escu3	ACTCTGACAGAGTTTTTCTGGT; TGCTTGTGGTTGTCAAGAT	TC	
	4/12	Escu1	TTCTCTTGGTCTATGGAAGGTG; GCTTGAAAGACAGTCACCAGG	CA	
	4/12	Escu5	ATGGGTGAACTCACTTGTTCG; TGCTAACGACCCTGGAGGT	CA	
	3/12	Escu4	TTCCCTCACAATTTTCCGAC; TATGTGCTGAAGTGAACCATCC	TG	
Reptiles					
Marine turtles: green, loggerhead, hawksbill, Pacific ridley, flatback, leatherback	6/6	Cm84	TGTTTTGACATTAGTCCAGGATTG; ATTGTTATAGCCTATTGTTCAGGA	CA	19
	6/6	Cc117	TCTTTAACGTATCTCCTGTAGCTC; CAGTAGTGTCAGTTCATTGTTTCA		
	5/6	Cm72	CTATAAGGAGAAAGCGTTAAGACA; CCAAATTAGGATTACACAGCCAAC		
	5/6	Ei8	ATATGATTAGGCAAGGCTCTCAAC; AATCTTGAGATTGGCTTAGAAATC		
	4/6	Cm3	AATACTACCATGAGATGGGATGTG; ATTCTTTTCTCCATAAACAAGGCC		
	4/6	Cm58	GCCTGCAGTACACTCGGTATTTAT; TCAATGAAAGTGACAGGATGTACC		
Timber rattlesnake	1/1	5A	CCAGAGCCATCAAGGCCCTT; TGCAGAGGCAGCACTTTGTTA	CA	20
	1/1	5-183	TTGTTGTAACCAGTGTGTGTGAT; CTGCAGACACTTATTATTATACC		
	1/1	7-144	CAGAGAAAAGGGAAGCATCAC; GCATACATGTATGTTTGTGTGCA		
	1/1	3-155	AAAAGTTAACAACTATGTAACCATT; TACATCACTTGCTGCTTCCTG		
			and two more		
Fish					
Stickleback, cod, lamprey, spotted ray, tope shark, lake sturgeon, conger eel, herring, pike, salmon, whiting	4/11	CIER11	TAGAGCGATCCGCAGCTGT; CTCGCTGCCTACTGCTTCC	CA	21, 22, 23
	7/11	CIER51	GCCAAAACACTGACGAGGTGA; TTTGCGCAAGCTTCAGGATGA		
	6/11	CIER62	GGTGCTGTCACTTTTGGCCAC; AACTCTGCTGGTCGCCACTCC		
	10/11	Gmo01	AAAAACAAAACGAGAGCGGT; GGAGTTGCTTGTGGTGGCAT	GA	
	9/11	Gmo02	CCCTCAGATTCAAATGAAGGA; GTGTGAGATGACTGTGTCG	CA	
	9/11	Gmo09	GCAGATTTGAATCAGCGTGT; CTCTGTGGGTATGTGCAAAG		
	7/11	Gmo123	GAGGGACATAAAGACACTT; AAACATGCATGCAGGCAAC		
	8/11	Gmo132	GGAACCCATTGGATTCAGGC; CGAAAGGACGAGCCAATAAC		
	4/11	Gmo145	GCATTGTAGGAACAACAATTAAC; GTGCATGTGCTCATTATAGC	GA	

Table 3. *Continued*

Species	P^a	Locus	Primers	Repeat	Ref.
Atlantic salmon, Pacific salmon, trout, char	7/10	Ssa4	ATTAGGCGGCAGCAGGCTC; TGTTCACTCACTGACACGCG	CA	24
	6/10	Ssa14	CCTTTTGACAGATTTAGGATTTC; CAAACCAAACATACCTAAAGCC		
	6/10	Ssa289	CTTTACAAATAGACAGACT; TCATCACGTCACTATCATC		
Insects					
Bees and bumblebees	6/6	A14	GTGTCGCAATCGACGTAACC; GTCGATTACCGATCGTGACG	CT, GGT	25
Subalpine ant	1/1	Myrt2	AAAGTATAATATGTAAGTAAATAA; GATCTGGGACATCTTGTG	GCGT	26
		Myrt3	GGTACTATCTAGTATCTAGTAATA; CGCGGCGGCACGTACCTA	GA,TG	
		Myrt4	CGTGCATGCGAGCATTCCGT; CGGCGATGCAAGTACGTC	CA	
Mosquito	1/1	AGXH7	CACGATGGTTTTCGGTGTGG; ATTTGAGCTCTCCCGGGTG	CA	27
		AGXH19	CTTTTTCTCCCCATTATCTC; CTGCAGTGTCCATTACGTAC		
		AGXH26	GGTTCCCTGTTACTTCCTGCC; CCGGCAACACAAACAATCGG		
		AGXH93	TTCCCAGCTCACCCTTCAAG; CCAACGTACAAACCTATCGC		
			and seven more		
Checkerspot butterfly	1/1	CINX1	ACATGAAACCGATTTTGAGG; CAGCTTAGTCCCCAAGTCA	CA	28
		CINX4	GGTCAGCGACTAAGAGTTAT; TATCCTTCAGACCAGCAGTT	AT	
Wasp	1/1	PACO3GT	CGCACATACATATACGTATAC; CGTTCCTTTGCATTTTCTCTGATTTACGG	CA	29
		PACO13	GAAAGAAGAAAAATACGGTAG; GCTTATCATCACACG	AAT	
			and three more		
Mollusc					
Freshwater snail	1/1	3	CAATCTTGTATCTATAATCCG; CCACTCCAGTAAGAAACAAAC	CA	30
		5	GAACTCGGTCTATCCACC; CTGCACATTAGGGATCA		
		11	TGAAACATGTTTTACGCATTG; ACATACGGCTAACAATTTGTATTAC	GATA	
			and three more		
Plants					
Pithecellobium elegans (a tropical tree)	1/1	Pel2	TAACGCAATCAGTTTATCAA; CACACTATTTATGTTCAAGA	GT,GA	31
		Pel3	AGAGATGGACTGGAAACTTC; CCTCCTTAGATTTCTTGTCT	TC,TA	

		Pel5	TCTCTGCACACAGGAACCCTTTTGC; CCAGAAATAAGGCTCTTTTGCACA and two more	AAAG	
Bur oak	1/1	MSQ3	CCCYYCYGACATTGCATATTCGA; CCAATTCGACAATTTCTTAGTGCA	GA	32
		MAQ4	TCTCCTCTCCCCATAAACAGG; GTTCCTCTATCCAATCAGTAGTGAG	AG	
		MSQ13	TGGCTGCACCTATGGCTCTTAG; ACACTCAGACCCACCATTTTTCC	TC	
Paspalum vaginatum (a seashore grass)	1/1	Pv35	TCGAAATCGAAAAAGAAGATCGTTC; GGCGCCAGCTACAAGGTTAG	GA	33
		Pv51	TCCCATCATCATCAGTTCTTCCAATC;GCCCTGTGCTATTATTCATCATCTT		
		Pv53	CTCGGAAACCGCAGCTCA; GCTCCGCCTCCTCTATTCCA		

[a]This is not an inclusive list. Instead it is intended to be representative of what has been published to date. For further references and a web-site address see Chapter 7.
[b]The proportion of species tested that were found to be polymorphic.

References

1. Ranjan, D., Shriver, M.D., Yu, L.M., Jin, L., Aston,E., Chakraborty, R., and Ferrell, R.E. (1994) *Genomics* **22**, 226.
2. Coote, T. and Bruford, M.W. (1996) *J. Hered.* **87,** 406.
3. Engel, S.R., Linn, R.A., Taylor, J.F., and Davis, S.K. (1996) *J. Mammal.* **77**, 504.
4. Menotti-Raymond, M.A., and O'Brien, S.J. (1995) *J .Hered.* **86**, 31.
5. Roy, M.S., Geffen, E., Smith, D., Ostrander, E.A. and Wayne, R.K. (1994) *Mol. Biol. Evol.* **11**, 553.
6. Ostrander, E.A., Sprague, G.F., and Rine, J. (1993) *Genomics* **16**, 207.
7. Coltman, D.W., Bowen, W.D., and Wright, J.M. (1996) *Mol. Ecol.* **5**, 161
8. Valsecchi, E. and Amos, W. (1996) *Mol. Ecol.* **5**, 151.
9. Paetkau, D., Calvert, W., Stirling, I., and Strobeck, C. (1995) *Mol. Ecol.* **4**, 347.
10. Paetkau, D. and Strobeck, C. (1994) *Mol. Ecol.* **3**, 489.
11. Craighead, L., Paetkau, D., Reynolds, H.V., Vyse, E.R., and Strobeck, C. (1995) *J. Hered.* **86**, 255.
12. O'Connell, M., Wright, J.M., and Farid, A. (1996) *Mol. Ecol.* **5**, 311.
13. Dallas, J.F., Dod, B., Boursot, P., Prager, E.M. and Bonhomme, F. (1995) *Mol. Ecol.* **4**, 311.
14. Dietrich, W., Katz, H., Lincoln, S., Shin, H.-S, Friedman, J., Dracopoli, N., and Lander, E. (1992) *Genetics* **131**, 423.
15. Ishibashi, Y., Saitoh, T., Abe, S., and Yoshida, M.C. (1995) *Mol. Ecol.* **4**, 127.
16. Primmer, C.R., Moller, A.P., and Ellegren, H. (1995) *Mol. Ecol.* **4**, 493.
17. Primmer, C.R., Moller, A.P., and Ellegren, H. (1996) *Mol. Ecol.* **5**, 365.
18. Hanotte, O., Zanon, C., Pugh, A., Greig, C., Dixon, A., and Burke, T. (1994) *Mol. Ecol.* **3**, 529.
19. Fitsimmons, N.N., Moritz, C., and Moore, S.S. (1995) *Mol. Biol. Evol.* **12**, 432.
20. Villarreal, X., Bricker, J., Reinert, H.K., Gelbert, L., and Bushar, L.M. (1996) *J. Hered.* **87**, 152.
21. Rico, C., Rico, I., and Hewitt, G. (1996) *Proc. R. Soc. Lond B.* **263**, 549.
22. Brooker, A.L., Cook, D., Bentzen, P., Wright, J.M., and Doyle, R.W. (1994) *Can. J. Fish. Aquat. Sci.* **51**, 1959.
23. Rico, C., Zadworny, D., Kuhnlein, U., and Fitzgerald, G.J. (1993) *Mol. Ecol.* **2**, 271.
24. McConnell, S.K., O'Reilly, P., Hamilton, L., Wright, J.M., and Bentzen, P. (1995) *Can. J. Fish. Aquat. Sci.* **52**, 1863.
25. Estoup, A., Solignac, M., Harry, M., and Cornuet, J.-M (1993) *Nucleic Acids Res.* **21**, 1427.
26. Evans, J.D. (1993) *Mol. Ecol.* **2**, 393.
27. Lanzaro, G.C., Zheng, L., Toure, Y.T., Traore, S.F., Kafatos, F.C., and Vernick, K.D. (1995) *Insect Mol. Biol.* **4**, 105.
28. Palo, J., Varvio, S.-L, Hanski, I., and Vainola, R. (1995) *Heraditas* **123**, 295.
29. Choudhary, M., Strassmann, J.E., Solis, C.R., and Queller, D.C. (1993) *Biochem. Genet.* **31**, 87.
30. Jarne, P., Viard, F., Delay, B., and Cuny, G. (1994) *Mol. Ecol.* **3**, 527.
31. Chase, M., Kesseli, R., and Bawa, K. (1996) *Am. J. Bot.* **83**, 51.
32. Dow, B.D., Ashley, M.V., and Howe, H.F. (1995) *Theor. Appl. Genet.* **91**, 137.
33. Liu, Z.-W, Jarret, R.T., Kresovich, S., and Duncan, R.R. (1995) *Theor. Appl. Genet.* **91**, 47.

A5

Computational resources for population analysis

J. CLAIBORNE STEPHENS and JILL PECON SLATTERY

1. Background and introduction

Concomitant with the rapid infiltration of DNA-based technology into population genetics laboratories has been the infusion of computers, software, and on-line databases into the daily activities of many population geneticists. The identification of abundant, highly variable loci, detectable by a wide variety of techniques has led to a revolution in the way researchers view and analyse levels of variation. Evaluation of genetic variation ranges from detection of allelic associations within and between loci, through sub-populations, populations, sub-species, up to inter-specific phylogenetic comparisons. Novices in this area would do well to consult any of several excellent texts (1–4) that address the analyses appropriate for these different levels of variation as well as the underlying theoretical issues. We have, somewhat arbitrarily, divided our coverage of current computer programs in this area to focus on two primary levels of enquiry: intra- and inter-population structure and phylogenetic inference.

The primary focus of population genetics is on assessment of diversity within or among populations and determination of whether or not allelic associations exist at several different levels. At a genotypic level, at individual loci, one can test for departure from Hardy–Weinberg equilibrium. At two or more loci, 'linkage disequilibrium' is tested as the departure of gametic frequencies from expectation under allelic independence among the loci. F-statistics for analysing the distribution of genetic diversity within the data are used to test the null hypothesis that the presumptive populations are really samples from one large panmictic population. One can then base 'genetic distance' estimates on the allele frequency distribution(s) to quantify any observed patterns of variation. Generally underlying all of these analyses is the characterization of the alleles at a given locus, or more often, multiple loci, followed by the estimation of the frequency distribution of these alleles among populations (1–3). We will discuss several appropriate programs in current use and provide web site addresses for current servers that deal with this area.

In contrast, phylogenetic reconstruction programs are applied to molecular data to estimate the evolutionary history of:

(a) alleles within a species;
(b) homologous loci among several related species;
(c) homologous and often paralogous copies of loci belonging to gene families; or even
(d) entire genome comparisons (e.g., DNA–DNA hybridization).

From these, inferences are often made about the evolutionary relationships among the species or populations from which these data were obtained, resulting in a phylogenetic tree. Phylogenetic reconstruction of taxa is generally most powerful in determining patterns of associations among taxa at the sub-species level or higher, although we note several powerful applications to non-traditional taxomic questions such as allelic and viral phylogenies.

Regardless of the specific question, three major methods are commonly employed in phylogenetic analyses: distance-based, maximum parsimony, and maximum likelihood. Each of these is derived by differing evolutionary assumptions that are not necessarily compatible. Common to all phylogenetic methods is an optimality criterion, employed to evaluate objectively how well alternative trees (including those inferred by the programs) fit the actual data.

For comparison of the three methods, consider a data set composed of nucleotide sequences from a particular gene amplified from different individuals. With a distance-based analysis, the initial step is to compute genetic distances among all pairs of sequences. Each distance estimate represents an average substitution rate per site since the pair of sequences last shared a common ancestor. By computing an average value, all sites, even those that exhibit no substitutions in any of the sequence comparisons, are included in the analysis. Choice of the appropriate distance measure depends on many factors:

- the type of gene (e.g., its function, whether it is mitochondrial or nuclear)
- frequency distribution of the four nucleotides within the data
- transition:transversion ratio
- mutation rate heterogeneity among sites.

These pairwise estimates are used as input for a clustering algorithm such as neighbour-joining or least-squares regression (Fitch–Margoliash). An optimality criterion for distance-based methods is that of minimum evolution whereby differences between the genetic distance matrix and those represented by the branch lengths of the resultant tree are minimized. Data sets consisting of frequency distributions of alleles or genotypes within a species or among closely related species are more appropriately addressed with distance-based analyses, and distance methods are the only choice for data that exists only in the form of summary pairwise distances (e.g., albumin immunological distances, DNA–DNA hybridization).

In contrast, maximum parsimony analysis requires that the data be considered as character states. Thus, for our example data set of molecular sequences, each site of the sequence is a distinct character with A, G, C, and T as possible character states. A tree is constructed by considering only those sites that are polymorphic with the ideal that each mutation arose only once within the phylogeny. Conflicting character changes within the topology are considered a measure of homoplasy (substitutions that are attributed to convergence, parallel evolution, or back-mutation). In searching for the best tree, maximum parsimony tests alternative trees under the optimality criterion that the tree of the smallest total number of steps (most parsimonious) is the best estimation of the true phylogeny.

The third method, maximum likelihood, is a statistically rigorous analysis of genetic data under a specific evolutionary model. This method is computer intensive; it searches for the most probable tree given the data and a specified model of evolution that determines mutation within the genetic marker(s). For a given tree, the likelihood of each substitution at each site within the sequence is computed under the specified model of mutation. Note that parametric values for the evolutionary model can also be tested and optimized with this approach. Using the optimality criterion that the tree of the greatest combined likelihood reflects the true phylogeny, the final tree is selected from all those possible. Because the probability of each mutation is computed for each tree considered, this can require prohibitively extensive computer resources with large data sets.

An important adjunct to phylogenetic analysis is bootstrap analysis, often used to evaluate trees obtained with maximum parsimony and distance-based methods. A bootstrap analysis is an iterative resampling method which tests how reliably the same topology is consistently derived from the data. It involves the creation of at least 100 artificial data sets by sampling with replacement from the original data. These artificial data sets are analysed and a consensus of the resultant trees is viewed as support for the nodes within the original tree.

Finally, since all phylogenetic trees are derived, either implicitly or explicitly, based on the pattern of mutations exhibited by a given genetic marker, many programs address the pattern(s) of molecular change such as that within a taxon or the evaluation of the nature of observed changes (e.g., transition/transversion bias, position within codon of protein coding genes). We will cover examples of these programs too. Our strategy is to focus on the programs that we know best or that have potential for the widest appeal, but also to try to provide enough information in short tabular format that the reader can easily pursue their own particular interests. There is considerable evolution of computer programs, both anagenic and cladogenic. So as not to unwittingly lead the reader down the road to extinction, we will begin generically, highlighting higher level resources such as servers and databases that we feel are likely to respond more dynamically to each researcher's interests.

2. Hardware, web sites, and databases

2.1 Supercomputer centres

Many facilities have supercomputers and staff whose mandate is to augment research, including population genetics and phylogenetics, by collaborating on analysis and novel algorithms in those areas. In particular, the National Science Foundation of the United States has a Supercomputers Centers Program, whose mission, in part, is the facilitation of scientific research that requires high performance computing such as phylogenetic inference. A basic site where many of these are listed is the 'Science and Supercomputers' site (see *Table 1* for URL), which has direct links to many sites worldwide. The National Center for Supercomputing Applications is another excellent place to start.

2.2 Web sites and databases

As mentioned earlier, computer programs often change, as do internet web sites. One of our main goals in this section is to provide links, either as e-mail, FTP addresses, or URLs, to current resources that are available by computer. In addition to truly useful programs, many such sites have additional features such as examples, FAQs (frequently asked questions), citations, bug reports, and pointers or links to related material. Before the advent of the Internet, many of these functions were handled by newsletters and release notes. Although extremely useful to the end user, newsletters are often tedious to produce and inevitably dated by the time they are received.

3. Individual software packages

3.1 Population genetics and molecular evolution

3.1.1 AMOVA 1.55

AMOVA (Analysis of Molecular Variance) began as a program for estimating the variance of genetic diversity within a single species using mitochondrial DNA (5). The statistics are analagous to *F*-statistics for nuclear genotypes and can be partitioned into a hierarchical design: individuals within populations; populations within regions; and among regions. The significance of each statistic is estimated using a permutation analysis, which avoids assuming statistical normality of the data. The current program has been expanded to handle codominant data such as restriction fragment length polymorphsims (RFLPs) and microsatellites, and includes computation of variance components, *F*-statistics, co-ancestry coefficients, a heteroscedasticity index, and permutational tests of the above population statistics. For more information contact Laurent Excoffier (e-mail: `excoffie@sc2a.unige.ch`) or visit `http://anthropologie.unige.ch/ftp/comp/win/amova/`. This program is currently only available for Microsoft Windows platforms.

3.1.2 Arlequin

Arlequin is a relatively new entry into population genetics software and is billed as 'an exploratory population genetics software environment'. Many features are listed, including intra-population data analyses (e.g., diversity indices, distance estimates, maximum-likelihood estimation of allele and haplotype frequencies, several tests of selective neutrality) and inter-population tests of genetic structure and estimates of related fixation indices and coefficients. The web page for Arlequin (see *Table 1* for URL) includes FAQs, Bug Reports, and mechanisms for feedback and downloading the software. For more information contact Laurent Excoffier (see above). Arlequin is currently only available for Microsoft Windows platforms.

3.1.3 GDA

Many of the analytical goals of AMOVA are shared by GDA (Genetic Data Analysis), a series of programs based on Bruce Weir's book of the same name (1). These programs are currently available on a use-at-your-own-risk basis.

Table 1. Web sites and other addresses for computational resources for population analysis

Databases, servers, and useful home pages	
Australian Biotech. Society	`http://ba-itumac1.lib.unimelb.edu.au/ABA/zPhylogenetics`
Biosciences Index	`http://golgi.harvard.edu/htbin/biopages`
EMBL	`http://www.embl-heidelberg.de/`
EMBnet	`http://www.ie.embnet.org/brochure/`
European Bioinformatics Institute	`http://www.ebi.ac.uk/biocat/Phylogeny.html`
Evolution and Phylogeny Laboratory	`http://www.no.embnet.org/phylogeny.html`
Genetics Jump Station	`http://www.ifrn.bbsrc.ac.uk/gm/lab/docs/genetics.html`
GenLink	`http://www.genlink.wustl.edu/`
Houston	`ftp.bchs.uh.edu`
Indiana	`ftp.bio.indiana.edu`
Laboratory of Molecular Systematics...	`http://nmnhwww.si.edu/gopher-menus/LaboratoryofMolecularSystematicsFTPServer.html`
Molecular Ecology home page	`http://ocean1.msrc.sunysb.edu/molecol/index.html`

Table 1. *Continued*

Molecular evolution software...	http://evolve.zps.ox.ac.uk/PhySoft/PhySoft.html
NCSA	http://www.ncsa.uiuc.edu
NCBI	http://www.ncbi.nlm.nih.gov
Oak Ridge	http://www.ornl.gov/TechResources/Human_Genome/genetics.html
Population genetics/evolution e-mail list	http://life.biology.mcmaster.ca/~brian/evoldir.html
Quantitative genetics resources	http://nitro.biosci.arizona.edu/zbook/book.html
Science and supercomputers	http://coaps.fsu.edu/science_sites.html
Tree of life	http://phylogeny.arizona.edu/tree/phylogeny.html
UConn systematics server	http://darwin.eeb.uconn.edu/systematics.html
Willi Hennig Society	http://www.vims.edu/~mes/hennig/software.html
Population genetics programs	
AMOVA	http://anthropologie.unige.ch/ftp/comp/win/amova/
Arlequin	http://anthropologie.unige.ch/arlequin
GDA	http://biology.unm.edu/~lewisp/gda.html
GENEPOP	ftp.cefe.cnrs-mop.fr
GeneStrut	ftp from csuvax1.mudoch.edu.au
MicroSat	http://Lotka.Stanford.edu/microsat.html
RAPDistance	http://life.anu.edu.au/molecular/software/rapd.html
RSVP	http://oeb.harvard.edu/~rice
Alignment and phylogenetics programs	
CLUSTAL	ftp.embl-Heidelberg.de
FREQPARS	http://nmnhwww.si.edu/gopher-menus/LaboratoryofMolecularSystematicsFTPServer.html
Hennig86 v1.5	http://www.vims.edu/~mes/hennig/software.html
MacClade	http://phylogeny.arizona.edu/macclade/macclade.html
MacT	ftp.bio.indiana.edu
MEGA	imeg@psuvm.psu.edu
MOLPHY 2.2	ftp from sunmh.ism.ac.jp
ODEN	ftp from bioslave.uio.no
PAUP version 3.1.1	ftp from onyx.si.edu/pub
PAUP*	scavotto@sinauer.com
PHYLIP	http://evolution.genetics.washington.edu/phylip
TreeView	http://taxonomy.zoology.gla.ac.uk/rod/treeview.html
VOSTORG	ftp from hgc6.sph.uth.tmc.edu

All conventional modes of analysis (allelism and polymorphism characterization, heterozygosity, F-statistics, genetic distances, disequilibrium tests) are supported. Additionally, care has been taken to allow input and output in formats that are useful to other programs. More information on GDA can be obtained at the web site listed in *Table 1*, maintained by Paul O. Lewis (e-mail: `lewisp@unm.edu`). GDA is currently only available for Microsoft Windows platforms.

3.1.4 MicroSat 1.5

Analyses of microsatellite data consist of computing a distance measure among individuals across multiple loci. At present, MicroSat incorporates many different estimates of genetic distance, each marked by a specific model of evolutionary change. MicroSat computes estimates of linearity, and diversity measures such as F_{ST}, heterozygosity, and mean variance for each locus. The resultant distance matrices can be imported into PHYLIP (see below) as input for cluster analyses or phylogenetic reconstruction. MicroSat also performs a bootstrap analysis for the genetic distance data to test the consistency of the data to derive the same associations within and between populations. The web site (see *Table 1*) maintains a list of known available platforms, currently primarily Macintosh. MicroSat is written and maintained by Eric Minch (e-mail: `eric@Lotka.Stanford.edu`).

3.2 Phylogenetic and related programs

3.2.1 MacClade 3.06

Evaluation of character evolution on alternative phylogenies is one of MacClade's greatest virtues. The many capabilities of this program include: extremely simple and intuitive interactive manipulation of trees, facilities for editing character data, and visualization of levels and direction of change for each character. Wayne P. Maddison and David R. Maddison have developed MacClade for exploring data sets that are either phylogenetic or potentially phylogenetic in nature. A wide variety of examples, FAQs, and information related to MacClade or phylogenetic research in general can be obtained from the MacClade web site (see *Table 1* for URL).

3.2.2 MEGA 1.01

MEGA (Molecular Evolutionary Genetics Analysis) is a series of programs covering many aspects of molecular evolution. Part of this coverage includes phylogenetic inference using distance and parsimony methods. A major advantage of this program is the ability to use and evaluate alternative models of molecular evolution (e.g., Jukes–Cantor, Kimura 2 parameter, Tamura). Also, aspects of the data not normally required for phylogenetic inference are highlighted by MEGA analyses, such as transition:transversion ratios and position within codon of protein coding genes. The original version was developed by Sudhir Kumar, Koichiro Tamura, and Masatoshi Nei and is available by contacting imeg@psuvm.psu.edu. MEGA 1.01 is currently only available for Microsoft Windows platforms.

3.2.3 PAUP 3.1.1 and PAUP*

PAUP version 3.1.1 conducts phylogenetic analysis under the evolutionary assumptions of maximum parsimony. Data are treated as discrete characters with the assumption that the derived phylogenetic tree of the shortest (most parsimonious) length reflects the true phylogeny. Application of maximum parismony to population data consisting of allele frequencies is not robust. The power of this algorithm is in discerning those changes in the molecular data which reflect a shared, derived pattern of evolution. Most applications of maximum parsimony concern questions in higher levels of taxonomy. This phylogenetic program is not applicable to data consisting of frequency distributions of alleles among populations.

In contrast, PAUP* (some web sites list this as PAUP 4), still under revision, has been upgraded to perform maximum parsimony, distance-based analyses and maximum likelihood analyses. The program offers numerous distance measures which reflect the appropriate patterns of substitution present within the molecular sequence data. The Macintosh version of PAUP* is extremely flexible, permitting the user to select and modify variables concerning data input, search parameters, permutation testing, resampling methods, analysis of tree distributions, topologies, and rooting. It is clearly superior in maximum parsimony analysis with many options available for character data manipulation, weighting schemes, imposition of tree constraints, alteration of numerous search parameters, and in tree description. The maximum likelihood search is versatile, and designed so that the user must develop a specific model of evolution in accordance with each empirical data set. PAUP is written and maintained by David L. Swofford (e-mail: Swofford@onyx.si.edu).

Updates from version 3.1 to version 3.1.1 can be obtained by FTP from onyx.si.edu in the /pub directory. Versions are for MacIntosh computers. PAUP* is destined to become available from Sinauer Associates. Contact scavotto@sinauer.com for more information. The program (for Power PC native, DOS/Windows, Macintosh and PowerMac, UNIX-DEC Alpha workstation and UNIX-Sun Sparc workstation) and user's book are tentatively priced at $100.00.

3.2.4 PHYLIP

Joe Felsenstein's PHYLIP (Phylogeny Inference Package) is one of the earliest and remains one of the most useful computer packages ever produced for phylogenetic inference. This is a versatile, free, computer package consisting of over 30 programs which can analyse molecular sequence data (both DNA and protein), distance matrix data, gene frequencies, discrete or continuous character data, and programs for plotting phylogenetic trees.

For the analysis of population structure and evolution, data sets typically consist of sequences from multiple individuals of a given species. Phylogenetic analysis of these data will cluster these individuals under the hypothesis of a shared evolutionary history. If there is sufficient population sub-division, the individuals will cluster in accordance with putative populations.

A typical analysis begins by computing a distance measure using DNADIST among all pairs of individuals resulting in a distance matrix. DNADIST offers four estimates of genetic distance: Jukes–Cantor model, Jin–Nei model, Kimura two-parameter model, or maximum likelihood. The resultant 'outfile' of a distance matrix is then used as input for tree-building algorithms of using either least squares regression (FITCH and KITSCH), UPGMA (NEIGHBOR) or by neighbour-joining (NEIGHBOR) which clusters the individuals into related groups. The output files consist of a diagram of the tree followed by the estimated branch lengths in 'outfile' and a 'treefile' which can be imported into a graphics program for plotting.

To test the reliability of the data to obtain this tree, a bootstrap analysis can be performed. The steps required begin with creation of multiple bootstrap data sets using SEQBOOT. This program creates an artificial data set of the same dimensions as the original data set by resampling with replacement of sites from the orginal data. This outfile is then used as input following the sequence as with the original data analyses (DNADIST, NEIGHBOR) except that 'analyze multiple data sets' must be selected. The resultant 'treefile' is then analysed by CONSENSE to compute a consensus tree of the multiple trees derived from the bootstrap data sets.

The entire PHYLIP package is available free of charge in formats compatible with most desktop computers and workstations (UNIX, VAX systems; DOS, Windows 3.1, Windows 95, Windows NT with PC/AT, PC/XT, 386, 486, and Pentium PCs; Macintosh and PowerMAC). Additionally, PHYLIP's web site (see *Table 1* for URL) contains a wide variety of useful information about phylogenetic reconstruction and other programs related to phylogenetic inference. Joe Felsenstein, the architect of PHYLIP, can be reached at `joe@genetics.washington`.

3.2.5 TreeView (0.95)

TreeView is an extremely convenient program for viewing trees or making publication quality figures from standard output from such programs as PAUP or PHYLIP. It is free and available for most Macintosh and IBM PC compatible platforms. The program was developed by Roderic D. M. Page and is easily obtained from the web site listed in *Table 1*. This web site is another excellent example of how a well-designed web page can provide almost all aspects of support for a computer program by providing the program, manuals, bug fixes, tips, and related information.

4. Summary

Clearly, there is an ever-expanding wealth of programs, data, and related information pertinent to population biologists simply for the asking. Any charges involved are often minimal, and we are currently fortunate that finances are not the major hurdle to obtaining these resources. We have emphasized e-mail and the Internet, since these are extraordinarily simple and inexpensive mechanisms, and are available worldwide. Of course, researchers lacking Internet or e-mail connections can still generally acquire many of these programs on disk. In closing, we commend the research community, certainly those cited above, but many more left unmentioned, for their efforts in making so many programs and so much information useful and easily available.

References

1. Weir, B. S. (1996) *Genetic Data Analysis II.* Sinauer Associates, Sunderland, MA.
2. Nei, M. (1987) *Molecular Evolutionary Genetics*. Columbia University Press.
3. Li, W.-H. and Graur, D. (1991) *Fundamentals of Molecular Evolution*. Sinauer Associates, Sunderland, MA.
4. Hillis, D. M., Moritz, C., and Mable, B. K. (eds) (1996) *Molecular Systematics.* Sinauer Associates, Sunderland, MA.
5. Excoffier, L., Smouse, P., and Quattro, J. M. (1992). *Genetics* **131**, 479.

A6

List of suppliers

Advanced Genetic Technologies Corporation, 19212 Orbit Drive, Gaithersburg, MD 20879, USA.

Alconox Inc., 9 East 40th St., New York, NY 10016, USA.

Amersham International, Amersham Place, Little Chalfont, Bucks, HP7 9NA, UK.

Amersham

Amersham International plc., Lincoln Place, Green End, Aylesbury, Buckinghamshire HP20 2TP, UK.

Amersham Corporation, 2636 South Clearbrook Drive, Arlington Heights, IL 60005, USA.

AMF Cuno, Microfiltration Products Division, 400 Research Pky, Meriden, CT 06450, USA.

Amicon Inc., 72 Cherry Hill Drive, Beverly, MA 01915-9727, USA.

Anderman

Anderman and Co. Ltd., 145 London Road, Kingston-Upon-Thames, Surrey KT17 7NH, UK.

Applied Biosystems Division, Perkin–Elmer Corporation, 850 Lincoln Centre Drive, Foster City, CA 94404, USA.

Avitech Diagnostics (formerly AT Biochem, ATGC), 30 Spring Mill Drive, Malvern, PA 19355, USA.

Beckman Instruments

Beckman Instruments UK Ltd., Oakley Court, Kingsmead Business Park, London Road, High Wycombe, Bucks HP11 1J4, UK.

Beckman Instruments Inc., PO Box 3100, 2500 Harbor Boulevard, Fullerton, CA 92634, USA.

Beckman Instruments Inc., 2500 Harbor Blvd., PO Box 3100, Fullerton, CA 92634, USA.

Becton Dickinson

Becton Dickinson and Co., Between Towns Road, Cowley, Oxford OX4 3LY, UK.

Becton Dickinson and Co., 2 Bridgewater Lane, Lincoln Park, NJ 07035, USA.

Bio

Bio 101 Inc., c/o Statech Scientific Ltd, 61–63 Dudley Street, Luton, Bedfordshire LU2 0HP, UK.

Bio 101 Inc., PO Box 2284, La Jolla, CA 92038–2284, USA.

Bio-101, PO Box 2284, La Jolla, CA 92038, USA.

Biometra Inc., PO Box 1554, D-37005, Gottingen, Germany.

Bio-Rad Laboratories

Bio-Rad Laboratories Inc., 2000 Alfred Nobel Drive, Hercules, CA 94547, USA.

Bio-Rad Laboratories Ltd., Bio-Rad House, Maylands Avenue, Hemel Hempstead HP2 7TD, UK.

Bio-Rad Laboratories, Division Headquarters, 3300 Regatta Boulevard, Richmond, CA 94804, USA.

Boehringer Mannheim

Boehringer Mannheim UK (Diagnostics and Biochemicals) Ltd, Bell Lane, Lewes, East Sussex BN17 1LG, UK.

Boehringer Mannheim Corporation, Biochemical Products, 9115 Hague Road, P.O. Box 504 Indianapolis, IN 46250–0414, USA.

Boehringer Mannheim Biochemica, GmbH, Sandhofer Str. 116, Postfach 310120 D-6800 Ma 31, Germany.

British Drug Houses (BDH) Ltd, Poole, Dorset, UK.

Calbiochem, P.O. Box 12087, La Jolla, California 92039-2087, U.S.A.

CBS Scientific, PO Box 856, Del Mar, CA 92014, USA.

Clontech Laboratories, Inc., 1020 East Meadow Circle, Palo Alto, CA 94303-4230, USA.

Difco Laboratories

Difco Laboratories Ltd., P.O. Box 14B, Central Avenue, West Molesey, Surrey KT8 2SE, UK.

Difco Laboratories, P.O. Box 331058, Detroit, MI 48232–7058, USA.

Dupont, Concord Plaza, Wilminto, DE 19898, USA.

Du Pont

Dupont (UK) Ltd., Industrial Products Division, Wedgwood Way, Stevenage, Herts, SG1 4Q, UK.

Du Pont Co. (Biotechnology Systems Division), P.O. Box 80024, Wilmington, DE 19880–002, USA.

Dynal, PO Box 158, Skoyen, N-0212 Oslo 2, Norway.

Eastman Kodak, Rochester, NY 14650 USA.

European Collection of Animal Cell Culture, Division of Biologics, PHLS Centre for Applied Microbiology and Research, Porton Down, Salisbury, Wilts SP4 0JG, UK.

Falcon (Falcon is a registered trademark of Becton Dickinson and Co.).

Fisher Scientific Co., 711 Forbest Avenue, Pittsburgh, PA 15219–4785, USA.

Flowgen Instruments, Broadoak Enterprise Village, Broadoak Road, Sittingbourne, Kent ME9 8AQ, UK.

Flow Laboratories, Woodcock Hill, Harefield Road, Rickmansworth, Herts. WD3 1PQ, UK.

Fluka

Fluka-Chemie AG, CH-9470, Buchs, Switzerland.

Fluka Chemicals Ltd., The Old Brickyard, New Road, Gillingham, Dorset SP8 4JL, UK.

FMC BioProducts, 1 Risingeveg, DK-2665 Vallensbaek Strand, Denmark.

GATC, Fritz-Arnold Strasse 23, 78467 Konstanz, Germany.

Gene Codes Corp., 2901 Hubbard Rd., Ann Arbor, MI, 48105, U.S.A.

Gibco BRL

Gibco BRL (Life Technologies Ltd.), Trident House, Renfrew Road, Paisley PA3 4EF, UK.

Gibco BRL (Life Technologies Inc.), 3175 Staler Road, Grand Island, NY 14072–0068, USA.

Hamilton, 4970 Energy Way, Reno, NV 89502, USA.

Hitachi Software Engineerging America, Ltd, 111 Bayhill DR Suite 395, San Bruno, CA 94066, USA.

Hoefer Scientific Instruments, 654 Minnesota Street, Box 77387, San Francisco, CA 94107, USA.

Arnold R. Horwell, 73 Maygrove Road, West Hampstead, London NW6 2BP, UK.

Hybaid

Hybaid Ltd., 111–113 Waldegrave Road, Teddington, Middlesex TW11 8LL, UK.

Hybaid, National Labnet Corporation, P.O. Box 841, Woodbridge, NJ. 07095, USA.

HyClone Laboratories 1725 South HyClone Road, Logan, UT 84321, USA.

International Biotechnologies Inc., 25 Science Park, New Haven, Connecticut 06535, USA.

ICI, Cellmark Diagnostics, Blacklands Way, Abingdon Business Park, Abingdon, Oxfordshire OX14 1DY, UK.

Idaho Technology, PO Box 50819, Idaho Falls, ID 83405, USA

Invitrogen Corporation

Invitrogen Corporation 3985 B Sorrenton Valley Building, San Diego, CA. 92121, USA.

Invitrogen Corporation c/o British Biotechnology Products Ltd., 4–10 The Quadrant, Barton Lane, Abingdon, OX14 3YS, UK.

Japanese Cancer Research Resources Bank, Gene Respiratory, National Institute of Health, 10–35 Kamiosaki 2-chome, Shinagawa-ku, tokyo 141, Japan.

Joshua Meier, 7401 West Side Avenue, North Bergen, NJ 13904, USA.

Kodak: Eastman Fine Chemicals 343 State Street, Rochester, NY, USA.

Kontes, 1022 Spruce Street, Vineland, New Jersey 08360, U.S.A.

LI-COR Inc., 4421 Superior Street, PO Box 4000, Lincoln, NE 68504, USA.

Life Technologies Inc., 8451 Helgerman Court, Gaithersburg, MN 20877, USA.

Merck

Merck Industries Inc., 5 Skyline Drive, Nawthorne, NY 10532, USA.

Merck, Frankfurter Strasse, 250, Postfach 4119, D-64293, Germany.

Millipore

Millipore (UK) Ltd., The Boulevard, Blackmoor Lane, Watford, Herts WD1 8YW, UK.

Millipore Corp./Biosearch, P.O. Box 255, 80 Ashby Road, Bedford, MA 01730, USA.

Molecular Dynamics, 928 E. Arques Avenue, Sunneyvale, CA 94068, USA.

Nalge Company, Rochester, NY 14602-0365, USA.

National Diagnostics, 1013–1017 Kennedy Boulevard, Manville, NJ 08835, USA.

NEN Research Products, Wedgewood Way, Stevenage, Hertfordshire SG1 4QN, UK.

New England Biolabs (NBL)

New England Biolabs (NBL), 32 Tozer Road, Beverley, MA 01915–5510, USA.

New England Biolabs (NBL), c/o CP Labs Ltd., P.O. Box 22, Bishops Stortford, Herts CM23 3DH, UK.

Nikon Corporation, Fuji Building, 2–3 Marunouchi 3-chome, Chiyoda-ku, Tokyo, Japan.

Northfork Products, PO Box 4347. Tumwater, WA 98501, USA.

Owl Scientific Inc., P.O. Box 566, Cambridge, MA, 02139, USA.

Palliard Chemical Co., Palliard Farm, Twinstead, Sudbury, Suffolk CO12 7PD, UK.

Perkin-Elmer

Perkin-Elmer Ltd., Maxwell Road, Beaconsfield, Bucks. HP9 1QA, UK.

Perkin-Elmer Ltd., Post Office Lane, Beaconsfield, Bucks, HP9 1QA, UK.

Perkin-Elmer-Cetus (The Perkin-Elmer Corporation), 761 Main Avenue, Norwalk, CT 0689, USA.

Pharmacia Biotech Europe Procordia EuroCentre, Rue de la Fuse-e 62, B-1130 Brussels, Belgium.

Pharmacia Biotech Inc., 800 Centennial Avenue, PO Box 1327, Piscataway, NJ, USA.

Pharmacia Biosystems

Pharmacia Biosystems Ltd. (Biotechnology Division), Davy Avenue, Knowlhill, Milton Keynes MK5 8PH, UK.

Pharmacia LKB Biotechnology AB, Björngatan 30, S-75182 Uppsala, Sweden.

Philip Harris Scientific, 618 Western Avenue, Park Royal, London W3 0TE, UK.

Poly Tech Products, 95 Propezi Way, Somerville, MA 02143, USA.

Princeton Separations, PO Box 300, Adelphia, NJ 07710, USA.

Promega

Promega Ltd., Delta House, Enterprise Road, Chilworth Research Centre, Southampton, UK.

Promega Corporation, 2800 Woods Hollow Road, Madison, WI 53711–5399, USA.

Qiagen

Qiagen Inc., 9600 De Soto Avenue, Chatsworth, CA 91311-5012, USA.

Qiagen Inc., c/o Hybaid, 111–113 Waldegrave Road, Teddington, Middlesex, TW11 8LL, UK.

Qiagen Inc., 9259 Eton Avenue, Chatsworth, CA 91311, USA.

Robbins Scientific 814 San Aleso Avenue, Sunnyvale, CA 94086, USA.

Sarstedt, 68 Boston Road, Beaumont Leys, Leicester, LE4 1AW, UK.

Savant Instruments, Inc., 100 Colin Drive, Holbrook, New York, 11741-4306, U.S.A.

Schleicher and Schuell

Schleicher & Schuell, PO Box 4, D-3354 Dassel, Germany.

Schleicher and Schuell Inc., Keene, NH 03431A, USA.

Schleicher and Schuell Inc., D-3354 Dassel, Germany. Schleicher and Schuell Inc., c/o Andermann and Company Ltd.

Serva Biochemicals, 50 A&S Drive, Pararnus, NJ 07652, USA.

Shandon Scientific Ltd., Chadwick Road, Astmoor, Runcorn, Cheshire WA7 1PR, UK.

Sigma Chemical Company

Sigma Chemical Company (UK), Fancy Road, Poole, Dorset BH17 7NH, UK.

Sigma Chemical Company, 3050 Spruce Street, P.O. Box 14508, St. Louis, MO 63178–9916.

James Skare, 665 North Street, Tewksbury, MA 10876, USA.

Sorvall DuPont Company, Biotechnology Division, P.O. Box 80022, Wilmington, DE 19880–0022, USA.

Starch Art, PO Box 268, Smithville, TX 78947, USA.

Stratagene

Stratagene Ltd., Unit 140, Cambridge Innovation Centre, Milton Road, Cambridge CB4 4FG, UK.

Strategene Inc., 11011 North Torrey Pines Road, La Jolla, CA 92037, USA.

United States Biochemical (USB), P.O. Box 22400, Cleveland, OH 44122, USA.

UV Products Inc., Science Park, Milton Road, Cambridge CB4 4BN, UK.

V.W.R. Scientific Products, 1310 Goshen Parkway, West Chester, Pennsylvania 19380, U.S.A.

Wellcome Reagents, Langley Court, Beckenham, Kent BR3 3BS, UK.

Index

Index

Index

Index